OLIVER

SHOP MANUAL O-201

Series ■ 66 ■ Super 66 ■ 77 ■ Super 77 ■ 88 ■ Super 88 ■ 770 ■ 880

Series ■ 660 Supplement

Series ■ 770 & 880 Supplement

Models ■ 99 (6 Cyl.) 4 Speed ■ Super 99 (6 Cyl.) 6 Speed ■ Super 99 GM (3 Cyl.) 6 Speed

Series ■ Super 99 GMTC ■ 950 ■ 990 ■ 995

Series ■ Super 55 ■ 550

Information and Instructions

This shop manual contains several sections each covering a specific group of wheel type tractors. The Tab Index on the preceding page can be used to locate the section pertaining to each group of tractors. Each section contains the necessary specifications and the brief but terse procedural data needed by a mechanic when repairing a tractor on which he has had no previous actual experience.

Within each section, the material is arranged in a systematic order beginning with an index which is followed immediately by a Table of Condensed Service Specifications. These specifications include dimensions, fits, clearances and timing instructions. Next in order of arrangement is the procedures paragraphs.

In the procedures paragraphs, the order of presentation starts with the front axle system and steering and proceeding toward the rear axle. The last paragraphs are devoted to the power take-off and power lift systems. Interspersed where needed are additional tabu specifications pertaining to wear limits, torquing, et

HOW TO USE THE INDEX

Suppose you want to know the procedure for R (remove and reinstall) of the engine camshaft. Your step is to look in the index under the main heading ENGINE until you find the entry "Camshaft." Now re to the right where under the column covering the trac you are repairing, you will find a number which in cates the beginning paragraph pertaining to the ca shaft. To locate this wanted paragraph in the man turn the pages until the running index appearing on top outside corner of each page contains the num you are seeking. In this paragraph you will find the in mation concerning the removal of the camshaft.

I&T Shop Service

P.O. Box 12901, Overland Park, KS 66282-2901
Phone: 800-262-1954 Fax: 800-633-6219
itshopmanuals.com

June, 2002
August, 2004
July, 2007
November, 2010

Cover art done by Sean Keenan

OLIVER

Series ■ 66 ■ Super 66 ■ 77 ■ Super 77
■ 88 ■ Super 88 ■ 770 ■ 880

Previously contained in I & T Shop Service Manual No. 0-10

SHOP MANUAL

OLIVER

MODELS

66, 77, 88, SUPER 66, SUPER 77, SUPER 88, 770 AND 880

IDENTIFICATION

Tractor serial number plate is located on the rear panel assembly.
Engine serial number is stamped on right rear flange of engine.

BUILT IN THESE VERSIONS:
Models 66, 77, 88, Super 77, Super 88, 770 and 880.
Single and dual wheel tricycle types, and adjustable and non-adjustable axle types.

Super 66:
Single and dual wheel tricycle types, and adjustable axle types.

IDENTIFICATION DATA

Series	Serial Nos.	Fuel	Cylinders Bore & Stroke
66 Row Crop	420001 UP	HC, CDF	4–3 5/16 x 3 3/4
		KD	4–3 1/2 x 3 3/4
66 Standard	470004 UP	HC, CDF	4–3 5/16 x 3 3/4
		KD	4–3 1/2 x 3 3/4
Super 66		HC, CDF	4–3 1/2 x 3 3/4
77 Row Crop	320001 UP	HC, LP-Gas	*6–3 5/16 x 3 3/4
		KD	6–3 1/2 x 3 3/4
		CDF	6–3 5/16 x 3 3/4
77 Standard	269001 UP	HC, LP-Gas	6–3 5/16 x 3 3/4
77 Standard	269001 Up	KD	6–3 1/2 x 3 3/4
		CDF	6–3 5/16 x 3 3/4
Super 77		HC, CDF, LP-Gas	6–3 1/2 x 3 3/4
770		HC, CDF, LP-Gas	6–3 1/2 x 3 3/4
88 Row Crop	120001 Up	HC, LP-Gas	6–3 1/2 x 4
		KD	6–3 3/4 x 4
		CDF	6–3 1/2 x 4
88 Standard	820001 Up	HC, LP-Gas	6–3 1/2 x 4
		KD	6–3 3/4 x 4
		CDF	6–3 1/2 x 4
Super 88		HC, CDF, LP-Gas	6–3 3/4 x 4
880		HC, CDF, LP-Gas	6–3 3/4 x 4

*Some of this group were originally equipped with gasoline burning engines having a 3 3/16 inches bore.

HC=Gasoline   CDF=Certified Diesel Fuel   KD= Distillate or tractor fuel   LP-Gas=Propane or propane butane

INDEX (By Starting Paragraph)

CONDENSED SERVICE DATA (For Non-Super Models)

GENERAL

Tractor Model	66HC	66KD	66D	77HC 77LPG	77KD	77D	88HC 88LPG	88KD	88D
Cylinders	4	4	4	6	6	6	6	6	6
Bore—Inches	3 5/16	3½	3 5/16	3 5/16	3½	3 5/16	3½	3¾	3½
Stroke—Inches	3¾	3¾	3¾	3¾	3¾	3¾	4	4	4
Displacement—Cubic Inches	129	144	129	194	216	194	231	265	231
Compression Ratio—Later Diesels	6.75	4.75	15.5	6.75	4.75	15.5	6.75	4.75	15.5
Compression Ratio—Early Diesels	——	——	15.0	——	——	15.0	——	——	15.0
Pistons Removed From:	Above	Above	Above	Above	Above	Above	Above	Above	Above
Main Bearings, Number of	3	3	3	4	4	4	4	4	4
Main Bearings Adjustable?	(1)	(1)	No	(1)	(1)	No	(1)	(1)	No
Rod Bearings Adjustable?	(1)	(1)	No	(1)	(1)	No	(1)	(1)	No
Cylinder Sleeves	Wet	Wet	Wet	Wet	Wet	Wet	Wet	Wet	Wet
Forward Speeds	(2)6	(2)6	(2)6	(2)6	(2)6	(2)6	(2)6	(2)6	(2)6
Forward Speeds—Optional	——	——	——	(3)4	(3)4	(3)4	(3)4	(3)4	(3)4
Generator and Starter Make	D-R	D-R	D-R	D-R	D-R	D-R	D-R	D-R	D-R

TUNE-UP

Tractor Model	66HC	66KD	66D	77HC 77LPG	77KD	77D	88HC 88LPG	88KD	88D
Firing Order	1, 2, 4, 3			1, 5, 3, 6, 2, 4					
Valve Tappet Gap—Inlet	.009C	.009C	.009C	.009C	.009C	.009C	.009C	.009C	.009C
Valve Tappet Gap—Exhaust	.016C	.016C	.016C	.016C	.016C	.016C	.016C	.016C	.016C
Inlet and Exhaust Valve Face Angle	45	45	45	45	45	45	45	45	45
Inlet and Exhaust Valve Seat Angle	45	45	45	45	45	45	45	45	45
Ignition Distributor Make	D-R	D-R	None	D-R	D-R	None	D-R	D-R	None
Ignition Distributor Model	1111704	1111704	——	1111702	1111702	——	1111702	1111702	——
Ignition Magneto Make	Wico	Wico	——	Wico	Wico	——	Wico	Wico	——
Ignition Magneto Model	XV4	XV4	——	XV6	XV6	——	XV6	XV6	——
Distributor Breaker Gap	.022	.022	——	.022	.022	——	.022	.022	——
Magneto Breaker Gap	.015	.015	——	.015	.015	——	.015	.015	——
Distributor Timing Retard Degrees	TDC	5°B	——	TDC	5°B	——	TDC	5°B	——
Distributor Timing Full Advance Degrees	22	27	——	22	27	——	22	27	——
Magneto Impulse Trip Point	5°A	5°A	——	5°A	5°A	——	5°A	5°A	——
Magneto Lag Angle Degrees	20	20	——	20	20	——	20	20	——
Magneto Full Advanced Timing Degrees	15	15	——	15	15	——	15	15	——
Flywheel Mark Indicating:									
Retarded Timing Distributor	TDC	½" B	——	TDC	½"B	——	TDC	½"B	——
Full Advanced Timing—Distributor	HC/IGN	KD/IGN	——	HC/IGN	KD/IGN	——	HC/IGN	KD/IGN	——
Full Advanced Timing—Magneto	Mag.	Mag.	——	Mag.	Mag.	——	Mag.	Mag.	——
Distributor Governor Advance	Refer to paragraph 529 in this manual								
Spark Plug Make	AC	AC	None	AC	AC	None	AC	AC	None
Model for Gasoline and LP-Gas	83S Com	85S Com	——	83S Com	85S Com	——	83S Com	85S Com	——
Electrode Gap	.025	.025	——	.025	.025	——	.025	.025	——
Carburetor Make, Model and Float	Refer to paragraph 460 in this manual								
Injector Bosch Model	——	——	PSB	——	——	PSB	——	——	PSB
Nozzle Bosch Model	——	——	AKB 605	——	——	AKB 605	——	——	AKB 605
Injector Timing Mark on Flywheel	——	——	FP or 23°	——	——	FP or 23°	——	——	FP or 23°
Nozzle Opening Pressure—PSI	——	——	1750	——	——	1750	——	——	1750
Engine Low Idle RPM—Battery & Diesel	375	375	675	375	375	675	375	375	675
Engine Low Idle RPM—Magneto Ignition	500	——	——	500	——	——	500	——	——
Engine High Idle RPM	1750	1750	1750	1750	1750	1750	1750	1750	1750
Engine Loaded RPM	1600	1600	1600	1600	1600	1600	1600	1600	1600
Belt Pulley No Load RPM	1079	1079	1079	1085	1085	1085	1085	1085	1085
PTO No Load RPM	582	582	582	582	582	582	582	582	582

SIZES-CAPACITIES-CLEARANCES
(Clearances in Thousandths)

Tractor Model	66HC	66KD	66D	77HC 77LPG	77KD	77D	88HC 88LPG	88KD	88D
Crankshaft Journal Diameter	2.2495	2.2495	2.2495	2.2495	2.2495	2.2495	2.6245	2.6245	2.6245
Crankpin Diameter	1.9995	1.9995	1.9995	1.9995	1.9995	1.9995	2.2495	2.2495	2.2495
Rod C to C Length	6.75	6.75	6.75	6.75	6.75	6.75	6.75	6.75	6.75
Camshaft Journal Diameter, Front	1.7495	1.7495	1.7495	1.7495	1.7495	1.7495	1.7495	1.7495	1.7495
Camshaft Journal Dia., 2nd, 3rd and 4th	1.7485	1.7485	1.7485	1.7485	1.7485	1.7485	1.7485	1.7485	1.7485
Piston Pin Diameter	1.2495	1.2495	1.2495	1.2495	1.2495	1.2495	1.2495	1.2495	1.2495
Valve Stem Diameter—Inlet	.372	.372	.372	.372	.372	.372	.372	.372	.372
Valve Stem Diameter—Exhaust	.371	.371	.371	.371	.371	.371	.371	.371	.371
Compression Ring Width	.124	.124	.124	.124	.124	.124	.124	.124	.124
Oil Ring Width	.249	.249	.249	.249	.249	.249	.249	.249	.249
Main Bearings, Diameter Clearance	.5-3.5	.5-3.5	.5-3.5	.5-3.5	.5-3.5	.5-3.5	.5-3.5	.5-3.5	.5-3.5
Rod Bearings, Diameter Clearance	.5-2.5	.5-2.5	.5-2.5	.5-2.5	.5-2.5	.5-2.5	.5-2.5	.5-2.5	.5-2.5
Piston Skirt Clearance	Refer to paragraph 443 in this manual								
Crankshaft End Play	4.5-10	4.5-10	4.5-10	4.5-10	4.5-10	4.5-10	4.5-10	4.5-10	4.5-10
Camshaft End Play	Controlled by spring pressure and thrust button								
Camshaft Bearing, Diameter Clearance	1.5-5	1.5-5	1.5-5	1.5-5	1.5-5	1.5-5	1.5-5	1.5-5	1.5-5
Cooling System—Gallons	3.5	3.5	3.5	4.5	4.5	4.5	4.5	4.5	4.5
Crankcase Oil—Quarts	4.0	4.0	4.0	5.0	5.0	5.0	6.0	6.0	6.0
Transmission and Differential—Quarts	18.0	18.0	18.0	18.0	18.0	18.0	18.0	18.0	18.0
Hydraulic System—Quarts	Refer to paragraph 604 in this manual								
PTO—Quarts	0.5	0.5	0.5	0.5	0.5	0.5	0.5	0.5	0.5

(1) Early production Non-Super models were provided with shims which only control crush of shell insert. Later and current production engines are equipped with non-adjustable shell inserts. (2) Two reverse speeds. (3) Four reverse speeds.

CONDENSED SERVICE DATA (Super, 770 & 880)

GENERAL

Tractor Model	Super 66HC	Super 66D	Super 77HC 770HC	Super 77D 770D	Super 88HC 880HC	Super 88D 880D
Cylinders	4	4	6	6	6	6
Bore—Inches	3½	3½	3½	3½	3¾	3¾
Stroke—Inches	3¾	3¾	3¾	3¾	4	4
Displacement—Cubic Inches	144	144	216	216	265	265
Compression Ratio (Prior to 1957)	7.0	15.5	7.0	15.5	7.0	15.5
Compression Ratio (After 1957)	7.3	15.5	7.3	15.5	7.3	15.5
Pistons Removed From:	Above	Above	Above	Above	Above	Above
Main Bearings, Number of	3	3	4	4	4	4
Main Bearings Adjustable?	No	No	No	No	No	No
Rod Bearings Adjustable?	No	No	No	No	No	No
Cylinder Sleeves	Wet	Wet	Wet	Wet	Wet	Wet
Forward Speeds	6	6	6	6	6	6

TUNE-UP

	Super 66HC	Super 66D	Super 77HC 770HC	Super 77D 770D	Super 88HC 880HC	Super 88D 880D
Firing Order	1, 2, 4, 3		1, 5, 3, 6, 2, 4			
Valve Tappet Gap—Inlet	.010C	.010C	.010C	.010C	.010C	.010C
Valve Tappet Gap—Exhaust	.016C	.020C	.016C	.020C	.016C	.020C
Inlet and Exhaust Valve Face and Seat Angle	45	45	Refer to 770 & 880 Supplement paragraph 1.			
Ignition Distributor Make	D-R	None	D-R	None	D-R	None
Ignition Distributor Model	1112555	—	1111750	—	1111750	—
Distributor Breaker Gap	.022	—	.022	—	.022	—
Distributor Timing Retard Degrees	TDC	—	TDC	—	TDC	—
Distributor Timing Full Advance Degrees	28	—	22	—	22	—
Flywheel Mark Indicating:						
Retard Timing Distributor	TDC	—	TDC	—	TDC	—
Full Advanced Timing—Distributor	HC/IGN	—	HC/IGN	—	HC/IGN	—
Spark Plug Electrode Gap	.025	—	.025	—	.025	—
Injector Bosch Model	—	PSB	—	PSB	—	PSB
Injector Timing Mark on Flywheel	—	FP or 26°	—	FP or 23°	—	FP or 23°
Nozzle Opening Pressure—PSI	—	1750	—	1750	—	1750
Engine Low Idle RPM—Battery and Diesel Ignition	375	675	375	675	375	675
Engine High Idle RPM	2200	2200	1750	1750	1750	1750
(770, 880)	—	—	1925	1925	1925	1925
Engine Loaded RPM	2000	2000	1600	1600	1600	1600
(770, 880)	—	—	1750	1750	1750	1750
Belt Pulley No Load RPM	1357	1357	1085	1085	1085	1085
(770, 880)	—	—	1193	1193	1193	1193
PTO No Load RPM	733	733	582	582	582	582
(770, 880)	—	—	597	597	597	597

SIZES-CAPACITIES-CLEARANCES

(Clearances in Thousandths)

	Super 66HC	Super 66D	Super 77HC 770HC	Super 77D 770D	Super 88HC 880HC	Super 88D 880D
Crankshaft Journal Diameter	2.2495	2.2495	2.2495	2.2495	2.6245	2.6245
Crankpin Diameter	2.250	2.250	2.250	2.250	2.4375	2.4375
Rod C to C Length	6.75	6.75	6.75	6.75	6.75	6.75
Camshaft Journal Diameter, Front	1.7495	1.7495	1.7495	1.7495	1.7495	1.7495
Camshaft Journal Diameter, 2nd, 3rd and 4th	1.7485	1.7485	1.7485	1.7485	1.7485	1.7485
Piston Pin Diameter	1.2495	1.2495	1.2495	1.2495	1.2495	1.2495
Valve Stem Diameter—Inlet	.372	.372	.372	.372	.372	.372
Valve Stem Diameter—Exhaust	.371	.371	.371	.371	.371	.371
Compression Ring Width	.124	.124	.124	.124	.124	.124
Oil Ring Width	.187 or .250	.187 or .250	.187 or .250	.187 or .250	.187 or .250	.187 or .250
Main Bearings, Diameter Clearance	.5-3.5	.5-3.5	.5-3.5	.5-3.5	.5-3.5	.5-3.5
Rod Bearings, Diameter Clearance	.5-2.5	.5-2.5	.5-2.5	.5-2.5	.5-2.5	.5-2.5
Piston Skirt Clearance	Refer to paragraph 443 in this manual					
Crankshaft End Play	2.5-10	2.5-10	2.5-10	2.5-10	2.5-10	2.5-10
Camshaft End Play	Controlled by spring pressure and thrust button					
Camshaft Clearance In Bushing—Front	1.5-3.5	1.5-3.5	1.5-3.5	1.5-3.5	1.5-3.5	1.5-3.5
Camshaft Clearance—In Unbushed Bores	2.5-6.5	2.5-6.5	2.5-6.5	2.5-6.5	2.5-6.5	2.5-6.5
Cooling System—Gallons	3.5	3.5	4.5	4.5	4.5	4.5
Crankcase Oil—Quarts	4.0	4.0	5.0	5.0	6.0	6.0
Transmission and Differential—Quarts	18.0	18.0	18.0	18.0	18.0	18.0
Hydraulic System—Quarts	Refer to paragraph 604 in this manual					
PTO—Quarts	0.5	0.5	0.5	0.5	0.5	0.5

FRONT SYSTEM (TRICYCLE TYPE)

MANUAL STEERING SYSTEM

The single and dual wheel versions are shown in Figs. O300, O301 and O303. On single wheel versions, the wheel fork is supported in two taper-roller bearings. On dual wheel versions, the vertical spindle is suported in a taper-roller bearing (13—Fig. O303) at the lower end and at the upper end by a bushing (33) which also controls the steering gear sector mesh adjustment.

The steering system, Fig. O300, consists of a worm and sector mounted in the front axle pedestal (support). The sector (19) is splined to the vertical spindle or wheel fork upper end.

401. **VERTICAL SPINDLE END PLAY.** On dual wheel models without power steering, adjust vertical spindle end play to 0.001-0.002 by adding or removing shims (25—Fig. O300) located between sector gear hub lower face and eccentric bushing (33). Remove the radiator grille, steering gear housing cover, and sector to gain access to the shims.

On single (fork mounted) wheel models without power steering, the vertical spindle shaft bearings are adjusted to zero end play by means of upper bearing cone retaining nut. Remove radiator grille and steering gear housing cover to gain access to retaining nut.

402. **WORMSHAFT.** Wormshaft bearings can be adjusted without disturbing the wormshaft and/or front support.

To adjust, disconnect front universal joint. Add or remove shims (S—Fig. O300) located between bearing retainer (21) and pedestal to eliminate all wormshaft bearing play but permitting shaft to rotate without binding.

To remove wormshaft assembly, disconnect front universal joint. Remove worm shaft bearing retainer cap screws; then, rotate wormshaft, to thread it out of mesh with sector and thus force worm shaft bearing cup (23) out of pedestal.

403. **GEAR MESH.** This adjustment applies to dual wheel models without power steering as the single (fork mounted) wheel models without

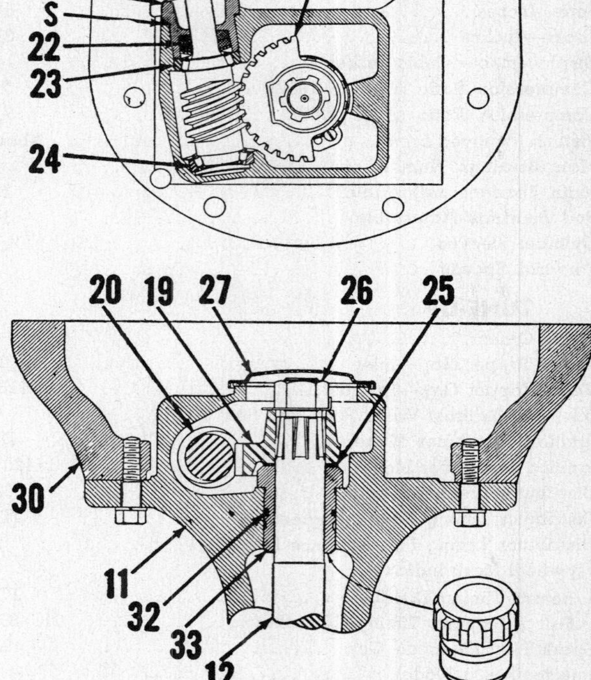

Fig. O300—Steering gear assembly used on dual wheel tricycle type models. Shims (S) control wormshaft bearings adjustment. Eccentric bushing (33) controls sector mesh adjustment. Oliver steering gear units used on adjustable axle models are similar.

11. Vertical spindle & steering gear housing
12. Vertical spindle
19. Sector
20. Wormshaft
21. Bearing retainer
22. Seals
23. Bearing cup
24. Bearing cup
25. Shims
26. Sector retaining nut
27. Gear housing cover
30. Main frame
32. Sealing "O" ring
33. Eccentric bushing

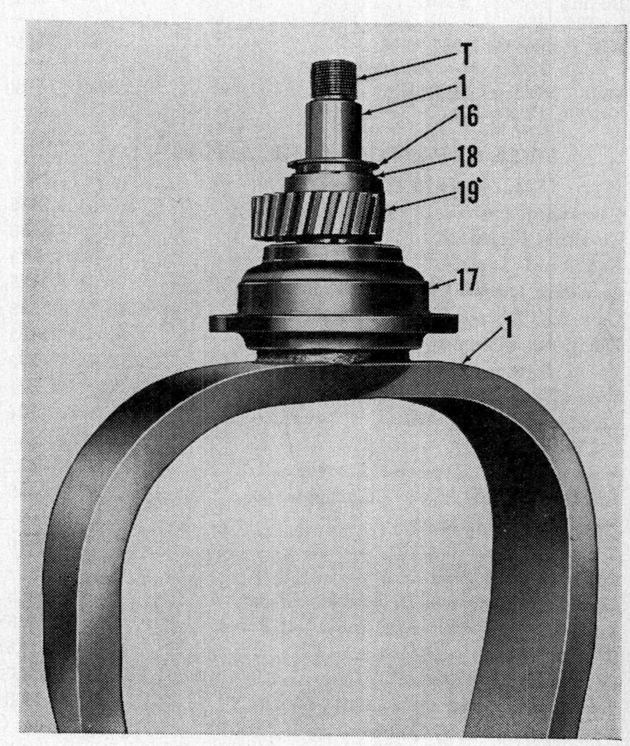

Fig. O301 — Wheel fork unit used on single wheel models. A nut on threads (T) locates the upper roller bearing and controls bearing adjustment. No sector mesh adjustment is provided for the steering gear.

1. Fork and spindle
16. Felt seal retainer
17. Bearing cage
18. Snap ring
19. Sector

power steering make no provision for gear mesh adjustment. The adjustment is controlled by the eccentric bushing (33).

To adjust, first remove radiator grille and gear housing cover. Turn steering wheel until wheels are in the straight ahead position. Working through gear housing cover opening, remove lock screw from eccentric bushing; then, rotate bushing in its bore until all backlash between sector and worm is removed. Temporarily lock bushing in this position; then, turn steering wheel from full left to full right. If gear does not bind when passing from right to left, tighten lock screw. If gear binds or drags heavily at any position, rotate bushing in its bore in an opposite direction, just enough to remove binding condition.

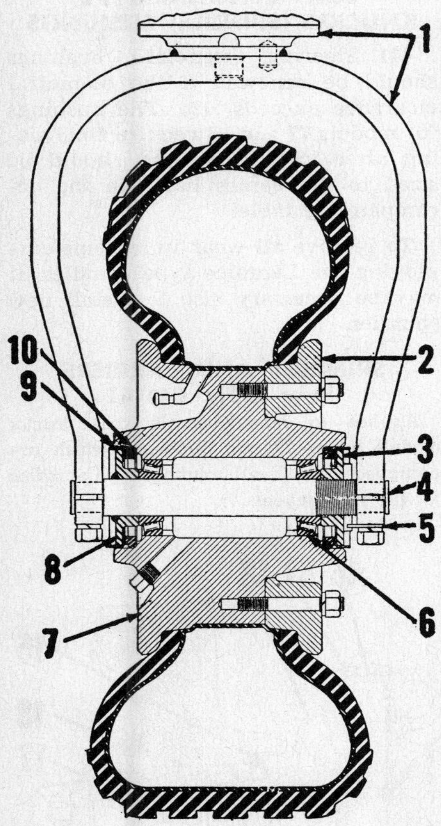

Fig. O302—Wheel fork and assembly as used on single wheel tricycle type models. Nut (3) controls wheel bearings adjustment.

1. Wheel fork & vertical spindle
2. Hub flange
3. Bearing adjusting nut
4. Wheel spindle
5. Nut lock
6. Bearing cup & cone
7. Wheel & hub
8. Bearing retainer (welded to spindle)
9. Seal cup
10. Seal

FRONT SUPPORT ASSEMBLY MANUAL TYPE STEERING

404. **WHEEL FORK R & R OR RENEW.** To remove wheel fork for renewal of bearing or seal, remove hood and radiator grille, and support front portion of tractor. Remove horizontal wheel axle and wheel assembly from fork. Remove steering gear housing cover, nut from threads (T—Fig. O301) and cap screws from lower bearing retainer (17). Bump fork spindle down and out of steering gear housing. At this time, lower bearing cup and cone, and oil seal can be renewed, after removing sector from spindle.

Adjust fork spindle bearings by means of upper nut to provide zero end play.

405. **VERTICAL SPINDLE R & R OR RENEW.** To R & R or renew the vertical spindle (12—Fig. O300 or O303), eccentric bushing (33) and/or lower bearing (13), first support the front end of tractor. Remove the hood, radiator grille, both wheel and hub units, and steering gear housing cover (27). Working through steering gear housing cover opening, remove sector retaining nut (26). Using a suitable puller attached to the lower end of the vertical spindle, pull spindle from sector and front support. Eccentric bushing, lower bearing and/or felt seal can be renewed at this time.

Renew eccentric bushing if same is out-of-round more than 0.006 and/or if it has more than 0.016 diametral clearance on vertical spindle.

Adjust vertical spindle bearing by adding or removing shims (25) located between sector gear hub and eccentric bushing to provide a 0.001-0.002 end play to vertical spindle.

406. **HORIZONTAL AXLE.** On models 77 and 88, the front wheels axle (horizontal spindle) is bolted to the vertical spindle as shown in Fig. O303. The axles can be removed or renewed by removing bolt (31) and bumping the axle in a downward direction and off the key. On models 66, the horizontal axle is in effect integral with the vertical spindle.

407. **SUPPORT ASSEMBLY R & R.** To remove front support as an assembly (assembly contains wheel fork or vertical spindle, and steering gear unit), proceed as follows: Support front of tractor and remove hood, radiator grille and radiator. Disconnect steering shaft front universal joint and remove cap screws retaining support to tractor frame. Raise front of tractor and remove support assembly.

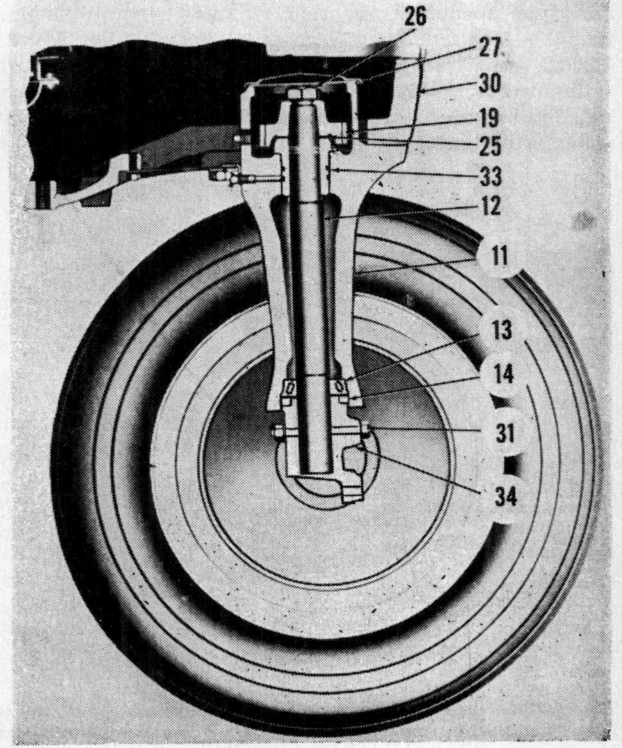

Fig. O303 — Non-power type steering system on dual wheel tricycle type tractors.

11. Vertical spindle & steering gear housing
12. Vertical spindle
13. Bearing cup & cone (lower)
14. Felt seal
19. Sector
25. Shims
26. Sector retaining nut
27. Gear housing cover
30. Main frame
31. Horizontal spindle retaining bolt
33. Eccentric bushing
34. Horizontal spindle

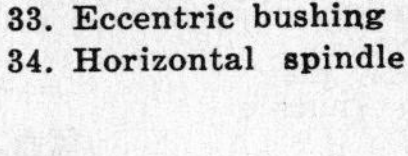

FRONT SYSTEM (Axle Type)

KNUCKLES

The non-adjustable type axles as used on models 66, 77 within the serial range of 269001-15970 and 88 within the serial range of 820001-10074 are equipped with reverse Elliot type steering knuckles, as shown in Figs. O304 and O305. Non-adjustable type axles used on models 77 after serial 15970, and 88 after serial 10074 are equipped with "live" (rotating) spindle for the front wheels as shown in Fig. O305A. The adjustable type axles as used on models 66, 77 and 88 are equipped with Lemoine type knuckles.

AXLE ASSEMBLY

408. The axle main member complete with wheels and knuckles can be removed as an assembly after disconnecting both tie rods from the steering gear sector arm, and the radius rod at the rear pivot. Raise and support the front end of the tractor; then roll the axle assembly rearward on non-adjustable types or forward on adjustable axle types and off the axle main member pivot pin.

408A. Check and adjust the front wheel toe-in of 3/16 inch by varying the length of the tie rods.

RADIUS ROD

409. On models 77 within the serial range of 320001-353582 and 88 within the serial range of 120001-147071 and all 66 adjustable and non-adjustable axle models, the radius rod pivots in a ball type socket at the rear. On adjustable axle models 77 after serial 353582, and 88 after serial 140171, the radius rod pivots in a renewable bushing. The bushing, which requires final sizing after installation to provide a

NON-ADJUSTABLE TYPE AXLE WITH REVERSE ELLIOT TYPE KNUCKLES

Models	66	77 & 88
Knuckle Pin Diameter	1.123	1.123
Knuckle Pin Bushing Inside Diameter	1.125 –1.126	1.125 –1.126
Pivot Pin Diameter	0.997 –1.000	1.245 –1.246
Pivot Pin Bushing Inside Diameter	None	1.247 –1.248

ADJUSTABLE TYPE AXLE

Models	66	77 & 88
Knuckle Pin Diameter	1.369 –1.370	1.497 –1.498
Knuckle Pin Bushing Inside Diameter	1.3735–1.3745	1.5015–1.5030
Pivot Pin Diameter	0.997 –1.000	1.497 –1.498
Pivot Pin Bushing Inside Diameter	None	1.5015–1.5030

.002-.003 diametral clearance, can be renewed after removing the pivot support bracket.

AXLE PIVOT PIN & BUSHING

410. Axle main member pivot pin bushings (41—Fig. O307) are used only on adjustable axle models 77 and 88, and non-adjustable axle models 77 and 88. Pivot pin bushings can be renewed after removing the axle main member as in paragraph 408. Bushings will require final sizing after installation to the values as listed in the accompanying table to provide a clearance of 0.001-0.003 or 0.0035-0.006 as listed.

410A. To renew the axle main member pivot pin, first remove the tractor grille. Working through the top of the axle bolster (support), remove the pivot pin retaining pin (39) by drilling off the riveted portion and bumping the retaining pin downward. Bump the pivot pin rearward and out of the axle bolster (support).

CONVENTIONAL TYPE KNUCKLE (SPINDLE) BUSHINGS

411. Steering knuckle bushings should be renewed if the diametral clearance exceeds .020. The bushings for models 77 and 88 require final sizing after installation and should be sized to the values listed in the accompanying table.

To remove all wear in systems employing the Lemoine type spindles, it may be necessary also to install new spindles.

SPINDLES AND CARRIERS
(Refer to Fig. O305A)

Applies to non-adjustable axle tractor models Super 77 and Super 88 which are equipped with "live" (rotating) type spindles for the front wheels.

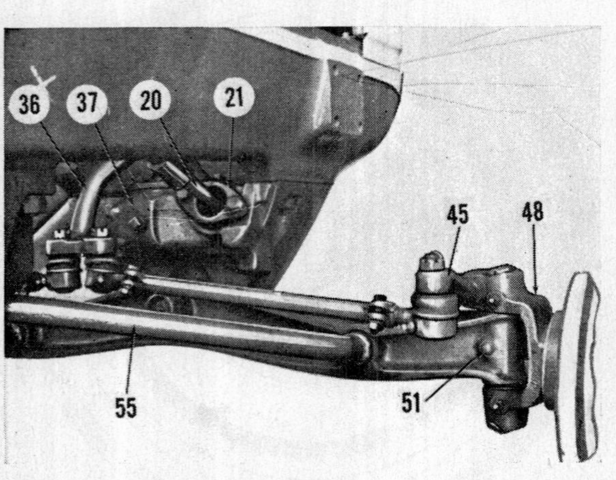

15. Seal
17. Bearing cup & cone (outer)
18. Bearing cup & cone (inner)
20. Wormshaft
21. Bearing retainer
36. Sector shaft arm
37. Axle bolster & gear housing
45. Steering knuckle arm
46. Cotter key
47. Expansion plug
48. Steering knuckle
49. Bushing
50. Thrust washers
51. Knuckle pin retaining pin
55. Radius rod

Fig. O304—Non-adjustable axle main member, and steering gear installation on model 88. Other non-adjustable axle models are similar.

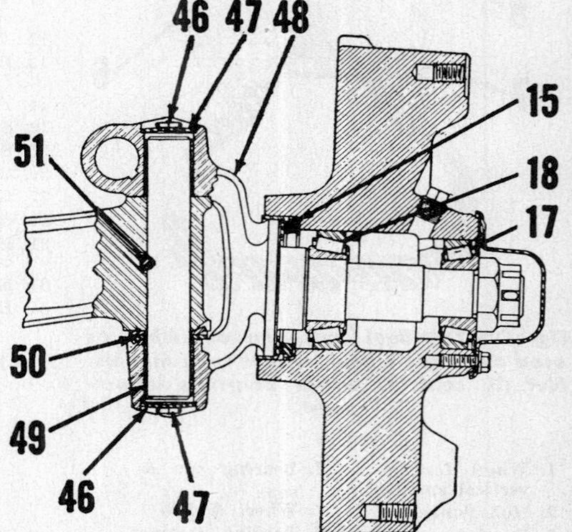

Fig. O305—Reverse Elliot type steering knuckle as used on non-adjustable axle models.

412. **SPINDLES.** Front wheels are carried on "live" (rotating) spindles which revolve in tapered roller bearings mounted in the spindle carriers (21—Fig. O305A). The spindle carriers are supported by stationary pivots (33 and 35), anchored in the axle yokes and provided with needle type roller bearings (65) and separate thrust bearings (66).

412A. Procedure for adjustment or renewal of front wheel bearings is conventional except that the knuckle arms (22) must be removed to obtain access to spindle nuts. Cups and cones may be renewed without removing the spindle carrier from the axle.

413. **SPINDLE CARRIERS.** To remove or renew the spindle carrier pivot pins, proceed as follows: Remove front wheels and bump out the Roll pins (34—Fig. O305A) which retain the spindle carrier pivot pins in the axle yoke. Disconnect knuckle arm

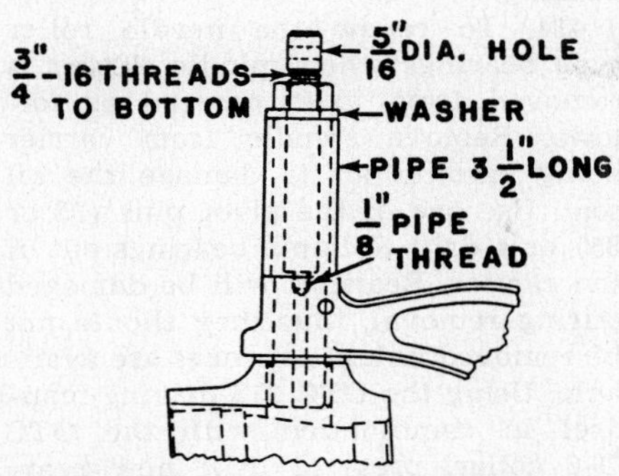

Fig. O305B—Using a screw, nut and 3½-inch length of pipe to extract pivot pins from "live" spindle type front axle. Pins have ⅛-inch pipe threads.

(22 or 25) from tie rod, or remove arm from carrier. Remove grease fittings (4) from pivot pins (33 and 35), and extract the pivot pins using a screw, nut and pipe as a puller as shown in Fig. O305B. A slide hammer type of puller, attached to a screw or rod having a ⅛ inch pipe thread, may be used also to remove the pivot pins.

The longer pivot pin installed in the top of the axle should engage the shorter pivot boss on the spindle carrier.

Spindle bearings are lubricated with ½ pint of engine oil which is introduced through the pipe plug opening in the carrier.

To renew the needle type bearings for the pivot pins, refer to paragraphs 413B and 414.

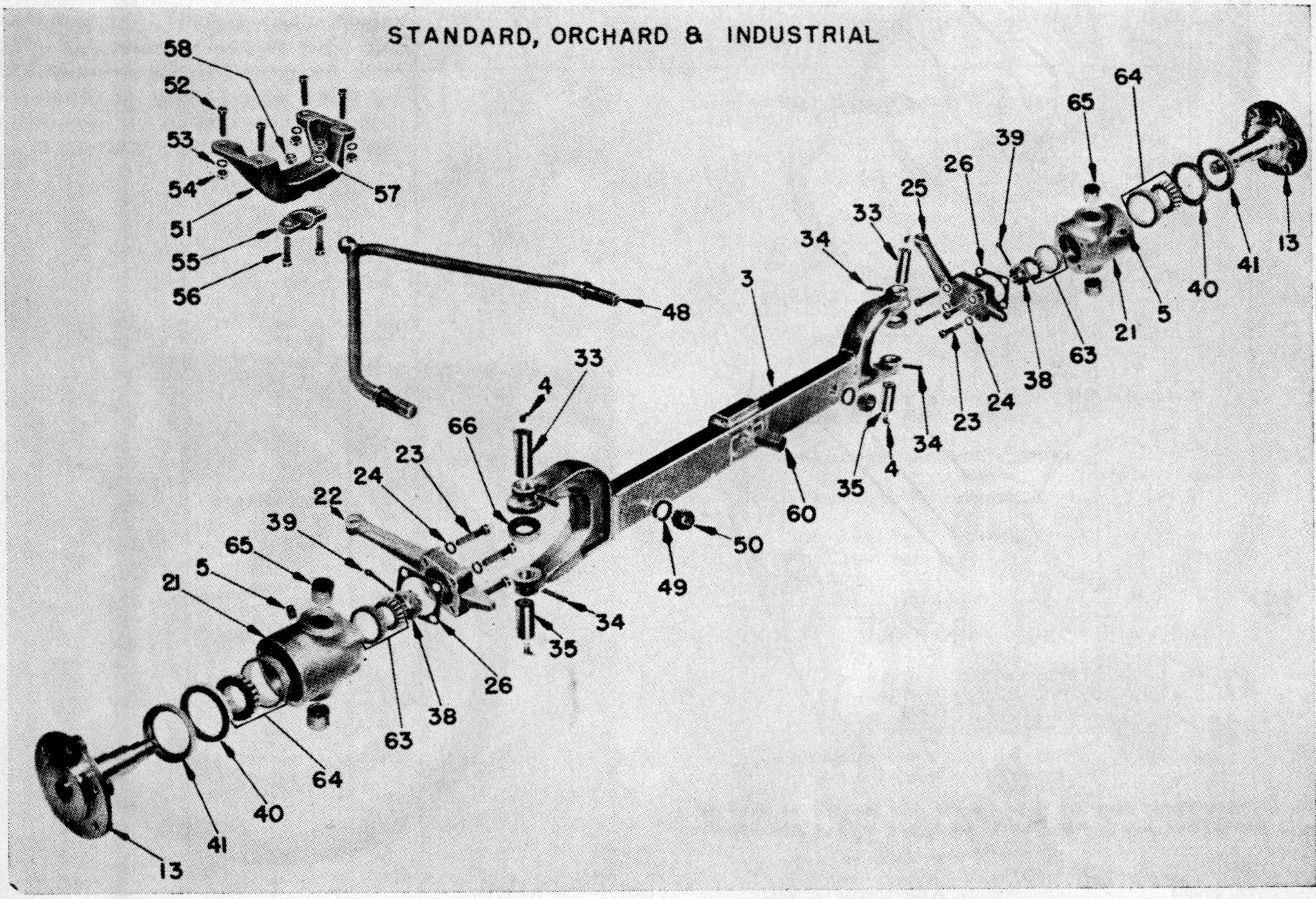

STANDARD, ORCHARD & INDUSTRIAL

Fig. O305A—Live spindle type front axle as used on series 77 after serial 15971, series 88 after serial 10075 also 770 and 880. Front wheel attaches to rotating (live) spindle (13).

3. Axle with bushing
13. Live spindle
21. Spindle carrier
22. Steering knuckle arm
25. Steering knuckle arm
33. Carrier pivot, upper
34. Roll pin
35. Carrier pivot, lower
40. Wheel seal
41. Seal wearing cup
48. Axle stay rod
55. Ball socket cap
51. Rod ball socket
60. Axle pivot bushing
63. Spindle bearing, inner
64. Spindle bearing, outer
66. Roller thrust bearing

413A. **SEALS, WEAR CUPS & WHEEL BEARINGS.** These parts can be renewed after removing the wheel spindles from the carriers.

It is important that the wearing surface of seal cup (41—Fig. O305A) be smooth and square with seal. Use crocus cloth or very fine sandpaper to remove slight nicks or foreign matter from seal cup. Apply a thin coat of shellac or gasket cement to outer edge of metal portion of each grease seal, and to the inner surface of the seal cup.

413B. **PIVOT PIN NEEDLE BEARINGS.** These bearings are of the closed end type, and the Oliver Corp. recommends the use of an Owatonna Driving Mandrel No. 815 on which is placed the Owatonna No. O-6 Driving Collar so as to prevent damaging the bearing. The recommended driver and collar combination applies pressure simultaneously to top and sides of bearing.

414. To renew the needle roller type bearings when spindle carrier is removed from axle, proceed as follows: Remove spindle from carrier being careful not to damage the oil seal. Use one of the pivot pins (33 or 35) or a drift to bump bearings out of the carrier. Bearings will be damaged during removal, thus they should not be removed unless new ones are available. Using the OTC 815 driving mandrel in combination with the OTC O-6 collar, press or drift new bearings into carrier until top of each bearing is flush with carrier.

MANUAL STEERING GEAR

Axle Type Tractors

Early production non Super models 66, 77 and 88 non-adjustable and adjustable axle models are equipped with a worm and sector type Oliver gear which is housed in front bolster (axle support). Later production non Super and all Super axle models are equipped with a Saginaw recirculating ball nut type steering gear which is contained in a separate housing and mounted on top of the front bolster (support).

Oliver field conversion kits are available for converting the Oliver gear installations to the Saginaw gear type. When making this change, it will be necessary, however, to rework the front frame on all models except the 66 adjustable axle model.

Models With Oliver Gear

(See Paragraph 421 for Saginaw)

415. **WORMSHAFT.** Wormshaft bearings on all axle types can be adjusted without disturbing wormshaft and/or front support. To adjust, disconnect steering shaft front universal joint. Add or remove shims (S—Fig. O306) between bearing retainer (21) and front axle bolster to eliminate all worm shaft end play but permitting shaft to rotate without binding.

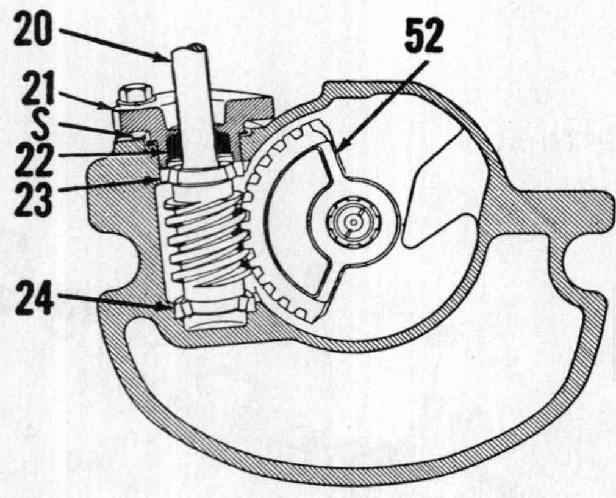

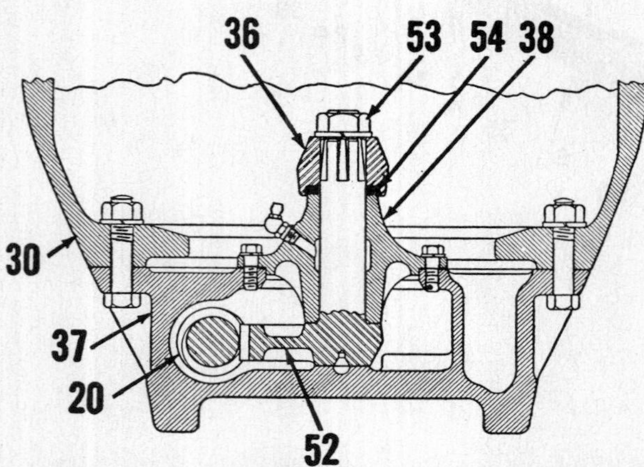

Fig. O306—Oliver steering gear (worm and sector) as used on early production non-Super models equipped with a non-adjustable axle.

S. Shims
20. Wormshaft
21. Bearing retainer
22. Seals
23. Bearing cup
24. Bearing cup
30. Main frame
36. Sector shaft arm
37. Axle bolster & gear housing
38. Sector shaft bearing & housing cover
52. Sector
53. Steering gear arm retaining nut
54. Felt seal

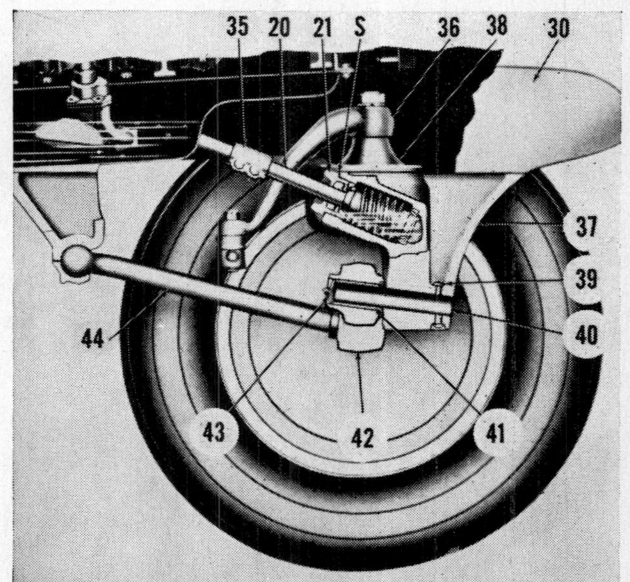

Fig. O307—Non-adjustable axle models, showing the axle assembly and Oliver steering gear unit.

S. Shims
20. Wormshaft
21. Bearing retainer
30. Main frame
35. Steering shaft coupling
36. Sector shaft arm
37. Axle bolster & gear housing
38. Sector shaft bearing & housing cover
39. Pivot pin retaining pin
40. Axle pivot pin
41. Bushing
42. Axle main member
43. Expansion plug
44. Radius rod

416. **GEAR MESH (NON-ADJUSTABLE AXLE MODELS).** This adjustment applies to non-adjustable axle type tractor models. Sector mesh adjustment is controlled by the gear housing cover (38—Fig. O306, O307 or O309), the shaft bore of which is located slightly off-center to the cover bolt hole circle. To adjust, turn steering wheel until front wheels are in the straight ahead position. Remove hood, grille and cap screws from gear housing cover (38), and rotate cover until all backlash between sector and worm is removed. Temporarily lock cover in this position; then, rotate steering wheel from full left to full right. If gear binds or drags heavily at any position, rotate cover in opposite direction just enough to remove binding condition, and reinstall the cover cap screws. Cap screws will enter different cover holes than previously.

417. **GEAR MESH (ADJUSTABLE AXLE MODELS).** This adjustment applies to adjustable axle type models. Sector gear mesh adjustment is controlled by an eccentric bushing (33—Fig. O308) located beneath the sector gear. To adjust, first remove radiator grille and gear housing cover. Turn steering wheel until front wheels are in straight ahead position. Remove lock screw from eccentric bushing; then, rotate bushing until all backlash between sector and worm is removed. Temporarily lock bushing in this position; then, rotate steering wheel from full left to full right. If gear binds or drags heavily at any position, rotate bushing in an opposite direction just enough to remove binding condition.

418. **SECTOR END PLAY.** The adjustable axle models have an end play adjustment which is controlled by a steel ball and adjusting screw located in gear housing cover. Adjust to zero end play of sector shaft by rotating the screw.

419. **OVERHAUL GEAR ASSEMBLY (NON-ADJUSTABLE AXLE).** Gear unit can be overhauled without removing steering gear housing (axle bolster) from tractor.

To remove wormshaft (20—Fig. O306), disconnect steering shaft sleeve coupling. Remove cap screws from wormshaft bearing retainer (21). Rotate wormshaft to thread worm out of mesh with sector and thus force rear bearing cup (23) out of housing. Examine roller bearing contacting portions of worm, and renew worm if surfaces are pitted, or worn. Method of renewal of wormshaft bearing cups and cones is self-evident. Adjust wormshaft bearings to zero end play with shims (S).

To remove sector and shaft (52), first remove hood, radiator grille and radiator. Disconnect tie rods from steering sector arm. Using a suitable puller, remove steering sector arm. Unbolt and remove steering gear case cover (cover also controls sector mesh adjustment) from housing. Correct excessive looseness of sector shaft bearing by renewing the unbushed cover (38) and/or sector and sector shaft (52). Adjust sector mesh as outlined in paragraph 416.

420. **OVERHAUL GEAR ASSEMBLY (ADJUSTABLE AXLE).** First disconnect steering shaft sleeve coupling. Remove cap screws from wormshaft bearing retainer. Rotate wormshaft to thread worm out of mesh with sector, and thus force rear bearing cup out of housing. If roller bearing contacting surfaces of worm are pitted or worn, renew the worm. Method of renewal of bearing cups and cones is self evident. Adjust worm shaft bearings to zero end play with shims (S—Fig. O308).

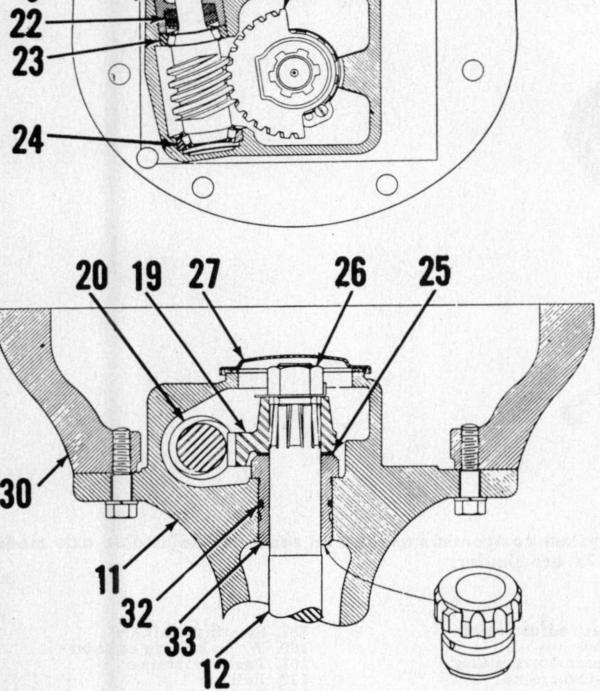

Fig. O308—Steering gear assembly typical of that used on adjustable axle models. Shims (S) control worm shaft bearings adjustment. Eccentric bushing (33) controls sector mesh adjustment.

Fig. O309—Worm and sector gear (Oliver) installation on non-adjustable axle models. Shown is non-Super model 88. Others non-Super models are similar.

20. Wormshaft
21. Bearing retainer
36. Sector shaft arm
37. Bolster and gear housing
38. Sector shaft bearing and housing cover

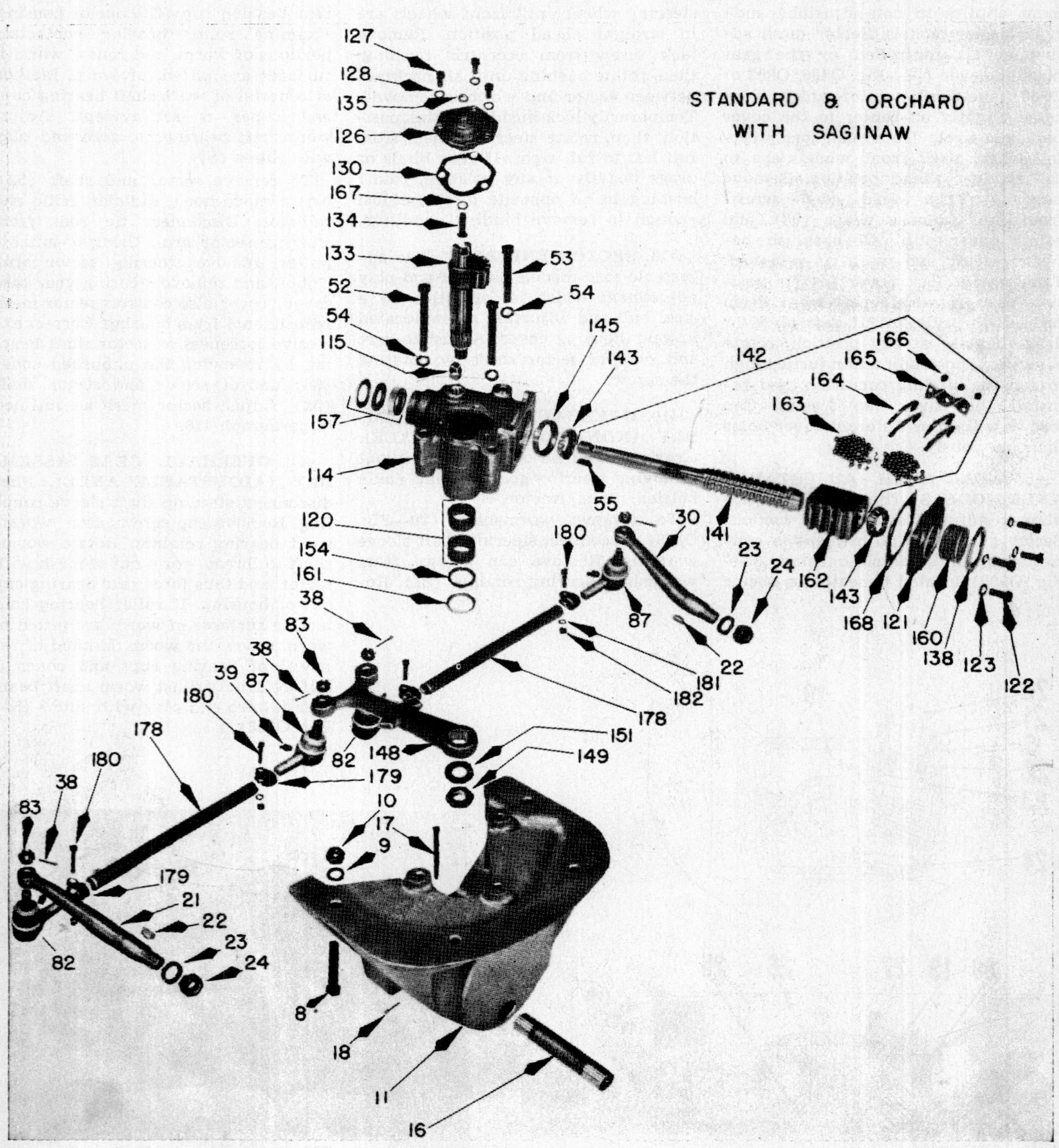

Fig. O310—Saginaw steering gear (recirculating ball type) and steering system components as used on some non-adjustable axle model 88 tractors. Models 66 and 77 are similar.

11. Front axle bolster
16. Axle pivot pin
17. Retaining pin
18. Cotter pin
21. Steering knuckle arm
22. Woodruff key
30. Steering knuckle arm
55. Woodruff key
82. Tie rod socket
87. Tie rod socket
114. Gear housing
120. Bushings
121. Housing cover
126. Gear housing cover
130. Gasket
133. Shaft and gear
134. Lash adjuster
135. Lash adjuster nut
138. Lock nut
141. Shaft & worm assy.
143. Bearing cone
145. Bearing cup
148. Pitman arm
151. Lock washer
154. Packing
157. Steering shaft seal
160. Worm bearing adjuster
161. Packing retainer
162. Ball nut
164. Ball guide
165. Guide clamp
167. Shims
168. Gasket
178. Tie rod (intermediate)

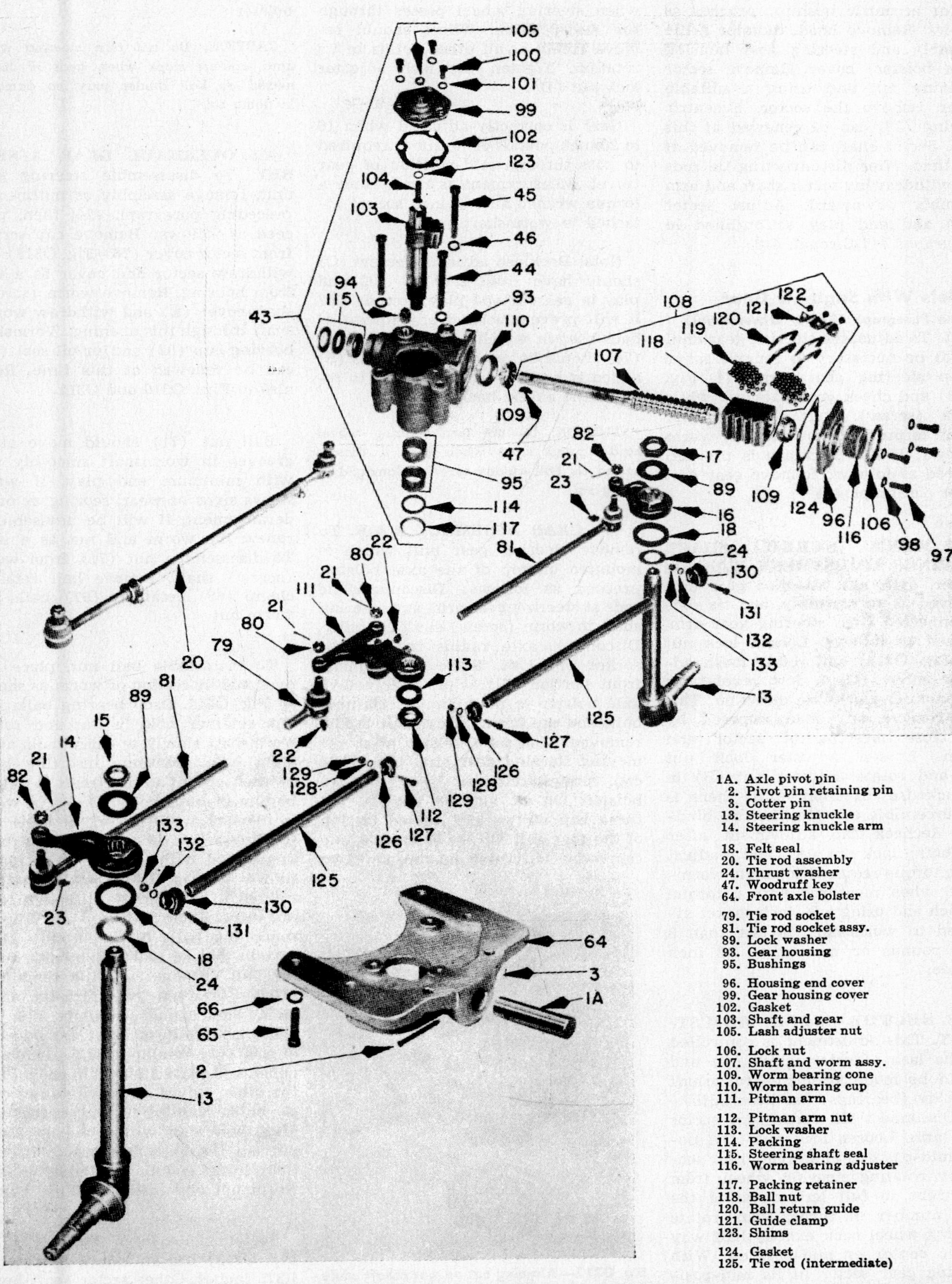

Fig. O311—Saginaw steering gear (recirculating ball type) and steering system components as used on some adjustable axle model 77 tractors. Models 66 and 88 are similar.

To remove sector, sector shaft and/or eccentric bushing, proceed as follows: Remove hood, radiator grille assembly and steering gear housing (axle bolster) cover. Remove sector retaining nut and using a suitable puller remove the sector. Eccentric bushing (33) can be renewed at this time. Sector shaft can be removed at this time, after disconnecting tie rods and withdrawing sector shaft and arm assembly downward. Adjust sector mesh and end play as outlined in paragraphs 417 through 418.

Models With Saginaw Gear

(See Paragraph 415 for Oliver Gear)

421. To adjust the steering gear unit, it will be necessary to remove same. Grasp steering gear arm (74—Fig. O313) and check gear backlash while worm (screw) shaft (68) is held steady to prevent movement of worm (screw). If any backlash is present, proceed as follows: Remove gear unit as per paragraph 424.

422. **WORM (SCREW) SHAFT BEARINGS ADJUSTMENT.** With axle bolster, axle, and attached gear unit removed as an assembly, and tie rods disconnected from steering gear arm, proceed as follows: Loosen lock nut (D—Fig. O313) and rotate mesh adjuster screw (C) a few revolutions in a counter-clockwise direction. This will remove any load imposed by the close meshing of sector and worm. Loosen adjuster lock nut (A) and rotate adjuster nut (B) in a clockwise direction, until there is no perceptible end play and no binding. Recheck the adjustment after tightening lock nut (A), and readjust if the torque required to rotate wormshaft, when measured with a torque wrench and using a ⅞ inch socket attached to wormshaft, is less than 8 inch pounds or more than 12 inch pounds.

423. **SECTOR MESH ADJUSTMENT.** This adjustment is controlled by the lash adjuster screw (C) and should be made after the wormshaft end play (bearings) has been adjusted. Disconnect tie rods from sector shaft arm. Loosen lock nut (D). Locate mid-position of steering gear sector by rotating steering wheel from full right to full left, counting the total number of turns; then, rotate steering wheel back exactly half way to the center or mid-position. With steering gear sector in its mid-position of travel, rotate lash adjuster (C) in a clockwise direction until a very slight drag (zero backlash) is felt only when steering wheel passes through the mid-position. Wheel should revolve freely at all other points in its rotation. Tighten the lash adjuster lock nut (D).

Gear is correctly adjusted when 16 to 20 inch pounds of torque is required to pass through mid-position of gear travel. Measurement is made with a torque wrench and ⅞ inch socket attached to wormshaft.

Note: Backlash adjusting screw (C) should have from zero to .002 end play in gear. If end play exceeds .002 it will prevent correct adjustment of backlash; in which case, sector cover (76) should be removed and shims (S) added at head of adjuster screw to remove the excess backlash.

CAUTION: Do not turn steering wheel hard against stops when gear is disconnected as ball guides may be damaged in doing so.

424. **GEAR ASSEMBLY R & R.** To remove steering gear unit which is mounted on top of the axle bolster, proceed as follows: Disconnect tie rods at steering gear arm, and steering shaft to worm (screw) shaft coupling. Disconnect axle radius rod at rear socket or pivot. Raise and support front portion of tractor, and remove axle bolster to main frame retaining bolts and cap screws. Gear unit can be removed from axle bolster after removing steering gear arm, and three cap screws retaining gear unit to bolster. On 77 and 88 models, the three cap screws are located on top of the gear unit. On the 66 models, one cap screw is located on top, and two are on the lower side of the axle bolster.

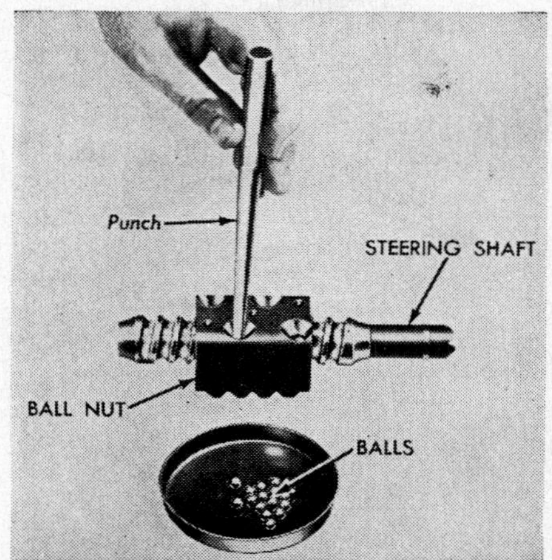

Fig. O312—Aligning nut on wormshaft while inserting balls in ball circuit. Insert one-half of the total number of balls in each circuit and guide.

CAUTION: Do not turn steering wheel hard against stops when gear is disconnected as ball guides may be damaged in doing so.

425. **OVERHAUL GEAR ASSEMBLY.** To disassemble steering gear unit, remove assembly as outlined in preceding paragraph 424; then, proceed as follows: Remove cap screws from sector cover (76—Fig. O313) and withdraw sector and cover as a unit from housing. Remove worm (screw) shaft cover (E) and withdraw wormshaft through this opening. Wormshaft bearing cup (62) and/or oil seal (61) can be renewed at this time. Refer also to Figs. O310 and O311.

Ball nut (71) should move along grooves in wormshaft smoothly and with minimum end play. If worm shows signs of wear, scoring or other derangement, it will be advisable to renew the worm and nut as a unit. To disassemble nut (71) from worm (screw) shaft, remove ball retainer clamp (66), retainers (67), balls and worm nut.

To reassemble ball nut, place nut over middle section of worm as shown in Fig. O312. Drop bearing balls into one retainer hole in nut and rotate wormshaft slowly to carry balls away from hole. Continue inserting balls in each circuit until circuit is full to bottom of both holes. If end of worm is reached while inserting balls and rotating worm in nut, hold the balls in position with a clean blunt rod as shown in Fig. O312 while shaft is rotated in an opposite direction for a few turns. Remove the rod and drop the remaining balls in the circuit. Make certain that no balls are outside regular ball circuits. If balls remain in groove between two circuits or at ends, they cannot circulate and will cause gear failure. Next, lay one-half of each split retainer, Fig. O314, on the bench and place 13 balls in each. Place the other halves of each retainer over the balls. Holding the halves together, plug their ends with heavy grease to prevent the balls from dropping out; then, insert complete retainer units in worm nut and install retainer clamp.

Sector shaft large bushing (72—Fig. O313) has an inside diameter of 1.375 inches. Other sector shaft bushing (75) has an inside diameter of 1.0625 inches. I&T suggested clearance of shaft in bushings is .0015-.003.

Select and insert shims (S) on sector mesh adjusting screw (C) to provide zero to .002 end play, before reinstalling sector and shaft in gear housing. Adjust worm (screw) shaft bearings and sector mesh as outlined in paragraphs 422 and 423.

CAUTION: Do not turn steering wheel hard against stops when gear is disconnected as ball guides may be damaged in doing so.

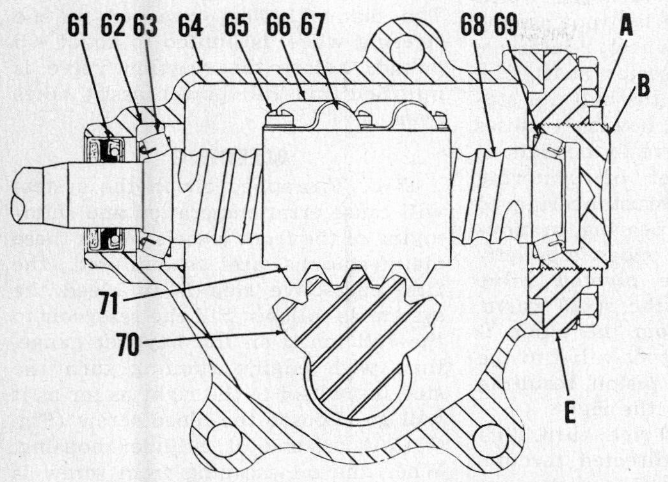

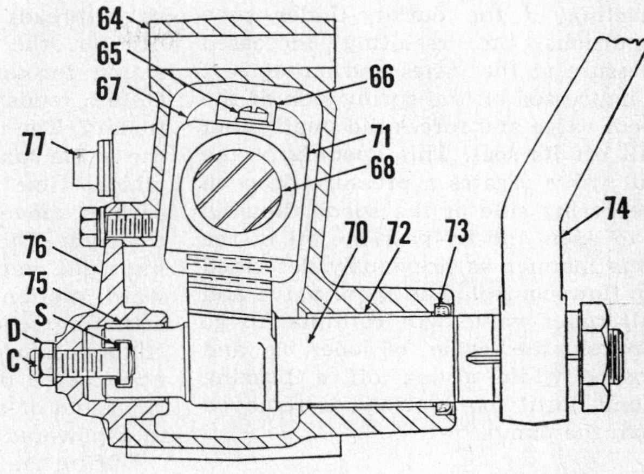

Fig. O313—Saginaw gear (recirculating ball type) used on some adjustable and non-adjustable axle models.

A. Lock nut
B. Wormshaft bearing adjuster nut
C. Lash adjuster
D. Lock nut
E. Wormshaft cover
S. Shim washers
61. Seal
62. Bearing cup
63. Bearing cone
64. Gear housing
65. Clamp retaining screw
66. Retainer clamp
67. Retainer
68. Wormshaft
69. Bearing cone
70. Sector & shaft
71. Ball nut
72. Bushing (large)
73. Seal
74. Sector shaft arm
75. Bushing (small)
76. Sector cover
77. Lubricant filler plug

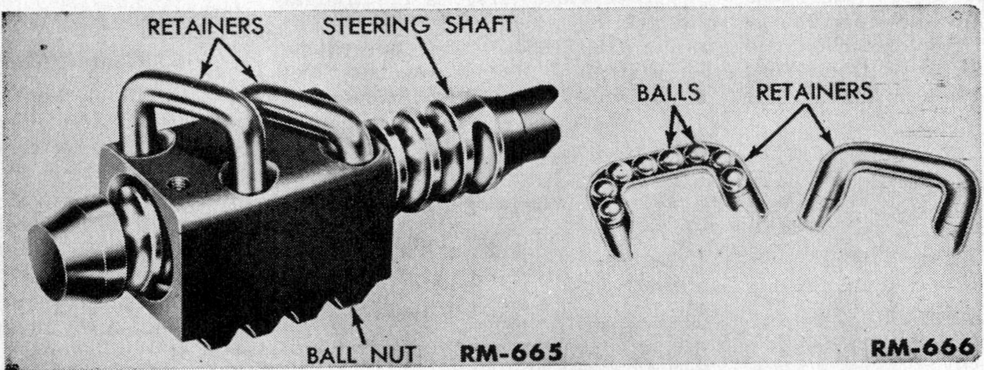

Fig. O314—Saginaw gear (recirculating ball type) showing ball nut and ball circuit retainers.

POWER STEERING

Super 77, 88, 770, 880

Power steering is an available option on these models. The power steering of Gemmer design utilizes a Vickers vane type pump as the energy source for the combination piston and rack and sector gear unit. Steering wheel effort required to actuate the "feel" type valving is 4-5 pounds. The steering pump is mounted on the end of the electric generator and is directly driven by the armature shaft.

PUMP OPERATION

425A. The flow of oil through the vane type pump is briefly as follows: Refer to Fig. O315. Oil returning from the steering gear at low pressure enters the inlet port and flows to the pump oil reservoir via a hole in the pump ring and the filter. Oil enters the pump from the reservoir through two drilled holes in the side of the cover and is drawn into the pumping chamber at both the body and the pressure plate sides through oppositely located passages in the rotor.

The pressurized oil leaves the pump chamber by two routes. One path is diverted through the small pressure sensing port to the pump outlet and steering gear and to the back of the flow control valve. The other path is to the inner end of the flow control valve to open the valve, and to the pressure plate chamber to seal the plate and assist in forcing the rotor vanes outwardly. Both the quantity of oil and its pressure are controlled by the spool type flow valve and the ball type relief valve contained inside it.

The need for control of the quantity of oil delivered arises when the pump is revolving at very high speeds. When the oil delivery quantity increases beyond a certain amount the oil pressure on the spring (back) side of the flow control valve spool decreases due to the increased pressure drop through the restriction (marked "pressure drop" in illustration) at the pump outlet. The higher pressure on the vane side of the spool then overcomes the spring and moves the spool to the left, thus by-passing the excess quantity of oil back to the reservoir as shown in the inset of the illustration. The valve will stay open until the pump rpm decreases to the designed maximum.

Control of pump maximum pressure is needed in case of an overload condition in the steering gear or a restricting of the outlet. Under such conditions the resulting increased pressure at the vanes and outlet port is impressed on the spring side of the spool valve and forces the small relief ball off its seat. This opening of the ball valve creates a pressure drop on the spring side of the spool allowing it to open and by-pass the oil in the same manner as previously described for flow control. The spool valve and ball relief valve will continue to go through the cycle of opening and closing while giving off a buzzing sound, until the overload is removed from the pump.

GEAR UNIT OPERATION

425B. Transfer of the manual effort of turning the steering wheel (to right or left) into power operation of the gear is controlled by the 8 centering springs built into the worm thrust bearing. The thrust bearing is shown at the left end of the two drawings in Fig. O315B and the centering springs are titled "riveted spring" and "free spring" respectively. One set of 4 springs is compressed during a right turn and the other set of four is compressed during a left turn. In neutral the spring pressure is balanced.

In making a right turn the worm shaft threads into the ball nut assembly but the front wheel resistance, acting through the sector shaft Fig. O315B, tends to hold the ball nut stationary. The reaction, however, causes the worm shaft to move to the right at which time one set of centering springs moves the thrust bearing to the right. The thrust bearing, in moving right, imparts an opposite or leftward motion to the pivoted valve actuating lever and the spool valve. High pressure oil from the pump is now routed by the spool valve to the right end of the rack piston, resulting in a powered turn to the right.

During the powered right turn, high pressure oil is also directed through the reaction control valve to the reaction chamber within the spool valve. Increasing pressure in this chamber assists the centering spring in trying to neutralize the spool valve and shut off the power. To prevent this occurence more effort is required at the steering wheel and it is this opposing pressure in the reaction chamber that gives the operator the "feel of the wheel" to avoid over-steering.

Operation of the gear in a left turn is similar except that the opposing set of valve centering springs activates the valve and rack piston which will travel in the opposite direction. The manual effort required at the steering wheel is limited to about 4-5 pounds before the reaction valve is nullified and full power assist takes over.

BLEEDING

425C. Entrapped air in the system will cause erratic operation and shimmying of the front wheels. When these manifestations are encountered, the first corrective step is to bleed the system as follows: Fill the reservoir to the full mark on the bayonet gauge, then with engine running turn the steering wheel to the right as far as it will go. Loosen the bleed screw (Fig. O315A) on end of cylinder housing. When the oil escaping from screw is free of air bubbles, tighten the screw and refill the reservoir.

Bleeding can be accomplished without removing the radiator by reaching in from behind. In doing so be very careful so as to avoid injury from contact with the fan drive pulley.

SYSTEM PRESSURE TEST

425D. If steering performance is not normal, first bleed the system. If bleeding does not correct the trouble, check the system pressure to determine the location of the trouble. To make the test disconnect the high pressure line at the oil pump and install a pressure gauge capable of covering the pressure range of 800-1100 psi. Install also, on the down stream side of the gauge, a shut-off valve of needle or globe type as shown in Fig. O315D.

With engine and oil warm and running at low idle and shut-off valve

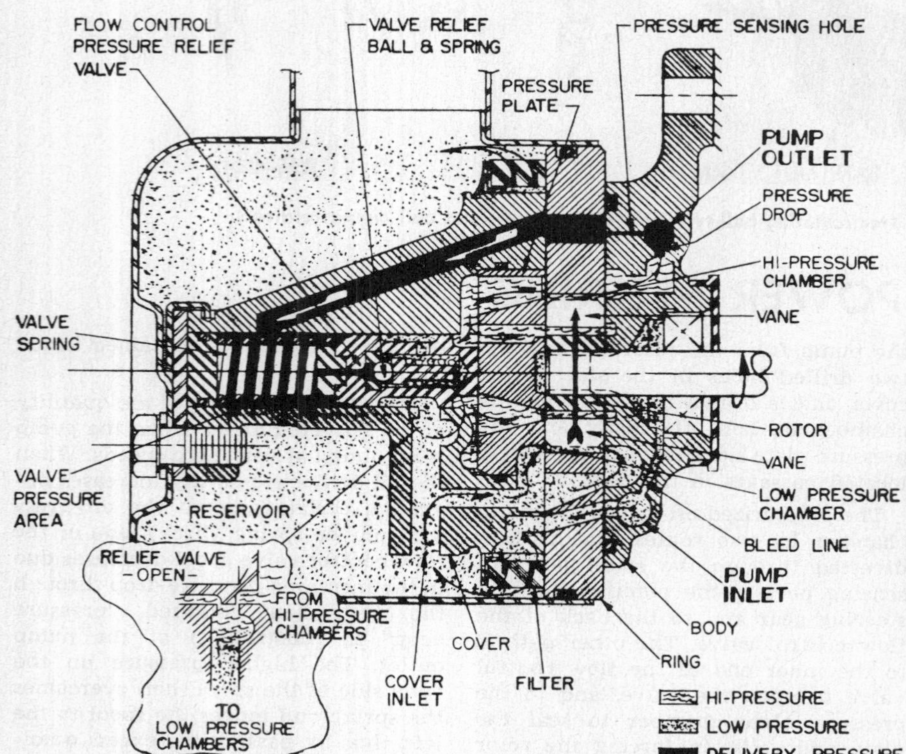

Fig. O315—Section through Vickers vane type pump used with power steering gear unit. Pump is equipped with a combination flow control and pressure relief valve.

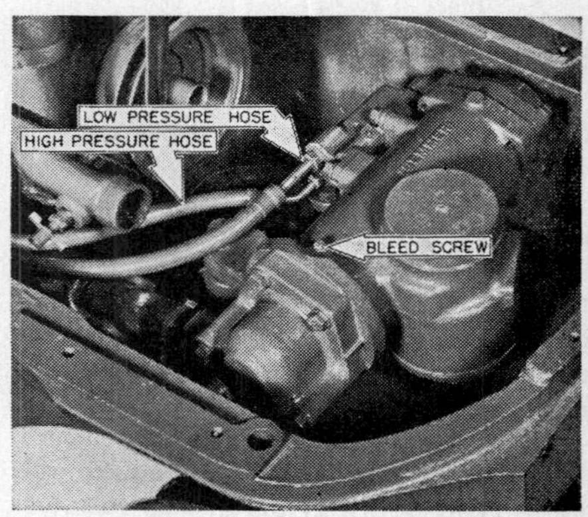

Fig. O315A — Location of bleed screw. Screw can be reached from behind without removing radiator but care must be exercised to avoid injury from contact with revolving fan drive pulley.

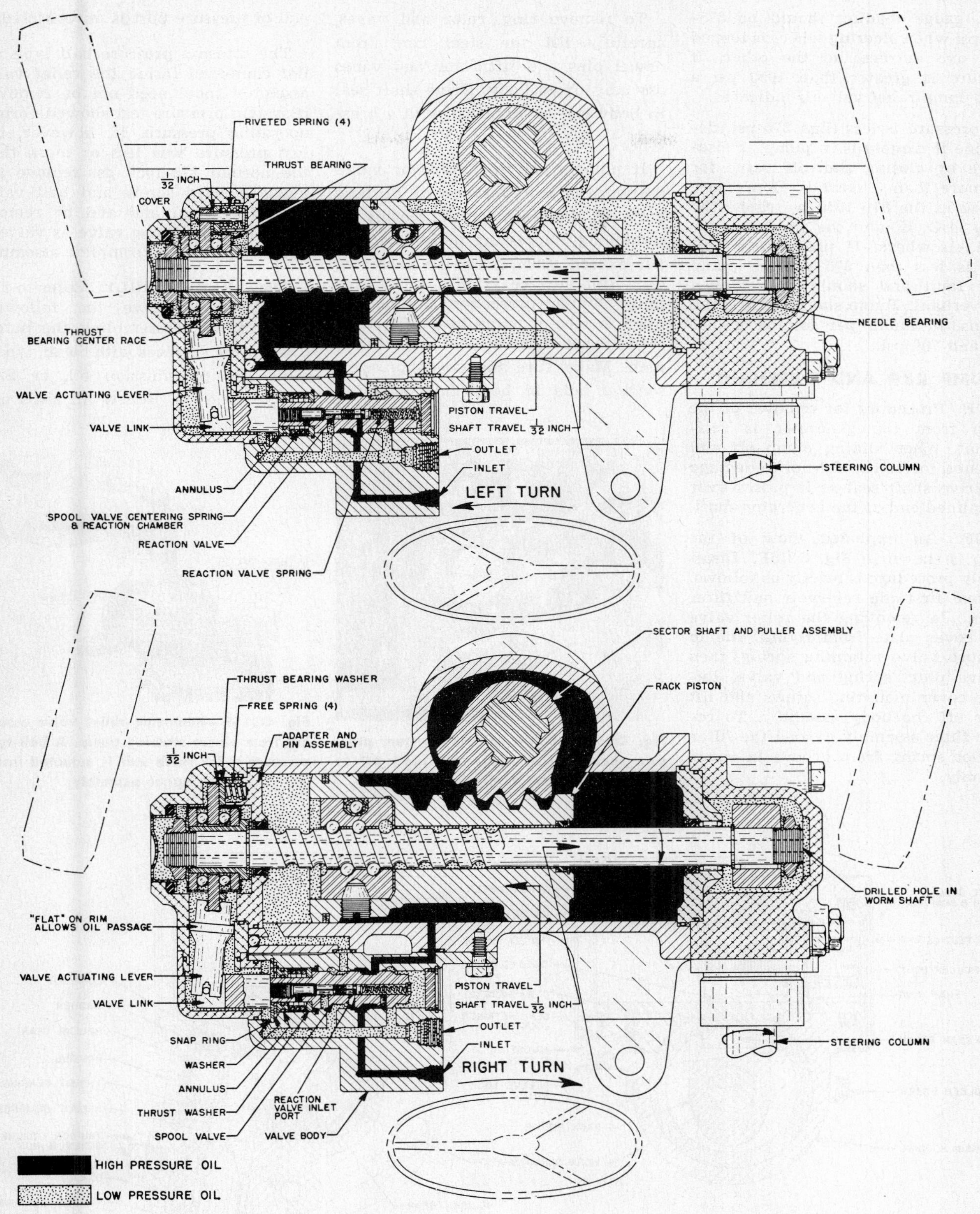

Fig. O315B—Sectional view of Oliver-Gemmer power steering gear showing positions of component parts when making a power assisted turn. Transfer of manual effort into power operation is controlled by 8 springs in the thrust bearing.

open, gauge reading should be 875-1050 psi when steering wheel is turned from one extreme to the other. If pressure is greater than 1050 psi a stuck pump relief valve is indicated.

If pressure is less than 875 psi, determine if trouble is in pump or elsewhere by closing shut-off valve for not more than 3 seconds. If pressure is now in the 875-1050 psi range the pump is O. K. and the trouble is located elsewhere. If pressure at this time is less than 875 psi the pump is at fault and should be removed for overhaul. Pump should deliver 1.3 gallons minimum per minute at 850 rpm and 700 psi.

PUMP R&R AND OVERHAUL

425E. Procedure for removal of the pump from the generator is self-evident. When sliding pump off and on generator be careful not to damage the drive shaft seal as it passes over the splined end of the generator shaft.

425F. An exploded view of the pump is shown in Fig. O315F. Disassembly procedure is briefly as follows: Remove first the reservoir and filter screen. In removing the relief valve hold cover plate from flying, due to pressure valve retaining spring, then remove plug, spring and valve. Remove cover mounting screws and lift cover off the body assembly. To remove filter assembly extract the filter retainer spring from groove in cover assembly.

To remove ring, rotor and vanes, carefully lift the steel ring from dowel pins and lift rotor and vanes assembly from body. Drive shaft seal in body can be removed with a brass rod.

If pump steel ring, rotor or vanes are worn or scored, renew all of these parts as an assembly. If scoring or scratching of pressure plate face does not exceed 0.005 it can be corrected on a fine surface grinder or by lapping. Similar correction can be applied to face of pump body if irregularities are not deeper than 0.0015 inch. Make sure the small pressure sensing hole in body face at inner end of pressure port is unrestricted.

The internal pressure ball type relief contained inside the relief valve assembly spool need not be removed if system pressure test showed normal operating pressure. If, however, the test pressure was less or more than the normal 875-1050 psi remove the hexagon head screw and ball valve from the spool and add or remove shims or install new valve as valve is available only as a complete assembly.

425G. **REASSEMBLY.** Refer to Fig. O315F and observe the following points when reassembling the pump: Coat inside surfaces with clean, type A automatic transmission oil, or SAE 10W engine oil. Renew all seals and

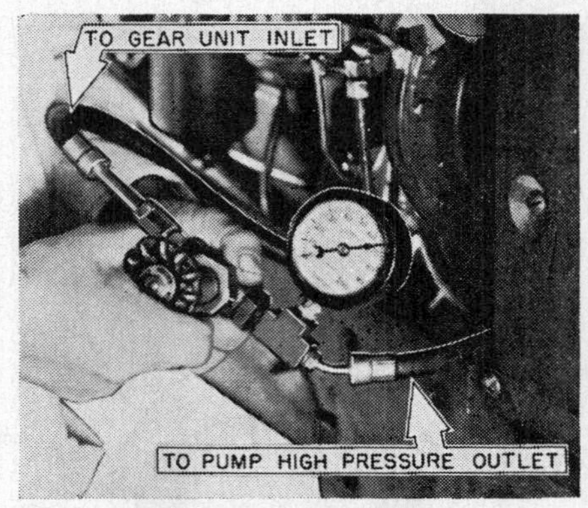

Fig. O315D—Power steering system pressure should be in range 875-1050 psi at extreme turn position.

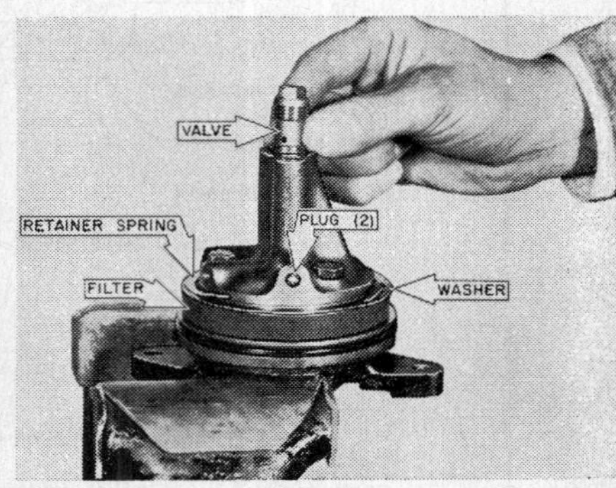

Fig. O315E—Removing relief valve assembly from power steering pump. A ball type pressure relief valve unit is mounted inside the spool assembly.

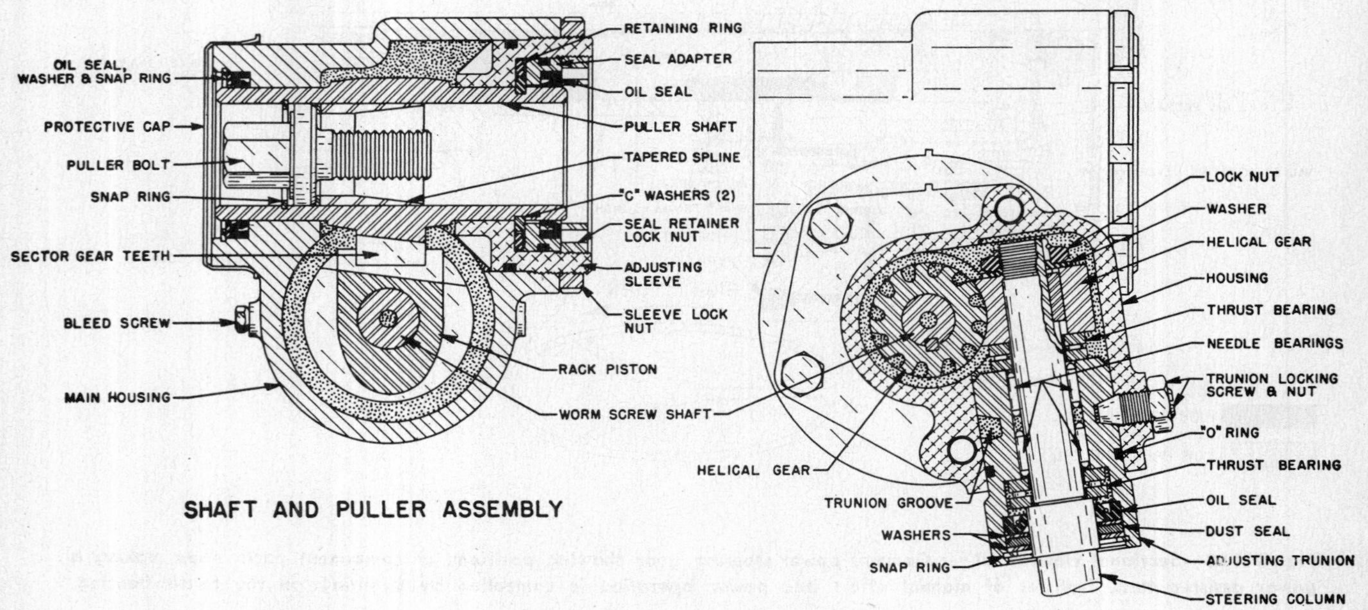

Fig. O315C—Additional sectional views of power steering gear unit. Refer also to Fig. O315B.

allow them to soak at least 20 minutes in the fluid before installing them. Lip of drive shaft seal in pump body should face reservoir end of pump. The small counterbore on rotor should face the mounting flange of the pump body. Rounded ends of rotor vanes should face outward. Apply a coating of light grease to rotor contacting surface of pressure plate to aid centering of rotor during pump installation. Install the alignment sleeve.

Refer to Fig. O315F. Install a new "O" ring in groove inside the cover. Add the filter retaining spring, filter retainer, flat side against spring, and the oil filter to outside of cover as shown. If relief valve spool does not move freely in bore of cover, polish it carefully with fine crocus cloth, being careful not to round the spool land edges. Hexagon end of valve faces outward. If a substitute drive shaft is available use it to rotate pump drive shaft. Pump rotor should rotate without binding. Correct the cause of binding before installing pump to tractor. Pump should deliver a minimum of 1.3 gallons per minute at 850 rpm and 700 psi.

GEAR ADJUSTMENT

425H. The gear portion of the steering system is correctly adjusted when not more than 3 inches of angular movement at the end of a 9-inch radius (18-inch diameter steering wheel rim) produces movement at the puller shaft (Fig. O315C) or moves the front wheels.

If arc of movement (at 9-inch radius) required to move the puller shaft or front wheels is more than 5 inches the gear needs either an adjustment or an overhaul. Before disassembling the gear for overhaul check to determine if the excess play is in the helical gears as follows:

Remove the steering gear assembly from the tractor as per paragraph 425I. Refer to Fig. O315G and with the gear completely assembled rotate the trunnion retainer screw out (counterclockwise) 3 full turns. While rotating the steering column in either direction, with one hand turn the knurled adjusting trunnion either way until an increase in drag is felt in rotating the column. Now back off the knurled trunnion until column drag is just removed. Tighten and lock the trunnion retainer screw. If this adjustment does not eliminate the excessive backlash the gear must be overhauled. Prob-

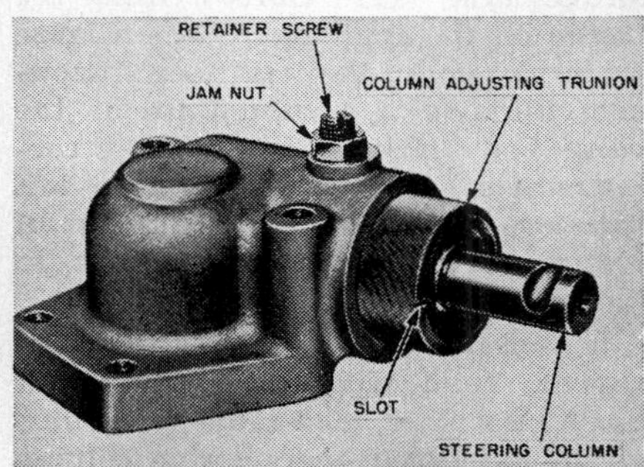

Fig. O315G—Helical gear housing shown here contains the only mechanical adjustment provided on the power steering gear.

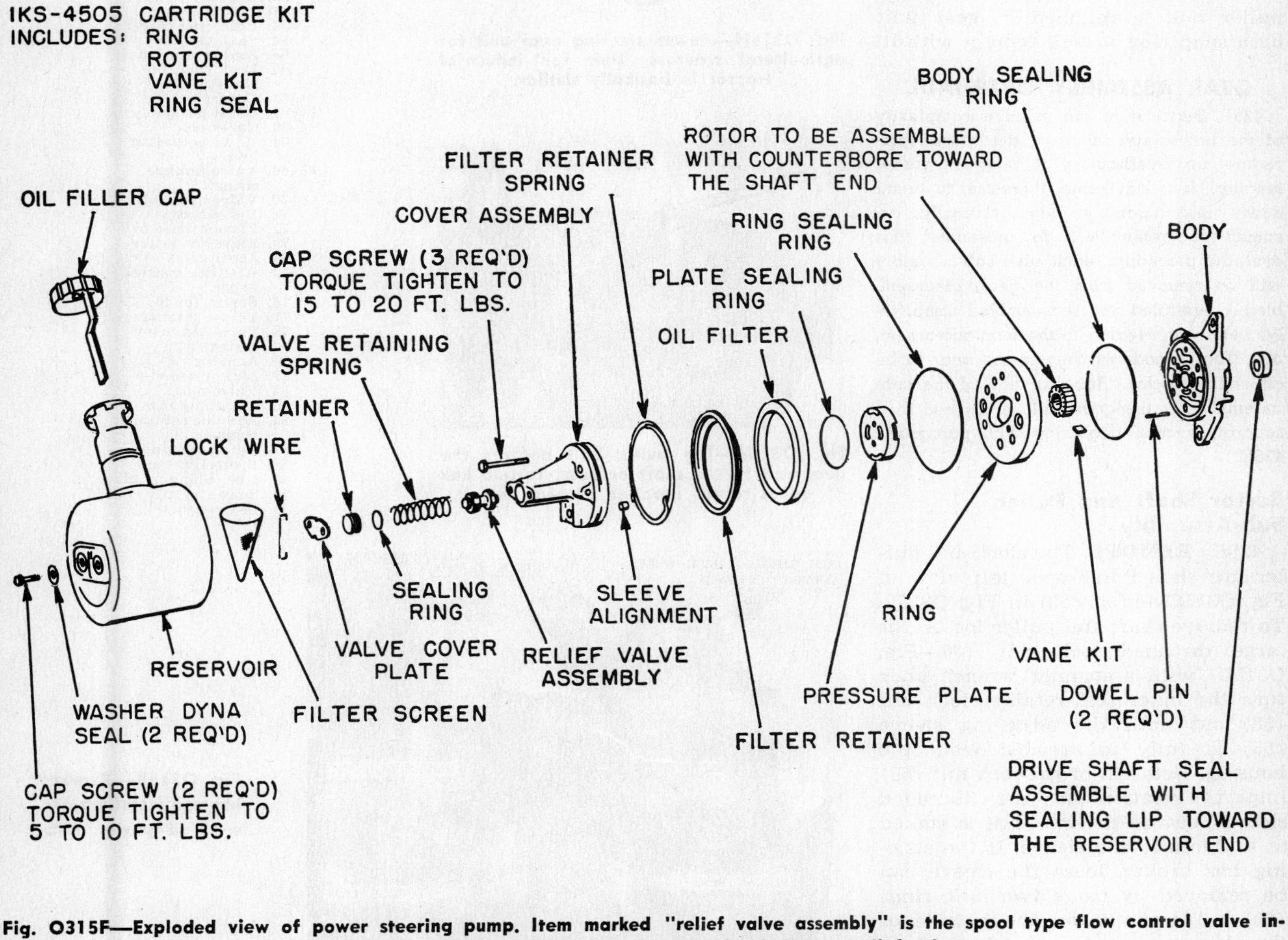

IKS-4505 CARTRIDGE KIT
INCLUDES: RING
ROTOR
VANE KIT
RING SEAL

OIL FILLER CAP
FILTER RETAINER SPRING
COVER ASSEMBLY
CAP SCREW (3 REQ'D) TORQUE TIGHTEN TO 15 TO 20 FT. LBS.
VALVE RETAINING SPRING
RETAINER
LOCK WIRE
ROTOR TO BE ASSEMBLED WITH COUNTERBORE TOWARD THE SHAFT END
BODY SEALING RING
RING SEALING RING
PLATE SEALING RING
OIL FILTER
BODY
SEALING RING
SLEEVE ALIGNMENT
RING
VANE KIT
RESERVOIR
VALVE COVER PLATE
RELIEF VALVE ASSEMBLY
PRESSURE PLATE
DOWEL PIN (2 REQ'D)
WASHER DYNA SEAL (2 REQ'D)
FILTER SCREEN
FILTER RETAINER
DRIVE SHAFT SEAL ASSEMBLE WITH SEALING LIP TOWARD THE RESERVOIR END
CAP SCREW (2 REQ'D) TORQUE TIGHTEN TO 5 TO 10 FT. LBS.

Fig. O315F—Exploded view of power steering pump. Item marked "relief valve assembly" is the spool type flow control valve inside of which is mounted the spring loaded ball type relief valve.

able internal causes of excess backlash are worn helical gear shaft keys, loose worm shaft lock nuts or damaged thrust bearing.

GEAR UNIT REMOVE AND REINSTALL

425I. Removal of power steering gear unit from tractor is accomplished as follows: Remove hood side panels, doors and radiator. Disconnect the torque link (Fig. O315J) from the flange on the gear unit. Place a large drain pan under the unit, then disconnect the two hydraulic lines at the valve body hose ports. Rotate steering wheel slowly from right to left extreme positions to force oil out of system. Plug the hose openings in the unit valve body.

Loosen or remove the steering shaft support from right side of cylinder block rear plate, then remove the universal joint from the steering column of the gear unit. Remove the protective cap from the gear housing and using a ⅞-inch socket turn the puller bolt (Fig. O315J) counter-clockwise until the gear unit can be lifted from the post, as shown in Fig. O315K. The puller bolt is retained in gear unit by a snap ring so will remain with it.

GEAR ASSEMBLY OVERHAUL

425J. Because of the relative complexity of the power steering gear the written procedure for overhauling it becomes rather lengthy. It is advisable, therefore, to break down the assembly into arbitrarily assumed sub-assemblies. In presenting this overhaul procedure, each such sub-assembly will be removed from the gear, disassembled, overhauled and reassembled completely, before proceeding to the next sub-assembly. The procedures assume the gear to be off of the tractor. Reinstallation of the sub-assemblies to the gear will be covered in a separate section beginning with paragraph 426E.

Sector Shaft and Puller Sub-Assembly

425K. **REMOVE.** The shaft and puller are shown in upper left view of Fig. O315C and at (52) in Fig. O315L. To remove shaft and puller loosen the large threaded lock ring (60—Fig. O315L) with a spanner wrench, then turn the inner seal retainer lock nut (58) out until the adjusting sleeve (53) is fully unthreaded from the housing. Rotation of the lock nut (58) imparts rotation to the threaded sleeve because the lock nut is staked at 4 places to the sleeve. If the staking has broken loose the sleeve can be removed by using two lock rings (60) as jam nuts on the outside of the sleeve.

425L. **OVERHAUL.** Extract the outer end oil seal by first carefully unstaking the lock nut (58) from the aluminum adjusting sleeve and turning the nut out of the sleeve. A pin spanner or equivalent is required for nut (58). It may be necessary to use two lock rings (60) on the sleeve and a spanner to hold sleeve from rotating while removing the seal lock nut. Bump end of shaft on wood block to dislodge seal and seal adaptor. Oil seal at opposite end can be removed at any time. Puller bolt is available only as an assembly with shaft, but can be removed after removing retaining and "C" washers. Install all new seals when reassembling. Be

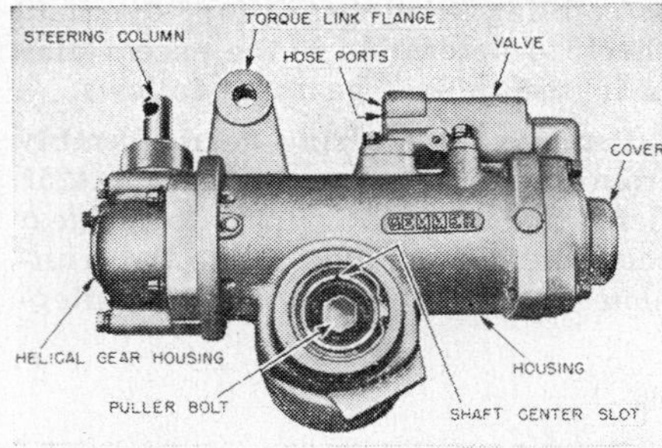

Fig. O315H—Power steering gear unit for agricultural tractors. Unit for industrial tractor is basically similar.

Fig. O315J—The puller bolt anchors the gear unit to the shaft or post. Torque link takes torque reaction of gear unit.

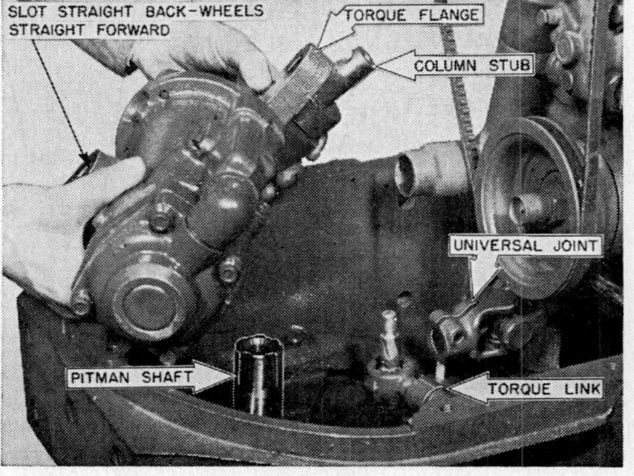

Fig. O315K—Removing and installing assembled power unit.

1. "O" ring
2. & 3. Housing
4. Shaft oil seal
5. Seal washer
6. Bleed screw
7. Worm screw
8. Worm & ball nut assembly
9. Ball nut
10. Ball
11. Ball return guide
12. Adapter and pin assembly
13. Seal retaining washer
15. Retaining ring
16. Seal adapter
17. Retaining ring
18. Thrust bearing bushing
19. Thrust bearing and spring assembly
20. Centering spring
22. 22A. 22B. Lock nut
23. Rack piston
24. Thrust ring & seal assembly
25. Seal spacer
26. Snap ring
27. Rack piston ring
29. Housing adapter
30. Needle bearing
32. Helical gear
34. & 38. Helical gear housing
39. Retainer screw
41. & 42. Column
44. Thrust washer
45. Needle thrust bearing
46. Trunnion
47. Needle bearing
48. Column oil seal
49. Dust seal
50. Seal washer
51. Snap ring
52. Sector shaft and puller assembly
53. Adjusting sleeve
54C. Washer
55. Retaining ring
56. Seal adapter
58. Seal retainer nut
60. Lock ring for No. 53
61. Cover
62. Cover assembly
66. Cover plug
68. Valve actuating lever
69. Valve linkage cover
70. Valve assembly
71. Valve body
72. Flare connector
73. Flare connector
74. Spool valve
75. Reaction control valve
76. Spring for No. 75
77. Thrust washer
78. Valve centering spring
80. Annulus
81. Link assembly
84. Valve plug
85. Retaining ring
86. Annulus retaining washer
87. Retaining ring
88. Protective cap
89. & 90. Shipping plug
92. Snap ring
94. Seal washer

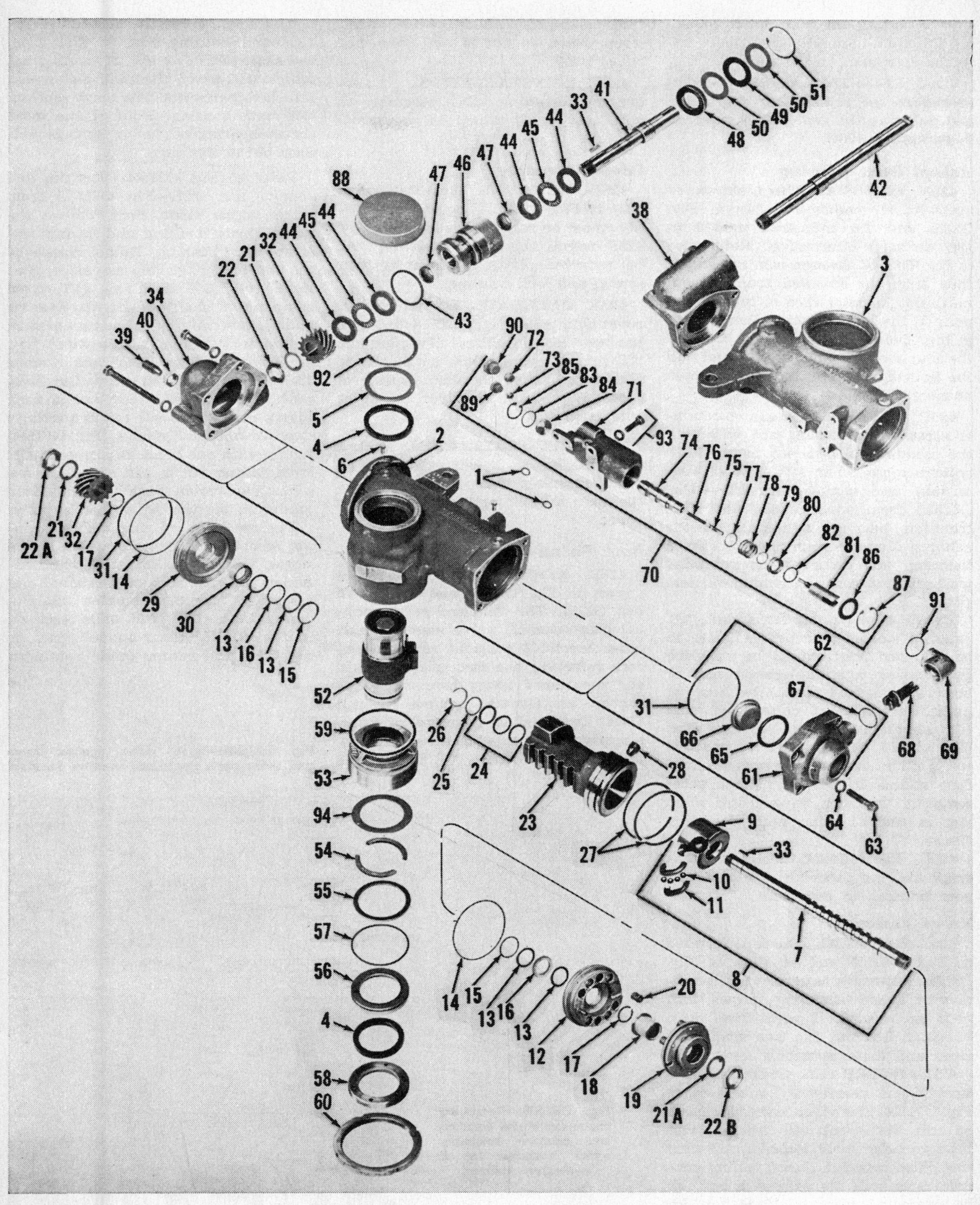

Fig. O315L—Exploded view of Oliver-Gemmer power steering gear unit. Refer to preceding page for parts legend.

sure to restake the lock nut by peening the aluminum adjusting sleeve into the four slots in the nut.

425M. **REINSTALLATION.** Correct procedure for reinstalling the sector and puller to the gear unit is covered in paragraph 426G.

Helical Gear Housing

425N. **REMOVAL.** This is shown for agricultural models at (34) in Fig. O315L and for industrial models at (38) in same illustration. Refer also to Fig. O315H. Loosen retainer screw, then turn the knurled trunnion out until slot in outer face of trunnion is pointed toward mounting flange or is at high point of eccentricity. Remove the four mounting cap screws and pull the helical gear housing off the main housing.

425P. **OVERHAUL.** Remove retainer screw from housing and withdraw the column and attached parts. With column mounted in soft jaws of vise, unstake and remove nut (22—Fig. O315L) from opposite end and slide trunnion, bearings and seals off the column. Remove snap ring (51) from trunnion and remove seal retaining washers, dust seal and column oil seal (48).

Install all new oil seals and "O" rings. If needle bearings (47) are to be renewed they should be installed by pressing against lettered end of bearing shell and should be flush at outer ends. Lip of column oil seal (48) should be toward bearings. Nut (22) should be tightened to produce just a slight drag on the bearings and then staked in position. When reassembling trunnion to housing align slot in end of same with mounting flange.

425R. **REINSTALL.** Refer to paragraph 426J for procedure for installing gear housing to gear unit.

Valve Assembly

425S. **REMOVAL.** This is illustrated in Fig. O315M and at (70) in Fig. O315L. Assembly is removed after removing three attaching screws. Lift body and contained parts from main housing. Lift out the two small "O" rings and valve actuating lever (68).

425T. **OVERHAUL.** Order of disassembly and reassembly is shown in Fig. O315M. The valve link is threaded into the spool; use wooden vise jaws to hold spool when unscrewing link. The inverted flared tubing connectors in body are pressed in but can be extracted using a 10-24 tap for small one and $\frac{5}{16}$-18 tap for the other. Run a nut well up on each tap, add a washer under it. Coat tap threads with grease, then after threading tap into connector about three turns, screw down on nut to pull connector out of body.

425V. **REINSTALLATION.** Procedure for installing valve assembly to gear unit is contained in paragraph 426H.

Cover Assembly

425W. **REMOVAL.** This is shown at (62) in Fig. O315L. The valve assembly must be removed as in paragraph 425S before the cover assembly can be removed. It is retained by four screws and lock washers.

425X. **OVERHAUL.** Do not remove cover plug (66—Fig. O315L) unless it has been leaking. Press new plug and "O" ring in evenly from inside but do not press against shoulder surface of plug. Plug must shoulder on cover counterbore.

426A. **REINSTALLATION.** Method of reinstallation is self evident. The valve assembly must be installed in the main housing before installing the cover.

Rack Piston Assembly

426B. **REMOVAL.** Rack piston is shown in Fig. O315P and at (23) in Fig. O315L. The shaft and puller, helical gear housing, valve assembly and cover assembly must be removed before removing the rack piston. After the mentioned parts have been removed, unstake and remove the nut (22A) from helical gear end of worm screw shaft. Remove gear, square key and retaining ring (17) from same end of shaft. Slide the needle bearing (30) and housing adapter (29) from the shaft. Push on end of worm shaft which will move the rack piston and attached parts out the cover end of the main housing. Four of the loose thrust bearing centering springs will drop out at this time.

Refer to Fig. O315N. Unstake and remove nut (22B—Fig. O315L) from worm screw shaft, then remove the worm thrust bearing and springs assembly and bushing. Do not disassemble the thrust bearing assembly. Remove retaining snap ring (17) from same end of shaft and lift the adapter and pin assembly from the rack piston. Mount rack piston in soft jawed vise as shown in Fig. O315P and remove the ball nut retaining screw. Carefully push worm screw to left out of rack piston and when ball nut is partially exposed hold ball guides (Fig. O315R) firmly with one hand, to prevent balls from falling out as nut and shaft are withdrawn from rack piston. Hold guides in position with tape or wire.

The nut balls are matched and cannot be obtained separately. The worm screw, ball nut, balls, guides and retainers are furnished only as a ball nut assembly. The disassembled ball nut containing a total of 23 balls is shown in Fig. 0315S. Early model gears do not have the return guide retainers.

Fig. O315M—Spool valve, reactor valve and other parts contained in valve housing.

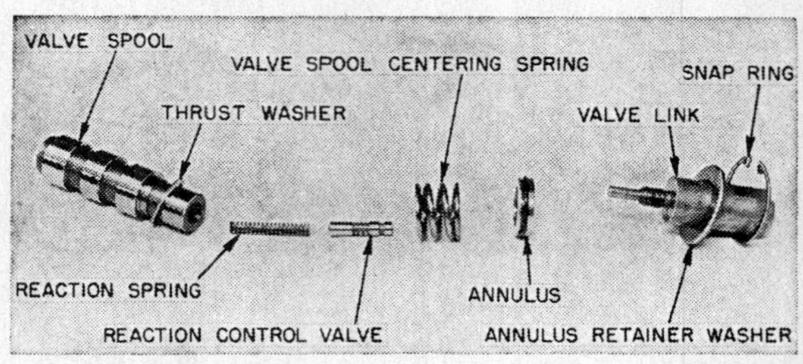

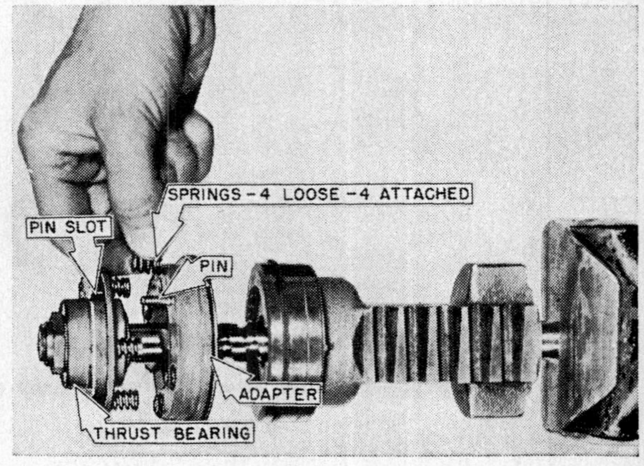

Fig. O315N—Removing the worm thrust bearing and adapter assembly which contains the 8 centering springs.

Remove the rack snap ring (26—Fig. O315L) spacer (25) and three-piece seal (24) from bore of rack piston. If piston rings are scored or otherwise damaged, use a suitable ring expander to install new rings.

426C. Install all new oil seals and "O" rings. If ball nut was disassembled, reassemble as follows: Slide ball nut over worm screw shaft with beveled side of nut facing long end of shaft. Position the nut near one end of the worm threads and drop 17 balls into it. Install remaining six balls into retainer and coat with Vaseline to hold them in place. If components of ball nut are in good condition and properly assembled, a pre-load of 2-6 inch pounds will be required for the ball nut to pass over the slightly larger diameter center point, on the worm screw shaft. Do not assemble the housing adapter or helical gear until rack is reinstalled to main housing.

Assemble the worm screw and nut, long end first, into the piston ring end of the rack piston, being extremely careful to avoid cutting the oil seal in opposite end of rack piston. Bring retaining screw hole in nut into register with mating hole in rack piston, then install a new retaining screw (28—Fig. O315L). Tighten the new screw to 30-35 foot pounds, then stake the piston into the screw slot at two places.

If needle bearing (30) in adapter housing (29) is worn install new one flush with counterbore, pressing on marked end of bearing cage. Remove snap ring (15) and install new seal assembly (13), rubber out, into adapter bore.

Reassemble the adapter and pin assembly, retaining ring, sleeve bushing and thrust bearing assembly to shaft as shown in Fig. O315T. Counterbore in flange end of bushing should cover the retaining ring. Reassemble washer and nut and with string wound around center race circumference of thrust bearing, tighten nut (22B—Fig. O315L) until a scale pull of 1½ to 3 pounds is required to rotate the bearing as shown in Fig. O315V. Stake lip edge of nut into keyway. Align pin in adapter with hole in thrust bearing washer and assemble the four loose springs in the adapter holes that are not aligned with the permanent springs. Temporarily secure adapter to thrust bearing with tape.

426D. **REINSTALL.** Procedure for reinstallation to the steering gear is contained in paragraph 426F.

REINSTALLATION OF SUB-ASSEMBLIES

426E. After the sub-assemblies have been removed, disassembled, overhauled and reassembled, as outlined in paragraphs 425J thru 426D they should be reinstalled to the main housing in the following order as indicated.

426F. **RACK PISTON, WORM AND BEARING.** Install this sub-assembly to main body as shown in Fig. O315W. Coat piston rings liberally with 10W oil and apply coating of light grease to main housing counterbore. Install a new "O" ring to adapter and pin assembly. Align flat on worm thrust

Fig. O315P — Removing the screw which retains the ball nut assembly inside the bore of the rack position.

Fig. O315R — Withdrawing ball nut assembly from rack piston and worm screw shaft.

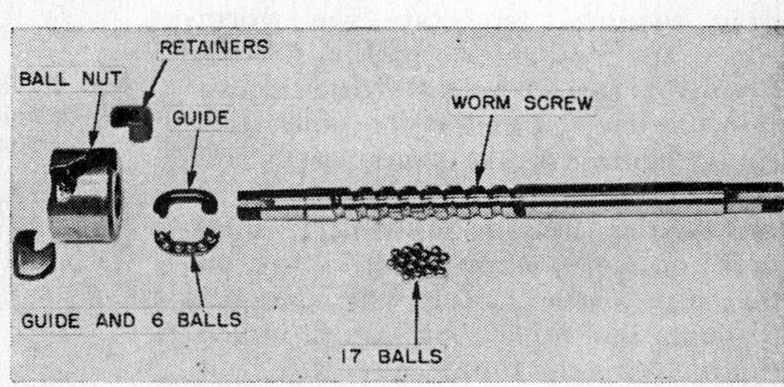

Fig. O315S—Worm screw shaft and ball nut disassembled.

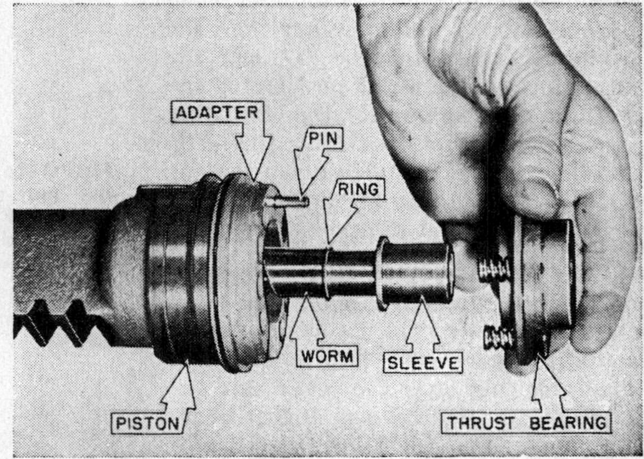

Fig. O315T—Assembling adapter and pin assembly, worm thrust bearing and related parts to short end of worm screw shaft.

bearing and spring assembly with the valve mounting surface of main housing, then gently push the assembly into the housing as shown.

The cover (61—Fig. O315L) can be installed only one way. Make sure that the pin on the flat surface of the bearing adapter enters the recess in the cover.

Place new "O" rings on outside of housing adapter, lubricate the housing bore, then carefully slide the adapter, hub end out, on to the worm screw shaft and into the housing. Install the snap ring (17) to shaft groove, then the gear key and helical gear (32) with counterbore of same over the snap ring. Mount assembly as shown in Fig. O316A, place brass rod through housing hole and wedge rack piston, then install lockwasher and nut. Tighten the nut to 25-30 foot pounds.

426G. **SECTOR SHAFT AND PULLER.** Position the rack piston with center tooth of same in register with center of remaining large hole in housing. The outside end of the sector shaft (52—Fig. O315L) has a slot or etched mark indicating center tooth of sector. Carefully enter marked sector tooth into mesh with marked tooth on piston, being extremely careful not to cut the housing oil seal. Start the adjusting sleeve (53) into the housing threads. If sector has been correctly installed the helical gear will rotate 5¾ revolutions in moving from one extreme to the other.

Using an inch-pounds torque wrench measure and record (on paper) the drag required to rotate the worm screw shaft as shown in Fig. O316B. Continue threading adjusting sleeve into the housing and at the same time rotate worm screw shaft back and forth through about 90 degrees, so as to move it through the center "high spot" on worm screw shaft. When an increase in drag is felt when passing through the high spot, again attach the inch-pounds torque wrench to nut and tighten adjusting sleeve until torque wrench registers an amount of drag 4½ to 9 pounds greater than the previously recorded drag. Lock the adjusting sleeve in this position by installing and tightening the adjusting sleeve ring type lock nut (60—Fig. O315L). Install protective cap locking it in position with 16-gauge wire or sealing compound if necessary.

426H. **VALVE AND BODY.** Assemble valve linkage cover (69—Fig. O315L) to body (71). Lower the valve actuating lever (68) into hole in end cover making sure the lever slot engages flat edge of worm thrust bearing shown in Fig. O315B. Install new "O" ring between actuating lever cover and housing end cover and two new small "O" rings into counterbores on mounting face of main housing. Place valve body on main housing with actuating lever arms engaged in slot on link in valve. Bolt the valve body to the main housing.

426J. **HELICAL GEAR HOUSING.** Bolt housing assembly (Fig. O315G) to main housing with slot in trunnion positioned as shown. Before installing the steering gear to tractor adjust the column as outlined in paragraph 425H.

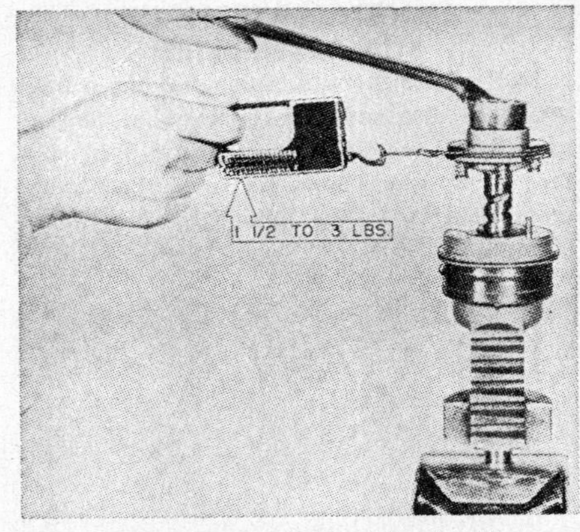

Fig. O315V—Adjusting preload of worm thrust bearing.

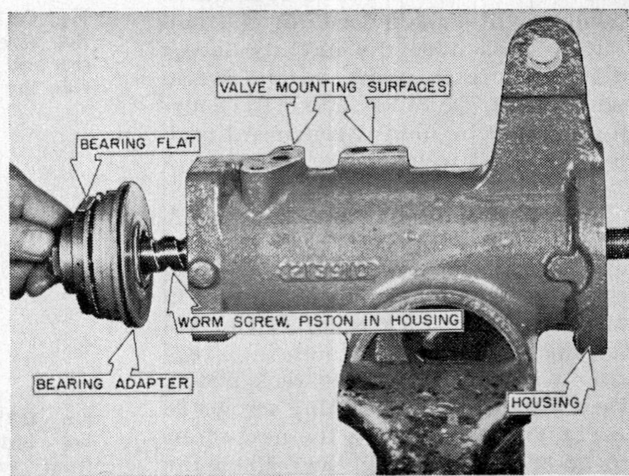

Fig. O315W — Installing rack piston, worm screw shaft and bearing assembly to main housing.

Fig. O316A — Installing nut to helical gear end of worm screw.

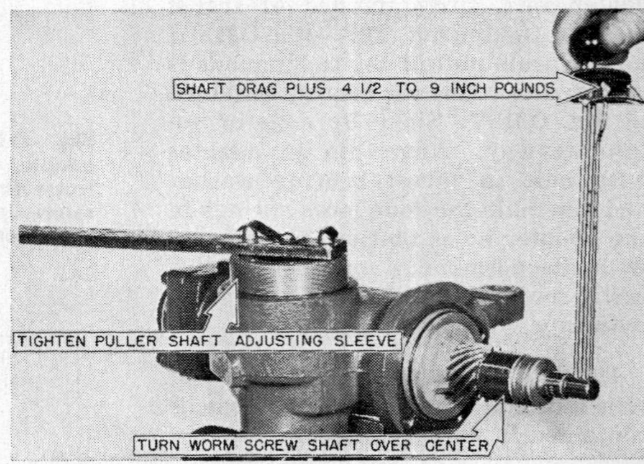

Fig. O316B — Adjusting the net shaft drag by means of the adjusting sleeve and an inch pounds torque wrench.

ENGINE AND COMPONENTS

R&R ENGINE

430. Remove tractor hood, fuel tank and batteries. If equipped with Power Booster remove shift lever retainer and control rod. Remove rear bolt from steering shaft on side of clutch housing. Unclip and disconnect wires at left and right front sides of rear panel. Unhook clutch return spring. Remove Diesel injection pump control rod or equivalent on gasoline models.

Remove the retaining capscrews from rear panel then lift same and move rearward far enough to pull steering shaft from its universal joint. Move rear panel assembly up and secure it to top of engine as in Fig. O-368C. Remove grille and radiator screen. Remove capscrews from radiator shield (baffle) and from radiator supports but do not remove radiator. Disconnect hoses from power steering pump and cover from Power Booster if so equipped.

Disconnect flexible lubrication tube for Power Booster at front attachment and pull tube out of Power Booster lower cover. Remove bolts from Power Booster clutch. NOTE: If paint marks are obliterated put an identification mark on Booster clutch parts so they can be reassembled to same relative position. This also applies to shims, the large lobe ends of which should point toward right side of tractor.

Remove main clutch rod. Disconnect front sections of steering shaft at clutch housing and front universal and remove shaft. Remove the engine support bolts. One front and one rear mounting bolt acts as a dowel and must be installed in proper holes. Remove PTO housing then remove retaining capscrews and pull overcenter clutch and PTO shaft assembly from tractor.

Install lifting eye to engine, attach hoist and lift engine slightly. Move assembly forward then lift from tractor. The Power Booster clutch if so equipped, can be removed at this time. Long hub of Booster front clutch disc faces forward and long hub of rear disc faces rearward.

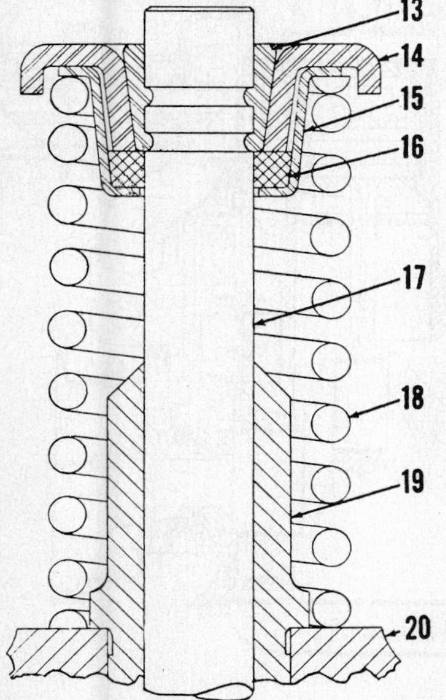

Fig. O318—Details of neoprene gasket and oil guard as installed on inlet valve stems.

13. Spring retainer lock
14. Spring retainer
15. Oil guard
16. Oil guard gasket
17. Inlet valve stem
18. Spring
19. Valve guide
20. Cylinder head

CYLINDER HEAD

431. To remove cylinder head, remove side panels, hood, and drain cooling system. Remove upper radiator hose, air cleaner, valve cover, valve rocker arm shaft oil line, and rocker arms and shaft assembly. On non Diesel models, remove carburetor from manifold but do not disconnect the air cleaner connection, fuel line, or controls from the carburetor. On Diesel engines, disconnect injector pump to nozzle fuel lines at nozzles, and cable to inlet manifold preheater. Remove head retaining screws and lift head from engine.

431A. On non Diesel engines, torque tighten all cylinder head retaining screws to 92-100 ft. lbs. On Diesel engines, torque tighten all cylinder head retaining screws except the drilled screw to 113-116 ft. lbs., and the drilled screw to 96-100 ft.-lbs.

Refer to paragraph 494 in this manual for data applying to the Diesel engine precombustion chambers (energy cells).

VALVES, GUIDES AND SEATS

Refer to 770 & 880 Supplement, paragraph 1 for Super 77 and Super 88 tractors.

432. Both inlet and exhaust valves have a face and seat angle of 45 degrees. Desired seat width is 3/64-1/16 inch. Seats may be narrowed, using 20 degree and 60 degree cutters.

Exhaust valve seat inserts are standard equipment on late production Super series HC and LP-Gas tractors. Inserts are available for all models and are installed with a 0.006 interference fit.

Inlet valves are provided with neoprene oil guards, as shown in Fig. O318, to prevent oil from passing into the combustion chamber via the valve stems. Install new oil guards each time the valves are reseated.

Adjust valve tappet gaps cold and to the following values:

Inlet 0.010
Non-Diesel, Exhaust 0.016
Diesel, Exhaust 0.020

432A. Total valve lift data below may be used for checking for worn cam lobes:

66-77 and Super 77 Total Lift
Intake Valve 0.312
Exhaust Valve 0.281
Super 66 Total Lift
Both Valves 0.360
88 and Super 88 Total Lift
Both Valves 0.344

Reject the camshaft when total valve lift is 0.030 less than shown in table.

433. The cast iron, shoulder type inlet and exhaust valve guides are not interchangeable and can be driven into position, using an Owatonna driver tool No. ST-106 or equivalent. Inlet guides are shorter in length. New presized guides have a bore diameter of .3745-.3755. Valve guides and/or valves should be renewed when the diametral clearance exceeds .0065 for inlet or .0075 for exhaust.

VALVE ROTATORS

Late production HC engines are factory equipped with Rotocap positive type valve rotators for the exhaust valves. Refer to Fig. O319. Kits containing shorter valve springs and rotators are available for servicing the engines not so equipped.

434. The positive type exhaust valve rotators require no maintenance but should be visually observed when the engine is running to make certain that each exhaust valve rotates slightly. Renew the rotator of any exhaust valve which fails to rotate.

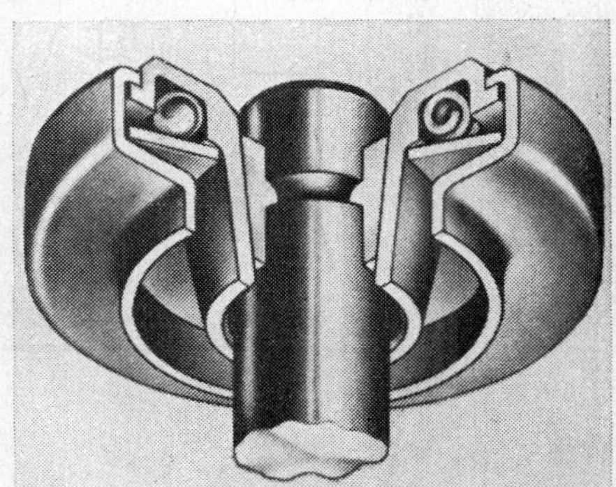

Fig. O319—Rotocap positive type valve rotators as installed on HC engines. Similar rotators are available for servicing engines not so equipped.

VALVE SPRINGS

435. Inlet and exhaust valve springs used on engines not equipped with the positive type valve rotators are interchangeable, whereas on engines equipped with valve rotators, the exhaust springs are shorter.

Springs with 10½ coils used on valves without rotators have a free length of $2\frac{25}{32}$ inches and should test: 44-52 lbs. @ $1\frac{15}{16}$ inches and 65-77 lbs. @ $1\frac{19}{32}$ inches.

Some engines without valve rotators are equipped with valve springs having a free length of 3¼ inches and 12 coils. Where these springs are encountered, it is recommended that they be changed to the latest 10½ coil type having a free length of $2\frac{25}{32}$ inches.

The valve springs used on valves equipped with positive rotator have a free length of $2\frac{5}{16}$ inches, 12 coils, and should require 43-47 lbs. to compress them to a height of 1 47/64 inches or 78-86 lbs. @ 1 25/64 inches.

VALVE TAPPETS

436. The mushroom type tappets (cam followers) operate directly in machined bores of the cylinder block. Tappets are supplied only in the standard size of 0.6240-0.6245 diameter (except early non-Super of 0.5605 diameter), and should have an operating clearance of 0.0005-0.002 in bore.

Renew tappets if their diameter is less than 0.619 late models, or 0.554 early models. If diametral clearance exceeds .007 with a new tappet, make up and install bushings in the block bores. Bushings should be final sized to a .624-.625 inside diameter.

Any tappet can be removed after removing the camshaft as outlined in paragraph 441. See paragraph 432 for tappet gap.

VALVE ROCKER ARMS

Rocker arms used in non Super engines prior to engine serial 939001 for the 66, 921910 for the 77, and 937157 for the 88 are equipped with wick type oilers as shown in Fig. O321. The rocker arms used in non Super engines after the aforementioned serial numbers and all later models are wickless. Refer to Fig. O322.

437. Rocker arms and shafts assembly can be removed for overhaul after removing the side panels, hood, valve cover and rocker arm shaft support retaining nuts.

437A. The rocker arm valve contacting surface can be refaced but the original radius of ⅜ inch must be maintained and the face must be parallel to the rocker arm shaft.

Desired diametral clearance between a new rocker arm and shaft is .0015-.0035 and with a maximum permissible clearance of .005. Semi-finished bushings (steel-backed, babbitt lined) are available for servicing the wick type rocker arms. Bushings should be bore sized to a .7445-.7455 diameter after installation. On wickless type rocker arms, it will be necessary to renew the rocker arm assembly to correct excessive running clearance as bushings are not available for service. Rocker arm shaft diameter is .742-.743.

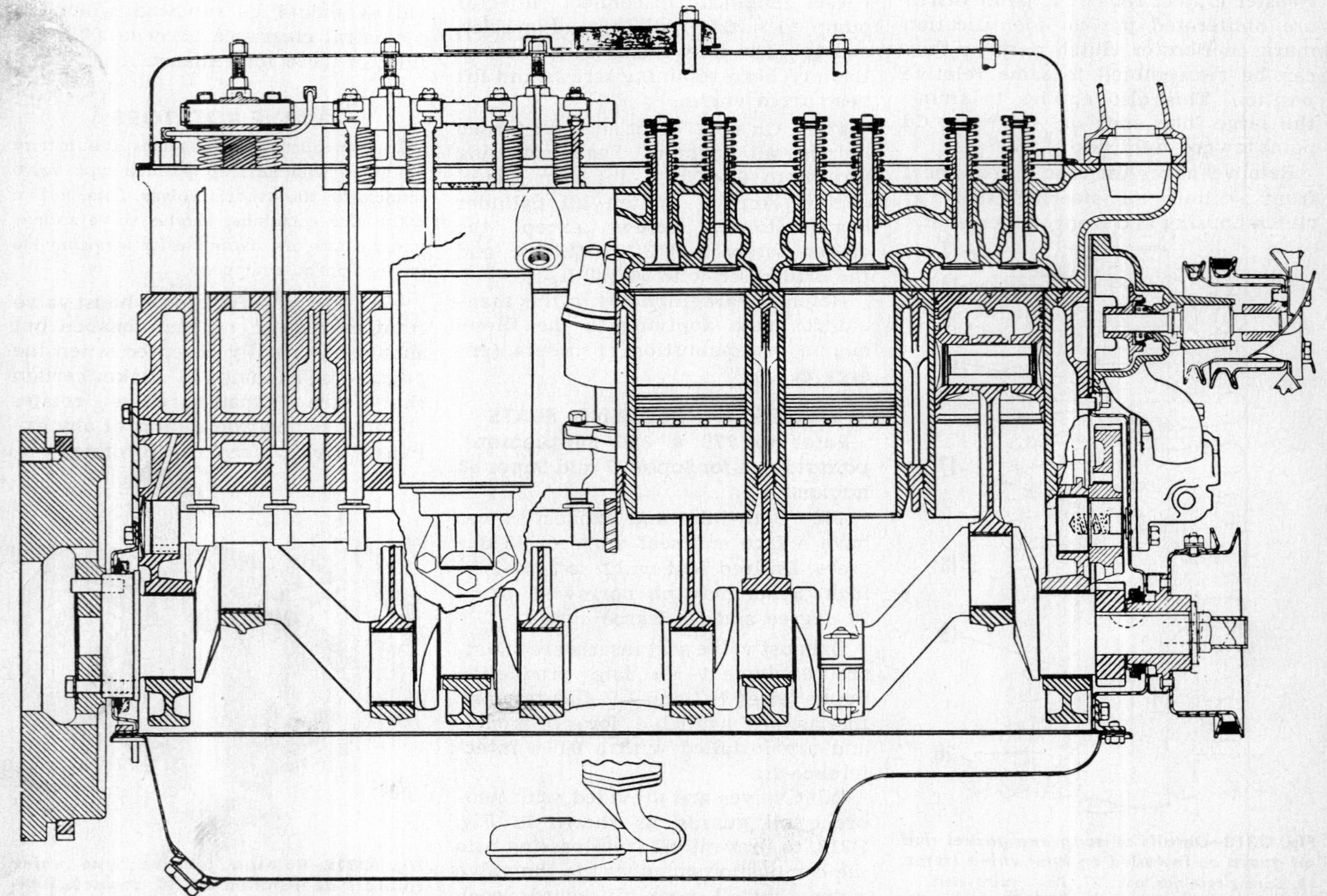

Fig. O320—Oliver 77 and 88 HC & KD engine. The four cylinder engine used in the model 66 is similar. Diesel engines are similar. On all models, rod and piston units are removed from above only.

CAUTION: The later type rocker arms (wickless type) are not interchangeable with the early wick type as the rocker arm shafts are different. However, the complete assembly of rocker arms and shafts are interchangeable providing the rocker arm shaft oil line is also changed to the later type.

Fig. O322A—Model 66 rocker arms (wick type) and shaft installation. Note offset rocker arms. Each rocker arm is marked "R" or "L" to identify its position on the shaft. Other models are similar.

21. Water outlet elbow
22. Rocker arm support
23. Spring
24. Rocker arm shaft
25. Right rocker arm
26. Left rocker arm
27. Special stud
28. Tappet adjusting screw
29. Oil line

437B. Each rocker arm is marked either "R" or "L" to identify its position on the shaft and same should be assembled to the shaft as shown in Fig. O322A. On wick type arm installations, install the shaft so that the oil holes face the valve stems; on the wickless type, install the shaft so the oil holes are facing up.

The G1 series of the non Super 88 engines are equipped with rocker arm shaft supports (Oliver No. K-203) which require the use of a washer

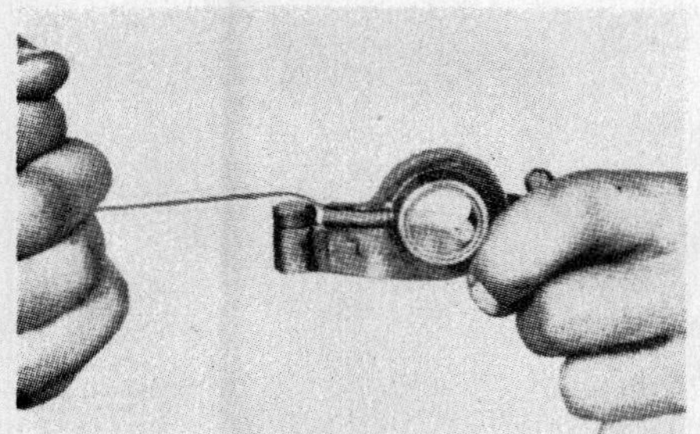

Fig. O321—Method of installing the oil wick on early production rocker arms.

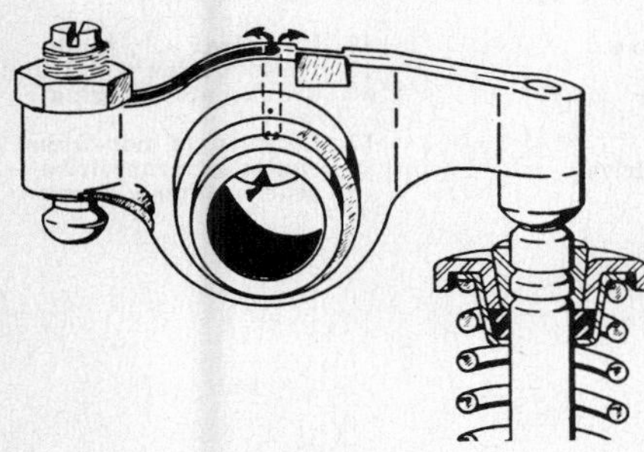

Fig. 0322—Wickless type rocker arms.

(Oliver No. K-210A) inserted between the head and support. These supports can be identified by the part number which is molded on the support. When reassembling these engines, be sure to install a washer under each support. Supports (Oliver No. K-203A) which are interchangeable with the K-203 supports are of the later type, and do not require the use of the washer.

Adjust valve tappets to value listed in paragraphs 432.

VALVE TIMING

438. Valves are correctly timed when the mark "C" on camshaft gear is meshed with an identical mark on the crankshaft gear, as shown in Fig. O324.

To check valve timing when engine is assembled, adjust number one cylinder inlet valve tappet gap to .005 more than normal operating clearance. Insert a .005 feeler gage between number one inlet rocker arm and valve stem. Crank engine over slowly until a slight resistance occurs when trying to withdraw feeler gage. At this time, inlet valve is just starting to open and flywheel mark "INT" (viewed through flywheel timing mark nspection port which is located on right side of engine) should be within ½-inch either way from index. After checking the valve timing, reset inlet valve tappet clearance to the value listed in paragraph 432.

TIMING GEAR COVER

439. To remove timing gear cover, proceed as follows: Remove engine side panels and hood. Drain engine coolant, and disconnect upper and lower radiator hoses. Remove radiator and grille assembly, generator belt, fan belt, and water pump. On non-Diesel engines, disconnect governor linkage at the governor and remove governor. Remove starting crank jaw. Using a suitable puller attached to two ⅜-16 cap screws which can be threaded into pulley, remove crankshaft pulley. Remove three oil pan to timing gear cover retaining cap screws and loosen all others. Remove timing gear cover retaining cap screws and remove the cover.

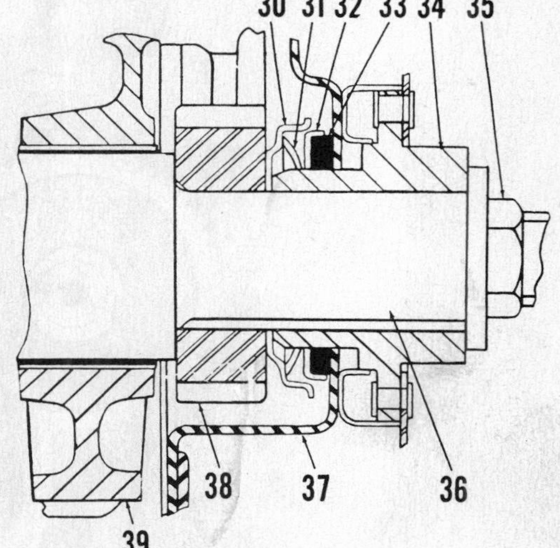

Fig. O323—Crankshaft front oil seal of treated cork (33) can be renewed after removing the timing gear cover. Parts (30), (31), (32) and (33) are available only as an assembly.

30. Seal housing
31. Spring
32. Seal retainer
33. Cork seal
34. Crankshaft pulley hub
35. Pulley retaining nut
36. Crankshaft
37. Timing gear cover
38. Crankshaft gear
39. No. 1 main bearing cap

TIMING GEARS

440. Timing drive on Diesel engines consists of three helical gears and the injection pump gear; on non-Diesel engines, two gears are used. Governor drive gear for non-Diesel engines is driven by camshaft gear. Refer to Figs. O324, O325 or O326.

Recommended gear backlash is .003-.005. Renew all gears when backlash between any pair exceeds .007.

The cylinder block for all models is designed for the installation of an idler gear. However, the gear is used only in the Diesel models. For all non-Diesel models, check to make certain that a plug (42—Fig. O324 or O325) is installed in what would normally be the idler shaft bushing bore on the Diesel engine.

440A. To remove camshaft gear, first remove timing gear cover. Remove camshaft gear from shaft by using a puller attached to two ⅜-16 cap screws which can be threaded into gear. Crankshaft gear can be removed in a similar manner as shown in Fig. O328. Avoid pulling the gears with pullers which clamp or pull on the gear teeth.

440B. The camshaft and crankshaft gears are stamped either "S" (standard), "O" (oversize), "U" (undersize) and amount of undersize or oversize. Original installation size is stamped

Fig. O324—Crankshaft and camshaft gear installation on non-Diesel engines. Mesh camshaft gear mark "C" with an identical mark on crankshaft gear. Diesel engines are similar. Plug "42" is installed on all non-Diesel engines.

38. Crankshaft gear
40. Governor shaft sleeve
41. Governor shaft bushing
42. Plug (non-Diesel)
43. Oil pressure relief valve
44. Camshaft gear
45. Thrust button
46. Flywheel timing port

Fig. O326—Timing gear installation on Diesel engines. Mesh camshaft gear mark "C" with an identical mark on crankshaft gear. Note call-out "12" indicates Diesel injection pump outlet for No. 1 cylinder. Model 77 is shown.

1. Nozzle return line
10. Timing port
38. Crankshaft gear
44. Camshaft gear
45. Thrust button
47. Injection pump drive gear
48. Idler gear
49. Thrust button
50. Crankshaft engaging ratchet
51. Location of non-Diesel engine governor drive gear bushing & sleeve

Fig. O325 — Crankshaft and camshaft gear installation on non-Diesel engines. The original gear installation size is stamped on timing gear cover gasket surface, shown at "F", and gear size is stamped as shown at "44". Diesel engines are similar.

on timing gear gasket surface of cylinder block in the vicinity of the cam gear. When renewing gears on an engine stamped "+ 1", use one standard gear and one 0.001 oversize gear.

440C. When reinstalling the cam gear, remove the oil pan and buck-up the camshaft at one of the lobes near the front end of the shaft with a heavy bar. Both the camshaft gear and crankshaft gear should be heated in oil to facilitate installation.

Reinstall gears with the "C" mark on crankshaft gear meshed with a similar mark on camshaft gear as shown in Fig. O324.

440D. The idler gear (48—Fig. O326) is installed only on Diesel engines and can be removed after removing timing gear cover. Running clearance of gear spindle in bushings should not exceed 0.004. When gear is in contact with front face of sleeve, idler gear face should be flush with crank gear face.

If clearance is greater than 0.004, or if flush condition is not obtained, renew the sleeve and bushings assembly as neither part is available separately. Bushings in new sleeve may require final sizing to provide the desired .0015-.002 running clearance.

In rare cases, it may be necessary to reface the thrust face of a new sleeve assembly in order to obtain the flush setting.

CAMSHAFT

441. To remove camshaft, first remove timing gear cover as outlined in paragraph 439. Remove ignition distributor from non-Diesel engines. On Diesel engines, remove fuel pump from right side of engine. On all engines remove rocker arms and shaft assembly, engine oil pan, and oil pump. Remove push rods and block up or support tappets (cam followers). Thread shaft and gear forward out of block bores.

Cam gear can be removed before or after camshaft has been removed from engine by using a suitable puller attached to the gear with two 3/8-16 cap screws.

441A. The front journal on all models rotates in a steel-backed, babbitt lined bushing. The two remaining journals on 4 cylinder models; 3 on 6 cylinder models, rotate in machined bores in the cylinder block.

Camshaft journal sizes are: Front, 1.7495-1.750; second, third and fourth, 1.7485-1.7495. Recommended running clearance of number one journal is .0015-.003 with a maximum of .005; all others, .0025-.0065 with a maximum permissible clearance of .007.

Camshaft lobe lift is .246-.250 inlet and exhaust for Super 66, 88 and Super 88. On 66, 77 and Super 77 intake lobe lift is .224, exhaust .204. Renew the camshaft if any lobe lift is .020 less than specified.

When the diametral clearance of machined bores exceeds .007, correction can be made by renewing the camshaft and/or cylinder block, or by reboring the bores in cylinder block to a diameter of 1.8745-1.8755 inches and installing bushings. Pre-sized service bushings as supplied for number one journals can be used and should be installed using a close fitting piloted driver.

In first production engines, the camshaft bushing was pinned. For repairs, it is recommended that this pin be removed by cutting same flush with machined bore in cylinder block.

Camshaft end play is controlled by a spring loaded thrust button. Thrust button spring has a free length of 1 3/16 inches and should have a pressure of 15.5-18.5 lbs. when compressed to a height of 25/32 inch.

441B. If for any reason the cam gear was removed from the shaft, the gear should be pressed on the shaft before installing shaft in engine.

ROD AND PISTON UNITS

442. Piston and connecting rod assemblies are removed from above only, after removing cylinder head and oil pan. Later series rods are offset and narrow side must be installed facing the nearest main bearing. Non-Super rods are installed with rod mark "Front" facing toward the timing gear end of the engine.

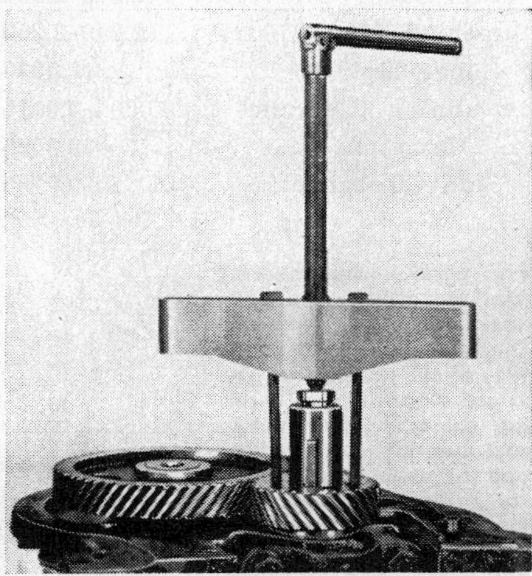

Fig. O328—Removing crankshaft gear on non-Diesel engines by using a puller attached to two 3/8-16 bolts which can be threaded into gear.

Tighten connecting rod bolts to 44-46 ft. lbs. torque on non-Diesel engines, and 46-50 ft. lbs. torque for Diesel engines.

PISTONS, SLEEVES AND RINGS

Engines as used in the 77 series, serial numbers Row Crop 320001-320145, and Standard 269001-269110 were originally equipped with sleeves having a 3.187 bore diameter. Later non-Super engines are equipped with sleeves having a bore diameter of 3.3125. It is recommended that engines with a 3.187 bore diameter be changed over to the 3.3125 bore diameter whenever the engine is overhauled, by installing the 3.3125 sleeves, pistons and cylinder head.

443. **PISTONS & SLEEVES.** Cast iron pistons were original equipment on non-Super 66, 77 and 88, and early production Super 88 engines. Aluminum alloy type pistons are installed on all later engines.

Desired piston skirt clearance is checked with a small spring scale using a ½-inch feeler gage as indicated in table. On later series with aluminum pistons clearance varies according to cast number on piston as indicated. Wear limit of pistons and sleeves is when a 0.006 x ½ inch feeler gage requires less than a 5-10 lb. pull on a spring scale to withdraw it. Pistons are available in standard size only.

Skirt clearance scale pull, ½-inch feeler:

Iron Pistons

Model	Pull	Feeler Thickness and Number of
66-77	3-6 lbs	Two 0.0015
88	3-6 lbs	One 0.002 and one 0.0015

Aluminum Pistons

Cast Number on Piston	Pull	Feeler Thickness and Number of
180604	3-6 lbs	One 0.002
180704	3-6 lbs	One 0.0015
192304	3-6 lbs	One 0.002 and one 0.0015
190304	3-6 lbs	One 0.0015

443A. Sleeves should be renewed when either of the following conditions exist: Taper, 0.008; out-of-round, 0.002; wear, 0.010.

Before installing wet type sleeves, clean all cylinder block sealing surfaces. The top surface of sleeve should extend 0.001-0.004 above top surface (gasket surface) of cylinder block. If this standout is in excess of 0.004, check for foreign material under sleeve flange. Excessive standout will cause water leakage at cylinder head gasket. Clearance between sleeve lower land and cylinder block bore should be 0.003-0.005. Use a white lead paste on the two neoprene sealing rings to facilitate installation of the sleeve and to help seal the sealing rings.

443B. **RINGS.** There are three compression rings and one oil control ring per piston. Factory supplied service rings are stamped "Top" and should be installed in this manner.

Recommended end gap for all compression rings is 0.012-0.023 with a maximum of 0.045 for service; oil control ring, 0.010-0.020 with a maximum of 0.045 for service. Recommended side clearance in grooves of all rings is 0.0015-0.003 with a reject value of 0.006.

PISTON PINS AND BUSHINGS

Refer to 770 & 880 Supplement, paragraph 2 for Super 77 and Super 88 tractors; to 660 Supplement for Super 66 tractors.

444. The 1.2495-1.2498 diameter full floating type piston pin is retained in piston bosses by snap rings and is available in oversizes of .005 and .010.

The split type graphite bronze piston pin bushings supplied for service require final sizing after installation. Bushings should be installed in the rod so that bushing outer edge is flush with outer edge of rod bore and the split side is at the top of the rod. Two bushings are installed in each rod. Piston pin bushing in connecting rod should be sized to provide a .0005-.001 clearance on pin, and the piston bosses sized to provide a .0002-.0004 clearance.

CONNECTING RODS AND BEARINGS

445. Connecting rod bearings as used in the first production non-Super, non-Diesel engines (engines prior to serial numbers: Model 66, 784312; model 77, 785090; model 88, 781958) were of the steel-backed, babbitt base renewable type fitted with shims for a limited range of adjustment. Refer to paragraph 446. Later tractors equipped with engines having serial numbers after the aforementioned serial numbers are equipped with shimless, non-adjustable, precision type bearings.

Engines equipped with the adjustable (shim) type connecting rod bearings can be changed over to use the precision, non-adjustable type bearings by changing connecting rod and cap assembly.

All Diesels are equipped with Tocco hardened crankshafts and micro copper-lead precision type bearing shells. Similar hardened shafts were installed on the non-Diesels beginning with Super 66 at engine serial 961991; Super 77 at 964959 and on Super 88 at 958338. Super and non-Super HC engines prior to adoption of the hardened shafts were equipped with micro lead-babbit bearing shells. The micro copper-lead shells should be used only with hardened crankshafts (all service shafts are hardened type) which can be identified by these stamped numbers on the crankshaft: 181411 for series 66; 186311 for series 77 and 191311 for series 88.

Micro babbit type bearing shells are supplied in undersizes of 0.003, 0.030 and 0.033. The micro copper-lead shells used with hardened shafts are supplied in undersizes of 0.003 and 0.020-inch.

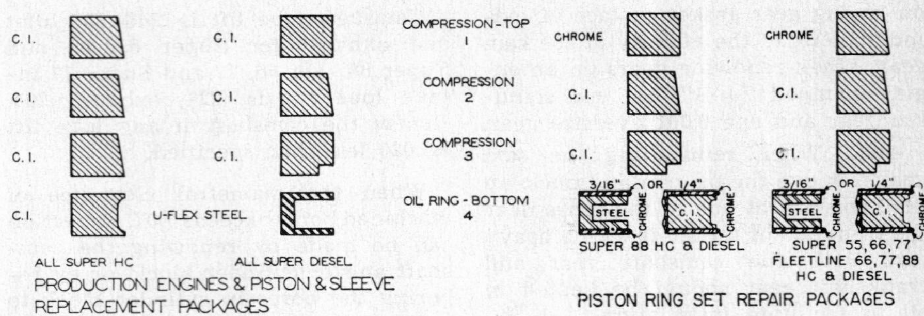

Fig. O328A — Configuration of the piston rings used in production and for service installation.

445A. Check crankshaft crankpins and bearings for wear, scoring and out-of-round condition.

Crankpin Diameter:
- Non-Super 66 & 77 ...... 1.999-2.000
- Non-Super 88 .... 2.249-2.250
- Super 66, 77 & 770 .... 2.249-2.250
- Super 88 & 880 .... 2.4375
- Running Clearance 0.0005-0.0015
- Maximum 0.0025
- Side Clearance 0.0075-0.0135

Bolt torque (Super 88 & 880) .... 56-58 ft.-lbs.
(Others) .... 46-50 ft.-lbs.

446. **BEARING ADJUSTMENT.** The following applies only to engines equipped with connecting rod bearings of the adjustable (shim) type. Connecting rod bearing shims which are provided for a limited range of adjustment do not fit between bearing shell parting surfaces, but only between bearing cap and its mating surface, as shown in Fig. O329. These shims control the amount of crush or pinch placed on shell and only indirectly control the bearing running clearance of .0005-.0015.

Individual shim thickness of the factory supplied shim pack of three shims is 0.002. In an emergency where new bearings are not available, the desired clearance can be obtained by reducing the height of the bearing shells. For each 0.002 reduction in height of a pair of shell inserts, remove a 0.002 shim from each side. Removal of metal from the parting surface of the shell (reducing) can be accomplished with fine emery paper and a flat surface, making an occasional check on the bearing shell height to prevent removal of too much metal.

For new bearing shells and crankshaft installation, use the standard

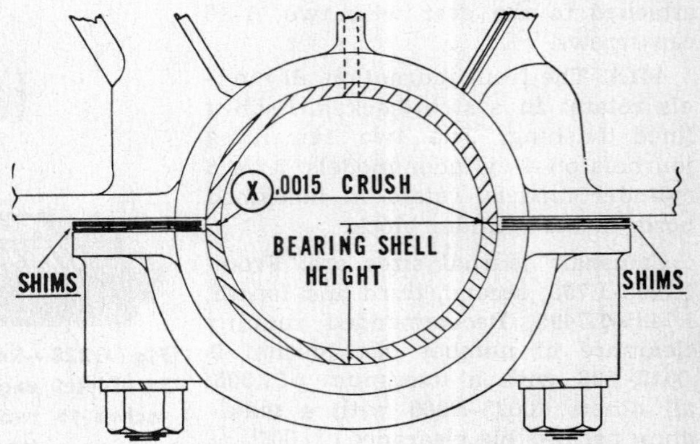

Fig. O329—Main bearing shell installation on early non-Super production non-Diesel engines. Note that the shims do not fit between bearing shell parting surfaces. Connecting rod bearing construction and adjustment is similar for early production models. Later non-Super and Super series production non-Diesel, and all Diesel engines use shimless, non-adjustable slip-in, precision rod and main bearings.

shim pack of three 0.002 shims which will automatically provide the correct running clearance of 0.0005-0.0015 and the correct bearing crush.

CRANKSHAFT AND BEARINGS

447. On non-Super model 66 non-Diesel engines prior to engine serial 783914; 77 non-Diesel engines prior to 785090, and 88 non-Diesels prior to 782665, the crankshaft is supported on slip-in type bearings provided with shims for a limited range of adjustment. Refer to paragraph 446. Main bearings used in all non-Diesel engines after the aforementioned serial numbers and all Diesel engines are of the shimless, non-adjustable precision type.

All Diesels are equipped with Tocco hardened crankshafts and micro copper-lead precision type bearing shells. Similar hardened shafts were installed on the non-Diesels beginning with Super 66 at engine serial 961991; Super 77 at 964959 and on Super 88 at 958338. Super and non-Super HC engines prior to adoption of the hardened shafts were equipped with micro lead-babbit bearing shells. The micro copper-lead shells should be used only with hardened crankshafts (all service shafts are hardened type) which can be identified by these stamped numbers on the crankshaft: 181411 for series 66; 186311 for series 77 and 191311 for series 88.

Micro babbit type bearing shells are supplied in undersizes of 0.003, 0.030 and 0.033. The micro copper-lead shells used with hardened shafts are supplied in undersizes of 0.003 and 0.020-inch.

447A. Recommended crankshaft end play of .0045 with a maximum of .010 for service is controlled by the flanged portion of the center main bearing shell on the 66, and by the number three main bearing shell on the others.

All main bearings can be renewed from below without removing the crankshaft.

In early production non-Super engines, only the upper half of the main bearing shell was flanged. On later and all Super production engines, both halves of the bearing are flanged. The later type of bearing shell is also used for servicing the earlier type.

447B. **BEARING ADJUSTMENT.** For adjustment procedure applying to adjustable (shim) type bearings, refer to paragraph 446.

448. Check crankshaft journals for wear, scoring and out-of-round condition.

Journal diameter,
  88 and 880 . . . . 2.624 -2.625
Journal diameter, others . 2.249 -2.250
Running clearance 0.0005-0.0035
  Maximum 0.0045
Renew or regrind if out-of-round more than 0.0015
Cap screw torque
  Super 88, 880 HC and Diesel . . . . 108-112 ft.-lbs.
  All others . . . . 87- 92 ft.-lbs.

CRANKSHAFT REAR OIL SEAL

449. The one piece treated cork oil seal, as shown in Figs. O330 and O332, can be renewed after removing the power take-off unit clutch and drive-shaft assembly on tractors so equipped as outlined in POWER TAKE-OFF section, paragraph 594. For tractor models equipped with hydraulic power lift but without continuous PTO, remove power lift driveshaft as outlined in POWER LIFT section, paragraph 607A. Remove engine clutch as outlined in paragraph 542 and flywheel. Cork oil seal and retainer assembly can be renewed at this time. Soak new seal assembly in oil before installing same.

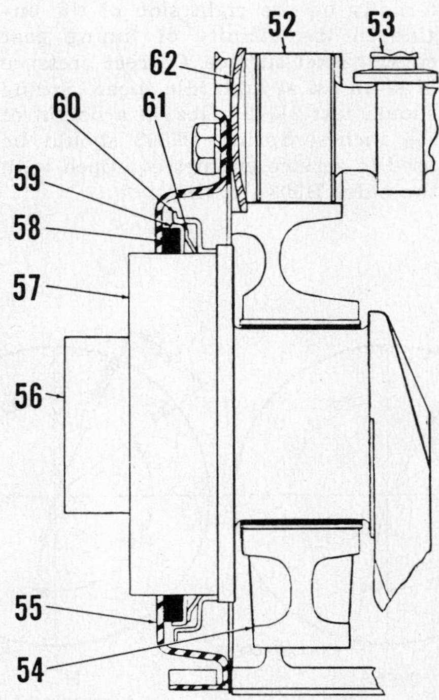

Fig. O330 — Cross-sectional view showing the crankshaft rear oil seal which is made of treated cork.

52. Camshaft
53. Cam follower
54. Rear main bearing cap
55. Oil retainer
56. Crankshaft
58. Cork seal
59. Seal retainer
60. Seal assembly retainer
61. Spring
62. Expansion plug

FLYWHEEL

450. **REMOVE & REINSTALL.** Remove power take-off clutch and drive-shaft assembly on tractors so equipped as outlined in POWER TAKE-OFF section, paragraph 594 or for tractor models equipped with hydraulic power lift but without continuous PTO, remove power lift drive shaft as outlined in POWER LIFT section, paragraph 607A. Remove engine clutch as outlined in CLUTCH section paragraph 542.

There is no need to mark the relative position of the flywheel to the crankshaft flange as same is doweled, and it can be installed in one position only.

Flywheel run-out when checked at rear face should not exceed .005. To install a new ring gear, heat same to 450 deg. F., and install the gear on the flywheel with beveled end of teeth facing the engine.

451. **TIMING MARKS.** Refer to Fig. O333. Flywheel marks are viewed through an inspection port located on right side of flywheel housing.

The non-Super 66, 77 and 88, and Super 77 and 88 non-Diesel engine flywheels are stamped as follows: TDC (indicating top center position of number one piston), INT (indicating inlet valve opens), IGN-HC (indicating gasoline and LP-Gas fully advanced timing).

The mark IGN-KD indicating the fully advanced timing point of 27 degrees ($2\frac{13}{16}$ inches) before top center for distillate engines is not always stamped on KD engine flywheels. To establish this mark, measure $2\frac{13}{16}$ inches BTDC.

The flywheel ignition timing mark "IGN-HC" on non-Super models 77

Fig. O332—Crankshaft rear oil seal and retainer (58) can be renewed after removing the engine flywheel.

62. Welch plug
63. Dowel pin

prior to engine serial 769409, and non-Super models 88 prior to tractor serial Row Crop-123301 and Standard-821086 was originally placed at 20 deg. (2⅛ inches) BTDC. Using this mark of 20 deg. BTDC to check fully advanced timing, required the engine to be operated at a rated load speed of 1600 rpm. Later non-Super 77 and 88, and all Super 77 and 88 models, including non-Super model 66, have the "IGN-HC" mark placed 22 deg. ($2\frac{9}{32}$") to permit the fully advanced timing to be checked at a no-load speed of 1750 rpm. All Super 66 non-Diesel engines have the IGN-HC mark located 28 deg. or $3\frac{7}{32}$ inches BTDC.

On engines equipped with magnetos, the flywheels are also stamped with a "MAG." mark which is located 15 degrees or $1\frac{9}{16}$ inches BTDC.

451A. Recommended injection pump timing for all Diesel engines except the Super 66, 770 and 880 is 23 degrees before TDC. On Super 66, 770 and 880 Diesel engines the FP mark is located 25 to 26 degrees BTDC. Some early production engines used in non-Super series tractors and Super 66 series tractors had the injection pump timing mark located either 23, 24 or 26 degrees BTDC.

OIL PUMP AND RELIEF VALVE

452. The vane type pump, Fig. O335, which is driven by the camshaft, is mounted on underside of cylinder block. Pump removal requires removal of oil pan and one cap screw attaching the pump body to the lower side of the cylinder block. It will be necessary to retime the ignition unit whenever the oil pump is removed.

452B. Disassembly is self-evident. Vanes should be installed to oil pump drive shaft so that flat sides of vanes will be facing the direction of normal rotation when viewed from the vane end, as shown in Fig. O335. The pump body and cover are assembled without a gasket. If the pump becomes worn do not attempt to repair it; renew the entire unit.

452C. Before installing oil pump to non-Diesel engines rotate engine crankshaft until number one piston is on compression stroke and flywheel mark "TDC" is indexed at inspection port. Install the oil pump so that the narrow side of the pump shaft, as divided by the slot (ignition unit drive slot) is on the crankshaft side and parallel to the crankshaft. Refer to Fig. O334 which shows the correct position of the ignition unit drive slot when viewed from above through ignition unit shaft hole in cylinder block. If drive slot is not in the position as shown, remove the oil pump and remesh the pump drive gear.

452D. The non-adjustable piston type oil relief valve is located externally on the right side of the engine in the vicinity of timing gear cover gasket surface. Correct pressure is 14-18 psi at high idle speed. Spring should test 2¼-2¾ lbs. at a height of $1\frac{1}{16}$ inches. Spring 1K303 should be used to service engines equipped with the older H303A spring.

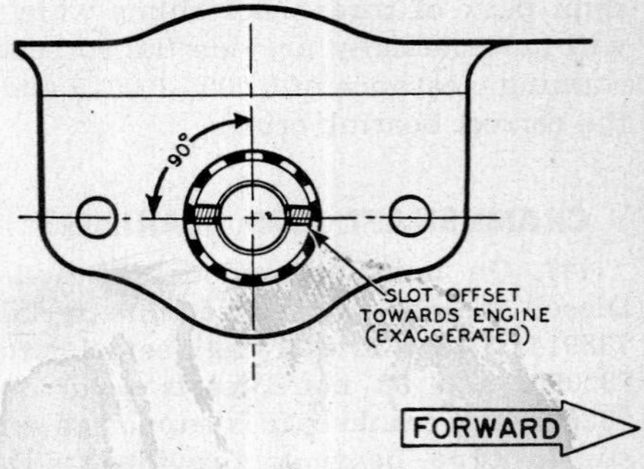

Fig. O334—Correct position of the ignition unit drive slot when viewed from the ignition unit mounting pad surface.

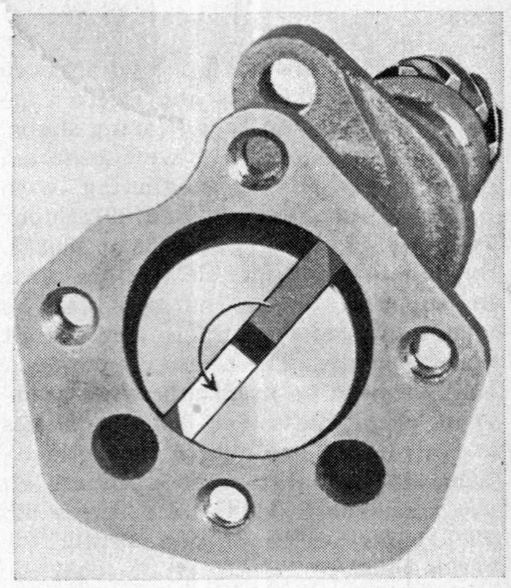

Fig. O335—Oil pump vanes should be installed so that the flat sides will face in the direction of rotation (counter-clockwise).

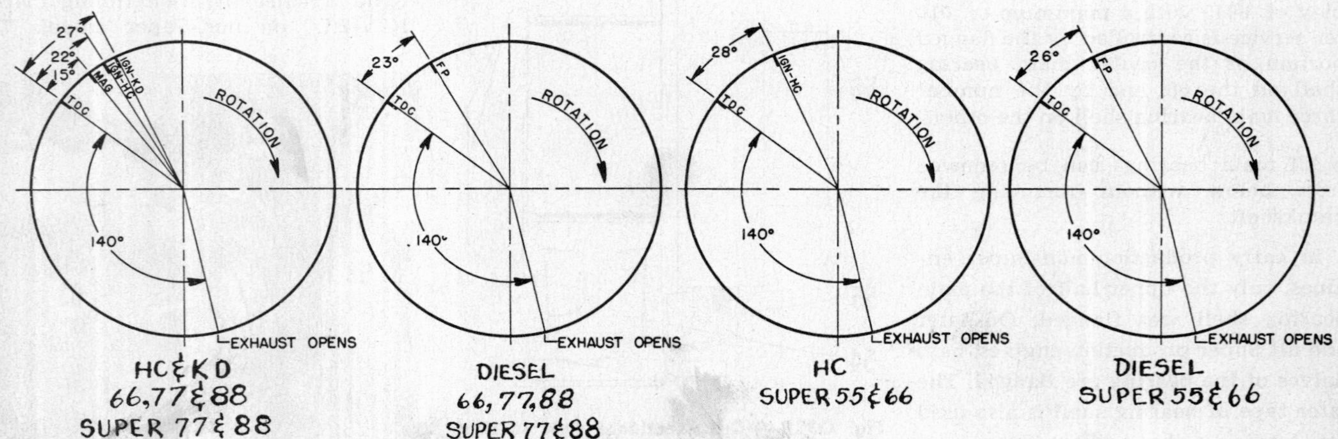

Fig. O333—Recommended flywheel timing marks on non-Super 66, 77 and 88, and Super 77 and 88 engines are as follows: IGN-HC is 22° or 2 9/32 inches BTDC; IGN-KD is 27° or 2 13/16 inches BTDC; FP is 23° or 2 41/64 inches BTDC; and MAG is 15° or 1 9/16 inches BTDC. Flywheel timing marks on Super 66 engines are: IGN-HC, 28° or 3 7/32 inches BTDC; and FP, 26° or 2 63/64 inches BTDC. If necessary, relocate flywheel marks to agree with foregoing data.

CARBURETOR

HC AND KD CARBURETOR

460. The carburetor application table applies to tractors equipped either with gasoline or distillate carbureted engines.

See accompanying table for float setting.

Clockwise rotation of idle adjusting needle richens the idle mixture; whereas similar rotation of the high speed needle leans the high speed (power) mixture.

CARBURETOR APPLICATION

Tractor Model	Carburetor Make	Carburetor Model	Float Setting
66 & 77 HC Prior 6502000	M-S	TSX-363	9/32
	Carter	UT-923S	9/32
66 HC After 6503000	M-S	TSX-603	9/32
	Carter	UT-2257S	17/64
66 KD	M-S	TSX-418	9/32
77 HC After 6503000	M-S	TSX-374	9/32
	Zenith	11580	1/4
77 KD	M-S	TSX-380	9/32
88 HC	M-S	TSX-374	9/32
88 HC Prior 6502000	Zenith	11580	1/4
88 HC After 6503000	M-S	TSX-610	9/32
	Zenith	11705	1/4
88 KD	M-S	TSX-221	9/32
770 HC	M-S	TSX-755	9/32
880 HC	Zenith	12259	1/4

CENTURY LP-GAS SYSTEM

LP-GAS CARBURETOR

Non-Super Models 77-88

Non-Super models 77 and 88 tractors are available with an LP-gas system designed and built by Century. Like other LP-gas systems, this system is designed to operate with the fuel supply tank filled to a level of not more than 80 per cent.

The LP-gas system consists of a model 3C-705A carburetor, series MO converter (regulator) equipped with a solenoid operated primer, a combined gas filter and fuelock (solenoid operated fuel shutoff), and a fuel tank equipped with a liquid filler valve, liquid outlet valve, vapor return valve, slip gauge, fixed level valve, and pop-off valve.

The Century model 3C-705A carburetor has 2 points of mixture adjustment, plus an idle stop screw. Refer to Figs. O336 and O337. This carburetor is designed for the engine to start with throttle valve in fully closed position.

470. **IDLE STOP SCREW.** Idle speed stop screw on the carburetor throttle should be adjusted to provide an idle speed of 375 crankshaft rpm (231 rpm at the belt pulley) for battery ignition equipped engines, or 500 crankshaft rpm (307 rpm at the belt pulley) for magneto ignition equipped engines.

471. **IDLE MIXTURE.** This adjustment synchronizes the carburetor throttle valve (9) to the gas valve (10). Make the adjustment as follows: With the engine warm, place the governor hand control lever (located on steering shaft column) in the idle position (top notch in quadrant). Remove the cotter pin from one end (A or B) of the carburetor drag link (7). Rotate the drag link adjusting screw (A or B) **in** or **out** until the engine runs smoothly. Rotating the adjusting screw clockwise (**in**) enriches the mixture. Adjust the idle stop screw to provide an idle speed of 375 crankshaft rpm for battery ignition equipped engines or 500 rpm for magneto ignition equipped engines.

Open the throttle to high idle speed of 1750 rpm and rotate the drag link adjusting screw (A or B) **in** or **out** until the engine runs smoothly. Recheck the low idle speed of 375 rpm (500 rpm for magneto equipped engines) and readjust if necessary; then, recheck the high idle speed drag link

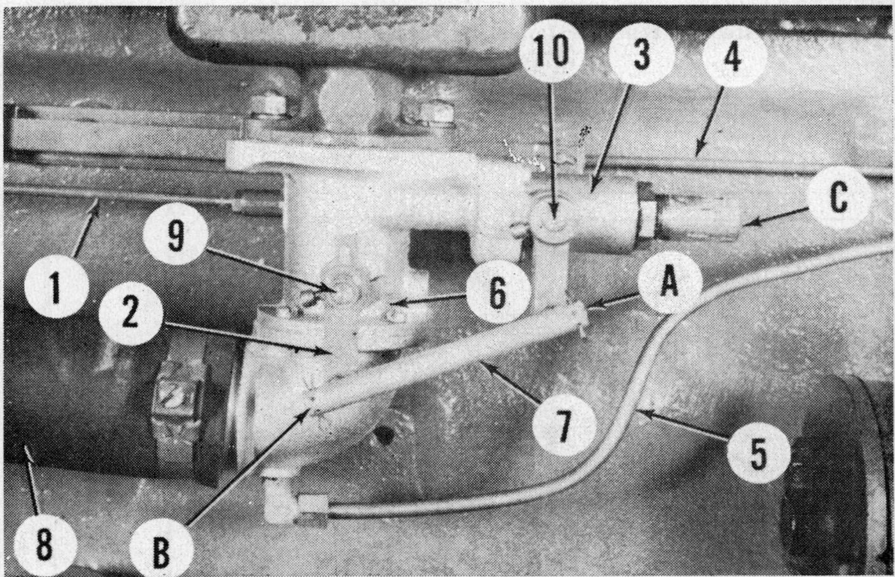

Fig. O336—Century model 3C-705A LP-Gas carburetor installation on non-Super models 77 and 88.

C. Carburetor fuel inlet
1. Governor to carburetor linkage
2. Throttle valve lever
3. Gas valve body
4. Hand control linkage
5. Balance line
6. Throttle body
7. Drag link
8. Air cleaner to carburetor hose
9. Throttle valve
10. Gas valve

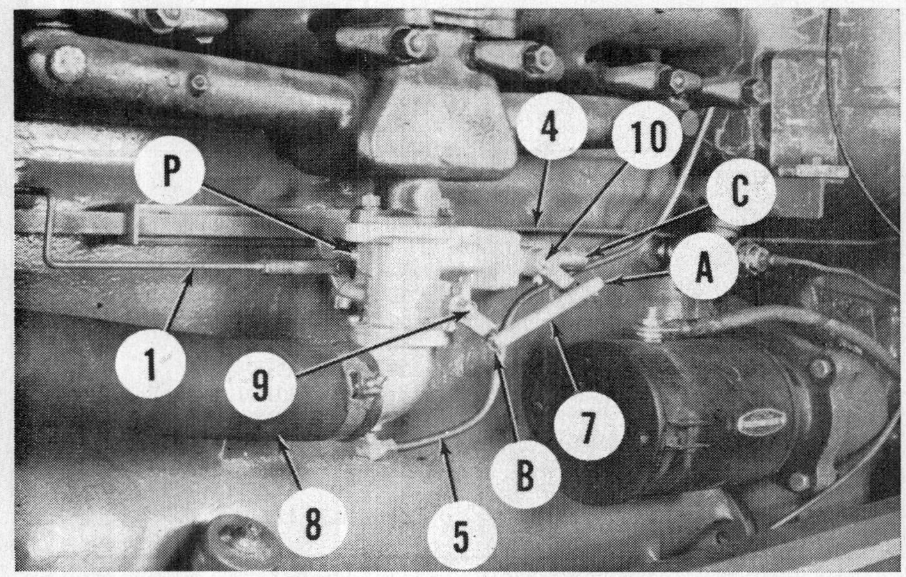

Fig. O337—Century model 3C-705A LP-Gas carburetor installation on non-Super models 77 and 88.

C. Carburetor fuel inlet
P. Spray bar (power adjustment)
1. Governor to carburetor linkage
4. Hand control linkage
5. Balance line
7. Drag link
9. Throttle valve
10. Gas valve

adjustment. Lock the adjustment by inserting a cotter pin in the drag link.

472. **POWER ADJUSTMENT.** In the model 3C-705A Century carburetor, the power mixture adjustment is controlled by the position of the carburetor spray bar (P—Fig. O337). The spray bar has no effect on the mixture except at wide open throttle.

472A. CENTURY METHOD WITHOUT ANALYZER. The spray bar is factory adjusted so that the holes in the bar are at approximately a 45 degree angle as shown in Fig. O339. With the engine running, loosen the spray bar set screw and rotate the spray bar (P—Fig. O337) slightly toward the "L" (lean) mark stamped on carburetor body. Open throttle valve quickly to the wide open position. If engine falters, rotate spray bar toward the "R" (rich) mark. Desired position of spray bar is where best acceleration at leanest setting is obtained. Tighten set screw to lock the adjustment.

472B. ANALYZER METHOD. In this method (Oliver's and Century's) the engine is operated with the carburetor throttle wide open and with sufficient load on the engine to hold the rpm to a maximum operating speed of 1600 rpm. Rotate the spray bar (P—Fig. O337) to give a reading of 12.5 on the exhaust analyzer with a gasoline scale or 14.0 on the analyzer with a LPG scale.

LP-GAS FILTER AND FUELOCK

473. The combined filter and fuelock are mounted on the tank unit. Refer to Fig. O340. The filter pack consists of felt pad, chamois and 2 brass screens. The fuelock is a solenoid operated fuel shut-off valve which operates on the ignition circuit.

473A. In most cases, the fuel strainer can be cleaned by opening the drain cock (located on lower side of filter) and then opening the fuel tank liquid valve (15) to blow out any accumulation of dirt or water.

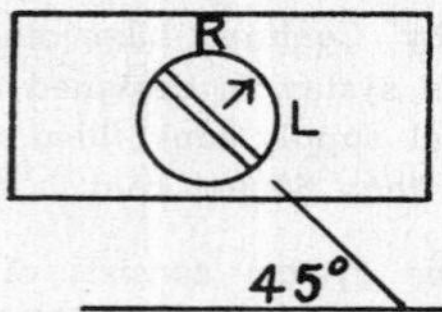

Fig. O339—The carburetor spray bar (power adjustment) is factory adjusted so that the holes are at a 45 degree angle.

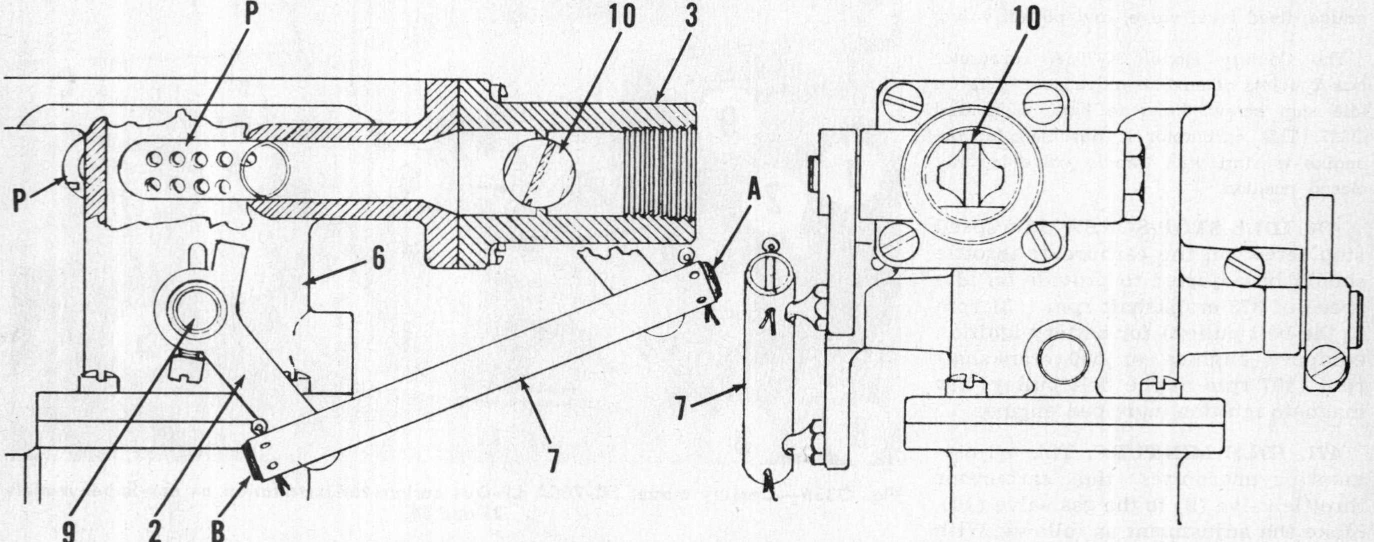

Fig. O338—Sectional view of a Century model 3C-705A LP-Gas carburetor showing the spray bar (P) and gas valve (10).

P. Spray bar (power adjustment)
2. Throttle valve lever
3. Gas valve body
6. Throttle body
7. Drag link
9. Throttle valve
10. Gas valve

Fig. O340—LP-Gas fuel filter and fuelock, tank, and regulator installation on non-Super models 77 and 88.

D. Drain cock
O. Water outlet
5. Balance line
12. Regulator cover
13. Regulator fuel inlet
15. Liquid valve
16. Solenoid (fuelock)
17. Solenoid to ignition coil wire
18. Fuelock
19. Fuse
20. Filter
22. Primer valve
23. Solenoid (primer)

473B. If for any reason the solenoid operated shut-off valve does not function properly, the valve can be disconnected from the fuel circuit by disconnecting the regulator to filter line (13); then remove the hexagon shaped valve body (18) and extract the shut-off valve. Reinstall the valve body and reconnect the filter to regulator line.

LP-GAS REGULATOR

474. **HOW IT OPERATES.** Refer to Figs. O340, O341 and O342. Fuel from the supply tank passes through the fuel filter (20) and the solenoid operated fuelock valve (18) and enters the regulating unit inlet (13) at a tank pressure of 20 to 175 psi and into the surge chamber where it is reduced from tank pressure to about 6 psi. Flow through the fuel inlet valve is controlled by the inlet pressure diaphragm (36). When the fuel enters the surge chamber, it expands rapidly and is converted from a liquid to a gas by heat from the engine coolant system. The vaporized fuel, then passes via the fuel outlet valve (43) into the low pressure chamber where it is drawn off at a pressure slightly below atmospheric to the carburetor via outlet (C). The fuel outlet valve is controlled by a large diaphragm (30) and spring (44). Unlike other regulating units, this unit has only the one fuel outlet which supplies all of the vaporized gas to meet every throttle opening. The other regulating units, usually have a separate line to supply the idling range demands of the engine.

A balance line (5) is connected to the carburetor air inlet horn so as to reduce the flow of fuel and thus prevent over richening of the mixture which would otherwise result when the air cleaner or the air inlet system becomes restricted.

A solenoid operated primer (23) which opens the fuel outlet valve is mounted on the front face of the regulating unit. The primer is used for starting purposes to fill the lines when the throttle is closed. The primer which can be operated manually is actuated by a separate switch located on the instrument panel.

Fig. O341—Left side installation view of the LP-Gas equipment used on non-Super models 77 and 88.

C. Regulator fuel outlet
D. Drain cock
O. Water outlet
P. Spray bar (power adjustment)
W. Water inlet
5. Balance line
7. Drag link
9. Throttle valve
10. Gas valve
13. Regulator fuel inlet
16. Solenoid (fuelock)
22. Primer valve
23. Solenoid (primer)
25. Solenoid to instrument panel switch

475. **REGULATOR OVERHAUL.** Remove the unit from the fuel tank and completely disassemble, using Fig. O342 as a reference. Thoroughly wash all parts and blow out all passages with compressed air. Inspect each part carefully and discard any which are worn.

475A. Before reassembling the unit, check the inlet pressure diaphragm for proper adjustment as follows: With the fuel inlet valve closed, the shoulder (X—Fig. O342) of the diaphragm lever link should be level or not more than 1/16 inch above the machined face of the main casting for the inlet pressure diaphragm. To obtain this adjustment, bend the valve lever (40) at the valve contacting end.

Continue to reassemble the unit, and before installing the fuel outlet diaphragm (30) and unit cover (12), note dimension (Y) which is measured from the face of the low pressure side of the casting to the diaphragm contacting surface of the fuel outlet valve operating lever (42) when the valve is in the closed position. Dimension (Y) should be ⅛-5/32 inch and can be obtained by bending the fuel outlet valve lever.

475B. After installing the regulator on the fuel tank, bleed the air from the regulator unit water circuit by opening the drain cock (D—Fig. O341) until there is a continuous flow of water without air bubbles.

GENERAL TROUBLE-SHOOTING

476. **SYMPTOM.** A sweating or frosting fuel filter.

CAUSE AND CORRECTION. Fuel filter is clogged with dirt which is restricting the flow of fuel. In most cases, the strainer can usually be cleaned by opening the drain cock located on the lower side of the filter bowl; then opening the fuel tank liquid valve so as to blow the dirt and water out of the filter.

477. **SYMPTOM.** Cold regulator shows moisture and frost after standing.

CAUSE AND CORRECTION. Trouble is due either to leaking valves or the valve levers are not properly set.

The fuel inlet valve can be cleaned and renewed as follows: Disconnect fuel filter to regulator line and remove the high pressure valve and body. The valve seat (39—Fig. O342) can be removed and turned over so that unused portion of the seat contacts the shoulder of the fuel inlet fitting.

The fuel outlet valve can be cleaned as follows: First remove the regulator cover (12). Open the fuel tank liquid valve and press down on the valve lever (42) to allow liquid fuel to wash over the seat.

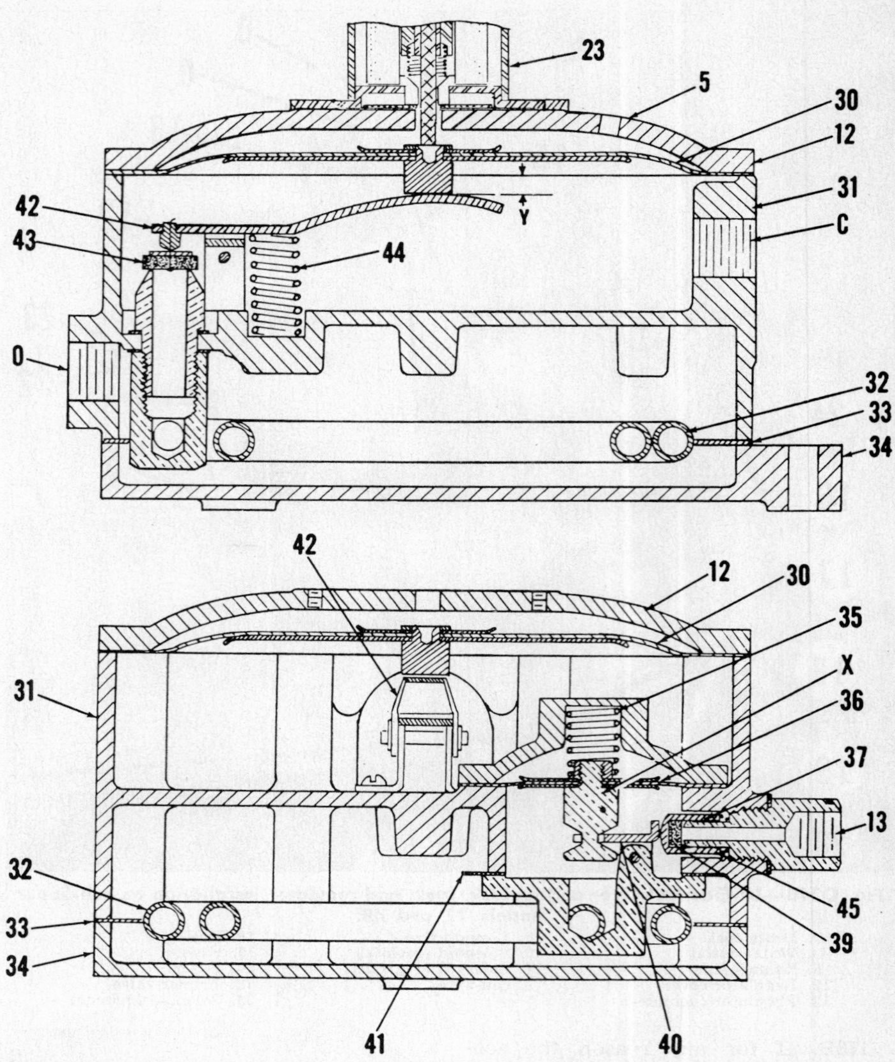

Fig. O342—Cross-sectional views of the LP-Gas regulator used on non-Super models 77 and 88. Top view is sectioned through the fuel outlet valve (low pressure) section. Lower view is sectioned through the fuel inlet valve (high pressure) section.

C. Fuel outlet
O. Water outlet
5. Balance line connection
12. Cover
13. Fuel inlet
23. Solenoid (primer)
30. Outlet valve diaphragm
31. Regulator body
32. Fuel vaporizer coil
33. Gasket
34. Regulator base
35. Spring
36. Inlet valve diaphragm
37. Inlet valve
39. Inlet valve seat
40. Valve lever
42. Valve lever
43. Outlet valve
44. Spring
45. Spring

Check the valve levers as per paragraph 475A.

478. **SYMPTOM.** Regulator body shows frost when engine is hot.

CAUSE AND CORRECTION. Poor water circulation through the heat exchanger is the cause of the trouble. To correct the trouble, it may be necessary to bleed the air from the regulator by opening the drain cock (D—Fig. O341), located on top of regulator, until there is a continuous flow of water without air bubbles. Also, check the water supply lines for restrictions.

479. **SYMPTON.** Engine fails to start.

CAUSE AND CORRECTION. Restricted fuel filter, inoperative fuelock (shut-off) valve, an overprimed engine or a low fuel tank can be the cause of the trouble.

Check and clean the fuel filter as per paragraph 473A.

The fuelock (shut-off) valve can be checked by turning the ignition switch to the on position and listening for a ping as the plunger valve is unseated. If the plunger valve ping cannot be heard, check the fuse which is inserted in the wire connecting the ignition switch to the fuelock. The valve can also be disconnected (removed) from the fuel circuit as per paragraph 473B.

DIESEL SYSTEM

480. Main components of the Diesel engine fuel system are: Refer to Fig. O343A. An American Bosch single 70MM plunger, constant stroke, sleeve control type model PSB4A fuel injection pump for series 66 engines or model PSB6A 70MM plunger fuel injection pump for series 77 and 88 engines. Injection pump is driven at crankshaft speed by an idler gear which meshes with the crankshaft gear. American Bosch or CAV closed pintle type fuel injectors are used.

Pre-combustion chambers (energy cells) are located in the combustion chambers of the cylinder head.

A flyweight type governor which is used to control the fuel delivery as a function of speed control is an integral part of the injection pump.

A camshaft operated, automotive diaphragm type primary fuel supply pump is used to supply fuel to the inlet side of the injection pump.

A primary fuel filter of the renewable cartridge type.

A secondary (final stage) fuel filter.

A heater plug (pre-heater) located in the inlet manifold elbow to assist engine starting in cold weather.

GENERAL TROUBLE-SHOOTING

481. The following data, supplied through the courtesy of American Bosch Company, should be helpful in shooting trouble on the Oliver Diesel engines.

481A. **SYMPTOM.** Engine does not idle well; erratic fluctuations.

CAUSE. Could be caused by faulty fuel injectors, also by a dirty overflow valve on pumps so equipped. The overflow valve should be removed and washed in cleaning solvent.

481B. **SYMPTOM.** Intermittent or continuous puffs of black smoke from exhaust.

CAUSE. Faulty fuel injector, also improper engine operating temperature can be the cause of the trouble.

481C. **SYMPTOM.** Fuel oil builds up (dilution) in the engine crankcase.

CAUSE. The trouble could be caused by a leaking gasket under the delivery valve, or badly worn plunger. The remedy for any of these conditions would be renewal of the complete hydraulic head as a unit as outlined in paragraph 485.

481D. **SYMPTOM.** Sudden heavy black smoke under all loads.

CAUSE. This calls for removal of the entire injection pump assembly for handling by competent personnel. The difficulty possibly is caused by a stuck displacer piston. Other possible causes are improperly adjusted smoke cam or dilution of the fuel by engine oil being by-passed by a damaged hydraulic distributor head filter.

481E. **SYMPTOM.** Poor fuel economy.

CAUSE. Water temperature too low. Check thermostat for proper functional control. Check for leakage.

481F. **SYMPTOM.** Engine low in power.

CAUSE. Filter between supply pump and injection pump may be clogged, or fuel supply is faulty. Due to type of fuel used, it may be necessary to advance the timing. Under no circumstances should the timing be advanced more than 4 degrees. Refer to paragraph 486.

481G. **SYMPTOM.** Engine rpm too low at full throttle position.

CAUSE. Could be caused by improper setting of the throttle linkage. Remove pump control lever cover and check if full travel is obtained at full load position of throttle control lever.

BLEEDING THE SYSTEM

482. Refer to Fig. O345. The Diesel fuel system should be bled or purged whenever the system has been disconnected or fuel tank emptied. To bleed the low pressure side of the system (fuel tank to injection pump), proceed as follows: Loosen bleed screws located on the primary fuel filter and the final stage filter. Operate priming lever (113—Fig. O347) on fuel supply pump with full strokes until clear fuel (free of air bubbles) flows past fuel filter bleed screw on the primary filters; then close the bleed screw. Repeat this procedure on the secondary filter; then bleed the high pressure system.

To bleed the high pressure system, loosen the fuel line connections at the fuel injectors; then using the starting motor, crank engine until clear fuel (free of air bubbles) flows past the fuel injector connections.

Fig. O343—Bosch model PSB6A injection pump as used on early production Diesel engine tractor models 77 and 88. Later pumps have hydraulic head shown in Fig. O343C but do not have the "PC" mark as shown. A similar pump is used on the model 66 tractor four cylinder engine.

A. Port closing line mark on drive hub
PC. Port closing line mark in pump timing port
4. Fuel inlet
7. Fuel shut-off arm
8. Lubrication line
9. Rated engine speed adjustment
12. No. 1 cylinder outlet
71. Governor linkage
72. Idle speed adjustment
73. Governor fork lever
74. Spring
75. Governor housing
76. Hub retaining screw
77. Drive gear hub
78. Pump cover
79. Plunger drive gear
80. Operating lever
82. Hydraulic head

INJECTION PUMP

483. Four cylinder series 66 diesels are equipped with Bosch pumps on which the first 8 digits in the model designation on the pump name plate are PSB 4A 70Z or PSB 4A 70Y. For the six cylinder 77 and 88 series the first 8 digits are PSB 6A 70Z or PSB 6A 70Y. For the 770 and 880 the first 8 digits are PSB 6A 70V.

New replacement pumps supplied by Oliver Corporation are calibrated for the 88 and 880 engine. When such a pump is installed on a 77 or 770 the smoke adjustment should be decreased one full turn to decrease the delivery for the smaller engine.

483A. **TROUBLE SHOOTING.** The following data, paragraphs 483B and 483C, should be helpful in shooting trouble on the injection pump.

483B. **FAULTY ACCELERATION.** If the engine fails to accelerate or to respond to the throttle, the trouble may be caused by excess friction in the pump control unit assembly 48—Fig. O343A. To check for excess friction in the pump control unit assembly, proceed as follows: With engine running, remove timing window from side of pump and observe action of the pump control arm which is connected to the governor control rod

If arm is sticking or moving erratically, stop the engine and disconnect the governor control rod. When arm is manually moved to either extreme of its movement, it should drop to neutral by its own weight. If movement is sticky, remove the pump control unit assembly by removing two attaching screws. Wash the control unit in an approved solvent. If this treatment does not produce free movement, disassemble the unit and clean the bearing surfaces by lapping with mutton tallow.

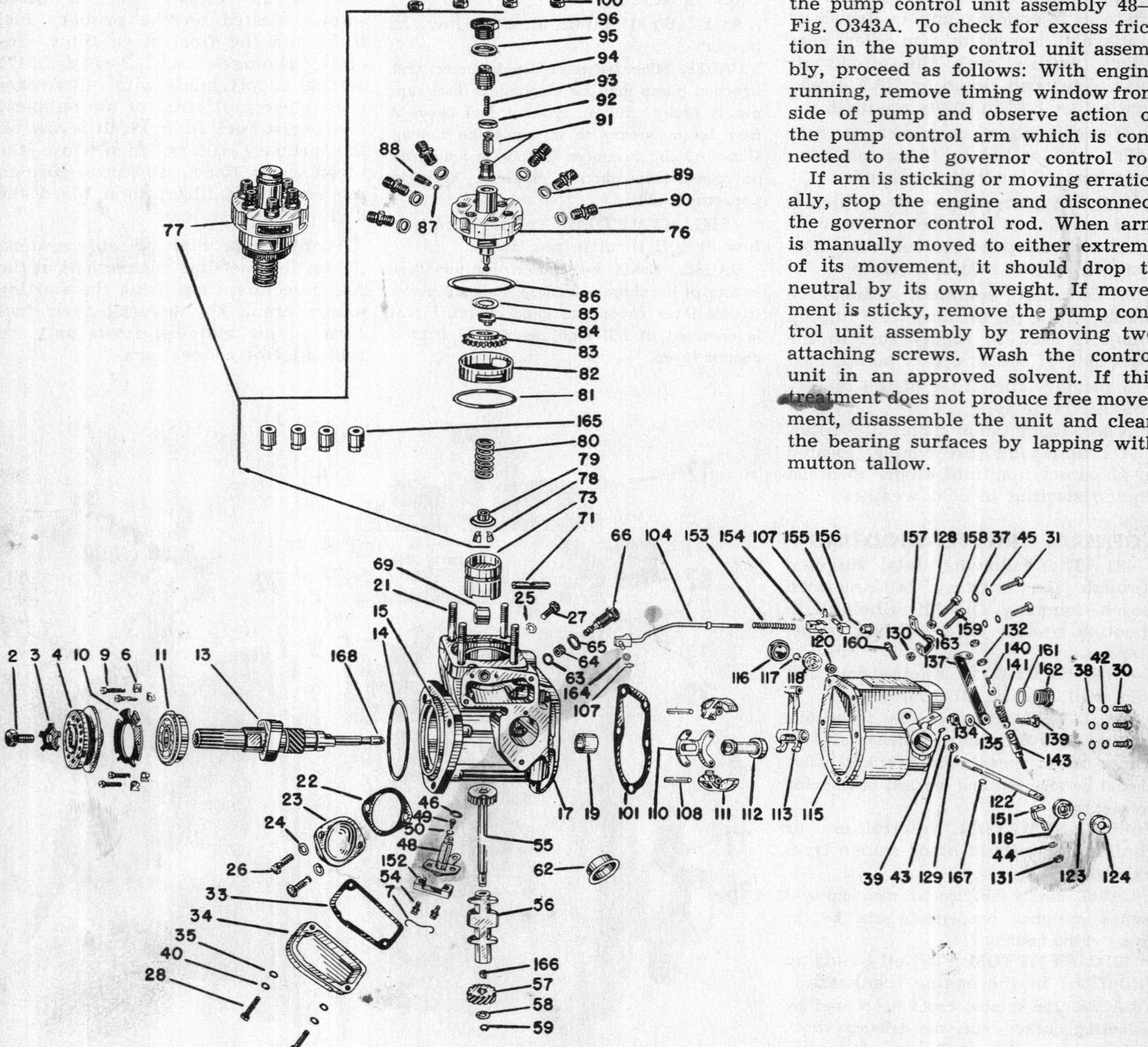

Fig. O343A—Components of American Bosch PSB type injection pump used on series 66, 77 and 88 diesel powered tractors. Field replacement pumps are basically the same for 77, 770, 88 and 880 except flow calibration.

3. Hub lockwasher
4. Drive hub
11. Camshaft bearing
13. Camshaft
15. Timing pointer
17. Pump housing
19. Camshaft bushing
23. Quill shaft pad
34. Timing window cover
48. Control unit
50. Snap ring
55. Quill shaft and gear
56. Quill shaft bushing
57. Camshaft driven gear
59. Snap ring
64. Filter screen
66. Filter screw
69. Tappet roller
71. Pin for roller 69
73. Tappet guide
76. Hydraulic head
77. Hydraulic head
78. Plunger lock
79. Spring seat, lower
80. Plunger inner spring
82. Gear retainer cover
83. Plunger drive gear
84. Plunger guide
85. Thrust washer
90. Discharge fitting
91. Delivery valve
93. Spring for valve 91
94. Holder for valve 91
100. Head nut
104. Control rod
108. Governor weight pin
110. Weight spider
111. Governor weight
112. Governor sleeve
113. Fulcrum lever
115. Governor housing
118. Bearing for shaft 122
120. Spacer for shaft 122
122. Fulcrum lever shaft
123. Snap ring
124. Seal for shaft 122
132. Spring adjusting nut
137. Operating lever
139. Pivot screw for 137
140. Spring adjusting screw
152. Control unit retainer
153. Control rod spring
156. Full load adjusting nut
157. Stop plate
158. Idle stop screw
159. Fuel shut-off screw
160. Full load stop screw
163. Stop screw spacer
165. Sleeve

If control unit arm swings freely when it is disconnected from the governor control rod, check the governor rod for full travel and smooth operation. If binding is encountered, remove governor housing from end of injection pump and clean or renew the parts as required to remove wear or eliminate sticking.

483C. **HARD STARTING, POOR PERFORMANCE.** Before condemning the pump assembly as the cause of these troubles, check condition of pump plunger as follows: Remove hydraulic head assembly as outlined in paragraph 485. Unload plunger spring (80—Fig. O343A) and remove split cone locks (78). Pry off gear retainer cover (82) and remove the plunger and control sleeve. Plunger, control sleeve and hydraulic head are matched parts.

Inspect these parts for scuff marks, scratches, and dull appearance of the lapped surfaces. A dull appearance indicates considerable wear. If any of these conditions exist, install a new hydraulic head assembly as per paragraph 485.

Install control sleeve with slot facing towards name plate side of pump head; then carefully install the plunger.

484. **DELIVERY CALIBRATION.** Fuel delivery data in accompanying table is for shops equipped to test and calibrate Bosch pumps.

FUEL DELIVERY CC PER NOZZLE FOR 500 TEST PLUNGER STROKES

Engine Rpm	66, 77	Super 66	Super 77	88	Super 88
200-400	15-16	17.5-18.5	17.5-18.5	18.5-19	22-23
1100	16.5-19	....	19-21.5	20-22.5	23.5-26
1350	....	19-21.5	....	....	....
1600	15-16	....	17.5-18.5	18.5-19.5	22-23
2000	....	17.5-18.5	....	....	....

Fig. O343C—American Bosch PSB injection pump after control arm unit has been removed. Engine oil filter (38) is being removed preparatory to removal of hydraulic head unit as per paragraph 485.

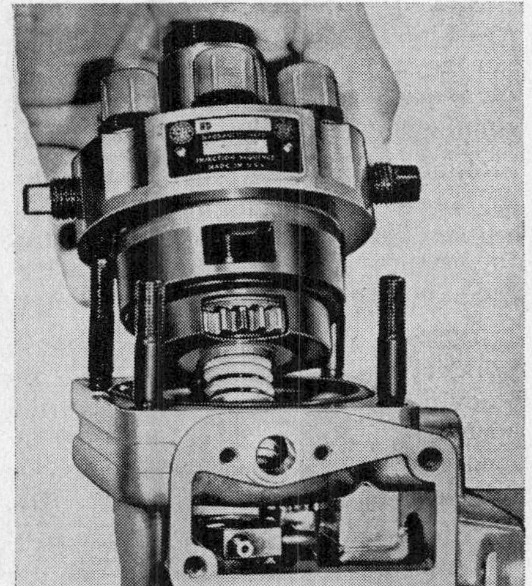

Fig. O343B—Injection pump hydraulic head assembly can be removed as shown after doing the preliminary work described in paragraph 485.

Fig. O344—Injection pump installation on model 77. Cap screw (72) controls engine idle speed, and nuts (9) control rated engine speed. Pump installation on other models is similar.

1. Fuel return
4. Fuel inlet
7. Fuel shut-off arm
8. Lubrication line
9. Rated speed adjustment
10. Pump timing port cover
11. Pump inspection port plug
12. No. 1 cylinder outlet
72. Idle speed adjustment
78. Pump cover
83. Hand-control linkage
85. Pump gear cover

When injection pump test stand equipment is not available, proceed as outlined in paragraph 484A.

484A. Load the engine until speed drops to 1200-1400 for Super 66, 770 and 880 engines, and 1000-1200 rpm for all other engines when the throttle hand control is in the wide open position. Using a ⅜-inch deep socket or a screwdriver, rotate the governor control rod adjusting nut, or stop screw Fig. O349 on early pumps, until the engine exhaust shows a slight indication of black smoke. Rotating the nut counter-clockwise lessens the smoke, but similar rotation of the stop screw increases the smoke.

485. **R & R HYDRAULIC HEAD UNIT.** Installation of a new or exchange hydraulic head unit is sometimes the indicated remedy when injection pump trouble is encountered. To remove the hydraulic head, proceed as follows: Remove pump timing window, and rotate the engine until the marked tooth of plunger drive gear (79—Fig. O343) is approximately in register either with stamped arrow head mark or stamped "O" mark. Remove pump control unit (80), and lubricating oil filter (38—Fig. O343C). Remove fuel lines, and nuts attaching hydraulic head to pump and lift off the head as shown in Fig. O343B. Do not attempt to lift off the hydraulic head without indexing the marked tooth of plunger drive gear with arrow head mark or "O" mark located in pump window. If the plunger drive gear is not indexed properly, a sheet metal plate on top of the quill shaft gear will prevent unmeshing of the plunger drive gear.

To reinstall the hydraulic head, reverse the removal procedure; and use new "O" rings.

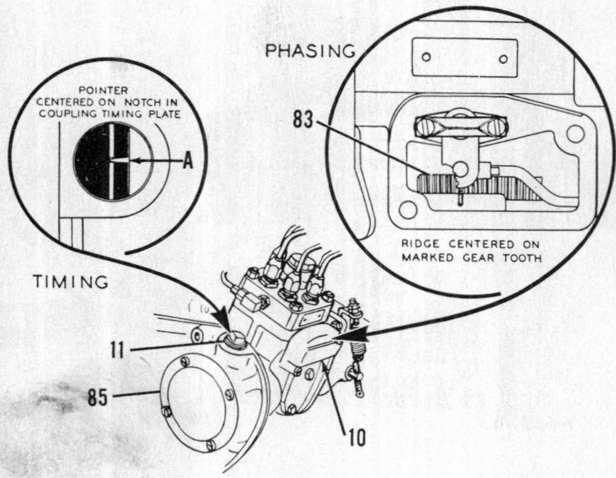

Fig. O344A—Reference points to be observed when timing and phasing the American Bosch PSB type pumps to Oliver Diesel engines as outlined in paragraphs 486 through 488.

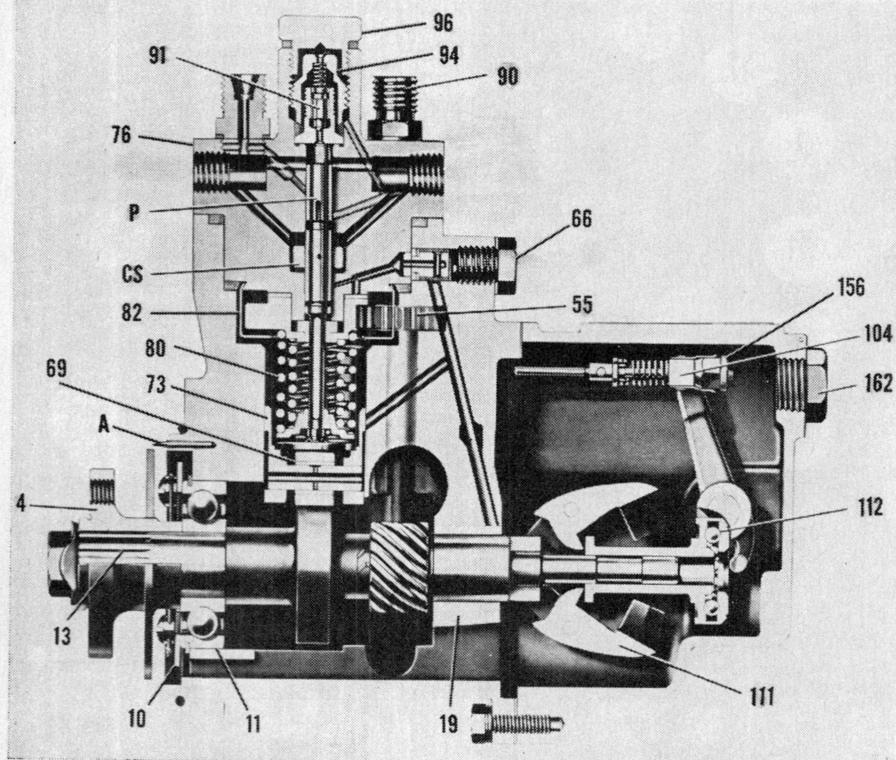

Fig. O344B—Sectionalized typical American Bosch type PSB injection pump. Pumps used on Oliver Diesel tractors are basically similar.

A. Timing pointer
CS. Plunger guide
P. Plunger
4. Drive coupling
10. Bearing retainer
11. Camshaft bearing
13. Camshaft
19. Camshaft bushing
55. Quill shaft and gear
66. Filter nut
69. Tappet roller
73. Tappet guide
76. Hydraulic head
80. Outer spring
82. Gear retainer cover
90. To nozzles (injectors)
91. Delivery valve
94. Holder for valve 91
96. Valve cap screw
104. Control rod pin
111. Governor weight
112. Governor sleeve
156. Load adjusting nut
162. Housing plug

486. **TIMING PUMP TO ENGINE.** Injection pump should be timed to engine so that closing of pump plunger port for number one outlet occurs when engine flywheel mark "FP" is indexed with flywheel inspection port notch. Timing mark FP for recommended injection pump timing on all but Super 66, 770 and 880 engines is located 23 degrees before TDC. Timing mark FP for Super 66, 770 and 880 is located 25-26 degrees before TDC. When it is known that the pump internal timing is O. K., and the flywheel mark FP is correctly located; the pump can be timed to the engine as follows:

486A. Remove inspection port plug (11—Fig. O344A) located in the timing gear cover directly above the pump drive gear. Also, remove the timing window cover (10) from side of pump.

Rotate engine crankshaft until number one piston is coming up on compression stroke; then slowly, until flywheel mark "FP" is aligned with flywheel housing inspection port notch.

NOTE: Compression stroke of No. 1 cylinder can be determined either by removing the pipe plug and cap from the No. 1 cylinder energy cell, or by removing the valve rocker arms cover and observing the closing of the No. 1 cylinder inlet valve.

At this time, the line mark on the injection pump coupling hub should be in register within $\frac{1}{32}$ inch with the timing pointer which extends from the front face of the pump. Refer to left circle in Fig. O344A. Both the line mark and timing pointer are viewed through the inspection port plug opening (11). If at this time, the marked tooth of the plunger drive gear (83) is registered within ¼ inch with the arrow or "O" mark stamped on timing window ledge the pump can be installed. If injection pump coupling hub mark is not in register with the timing pointer, as stated, it will be necessary to remesh the pump gear as outlined in paragraph 468B.

486B. **REMESHING PUMP GEAR.** To remesh pump gear for retiming pump to engine, proceed as follows: First position engine crankshaft so that flywheel mark "FP" is indexed with flywheel housing inspection port notch when No. 1 cylinder is on compression stroke. Refer to Flywheel Timing Marks, paragraph 451, to establish correct location of timing mark FP. Remove the pump gear cover which is located on front face of engine timing gear cover. Working through pump gear cover opening, remove two cap screws retaining pump drive gear (47—Fig. O346A). Using a socket wrench, rotate pump camshaft until pump coupling hub line mark registers with timing pointer (A—Fig. O344A); then reinstall pump drive gear retaining cap screws. Several trials of remeshing pump drive gear with idler may be necessary before finding the holes which will admit the cap screws without throwing hub line mark out of register with the timing pointer.

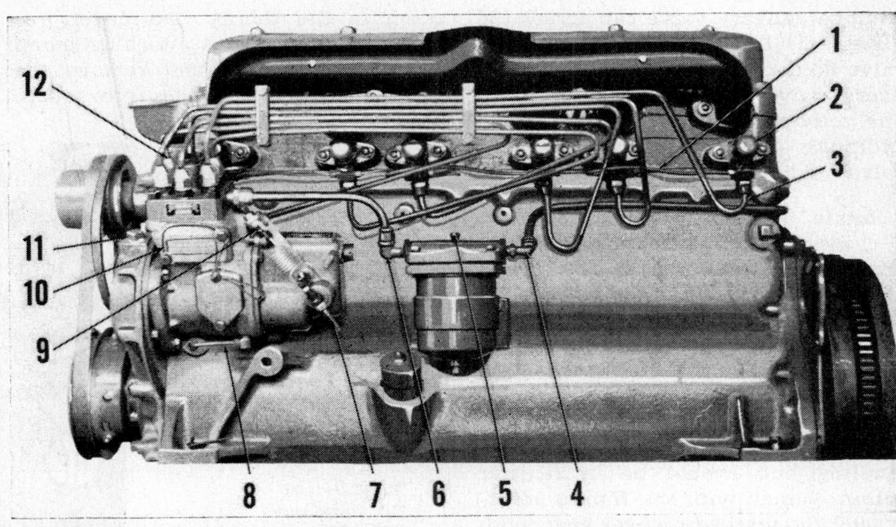

Fig. O345—Oliver 6 cylinder Diesel engine (left side). Model 66 engine is similar except that the primary fuel filter is located on the right side of the engine.

1. Fuel return line
2. Nozzle
3. High-pressure fuel line
4. Primary filter inlet
5. Bleed screw
6. Primary filter outlet
7. Hand control
8. Pump lubricating line
9. High speed no-load governor adjustment
10. Pump timing window
11. Pump timing plug
12. Location of No. 1 nozzle connection on pump head

Recheck injection pump to engine timing, and lock the pump drive gear cap screws with wire.

487A. **PUMP INTERNAL TIMING.** If pump has been disassembled, it should be timed internally at reassembly as follows: Refer to Figs. O344B and O344C.

Insert pump camshaft (13) into pump housing so that wide groove or keyway at drive end of camshaft is in register with the "CLW" mark on bearing retainer plate (10). Install the quill gearshaft (55) through bottom of pump housing so that when the spiral gear (located on lower end of quill gearshaft) is meshed with spiral gear on camshaft, the open tooth of the spur gear (located at upper end of quill gearshaft) will be in register with the drill mark which is located on the counterbore of the pump housing. Install hydraulic head assembly so that line marked tooth of plunger drive gear is in register with the stamped arrow or "O" mark which is located on the timing window ledge.

487B. PUMP INTERNAL PHASING. Injection pump is correctly phased internally when the flow of fuel from hydraulic head No. 1 outlet ceases immediately when pump coupling hub line mark is in register with the timing pointer. Phasing, which establishes the injection pump port closing point by using the flow method, is checked as follows:

Mount injection pump in a vise, and connect a fuel oil line from a gravity supply tank to the inlet side of the hydraulic head. Remove constant bleed (overflow) line fitting from hydraulic head and replace it with a ¼ inch pipe plug. Bleed pump of all air by loosening this pipe plug and rotating the pump camshaft. Next, place operating lever in full load position.

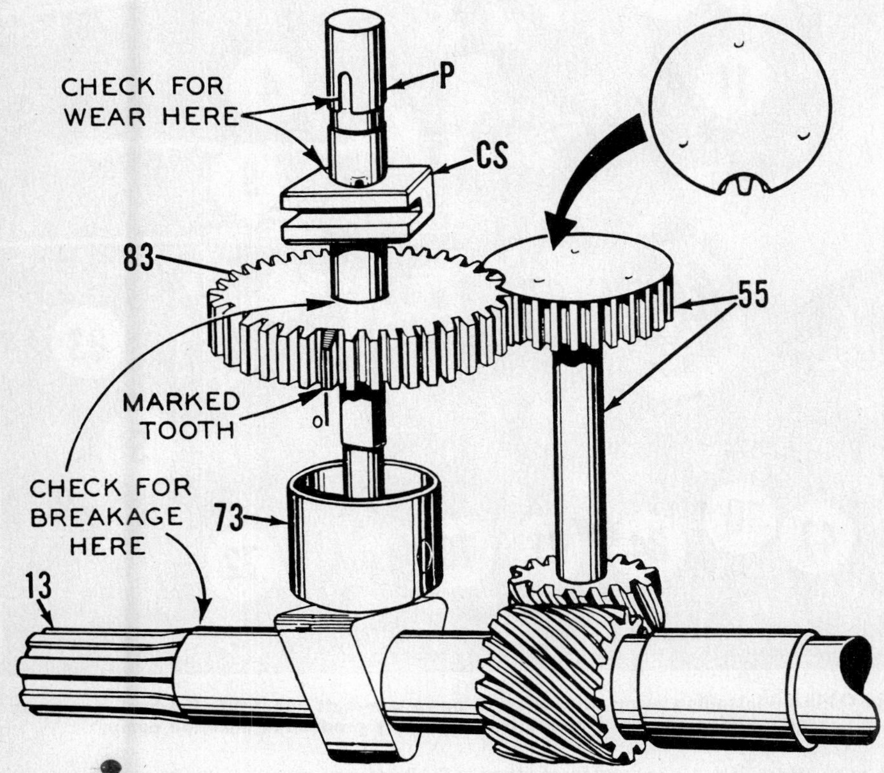

Fig. O344C—Schematic view of shafts and gears in type PSB American Bosch injection pumps showing likely wear points.

Remove delivery valve cap screw (96—Fig. O344B), unscrew the delivery valve holder (94) and lift out the delivery valve spring and valve (91). The delivery valve body is left in the hydraulic head. Reinstall delivery valve holder (94) and cap screw (96).

Rotate injection pump camshaft in a clockwise direction (viewed from drive end) until the marked tooth of plunger drive gear approaches the stamped arrow or "O" mark located on timing window ledge. Continue rotating the pump camshaft until the flow of fuel oil stops at the No. 1 outlet in the hydraulic head. At this time, the scribed line mark on drive coupling hub should be in register within $\frac{1}{32}$ inch with the timing pointer; and the marked plunger gear tooth will be about one tooth to the right of the stamped arrow or "O" mark located on timing window ledge.

If the drive coupling hub line mark and timing pointer are out of register more than ⅜ inch, the pump is assembled incorrectly. If the out-of-register is ⅛ inch or less, remove the old line mark and affix a new line mark on the drive coupling hub (4).

Repeat the phasing check until constant results (line mark and timing pointer register) are obtained. Reinstall delivery valve, spring and overflow valve which were previously removed.

488. **REMOVE AND REINSTALL PUMP.** Procedure for removal of injection pump is as follows: Shut off fuel supply at tank, and remove injection pump oil line. Disconnect fuel stop wire and governor control rod at the injection pump. Remove gear cover from front face of engine timing gear cover. Working through this opening in the timing gear cover, remove two cap screws attaching pump drive gear to hub; and remove the gear. Disconnect fuel lines from pump. Remove two cap screws and one bolt attaching injection pump to engine. Some mechanics prefer to disconnect the pump high pressure lines at the

Fig. O345B—View of later production American Bosch injection pump showing nuts (132) which control high idle engine speed and screw (158) which controls low idle speed.

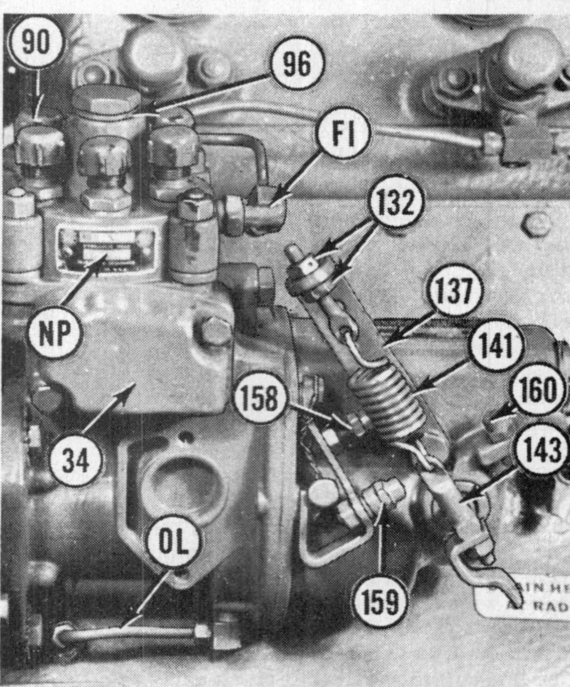

FI. Fuel inlet
NP. Name plate
OL. Oil line
34. Timing window cover
90. To injectors
96. Delivery valve cap
137. Operating lever
141. Governor spring
143. Shutoff screw
160. Full load stop screw

Fig. O346A—Installing pump drive gear to pump camshaft hub, while flywheel, and pump gear hub marks are properly indexed. Early production injection pump.

S. Smoke stop adjustment
1. Fuel return
4. Fuel inlet
9. Rated speed adjustment
10. Pump timing port
11. Pump timing port
47. Pump drive gear
71. Governor linkage
72. Idle speed adjustment
78. Pump cover
83. Hand-control linkage
84. Timing gear cover

Fig. O345A—Checking freedom of the injection pump delivery control unit as outlined in paragraph 483B.

injector nozzles; remove the pump with these lines attached; then, remove the injector nozzle lines from pump hydraulic head when pump is on the bench.

Reinstall and time injection pump to engine as outlined in paragraphs 486 thru 486B. After installing the pump, bleed the fuel system as outlined in paragraph 482. Adjust engine governed speed as outlined in paragraph 489A.

488A. **SMOKE STOP ADJUSTMENT.** Refer to Delivery Calibration, paragraphs 484 and 484A.

INJECTION PUMP GOVERNOR

The Diesel engine governor (flyweight type) is an integral part of the injection pump.

489. **LINKAGE FREEDOM.** Refer to Fig. O345A. The necessity for free movement of the governor linkage extends to the pump delivery control unit, located on the injection pump under the timing window. To check freedom of the pump control unit, disconnect unit arm from governor control rod. If arm shaft is tight in its sleeve, remove the pump control unit and free-up or renew the unit as outlined in paragraph 483B.

489A. **GOVERNOR ADJUSTMENT.** Engine idle no-load speed of 675 crankshaft rpm on all models is adjusted by means of screw (72 — Fig. O344) or screw (158—Fig. O345B) located on the outside of the injection pump housing. Rotate the adjusting screw in or out to decrease or increase the engine idle speed.

The high idle engine speed of 2200 crankshaft rpm (733 pto rpm) on Super 66; of 1925 (597 pto) on 770 and 880 and 1750 crankshaft rpm (582 pto rpm) on all other models, is adjusted by varying the tension of governor spring by means of adjusting nuts (132—Fig. O345B) or those on eye bolt (9—Fig. O346A).

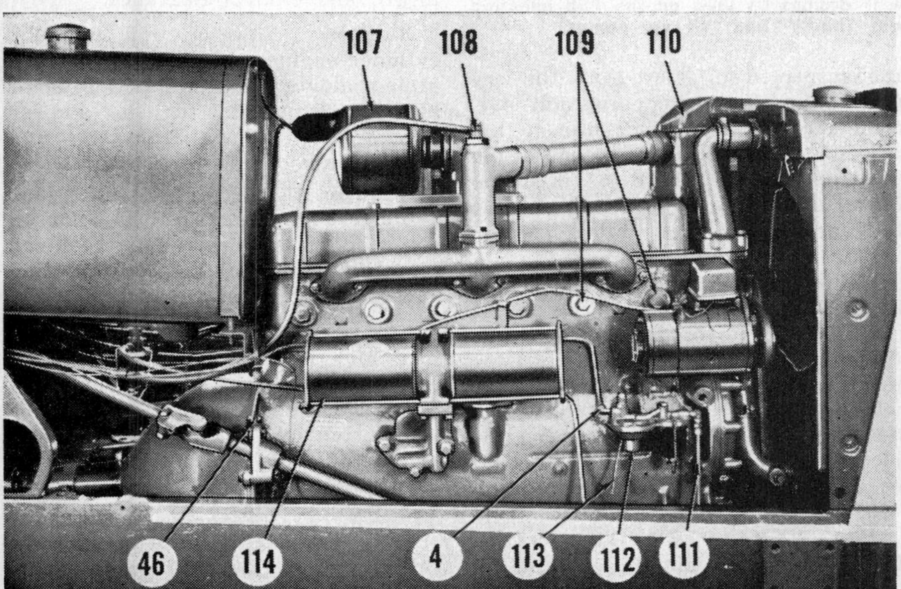

Fig. O347—Model 77 and 88 Diesel engine installation, right side view.

4. Fuel pump outlet
46. Flywheel timing port cover
107. Exhaust muffler
108. Manifold pre-heater
109. Energy cells
110. Air cleaner
111. Fuel pump inlet
112. Fuel pump
113. Priming lever
114. Oil filter

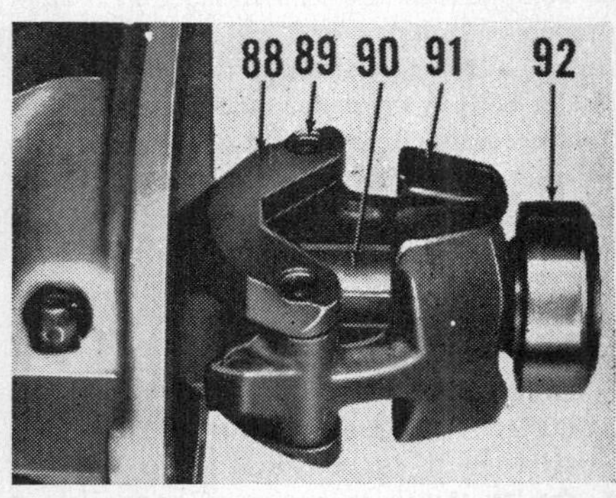

Fig. O348—Flyweight type governor, which is used to control the fuel delivery as a function of speed control, is an integral part of the injection pump.

88. Weight carrier
89. Pin
90. Pump camshaft
91. Weight
92. Bearing (ball)

Non-Super Model 88 Row Crop

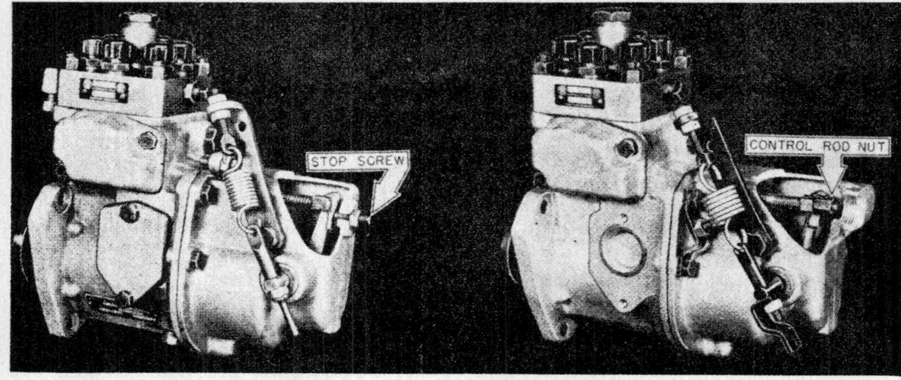

Fig. O349—Showing two smoke stop adjustment designs as used on the PSB injection pumps. Left: The "Z" type pump. Right: The "Y" and "V" type pumps.

FUEL INJECTORS (NOZZLES)

American Bosch or CAV closed pintle type fuel injectors are used. Furnished components of both nozzles are interchangeable. Later production nozzles and all service nozzles have the drip connection in the lower side of holder body instead of in the protection cap. This necessitates rework of drip lines or use of new lines. The extreme pressure of the injector nozzle spray is dangerous and can cause the fuel to penetrate the human flesh. Avoid this source of danger when checking the nozzles.

490. **LOCATING A FAULTY INJECTOR.** If one engine cylinder is misfiring, it is reasonable to suspect a faulty fuel injector. Generally, a faulty injector can be located by loosening the high pressure line fitting on each injector in turn; therby preventing fuel from entering the combustion chamber. As in checking spark plugs in a spark ignition engine, the faulty injector is the one which, when its fuel line is loosened, least affects the running of the engine. Remove the suspected injector from the engine as outlined in paragraph 491; then reconnect the fuel line to the injector. With the discharge end of the injector directed where it will do no harm, crank the engine and observe the spray pattern as shown in Fig. O350.

If the spray pattern is ragged, it is likely that the injector nozzle is the cause of the misfiring. To check the diagnosis install a new or rebuilt injector or an injector from cylinder which is firing regularly. If the cylinder fires regularly with the other injector, the condemned injector should be serviced as outlined in paragraph 492.

491. **R&R INJECTOR UNITS.** Before loosening any lines, wash all connections with fuel oil. Injector removal procedure which varies according to the cylinder location, is as follows:

Number 1 injector unit either on four or six cylinder engines can be removed after disconnecting all injector nozzle fuel lines and lifting the fuel line harness assembly off the engine.

Number 2 injector unit on four cylinder engine, or nos. 2, 3, 4 and 5 units on six cylinder engines can be removed after removing the two or three clamps from the fuel line harness and springing slightly one of the injector nozzle high pressure fuel lines to provide removal clearance.

Number 3 injector unit on four cylinder engines only can be removed after removing the number 1 injector nozzle high pressure fuel line and removing two clamps from the fuel line harness.

Number 4 injector unit on four cylinder engines or No. 6 injector on six cylinder engines can be removed after disconnecting the leak-off line and the high pressure fuel line.

After disconnecting the high pressure and leak-off fuel lines, cover open ends of lines and pump with composition caps to prevent entrance of dirt. Remove injector nozzle holder body stud nuts and carefully withdraw injector nozzle from cylinder head.

491A. Thoroughly clean injector recess in cylinder head before reinstalling injector. It is important that the seating surfaces of the recess be free

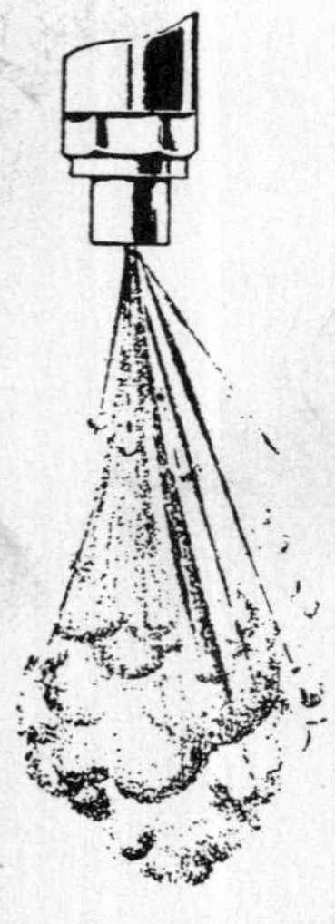

Fig. O350 — Spray patterns of a standard pintle type injector. Left: A poor spray pattern. Right: Ideal spray pattern.

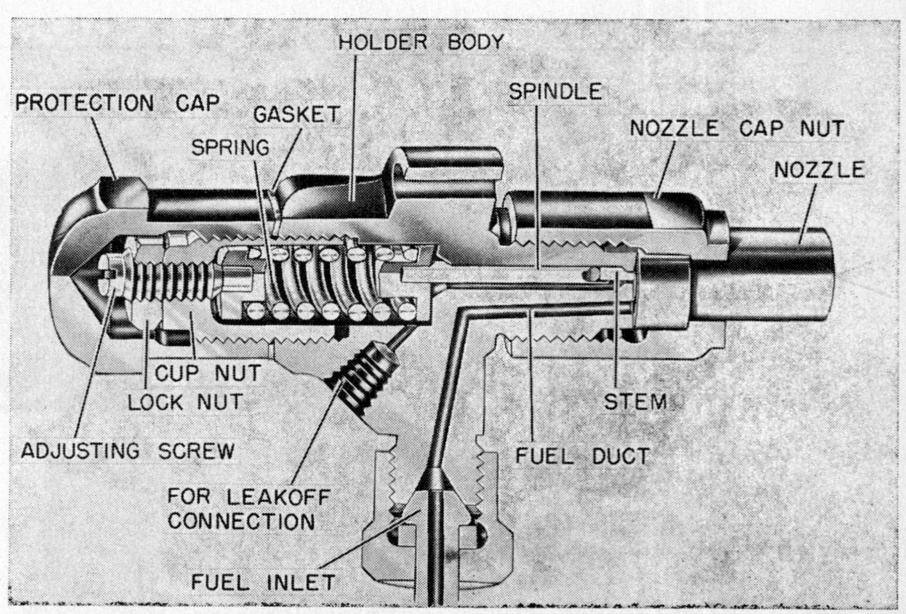

Fig. O351—Sectional view of the fuel injector used on the later Oliver Diesel engines. Nozzle opening pressure of 1750 psi is controlled by adjusting screw. Leak-off connection was located in protection cap on early injectors.

of even the smallest particles of carbon which could cause the injector to be cocked and result in blow by of hot gases. No hard or sharp tools should be used for cleaning. Bosch recommends the use of a wooden dowel or brass bar stock which can be shaped for effective cleaning. Do not re-use the copper ring gasket; install a new one. Tighten nozzle holder stud nuts to 14-16 foot pounds torque.

492. **SERVICING FUEL INJECTORS.** Hard or sharp tools, emery cloth, crocus cloth, grinding compounds or abrasives of any kind should **NEVER** be used in the cleaning of fuel injectors.

492A. DISASSEMBLY OF INJECTOR NOZZLE. Carefully clamp the nozzle holder body in a vise, and remove the injector nozzle unit consisting of the valve (V—Fig. O352) and the tip (T). If nozzle valve cannot be readily withdrawn from nozzle valve tip with the fingers, soak the nozzle unit in fuel oil, acetone, carbon tetrachloride or equivalent. Refer to paragraph 492B for cleaning and inspection procedures.

The nozzle valve and nozzle valve tip are a mated (non-innterchangeable) unit, and they should be handled accordingly as neither part is available separately.

492B. CLEANING AND INSPECTION. Soak all fuel injector parts in fuel oil, acetone, carbon tetrachloride or equivalent, being careful not to permit any of the polished surfaces to come into contact with any hard substance.

All surfaces of the nozzle valve should be mirror bright except the contact line of the beveled seating surface. Polish the valve with mutton tallow which is applied with a soft cloth or felt pad. The valve may be held by its stem in a revolving chuck during the polishing operation. A piece of soft wood well soaked in oil, or a brass wire brush will be helpful in removing carbon from the valve. If the valve shows any dull spots on its sliding surfaces, or if a magnifying glass inspection shows any nicks, scratches or etching, discard the valve and the nozzle valve tip or send them to an authorized service station for possible overhaul.

The inside surface of the valve tip can be cleaned with a piece of soft wood which is formed to a point that corresponds to the angle of the valve seat. The wood should be well soaked in oil. Some Bosch mechanics use an ignition distributor felt oiling wick instead of the soft wood for cleaning the seat in the valve tip. Delco-Remy distributor oiling wick, part DR804076, is suitable for this purpose. Shape the end of the wick to conform to the seat angle, and coat the formed end with tallow for polishing.

The orifice at the end of the valve tip can be cleaned with a wood splinter. Outer surfaces of the nozzle should be cleaned with a brass wire brush and a soft cloth soaked in carbon solvent. If there is any erosion or loss of metal at or adjacent to the valve tip orifice, discard both the valve tip and valve unit.

492C. Clean the exterior of the nozzle holder, except the lapped sealing surface, with a brass wire brush and carbon solvent.

The sealing surface of the valve tip, where it contacts the nozzle holder body, must be flat and shiny. Remove any discloration on these lapped surfaces by re-lapping with fine compound on a lapping plate using a figure 8 motion. This work requires skill and requires the use of an approved lapping outfit. If such equipment is not available send the nozzle holder and valve tip to an authorized service station for re-lapping.

Clean deposits from nozzle holder body cap nut with a brass wire brush and carbon solvent. Shallow irregularities which would provide a possible leakage path across the gasket contacting surface of cap nut can be removed by reduction lapping,

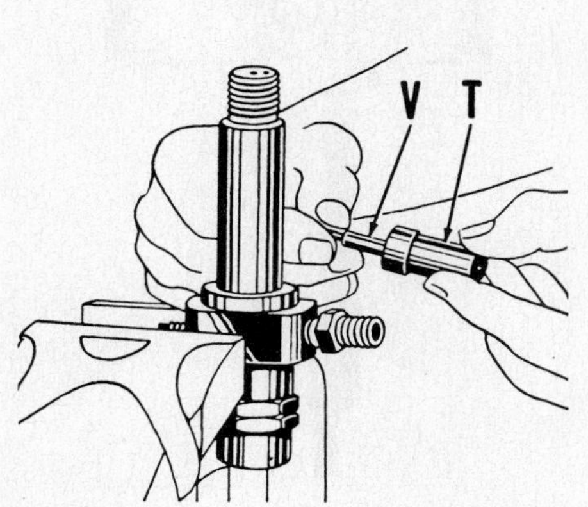

Fig. O352—Removing Oliver Diesel engine injector nozzle tip (T) and the valve (V).

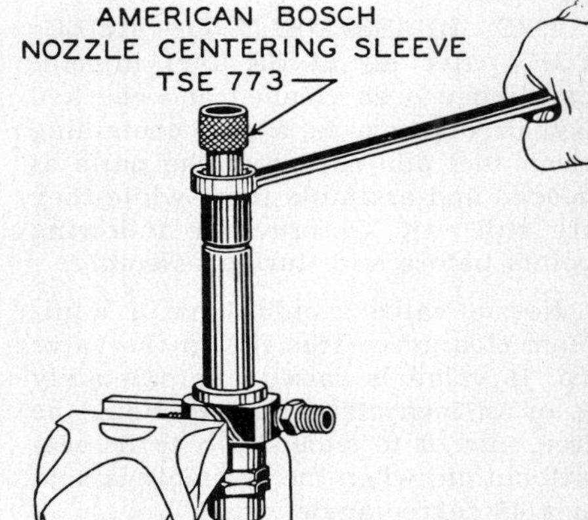

Fig. O353—Using a Bosch tool No. TSE773 to center the nozzle tip in the cap nut.

Non-Super Model 66 Row Crop

If irregularities cannot be corrected in a reasonably short time, renew the cap nut.

492D. DISASSEMBLY OF NOZZLE HOLDER BODY. If the shop is not equipped with a fuel injector tester, disassembly of the nozzle holder body should not be attempted. If a tester is available, proceed as follows: Remove the protection cap, spring retaining cap nut, pressure adjusting spring, spring upper seat, and lower seat and spindle.

Carefully check the lower spring seat and spindle. Renew the spindle if it is bent, if its end which contacts the nozzle valve is cracked or otherwize damaged, or if the spring seat shows any signs of wear.

Carefully check the pressure adjusting spring. Renew the spring if it is rusted, pitted or shows any surface cracks.

492E. REASSEMBLY OF INJECTOR. After all of the fuel injector parts have been cleaned and checked lay all of them in a pan containing clean fuel oil. Withdraw the parts as needed and assemble them while they are still wet. Observe the following points before and during assembly:

Nozzle valve should have a minimum clearance (free fit) in the valve tip. If valve is raised approximately ⅜ of an inch off its seat it should be free enough to slide down to its seat without aid when the assembly is held at a 45 degree angle.

If parts have been properly cleaned and polished, all sliding and sealing surfaces should have a mirror finish. Renew any part which does not meet this condition, or send the mating pieces to an authorized service station for possible reconditioning.

It is desirable that the nozzle valve tip be perfectly centered in the nozzle valve tip cap nut. A centering sleeve, American Bosch tool TSE773, is available and should be used for this purpose as shown in Fig. O353.

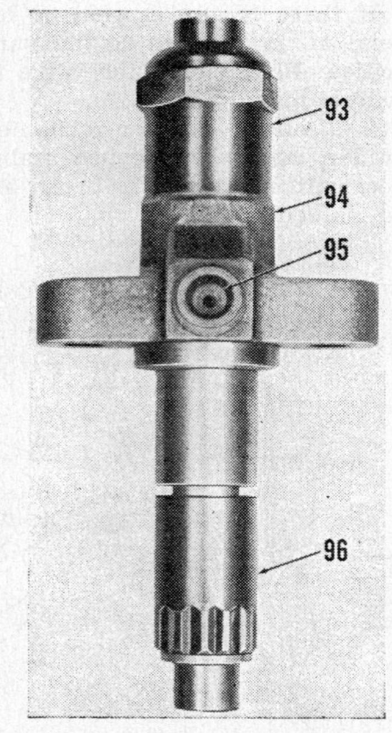

Fig. O354—Bosch early type closed pintle fuel injector.

93. Cap
94. Nozzle holder
95. Fuel inlet
96. Cap nut

Adjust the fuel injector opening pressure and check the condition of the fuel injector as outlined in paragraphs 493 through 493C.

493. **TEST & ADJUST FUEL INJECTORS.** The job of testing and adjusting the fuel injector requires the use of an approved nozzle tester. The fuel injector should be tested for leakage, spray pattern and opening pressure. Refer to Fig. O351.

493A. OPENING PRESSURE. Recommended fuel injector opening pressure of 1750 psi can be obtained by rotating the pressure adjusting screw. If a new pressure adjusting spring has been installed, adjust the opening pressure to 1770 psi or 20 psi higher to compensate for subsequent spring set.

While operating the injector tester handle, observe the gage pressure when the valve opens. The gage pressure drop when the valve opens should not exceed 300 psi. A greater pressure drop usually indicates a sticking valve.

493B. LEAKAGE. To check the fuel injector for valve leakage, actuate the tester handle slowly; and as the gage needle approaches 1750 psi, observe the valve tip orifice for drops of fuel. If drops of fuel are observed at pressures less than 1750 psi, the nozzle valve is not seating properly.

A slight overflow of fuel at the leak off port is normal. If the overflow is greater than that of a new fuel injector, it indicates excessive wear between the cylindrical surfaces of the nozzle valve and the valve tip.

Non-Super Model 88 Standard

493C. SPRAY PATTERN. Operate the tester handle slowly until nozzle valve opens, then stroke the lever with light quick strokes at the rate of approximately 100 strokes per minute. The correct spray pattern is shown at the right in Fig. O350. If the spray pattern is ragged, disassemble the nozzle valve tip and carefully check and recondition the parts as required.

DIESEL ENERGY CELLS

494. These assemblies are mounted directly opposite from the fuel injectors in the cylinder head, as shown in Fig. O356. Oliver Corp. catalogs the cell body (100) with orifice insert and cell cap (99) separately. It is suggested, however, that both of these parts be renewed if either one requires renewal.

In almost every instance where a carbon-fouled or burned energy cell is encountered, the cause is traceable either to a malfunctioning injector, incorrect fuel or incorrect installation of the energy cell. Manifestations of a fouled or burned unit are misfiring, exhaust smoke, loss of power or pronounced detonation (knock).

494A. REMOVAL. Any energy cell can be removed without removing any engine component. To remove an energy cell, first remove the threaded cell holder plug (97) and take out the cell holder spacer (98). With a pair of thin nosed pliers, remove cell cap (99). To remove the cell body screw a 3/4-16 NF bolt into the threaded end of the cell body. A nut and collar on the bolt will make it function as a puller. If puller is not available, remove the fuel injector; and use a brass rod to drift the cell body out of the cylinder head.

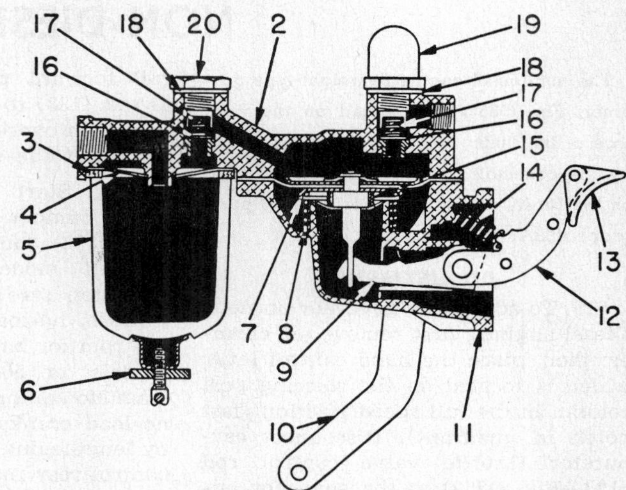

7. Diaphragm
8. Diaphragm spring
9. Pump link
10. Priming lever
12. Rocker arm
14. Rocker arm spring
15. Diaphragm spring
16. Valve
19. Air dome

Fig. O357—Sectional view of the Diesel primary fuel supply pump.

494B. CLEANING. Clean all carbon from front and rear crater of cell body using a brass scraper or a shaped piece of hard wood. Clean the exterior of the energy cell with a brass wire brush, and soak the parts in carbon solvent. Reject any part which shows signs of leakage or burning. If parts are not burned re-lap sealing surfaces (99 and 100) by using a figure 8 motion on lapping plate coated with fine compound.

STARTING AID (PREHEATER)

495. Diesel engines are equipped with separate inlet and exhaust manifolds. A heating coil (108—Fig. O347) located in the inlet manifold elbow is used to preheat the air and assist starting in cold weather.

Operation of the preheater can be checked by depressing the control switch which is located on the instrument panel, for approximately 30 seconds; then place your hand on the base of the preheater to see if it is warm.

PRIMARY FUEL SUPPLY PUMP

496. Diesel engines are equipped with an AC automotive, diaphragm type primary fuel supply pump, Fig. O357, flange-mounted on the right or left side of the engine. The pump, equipped with a hand operated priming lever for use in bleeding the low pressure side of the fuel system, is actuated by the engine camshaft. A satisfactory pump will show a 11-15 psi gage reading when checked at the outlet side.

496A. Removal and disassembly of the fuel supply pump is self-evident after an examination of the unit. To eliminate the possibility of overstretching the fabric material, when installing a new diaphragm, place the priming lever in a perpendicular (parallel with pump mounting pad surface) position when tightening the cover retaining screws.

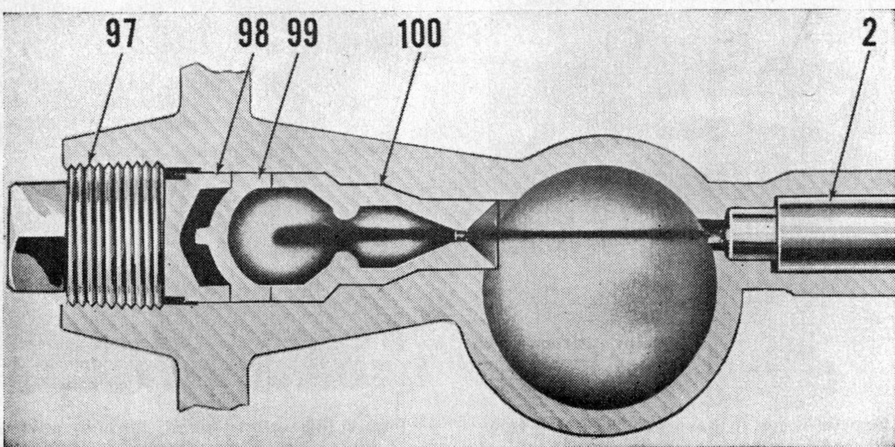

Fig. O356—Energy cell (pre-combustion chamber) is located in each cylinder and opposite to the nozzle.

2. Nozzle 97. Cell holder 98. Spacer 99. Cell cap 100. Cell body

NON-DIESEL GOVERNOR

The non-Diesel engine flyweight type governor, Fig. O358, is mounted on the front face of the timing gear cover, and is driven by the camshaft gear. For adjustment data on the Diesel engine governor, refer to paragraph 489A.

ADJUSTMENT

510. To adjust the governor on non-Diesel engines, first remove air cleaner; then, place the hand control lever which is located on the steering post column in the full speed position (last notch in quadrant). Disconnect carburetor throttle valve control rod (133—Fig. O359) at the governor operating arm (118). Place carburetor throttle valve in the wide-open position, and the governor unit lever in full forward position. Adjust length of rod (133) to provide approximately 1/16 inch over-travel, as shown; then, reconnect the rod.

510A. Start and warm up engine. Turn bumper spring screw (116—Fig. O358) "out" as far as it will go. On all models, adjust carburetor idle stop screw to provide a closed throttle, no-load crankshaft speed of 375 rpm for battery ignition equipped models, or 500 crankshaft rpm for magneto equipped models. High speed, no-load crankshaft rpm is controlled by lengthening or shortening the hand control rear rod (136—Fig. O360).

With the high speed adjustment checked and adjusted as outlined, rotate governor bumper spring (116—Fig. O358) inward to eliminate any tendency for the engine to surge but not enough to increase the engine speed.

If surging persists check to make sure bumper spring is of correct $\frac{7}{16}$-½ inch length.

The following table provides data for checking the governor adjustment.

NO-LOAD RPM

Models	770, 880	Super 66	All Others
Crankshaft	1925	2200	1750
Belt Pulley	1193	1357	1085
PTO	597	733	582

R&R AND OVERHAUL

511. Governor can be removed without disturbing the radiator by proceeding as follows: Remove the air cleaner. Disconnect governor linkage and remove governor to timing gear cover retaining cap screws. *Carefully withdraw the unit so as to prevent the loss of the drive gear bronze thrust washer (123—Fig. O358).*

511A. Governor drive gear (133) is a press fit and is keyed to the governor shaft (139). Gear can be removed with the use of a suitable puller. Governor weight carrier (122) is a press fit and keyed to the shaft. The weight carrier should be removed from the shaft by

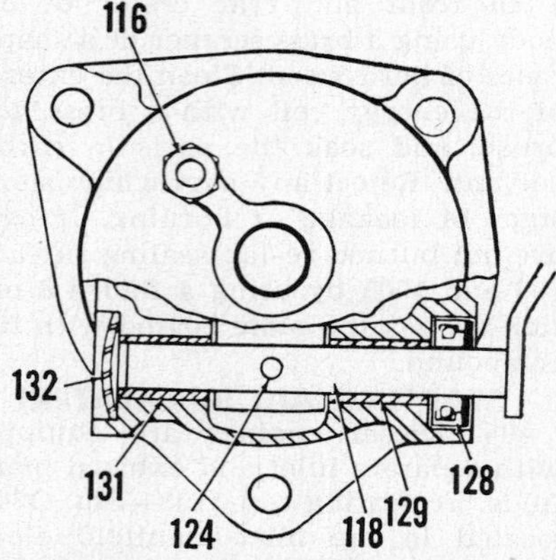

Fig. O358—Governor assembly used on non-Diesel engines, cross sectional view. Steel sleeve (40) and bushing (41) are located in the front face of the cylinder block.

40. Sleeve
41. Bushing
116. Bumper spring screw
117. Governor housing
118. Fork
119. Thrust bearing
120. Weight
121. Weight pin
122. Weight carrier
123. Thrust washer
124. Fork retaining pin
125. Expansion plug
126. Bushing
127. Expansion plug
128. Oil seal
129. Bushing
131. Bushing
132. Expansion plug
133. Drive gear
138. Snap ring
139. Governor shaft

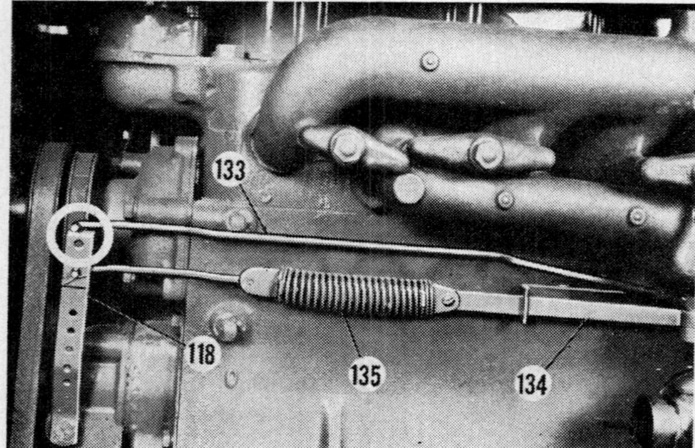

Fig. O359—Carburetor throttle valve to governor operating fork lever rod (133) is adjusted to have 1/16-inch over-travel.

118. Governor operating fork lever arm
134. Control rod (front)
135. Governor spring

Fig. O360—High speed, no-load engine rpm of 1750-1925 or 2200 is controlled by lengthening or shortening control rod (136).

M. Carburetor idle mixture adjustment screw
134. Control rod (front)
136. Control rod (rear)
137. Water temperature gauge sending unit

pressing on the gear end of the shaft. Disassembly of the weight unit is self-evident after an examination. The bushings (126 & 129 & 131) require final sizing after installation to provide a clearance of 0.0015-0.002.

Governor gear hub rotates in a steel backed babbitt bushing (41) and sleeve assembly which is pressed into the front face of the cylinder block. To renew the assembly, it will be necessary to remove timing gear cover and use a suitable puller to remove same as shown in Fig. O362. Renew bushing when diametral clearance exceeds .005. Pre-sized service bushings have a bore diameter of 1.002.

Correction of excessive wear on governor gear and thrust washer on early production model 88 tractors is corrected by installation of later governor shaft assembly Oliver No. IKSA-251A and lighter weights.

When later shafts and weights are installed on older 88 tractors the Oliver No. IH 309 Governor Spring Coil Cut-Out shown in Fig. O362A should be removed from the governor spring. After removing the IH-309 reset the IKS-277 Spring End to position shown in Fig. O362A. When this is done there should be 19 working coils between stops IKS-277 and IH-278.

COOLING SYSTEM

RADIATOR

520. To remove the radiator, first drain the cooling system and disconnect the upper and lower radiator hoses. Remove hood side panels, hood and radiator grille center clamp. Disconnect shutter control on models so equipped. Remove six cap screws retaining the radiator and shell assembly to the tractor front frame. Loosen air cleaner pipe clamp (located at top left side of radiator) and disconnect air cleaner inlet tee. Lift radiator and shell assembly off of tractor.

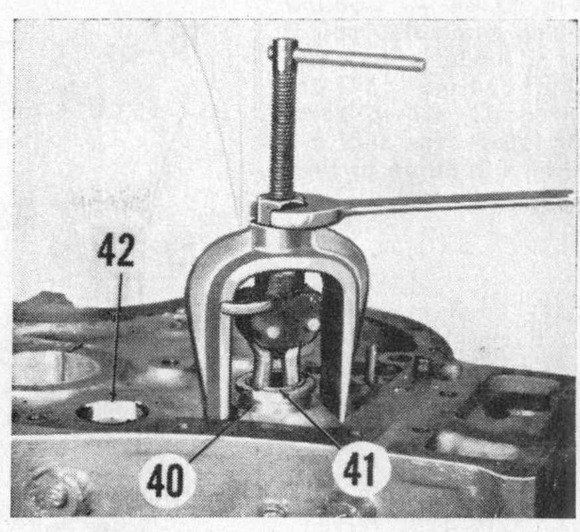

Fig. O362—Using a puller to remove the governor gear sleeve (40) and bushing (41) assembly on non-Diesel engines. Bore (42) is plugged on non-Diesel engines.

THERMOSTAT

521. Thermostat is located in engine water outlet elbow. Gasoline, LP-Gas and Diesel engine thermostats start to open at 170-176 deg. F., and are fully open (3/8 inch) at 196 deg. F. Distillate engine thermostats start to open at 180-185 deg. F., and are fully open (3/8 inch) at 205 deg. F.

WATER PUMP

522. **REMOVE & REINSTALL.** To remove the water pump, first remove the radiator as outlined in paragraph 520. Remove fan blades, fan belt, and generator belt from models so equipped. Remove pump to engine retaining cap screws and remove pump.

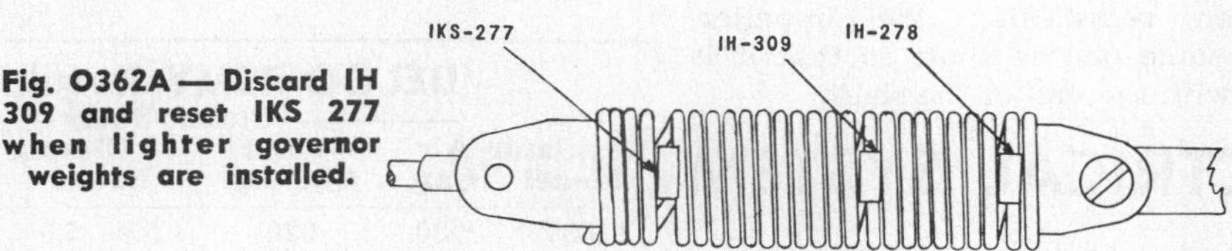

Fig. O362A — Discard IH 309 and reset IKS 277 when lighter governor weights are installed.

Fig. O363—Diesel engine timing gears and injection pump installation.

A. Port closing line mark on hub
10. Pump timing port cover
12. No. 1 cylinder outlet
38. Crankshaft gear
44. Camshaft gear
45. Thrust button
47. Pump drive gear
48. Idler gear
49. Thrust button
51. Governor shaft bushing & sleeve location (non-Diesel engines only)
76. Pump shaft cap screw
77. Drive gear hub
78. Pump cover
82. Hydraulic head
86. Gear retaining cap screw

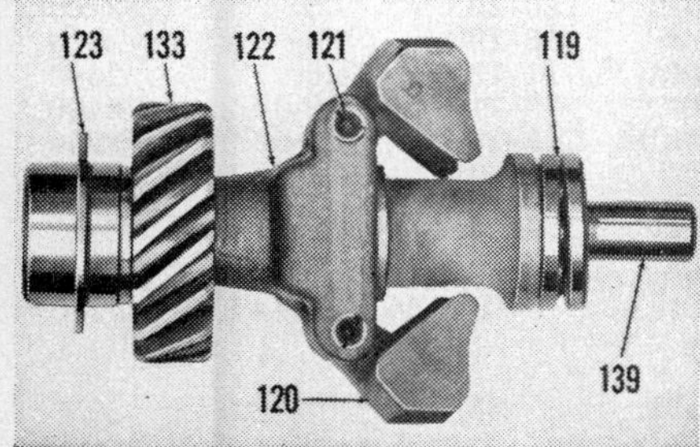

Fig. O361—Governor weight and shaft unit.

119. Thrust bearing
120. Weight
121. Weight pin
122. Weight carrier
123. Thrust washer
133. Drive gear
139. Governor shaft

522A. **OVERHAUL.** To disassemble the pump, proceed as follows: remove pump cover (151—Fig. O364). Remove snap ring (157) from front of water pump housing. Press shaft and bearing assembly toward fan blades end and out of the impeller and pump housing. Pump seal assembly can be renewed at this time.

Refer to 770 & 880 Supplement, paragraph 4 for Super 77 and Super 88 tractors; to 660 Supplement for Super 66 tractors.

Check condition of seal seat which is integral with pump body. If seat is rough or scored, renew pump body, or resurface seat by piloting the cutting tool from shaft bore. The shaft and prelubricated bearings are serviced as an assembly only.

When reinstalling the impeller, press same on the shaft so that it is flush with the end of the shaft.

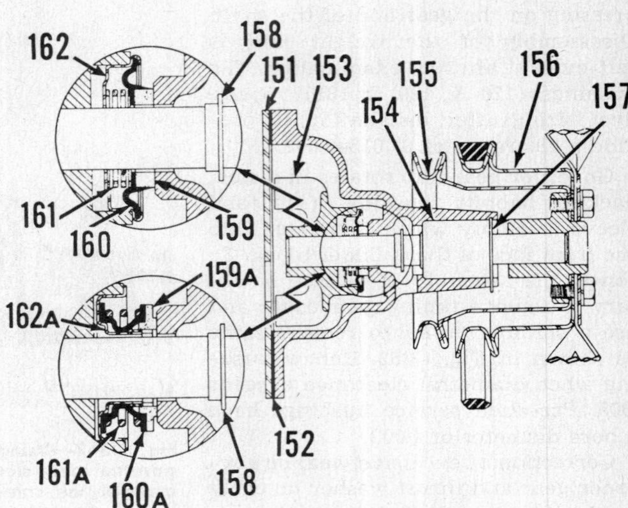

Fig. O364 — Coolant pump assembly, typical of all models. Shaft and shaft bearing (154) are integral. Components of latest type seal are shown in circle at lower left.

151. Cover
152. Gasket
153. Impeller
154. Shaft and bearing assembly
155. Generator belt drive pulley
156. Pump pulley
157. Snap ring
158. Slinger
159. Seal (carbon)
159A. Seal (carbon)
160. Seal (neoprene)
160A. Seal (neoprene)
161. Seal spring
161A. Seal spring
162. Seal retainer
162A. Seal retainer

ELECTRICAL SYSTEM

GENERATOR AND REGULATOR

525. Test specifications for the Delco-Remy generator and generator regulator equipment used on Oliver non-Super and Super models are listed in an accompanying table. Early spark ignition engines are equipped with 6 volt systems; Diesels and later HC engine have 12 volt systems.

STARTING MOTOR

526. Test specifications for the Delco-Remy starting motor equipment used on Oliver tractor engines are listed in an accompanying table.

DELCO-REMY Regulator Test Specifications

Regulator Model	Cutout Relay Air Gap	Cutout Relay Point Opening	Cutout Relay Closing Voltage Range	Cutout Relay Closing Voltage Adjust	Voltage Regulator Air Gap	Voltage Regulator Setting (Volts) Range	Voltage Regulator Setting (Volts) Adjust
1118305	.020	.020	5.9- 7.0	6.4	.075	6.6- 7.2	6.9
1118306	.020	.020	11.8-14.0	12.8	.075	13.6-14.5	14.0
1118790	.020	.020	5.9- 7.0	6.4	.075	6.8- 7.4	7.1
1118791	.020	.020	11.8-14.0	12.8	.075	14.0-15.0	14.4

5849—Refer to specifications listed below.

Model 5849 DELCO-REMY Step-Voltage Control Test Specifications

VOLTAGE CONTROL UNIT

Air Gap—Inches .... .040
Point Opening—Inches .... .015
Contact Spring Tension—Ounces 1.1
Armature Travel—Inches .... .040
Opening Range Volts 14.0-15.0
Adjust to 14.4
Closing Range Maximum Volts 12.5

CUTOUT RELAY

Air Gap—Inches .... .015
Point Opening—Inches .... .020
Closing Range Volts 12.6-14.0
Adjust to 13.3

DELCO-REMY Generator and Regulator Application and Generator Test Specifications

Tractor Model	Generator Model	Regulator Model	Brush Spring Tension Ounces	Field Draw @ 80 Degrees F. Volts	Field Draw @ 80 Degrees F. Amperes	Output Hot or Cold	Output Amperes	Output Volts	Output Generator R.P.M.
NON-SUPERS:									
66, 77 & 88 HC	1101391 (3 br.)	1118305	16	6	2.6 -2.9	Cold	15-17	6.9- 7.1	2000
						Hot	11-13	6.9- 7.1	2000
*66, 77 & 88 HC	*1100504 (3 br.)	1118305	16	6	2.5 -2.72	Cold	20-25	7.0	2400
						Hot	16-19	6.9- 7.1	2500
66, 77 & 88 D	1100953 (3 br.)	1118791	16	12	2.0 -2.14	Cold	11-13	14.0	2300
						Hot	9-11	13.8-14.2	2400
77 Diesel	1101779 (3 br.)	5849	16	12	1.6 -1.69	Cold	8-10	14.4-14.9	2200
						Hot	6- 8	14.1-14.5	2400
88 Diesel	1101779 (3 br.)	5849	16	12	1.6 -1.69	Cold	8-10	14.4-14.9	2200
						Hot	6- 8	14.1-14.5	2400
*77 & 88 D	*1101775 (3 br.)	1118306 or 1118791	16	12	1.6 -1.69	Cold	8-10	13.8-14.2	2000
						Hot	6- 8	13.8-14.2	2200
SUPERS:									
66, 77 & 88 HC	1100029 (2 br.) or 1100035	1118790	28	6	1.85-2.03	Cold	35	8.0	2650
66, 77 & 88 D & 770, 880 D & HC	1100314 (2 br.)	1118791	28	12	1.58-1.67	Cold	20	14.0	2300
All Power Steering	1100322 (2 br.)	1118791	28	12	1.58-1.67	Cold	20	14.0	2300

*Latest application for non-Super models.

DELCO-REMY Starting Motor Test Specifications

Tractor Model	Motor Number	Brush Spring Tension Ounces	No Load Test Volts	No Load Test Amperes	No Load Test R.P.M.	Lock Test Volts	Lock Test Amperes	Lock Test Torque Ft. Lbs.
66 HC	1107065	24	5.67	80	5000	3.37	525	12
66, 77 & 88 HC	1107939	24	5.67	80	5500	3.0	600	14
66D	1108610	24	11.25	75	6000	5.85	615	29
77 & 88 D	†1109225, 1109233	36-40	†11.7	95	8000	†2.8	†570	20
77 & 88 D	1109170	36-40	11.4	65	6000	5.0	725	44
77 & 88 D	‡1109225, 1109233	36-40	‡11.6	95	8000	‡2.2	‡600	20
SUPERS, 770 & 880								
*All HC	1107682	35	10.6	112	3240	3.5	350	....
*66 Diesel	1108648	24	11.8	40–70	6800 9200	5.85	615	29
*77, 770 & 880 D	1113088	48	11.5	50	6000	3.3	500	22
*88 Diesel	1113075 or 1113007	48	11.5	50	6000	3.3	500	22

*Latest application. †Starting motors with 15 slot wound armatures. ‡Starting motors with 25 slot lap wound armatures.

IGNITION SYSTEM

529. Test specifications for the Delco-Remy battery ignition distributors used on the Oliver non-Super and Super models are listed in an accompanying table.

529A. Non-Super 66 tractor engines can be equipped, as optional equipment, with a Wico XV1728 or Wico XVD2224 magneto. Non-Super 77 and 88 tractor engines can be equipped, as optional equipment, with Wico XV-6 or Wico XVD-2218 magneto. All magnetos have a 20 degree lag angle impulse coupling.

IGNITION TIMING

Battery Ignition

530. On 4 cylinder engines, the firing order is 1-2-4-3; on 6 cylinder units, 1-5-3-6-2-4. The battery ignition distributor rotates in an anti-clockwise direction when viewed from the drive end.

531. **GASOLINE LP-GAS ENGINES** To time the distributor on a gasoline or LP-Gas burning engine, crank engine until No. 1 piston (timing gear end) is on compression stroke; then, slowly rotate flywheel until flywheel mark "TDC" is indexed with notch at inspection port. Install distributor to engine with rotor positioned to fire No. 1 cylinder. Rotate distributor housing anti-clockwise until breaker contacts have just started to open; then, tighten the distributor clamping screws.

Use a timing light and check running timing at rated engine rpm (or higher) to the flywheel timing marks as tabulated below. Check location of the IGN-HC mark and relocate it, if necessary, to agree with latest recommendations. If running spark does not occur when flywheel mark "IGN-HC" registers with notch at inspection port, loosen distributor clamp screws, and rotate distributor body until registration is obtained.

Non-Super 66, 77 and 88 HC & LP-Gas, and Super 77 and 88 HC:

Retard Timing Mark....TDC
Fully Advanced Timing Mark....IGN-HC
Timing Mark BTDC Deg. 22
Timing Mark BTDC Inches....2 9/32

Non-Super 66, 77 and 88 KD:

Retard Timing Mark....½" BTDC
Fully Advanced Timing Mark....IGN-KD
Timing Mark BTDC Deg. 27
Timing Mark BTDC Inches....2 13/16

Super 66 HC:

Retard Timing Mark....TDC
Fully Advanced Timing Mark....IGN-HC
Timing Mark BTDC Deg. 28
Timing Mark BTDC Inches....3 7/32

770-880 HC and LP-Gas:

Retard Timing Mark....TDC
Fully Advanced Timing Mark....IGN-HC
Timing Mark BTDC Deg. 28

532. **DISTILLATE ENGINES.** Non-Super KD engines use the same distributor as is used for both gasoline and LP-Gas engines, but the static timing must be set 5 degrees earlier in order to make the full advanced spark occur 27 degrees before top center as recommended by Oliver Corp. First step in the timing procedure is to scratch or paint a mark on the flywheel at a point 5 degrees or ½ inch ahead of the stamped "TDC" mark for static timing. If there is no "IGN-KD" mark, paint the flywheel 5 degrees or ½ inch ahead of the stamped mark "IGN-HC".

Now crank engine until No. 1 piston is coming up on compression stroke; then, slowly rotate flywheel until the 5 degree or ½ inch mark before TDC is indexed with the notch at the inspection port. Install the distributor to the engine with the rotor positioned to fire No. 1 cylinder. Rotate distributor housing anti-clockwise until breaker points have just

DELCO-REMY Ignition Distributors Test Specifications

NOTE: Rotation is viewed from driving end. Advance data are listed in distributor degrees and R.P.M.

Tractor Model	Distributor Model	Rotation	Contact Gap Inches	Cam Angle Degrees	Start Advance Degrees @ R.P.M.	Intermediate Advance Degrees @ R.P.M.	Maximum Advance Degrees @ R.P.M.
Non-Super 66	1111704	CC	.022	25-34	0-2 @ 200	4-6 @ 500	9-11 @ 850
Super 66	1112555	CC	.022	25-34	0-2 @ 200	6-8 @ 600	13-15 @ 1075
Super 77 & 88	1111702	CC	.022	31-37	0-2 @ 200	4-6 @ 500	9-11 @ 850
*Super 77 & 88	*1111750	CC	.022	31-37	0-2 @ 200	4-6 @ 500	9-11 @ 850
770 & 880	1112587	CC	.022	31-37	0-2 @ 250	8-10 @ 750	12-14 @ 1000

*Latest application. Breaker arm spring tension 17-21 ounces.

started to open; then, tighten the distributor clamping screws. Use a timing light, and with the engine running at 1750 rpm (or higher), check to make sure that the spark occurs when the "IGN-KD" mark or 27 degree BTDC mark registers with the notch at inspection port.

Magneto Ignition

533. To time engines equipped with a Wico magneto, first adjust the breaker gap to .015. Magneto is timed to the full advanced timing mark "MAG" which is located 15 degrees or $1\frac{9}{16}$ inches before the flywheel mark "TDC".

Rotate engine crankshaft until the flywheel mark "MAG" (early production flywheels are not marked) which is located $1\frac{9}{16}$ inches BTDC, is indexed with flywheel inspection port notch. Offer magneto to the engine with rotor positioned to fire number one cylinder. On model 66, the magneto body should face rearward; on models 77 and 88, the magneto body should face forward. Next, rotate the magneto body in a counter-clockwise direction until the impulse coupling releases with a sharp snap. Rotate the magneto body back to its original position and until the breaker contacts have just started to open; then, tighten the magneto clamping screws.

Use a timing light to check the running timing of 15 degrees BTC at an engine speed of approximately 800 rpm or at a speed when the impulse coupling is not functioning.

If running spark does not occur at 15 degrees BTC (flywheel mark "MAG"), loosen magneto body clamp screws, and rotate magneto body until registration is obtained.

ENGINE CLUTCH

Power Take-Off Clutch data begins with paragraph 590

APPLICATION

540. Clutch model applications and specifications are listed in an accompanying table. It is recommended that the latest cover assembly as listed be used in servicing the early production models.

ADJUSTMENT

Series 66-Early Non Super 77 & 88

541. On series 66 tractors and early production non Super 77 and 88 models the clutch control linkage is of the long-travel-low-leverage-pedal type. On these installations one of which is shown in Figs. O365 and O366, lined plate facings can be worn down to the rivets before all of the inbuilt free pedal travel has been used. In other words unless linkage has been tampered with it will require no adjustment during the life of the tractor. To check and **permanently** adjust the linkage proceed as follows:

Disconnect pedal rod. Make sure that top of release fork (53—Fig. O366) is all the way back against clutch bell housing. Position the pedal with its stop (X—Fig. O365) against the bottom of the platform. Now without changing position of release fork or pedal adjust overall length of pedal rod until pedal rod will freely enter hole in pedal. On non-Super 77 and 88 the adjustment is by means of the nuts (57) at front end of rod as shown whereas on all series 66 tractors it is at the pedal end of the rod.

When the pedal free play on these hookups is less than ¾ inch the clutch should be relined, otherwise the release fork may foul on the clutch cover and damage result or if an at-

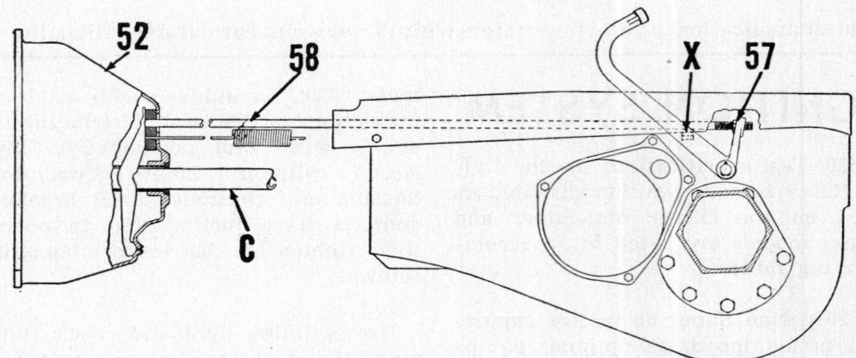

Fig. O365—On all 66 and on early production non-Super 77 and 88 the pedal linkage should not be adjusted to compensate for worn clutch facings. When pedal free travel on these models is less than ¾-inch remove the clutch for overhaul.

C. Clutch shaft
P. PTO unit drive shaft
49. Flywheel
50. Cover
51. Release bearing
52. Clutch housing
53. Fork
54. Felt retainer
55. Felt
56. Clutch pedal spring
57. Actuating rod adjusting nut
58. Actuating rod
59. Rod seal
60. Release lever
61. Eye bolt
62. Lock nut
63. Lined plate
64. Bushing
65. PTO drive shaft splined hub

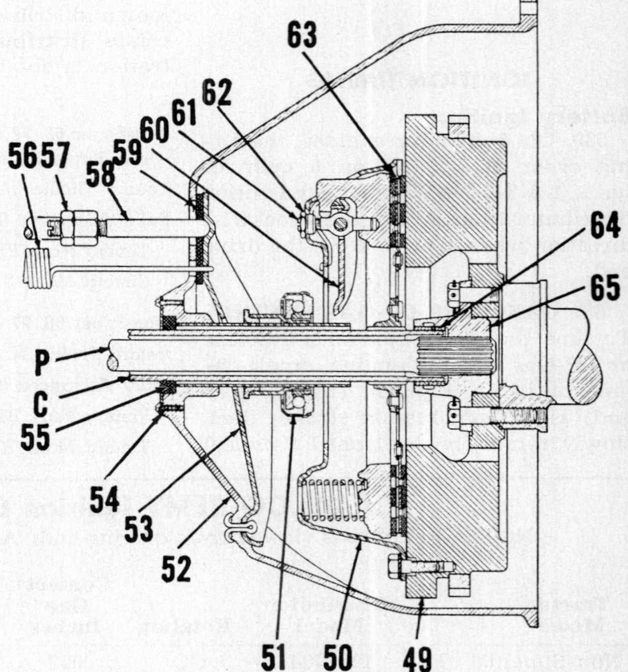

Fig. O366—Clutch installation on early production non-Super 77 and 88 tractors. Series 66 is similar except actuating rod length adjustment (57) is at pedal end of rod. See text for reason why actuating rod length should not be adjusted. Later production non-Super 77 and 88 and all Super models are equipped with a ball type clutch pilot bearing instead of the bushing (64) shown. Later 77 and 88 models are also provided with a pedal free travel adjustment.

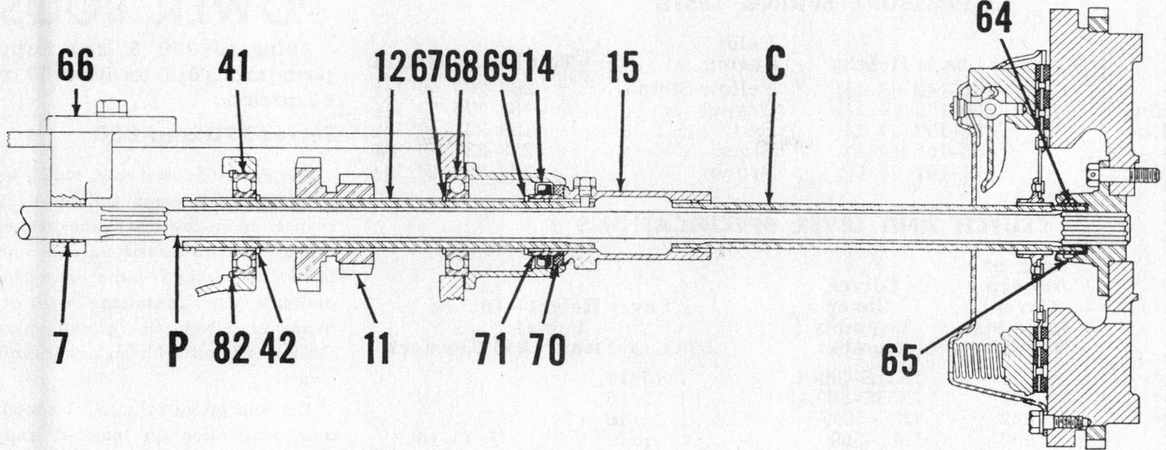

Fig. O367—Engine clutch shaft and continuous type pto shaft assembly. Note PTO shaft (P) is located inside hollow clutch shaft (C) and is splined to the flywheel hub. On models 77 and 88, PTO shaft bushing (7) is contained in a support bracket. On model 66 this bushing is located in transmission housing dividing wall. Later non-Super models 77 and 88 and all Super models are equipped with a ball type bearing instead of a bushing (64) as shown for the clutch shaft.

C. Clutch shaft
P. PTO unit drive shaft
7. Bushing
11. Mainshaft gear (608)
12. Transmission mainshaft
14. Oil seal
15. Coupling
41. Mainshaft bearing (rear)
42. Snap ring
64. Bushing
65. PTO drive shaft hub
66. PTO shaft support
67. Snap ring
68. Mainshaft bearing (front)
69. Snap ring
70. Felt seal
71. Sleeve
82. Snap ring

tempt is made to shorten the release rod it will drop out of position.

Late 77, 88, 770 and 880

541A. On the late production tractors a high leverage (softer pedal) limited travel linkage is used. This limited travel linkage and pedal can also be installed on the early production non-Super 77 and 88 when a softer pedal (less physical effort) action is required. Because of the limited release travel the linkage will stand two or three free pedal travel adjustments before the clutch facings are worn out.

On these hookups the pedal should always have a minimum of ¾ inch free travel. Obtain this minimum amount of free travel by adjusting the pedal rod at forward end of linkage shown in Fig. O367B. If clutch does not fully release when pedal is depressed to a point where its stop is within one inch of the bull gear cover as shown in Fig. O367A, the clutch should be removed for repairs.

REMOVE AND REINSTALL

542. The preferred method of removal of clutch is to first remove the engine and radiator as a unit as outlined in paragraph 430. With engine out of tractor, remove Power Booster clutch if so equipped as per paragraph 430, then unbolt clutch housing from engine and main clutch from engine flywheel.

A less efficient method of removing clutch is accomplished without removing the engine. This involves approximately the same engine panel work and requires removal of input shaft coupling and sliding the clutch bell housing and clutch shaft rearward after removing the housing screws.

OVERHAUL

543. Use the specifications as listed in the accompanying specifications table for adjusting the release lever height and checking the pressure springs.

A field change-over package, containing a new clutch shaft, ball bearing and PTO drive hub, is available for replacing the clutch shaft pilot bushing with a ball bearing.

Fig. O367A—On late production non-Super 77, 88 and all Super 77 and 88 tractors, the clutch should be fully released when pedal is depressed to point where its stop is 1-inch from bull gear cover as shown.

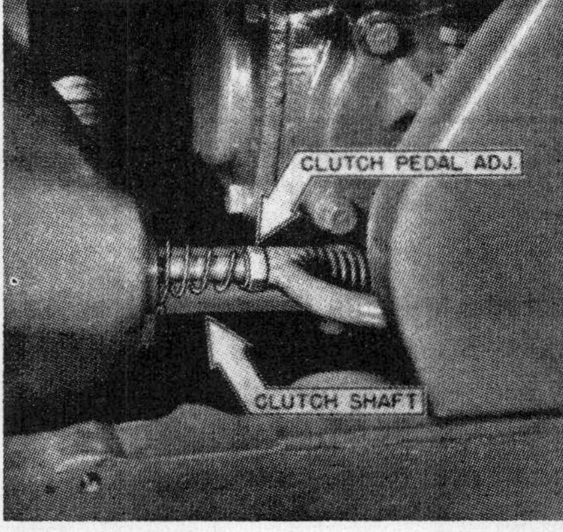

Fig. O367B — Showing location of clutch pedal rod length adjustment on later production series 77 and 88 tractors. Refer to text for reason why free travel should not be adjusted on early production tractors.

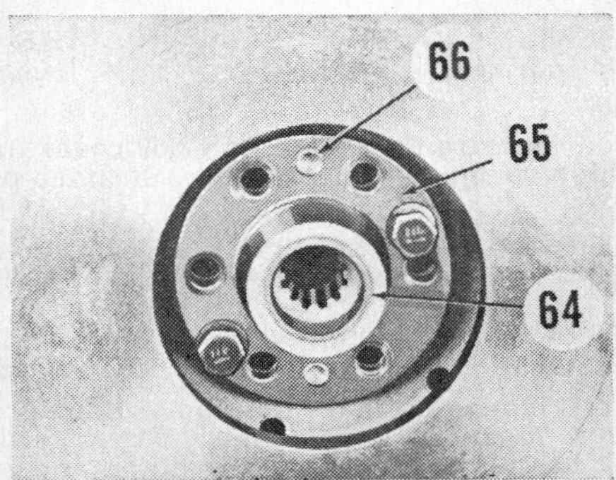

Fig. O368—Continuous type power take-off unit drive shaft splined hub and clutch shaft pilot bushing installation on flywheel. Later production non-Super models 77 and 88 and all Super models are equipped with a ball bearing instead of bushing (64).

PRESSURE SPRING TESTS

Color Spring	Test Lbs. & Height	Color Spring	Test Lbs. & Height
Purple	130-140 @ 1 11/16	Yellow Stripe	289-299 @ 1 13/16
Light Blue	160-170 @ 1 11/16	Orange	165-175 @ 1 1/2
Dark Blue	183-197 @ 1 11/16	Red	189-201 @ 1 1/2
Yellow	155-165 @ 1 1/2	Black	223-237 @ 1 1/2
Pink	177-187 @ 1 9/16	Brown	224-236 @ 1 11/16

CLUTCH AND LEVER SPECIFICATIONS

Clutch Model	Borg or Auburn Cover Assembly Number	Oliver Cover Assembly Number	Lever Height—Inches Using: .330 Keystock	Lever Height—Inches Using: .285 Keystock
B & B 11″	871	IKAS-569A	1 15/16	......
B & B 11″	879	IKBS-569A	1 15/16	......
B & B 11″	882	IKS-569A	1 15/16	......
B & B 9″	952	IMS-569	......	1 15/16
B & B 8″	989	ILS-569	......	1 13/16
B & B 8″	1315	ILAS-569	......	1 13/16
B & B 10″	992	IKS-569	1 15/16	......
B & B 9″	1329	IMBS-569	......	1 15/16
B & B 10″	1323	IKAS-569	1 15/16	......
Auburn 9″	3395 or 100009-1	IMAS-569	......	1 15/16
Auburn 10″	3281 or 100012-1	IKBS-569	1 15/16	......

Clutch lever height is the distance measured between friction face of flywheel and release bearing contacting surface of release levers.

CLUTCH APPLICATION AND SERVICE CHANGES

Tractor Model	Clutch Make and Size	Oliver Cover Assembly No.	Borg or Auburn Cover No.	Cover Pressure Springs
66 Early	B&B 8″	ILS-569	989	6 Orange
66 Later	B&B 8″	ILAS-569	1315	3 Red and 3 Black
66 Later	Auburn 9″	IMAS-569	3395	3 Tan

Service the above Borg & Beck ILS-569 cover into ILAS-569 by discarding the orange springs and installing 3 red springs Oliver No. IL559A and 3 black springs Oliver No. IL559B. The Auburn No. IMAS-569 cover assembly is recommended for service on Industrial 66 tractors.

Tractor Model	Clutch Make and Size	Oliver Cover Assembly No.	Borg or Auburn Cover No.	Cover Pressure Springs
77 Early	B&B 9″	IMS-569	952	9 Yellow
77 Later	B&B 9″	IMBS-569	1329	6 Red and 3 Yellow
77 Later	Auburn 9″	IMAS-569	3395	3 Tan
77 Later	Auburn 9″	IMAS-569	3395	3 Yellow Stripe

Service the above Borg & Beck IMS-569 cover into IMBS-569 type by discarding 6 of the 9 yellow springs and installing in their place 6 red springs Oliver No. IL-559 and one of the yellow springs at each third mounting point. Service the Auburn clutch by installing 3 heavier yellow striped springs No. IM-559C in place of tan springs.

Tractor Model	Clutch Make and Size	Oliver Cover Assembly No.	Borg or Auburn Cover No.	Cover Pressure Springs
77 Ind.	B&B 10″	IKS-569	992	9 Light Blue
77 Ind.	B&B 10″	IKAS-569	1323	3 Brown and 6 Light Blue

Service the above IKS-569 cover into IKAS-569 by discarding 3 of the light blue springs and installing 3 heavier brown springs Oliver No. IM-559A.

Tractor Model	Clutch Make and Size	Oliver Cover Assembly No.	Borg or Auburn Cover No.	Cover Pressure Springs
88 Early	B&B 10″	IKS-569	992	9 Light Blue
88 Later	B&B 10″	IKAS-569	1323	3 Brown and 6 Light Blue
88 Later	Auburn 10″	IKBS-569	3281	Pink
88 Later	Auburn 10″	IKBS-569	3281	Yellow

When servicing the above Borg & Beck covers proceed as indicated above for 77 Industrial models. When servicing the Auburn cover IKBS-569, if it has pink springs discard them and install the heavier yellow springs Oliver No. IK-559C.

Tractor Model	Clutch Make and Size	Oliver Cover Assembly No.	Borg or Auburn Cover No.	Cover Pressure Springs
88 Ind.	B&B 11″	IKS-569A	882	8 Dark Blue
88 Ind.	B&B 11″	IKAS-569A	871	12 Purple
88 Ind.	B&B 11″	IKBS-569A	879	12 Light Blue

When servicing the covers having dark blue or purple springs discard the springs and install 12 light blue springs Oliver No. IB-559A.

POWER BOOSTER

Refer to 770 & 880 Supplement, paragraphs 6-10 for Super 77 and Super 88 tractors.

Series 770 and 880

The Power Booster unit, which is available as optional equipment on series 770 and 880, is inserted in the drive line between engine clutch and transmission. The unit is comprised of a dual plate over-center clutch and a housing containing a set of reduction gears of which the output gear (R—Fig. O368B) is fitted with an over-running (Sprag) clutch.

The unit provides a 1.32:1 speed reduction and information on removal and overhaul follows.

544. REMOVE AND REINSTALL. To remove the Power Booster unit the tractor engine must be removed. To remove engine from tractor, proceed as follows: Remove hood side panels and hood, drain radiator but do not disconnect radiator hoses. Disconnect the excess fuel line and the wire from gasoline gage sending unit at top of fuel tank. Disconnect control rod and drain line from water trap, then disconnect fuel line. Remove the nuts from the tank retaining straps and remove the straps, then lift fuel tank from tractor.

Disconnect battery cables, battery retainer and remove batteries. Remove the capscrew retaining Power Booster shift lever so that lever can move outward, then disconnect the control rod between the shift lever and Power

Fig. O368A—Showing installation of coolant pump on non-Diesel engines. Pump installation on Diesel engines is similar.

Booster unit. Remove capscrew from rear yoke of steering universal located on right side of clutch housing. Disconnect the injection pump control rod at injection pump and bellcrank. Unclip and disconnect electrical wires at both left and right bottom sides of rear panel assembly and unhook the main clutch pedal return spring.

Remove the capscrews which secure the rear panel assembly, then lift the rear panel assembly enough to pull steering shaft from universal. Now carefully move rear panel assembly forward over engine and secure to engine as shown in Fig. O368C.

544A. Remove radiator grille and screen, then remove capscrews from radiator supports and the front frame shield. Disconnect hydraulic lines from the power steering pump and remove top cover from Power Booster unit as shown in Fig. O368D. Disconnect the flexible lubrication tube and fitting at its front attachment and unthread from hole in the Power Booster unit lower cover.

It should be noted that the Power Booster clutch is identified for reassembly by paint marks. However, if paint marks are gone, or not readily recognizable, use scribe marks to insure that the clutch, as well as the shims and shim blocks are assembled in their original position. Note also that the large lobe end of shim and shim block points toward right side of tractor.

Now remove capscrews from Power Booster clutch cover. Clutch can be rotated by disengaging engine main clutch. Remove engine clutch rod.

Disconnect front section of steering shaft from engine clutch housing and front universal and remove steering shaft. Remove the engine support bolts and note that left front and right rear bolts are dowel bolts. These must be reinstalled in their original positions.

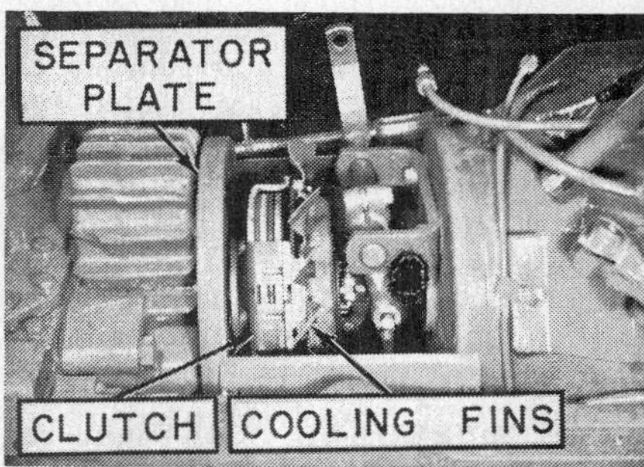

Fig. O368D—Power Booster unit with top cover removed. Note position of cooling fins.

Fig. O368C — View showing rear panel assembly removed and secured to top of engine. Rear panel assembly, engine and radiator assembly can be removed as a unit.

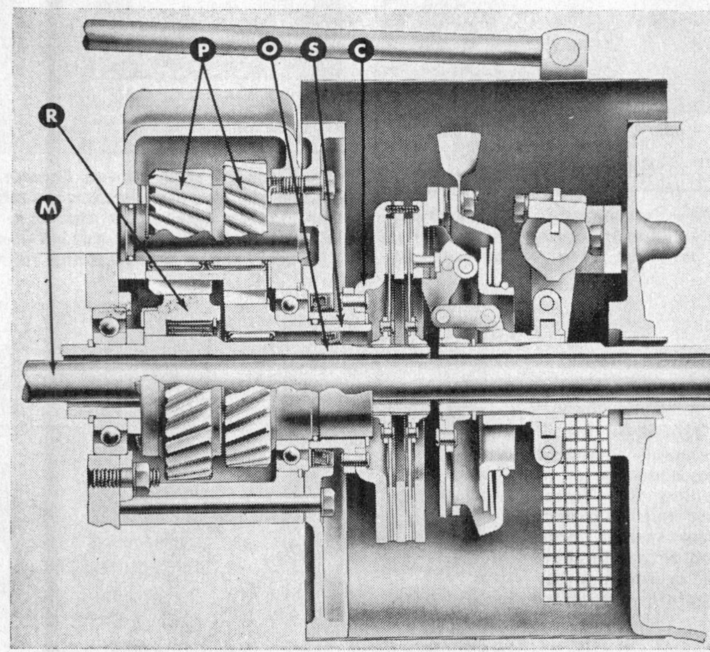

Fig. O368B — Cut-away view of the Power Booster unit used on series 770 and 880.

C. Clutch
M. PTO drive shaft
O. Output shaft
P. Counter-shaft gear
R. Output gear
S. Input gear

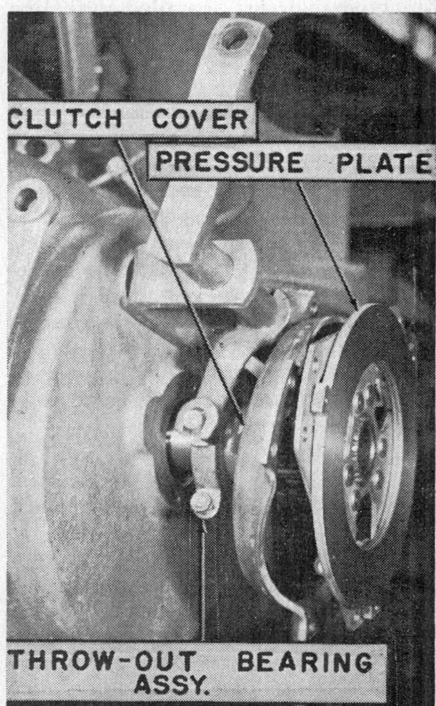

Fig. O368E—Rear of removed 770 engine used in tractor with transmission Power Booster showing a portion of the Power Booster clutch.

Remove the power take-off clutch housing, then unbolt and remove, as a unit, the power take-off clutch and shaft. Attach a lifting eye and hoist to engine and with a wire or short piece of rope tied from radiator to hoist to help support radiator assembly, lift engine, radiator, engine clutch and clutch shaft from tractor as a unit. See Fig. O368C.

544B. **OVERHAUL.** With engine removed as in paragraphs 544 and 544A, remove the Power Booster clutch discs and center plate. Note that long hub of front plate faces forward and the long hub of rear plate faces rearward as shown in Fig. O368F. Be careful not to lose or damage the three small pressure springs.

Remove capscrews from separator plate and pull separator plate and clutch housing from Power Booster housing. See Fig. O368G. Then remove snap ring from front of clutch housing and remove clutch housing and separator plate from input gear.

Drain Power Booster housing, unbolt and remove same from transmission housing then remove output gear from transmission input shaft as shown in Fig. O368J.

Remove countershaft by tapping on offset end of countershaft and catch the double gear and thrust washer as shaft comes out.

Inspect all parts for damage or undue wear and renew as necessary. Lip type seals are installed with lips facing toward rear of tractor. Install lip seal and needle bearing in input gear by using the following Oliver tools. ST-125-B Input Bearing and Seal Driver, ST-125-C Bearing Driver and ST-125-D Driving Mandrel. If these tools are not available, measure both seal and needle bearing prior to removal and install new parts to same measurement. Sprag (over-running) clutch can be removed from output gear by removing a snap ring and is available only as a unit. If oil slinger is removed from rear of output gear be sure that flutes align with oil holes in gear when reinstalling.

544C. When reassembling refer to Fig. O368B, and proceed as follows: Use new "O" ring and install shaft with offset end toward front and flat land of shaft at top, double countergear and thrust washer. Larger gear will be on forward side and locating tang of thrust washer will be located in the top capscrew hole.

With Sprag clutch and oil slinger installed in output gear, slide output gear on transmission input shaft. When holding the output gear, the transmission input shaft should turn only in a clockwise direction when viewed from front side.

Use a new gasket and install the Power Booster housing (with countershaft gear and shaft installed) and tighten retaining capscrews finger tight. Now install special aligning tool (Oliver No. ST125-A) as shown in Fig. O368H and alternately tighten capscrews to 27 ft.-lbs. torque. Remove aligning tool and mount a dial indicator to input shaft as shown in Fig. O368K. Input shaft and bore of housing MUST be concentric to within 0.005. If not, loosen and retighten housing retaining capscrews and recheck. If housing and shaft cannot be brought within limits by shifting housing, a new housing and/or input shaft will be needed.

Balance of reassembly is evident, keeping in mind that the top capscrew in separator plate must not be more than one inch long.

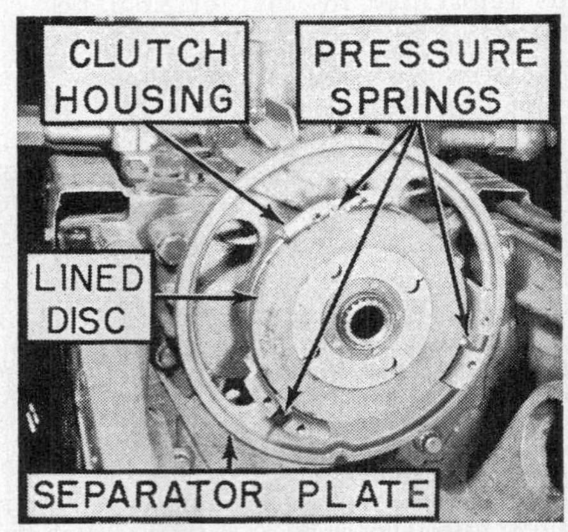

Fig. O368F — Power Booster clutch unit after engine has been removed. Long hub of front clutch disc faces forward. Long hub of rear clutch disc faces rearward.

Fig. O368J—View showing front face of transmission housing after Power Booster unit has been removed.

Fig. O368G—Power Booster unit with separator plate removed. Sprag (over-running) clutch is in inner bore of output gear.

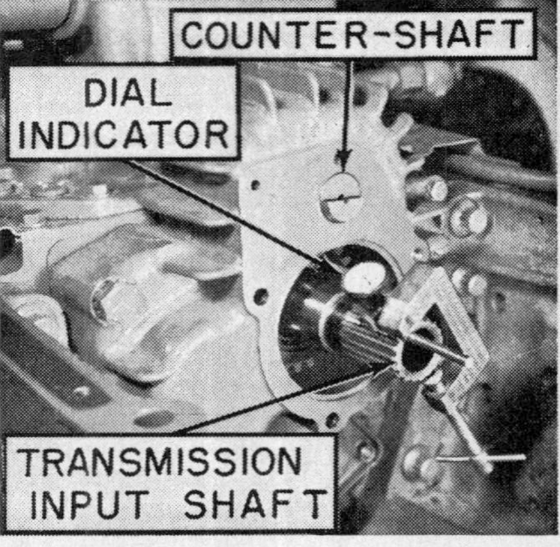

Fig. O368K—With Power Booster unit installed as shown, check concentricity of transmission input shaft and bore of housing with a dial indicator. Run-out must not exceed 0.005. Refer to text.

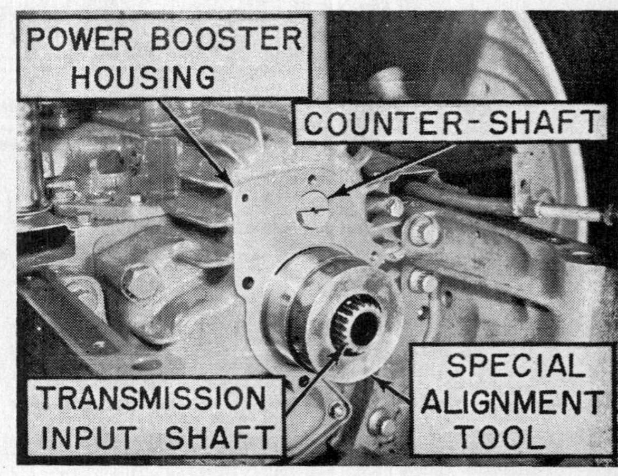

Fig. O368H — Install the special alignment tool as shown prior to tightening the Power Booster housing retaining cap screws. Note postion of countershaft which is retained by separator plate.

TRANSMISSION

Refer to 770 & 880 Supplement paragraphs 12-27 for Reverse-O-Torc transmission and to paragraphs 28-30 for Helical Gear Transmission.

Except for differences in size, number of forward and reverse speeds, and gear ratios, the transmissions for other models are similar. Differences, which affect disassembly, overhaul and/or reassembly, will be noted. Refer to Figs. 0378A, 0378B and 0387C.

BASIC INFORMATION

550. Transmission shafts, differential unit, final drive bevel pinion and ring gear, bull pinions, and bull gears are carried in the one housing (rear main frame), Fig. O370. Although most transmission repair jobs involve overhaul of the complete unit, there are infrequent instances where the failed or worn part is so located that the repair can be completed safely, and without complete disassembly of the transmission. In effecting such localized repairs, time will be saved by observing the following as a general guide.

Main (Input) Shaft. The main input shaft (12) rotates on ball type bearings which require no adjustment. Input shaft can be removed after removing belt pulley unit, power take-off unit drive shaft, or hydraulic lift pump drive shaft on models so equipped, center shifter rail fork, and transmission top front cover or lift unit if so equipped.

Shifter Rails and Forks. Shifter rails and forks, Fig. O372, are accessible for overhaul after removing transmission cover or lift unit if so equipped, belt pulley unit, power take-off unit drive shaft, or hydraulic lift pump drive shaft and transmission main (input) shaft. Refer to paragraph 554 for data on removing the gear shift lever.

Countershaft. Countershaft bearings can be adjusted with shims which are located between front face of transmission housing and bearing retainer (35—Fig. O370) to obtain .001-.003 end play of shaft. Bearings can be adjusted after removing the clutch compartment side panels, transmission top cover or lift unit and countershaft front bearing retainer.

The countershaft (30) can be removed after removing the transmission cover or lift unit, belt pulley unit, power take-off unit drive shaft, or hydraulic lift pump drive shaft, center shifter rail fork and transmission main (input) shaft (12).

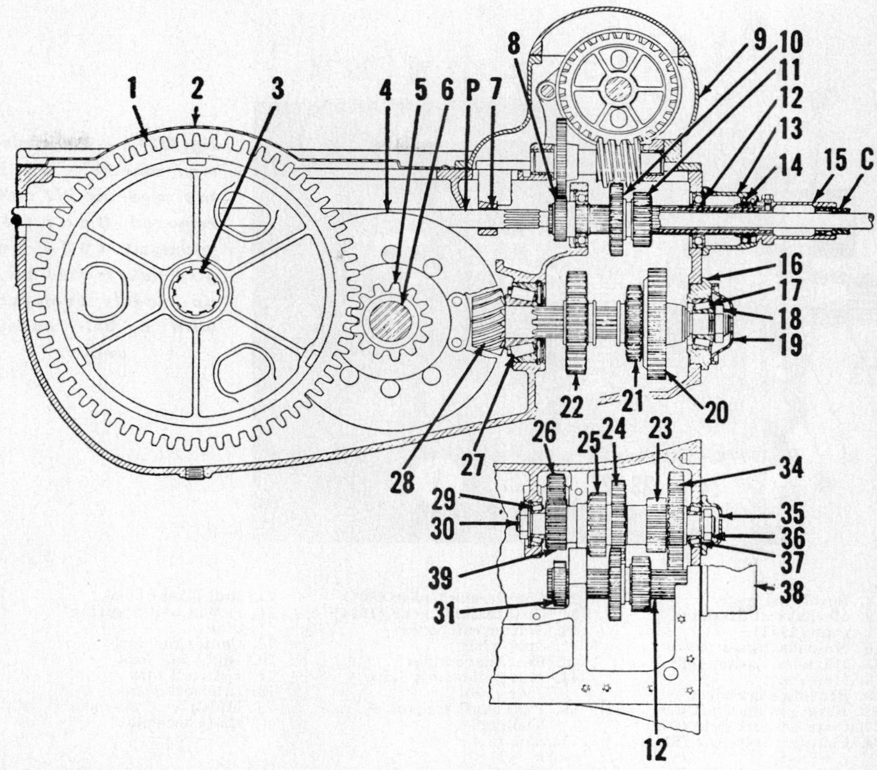

Fig. O370—Transmission shafts, bevel pinion and ring gear, differential unit, bull pinions, and bull gears are contained in the transmission housing (rear main frame). Except for differences in size, and gear ratios, all models are similar. To facilitate reassembly of transmission gears, the gear hubs are stamped with a number which is also indicated in the legend.

C. Clutch shaft
P. PTO unit drive shaft
1. Bull gear
2. Rear cover
3. Wheel axle shaft
4. Bevel ring gear
5. Bull pinion
6. Differential unit shaft
7. Bushing
8. Mechanical lift drive gear (1542)
9. Mechanical lift unit
10. Mainshaft gear (608)
11. Mainshaft gear (608)
12. Mainshaft
13. Bearing retainer
14. Oil seal
15. Coupling
16. Bearing cage
17. Bearing
18. Nut
19. Bearing cage cover
20. Bevel pinion shaft gear (1023)
21. Bevel pinion shaft gear (624)
22. Bevel pinion shaft gear (621)
23. Countershaft gear (609)
24. Countershaft gear (611)
25. Countershaft gear (607)
26. Reverse idler (1021)
27. Bearing
28. Bevel pinion shaft
29. Bearing
30. Countershaft
31. Snap ring
34. Countershaft gear (1024)
35. Bearing retainer
36. Nut
37. Bearing
38. Shifter rail cover
39. Countershaft gear (612)

10. Mainshaft gear (608)
12. Mainshaft
38. Shifter rail cover
43. Bearing retainer
44. Belt pulley unit drive gear (1004)
45. Snap ring
71. Sleeve
73. Rail detent block
74. Detent ball & spring plug
87. "O" ring seal

Fig. O371—Installation details of the belt pulley unit drive gear on models equipped with a belt pulley unit.

Bevel Pinion Shaft. Mesh position of bevel pinion shaft (28—Fig. O370) is

Fig. O372 — Transmission assembly with top cover or lift unit removed. Gear (44) (stamped 1004 on gear hub) is used only on models equipped with a belt pulley unit.

4. Bevel ring gear
8. Mechanical lift drive gear (1542)
10. Mainshaft gear (608)
11. Mainshaft gear (608)
12. Mainshaft
19. Bearing cage cover
21. Bevel pinion-shaft gear (624)
23. Countershaft gear (609)
24. Countershaft gear (611)
25. Countershaft gear (607)
34. Countershaft gear (1024)
38. Shifter rail cover
41. Snap ring
43. Bearing retainer
44. Belt pulley unit drive gear (1004)
66. PTO shaft support & bushing
71. Sleeve
73. Rail detent block
74. Detent ball & spring plug
75. Center rail fork
76. Right rail fork
77. Left rail fork
78. Interlock yoke
79. Multiple reverse idler shaft bore plate

non-adjustable. Bevel pinion shaft can be purchased separately from the bevel ring gear.

Bevel pinion shaft bearings can be adjusted with nut (18) located on forward end of shaft to provide a bearing preload required to rotate the shaft of 8-12 inch lbs. on the 66, 12-16 inch lbs. on the 77, and 16 to 20 inch lbs. on the 88; or a spring scale pull required to rotate the shaft of 2-3 lbs. on the 66, 3 to 5 lbs. on the 77, and 5 to 6 lbs. on the 88 when scale is attached to teeth of gear (20).

Bevel pinion shaft can be removed after removing transmission main shaft (12), the right bull gear, both bull pinions and differential side gears, and moving differential unit toward right side of housing to permit bevel pinion shaft to be withdrawn rearward.

Reverse Idler Shaft. Reverse idler shaft and gear (26) can be removed after removing the countershaft (30).

Multiple Reverse Idler. Multiple reverse idler shaft and gear (81—Fig. O374) can be removed after removing transmission cover or lift unit.

8. Mechanical lift drive gear (1542)
10. Mainshaft gear (608)
11. Mainshaft gear (608)
12. Mainshaft
16. Bearing cage
18. Nut
20. Bevel pinion shaft gear (1023)
21. Bevel pinion shaft gear (624)
22. Bevel pinion shaft gear (621)
23. Countershaft gear (609)
24. Countershaft gear (611)
25. Countershaft gear (607)
26. Reverse idler (1021)
27. Bearing
28. Bevel pinion shaft
29. Bearing
30. Countershaft
34. Countershaft gear (1024)
36. Nut
37. Bearing
39. Countershaft gear (612)
40. Reverse idler shaft
41. Snap ring
42. Bearing
43. Bearing retainer
44. Belt pulley unit drive gear (1004)
45. Snap ring
47. Bearing shield

Fig. O373—Transmission shafts and gears assembly typical of all models.

OVERHAUL

551. Procedure data on overhauling the various components which make up the transmission are outlined in the following paragraphs. The location of transmission gears on their respective shafts will be indicated in the text by figure call-out numbers as well as by their part number which is stamped on the gear flange.

552. PREPARATION. Overhaul of the complete transmission requires the following: Drain transmission and differential compartments. Remove belt pulley, hood side panels, clutch compartment panels, hood, battery, fuel tank, governor control rod from lower side of the battery carrier, head light wires at the fuse block, light wires from underside of the platform (transmission housing rear cover) and belt pulley shifter control. Disconnect steering shaft at the rear universal joint. Remove instrument panel retaining cap screws and lay the panel with remainder of wires still attached on top of engine.

An alternate method of access to the transmisison is to remove the engine. radiator and instrument panel assembly as a single unit as outlined in paragraph 430 then proceed as follows:

Block-up and support rear portion of tractor and remove both rear wheels and hubs.

From models so equipped, remove belt pulley unit, power take-off unit and drive shaft or hydraulic lift pump drive shaft, and transmission top cover

553. MAIN (INPUT) SHAFT. Transmission main (input) shaft (12—Fig. O374) can be removed as follows: First perform work as outlined in paragraph 552; then, remove engine clutch shaft to transmission main shaft coupling (15—Fig. O370), and main shaft front bearing retainer (13) on models without belt pulley unit. Remove clutch coupling spacer (71—Fig. O374), and "O" ring oil seal.

On models equipped with a belt pulley unit, remove clutch coupling spacer (71—Fig. O371), "O" ring oil seal (87), snap ring (45), belt pulley drive gear (44) (stamped 1004 on gear) and main shaft front bearing retainer (43).

On models equipped with mechanical type power lift unit, remove snap ring (83—Fig. O374) and power lift drive gear (8) (stamped 1542 on gear).

Use a small clamp to hold the main shaft rear bearing (42—Fig. O373) in housing bore. Bump main shaft toward front of transmission until front bearing is clear of case bore. Remove rear bearing positioning snap ring (41—Fig. O372) from shaft and slide gear (10 & 11) (stamped 608 on gear) off of shaft. Withdraw shaft and front bearing forward.

553A. Transmission main (input) shaft rotates on ball type bearings which require no adjustment. Reinstall shaft, gears and components in the order as in Figs. O378A, B and C.

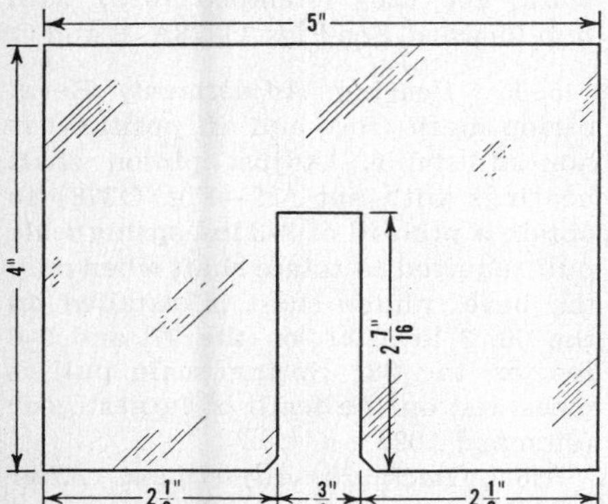

Fig. O373A—A sheet metal plate made to the dimensions as shown will prevent the loss of the gear shift lever springs.

Fig. O376—Reverse idler gear installation.

22. Bevel pinion shaft gear (621)
26. Idler gear (1021)
28. Bevel pinion shaft
88. Shaft retaining bolt

Fig. O374—Transmission mainshaft installation. Gear (8) is used only on models equipped with a mechanical type lift unit. Gear (44) is used only on models equipped with a belt pulley unit.

4. Bevel ring gear
8. Mechanical lift unit drive gear (1542)
10. Mainshaft gear (608)
11. Mainshaft gear (608)
12. Mainshaft
24. Countershaft gear (611)
38. Shifter rail cover
44. Belt pulley unit drive gear (1004)
71. Sleeve
76. Right rail fork
77. Left rail fork
78. Interlock yoke
81. Multiple reverse idler (1021)
83. Snap ring
84. Elastic stop nut

Fig. O375—Using Oliver special tool STS 100, shown in Fig. O377, to push countershaft forward and out of gears and transmission housing.

23. Gear (609)
24. Gear (611)
25. Gear (607)
28. Bevel pinion & shaft
34. Gear (1024)
39. Gear (612)
80. Special tool STS 100

554. SHIFTER RAILS AND FORKS. To prevent the gear shift lever springs from dropping out of the lever socket and into the transmission housing when only the gear shift lever is to be removed, make-up a sheet metal plate as shown in Fig. O373A. Insert the plate under the lever socket before lifting the lever and socket from the transmission. Also, the plate can be used when reinstalling the gear shift lever and socket.

554A. To remove the shifter rails and forks, first remove the transmission main (input) shaft (12—Fig. O372) as outlined in paragraph 553. Remove interlock yoke (78), shifter fork to rail retaining cap screws, shifter rail cover (38) located on forward outside wall of transmission housing, and remove the shifter rails and detent block (73) as an assembly from front of transmission. Further disassembly is self-evident.

555. COUNTERSHAFT. The countershaft unit (Fig. O375) can be removed after removing the transmission main shaft (paragraph 553) and shifter rails and forks (paragraph 554). Remove countershaft bearing retainer (35—Fig. O370) and shims from forward end of shaft, and nut from rear end of shaft. Using an Oliver STS-100 tool or an equivalent jack as shown in Fig. O377, and installed as shown in Fig. O375, push shaft forward and out of rear bearing cone; then, remove gears (39) (stamped 612), (25) (stamped 607), (24) (stamped 611), (23) (stamped 609), (34) (stamped 1024), out through top cover opening.

555A. Reinstall countershaft and gears by reversing the removal procedure. Install gears on countershaft in the following manner: (34) 1024 gear with bevel edge of teeth rearward, (23) 609 gear with hub rearward, (24) 611 gear with bevel edge of teeth forward, (25) 607 gear with hub toward rear, and (39) 612 gear with hub forward. See Figs. O378A, B and C.

Adjust countershaft bearings with shims inserted under front bearing retainer to provide the countershaft with a .001-.003 end play.

556. BEVEL PINION SHAFT. To remove the bevel pinion shaft (28—Fig. O378), perform work as outlined in paragraphs 552, 553 and 554, then proceed as follows: Remove right bull gear (by removing the axle shaft), both brake assemblies, both bull pinions and differential side gears and move the differential unit toward the right side of the housing to permit the bevel pinion shaft to be withdrawn rearward. On Super series be careful not to lose the differential shaft thrust bearing adjusting shims when removing the bull pinions.

After moving the differential unit toward the right side of the transmission housing, continue as follows: Remove bevel pinion shaft front bearing cap and the nut (18—Fig. O378). Bump pinion shaft rearward and out of front bearing cone. From the forward end of the shaft, remove the snap ring and spacer; then, remove gears (20, 21, 22—Fig. O373) out through transmission housing top cover opening.

556A. The bushing installed in the hub of the bevel pinion shaft idler gear (20) (stamped 1023 on gear) requires final sizing after installation to 1.1875-1.1885 on model 66; to 1.2500-1.2510 on model 77; and to 1.4375-1.4385 on model 88 so as to provide a .0025-.004 diametral clearance.

Bevel pinion shaft can be purchased separately from the bevel ring gear, or in a matched set with the bevel ring gear.

556B. Reinstall bevel pinion shaft, and gears by reversing the removal procedure. Install gears in the following manner: (22) (stamped 621) with fork groove forward, (21) (stamped 624) with fork groove rearward, and (20) (stamped 1023) with hub forward. See Figs. O378A, B and C.

556C. *Bearing Adjustment.* Bevel pinion mesh (fore and aft position) is non-adjustable. Adjust pinion shaft bearings with nut (18—Fig. O378) to obtain a preload of 2-3 lbs. spring scale pull required to rotate shaft when only the bevel pinion shaft is installed on the 66, 3 to 5 lbs. on the 77, and 5-6 lbs. on the 88. Spring scale pull is measured on the teeth of largest gear (stamped 1023).

556D. *Backlash Adjustment.* After installing the differential unit adjust differential carrier bearings (located on bull pinions) by means of shims interposed between the differential bearing retainers and the transmission housing to provide a .000-.003 end play.

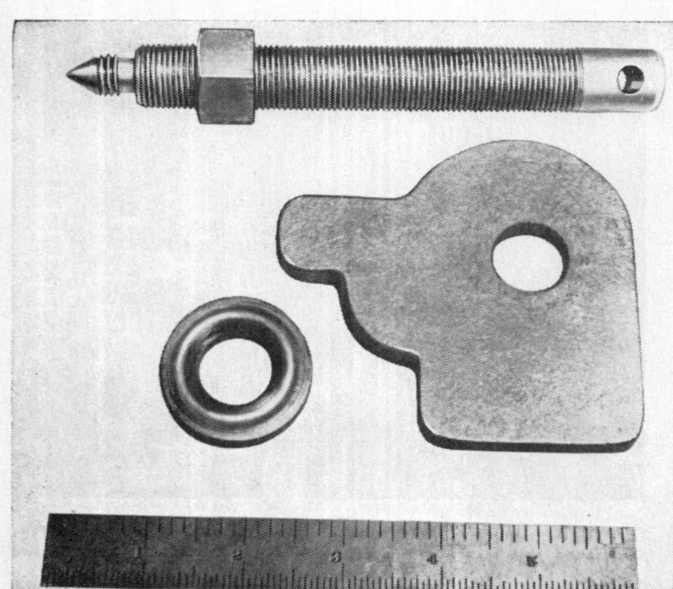

Fig. O377—Oliver special tool STS 100 which is used to remove countershaft as shown in Fig. O375.

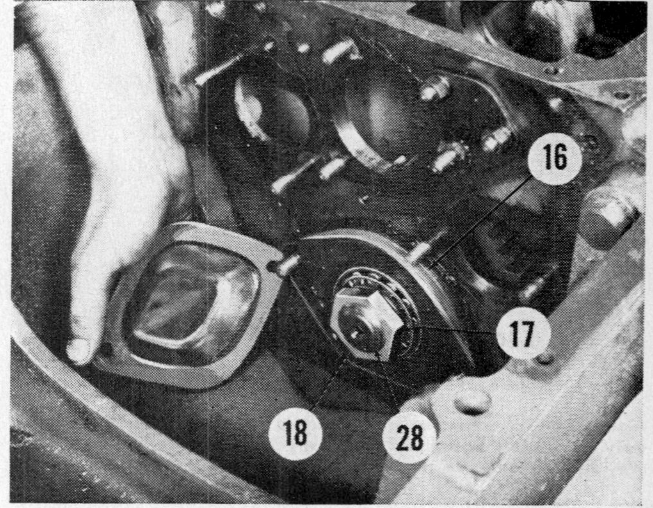

Fig. O378—Bevel pinion shaft with front bearing cage cover removed to show nut (18) which is used to adjust the shaft bearing to a slight preload.

16. Bearing cage
17. Bearing
18. Adjusting nut

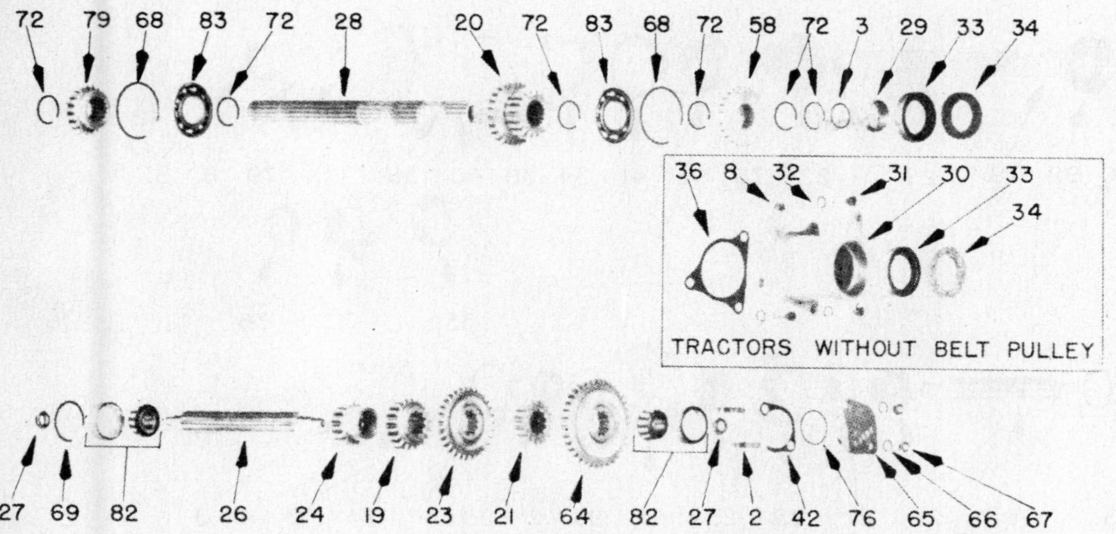

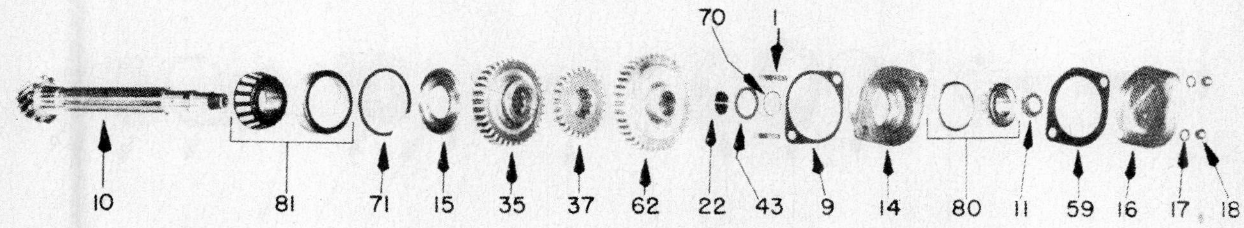

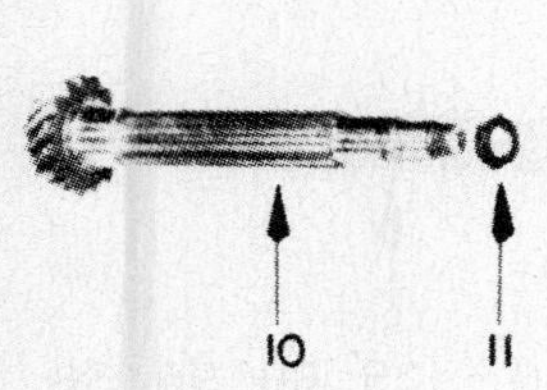

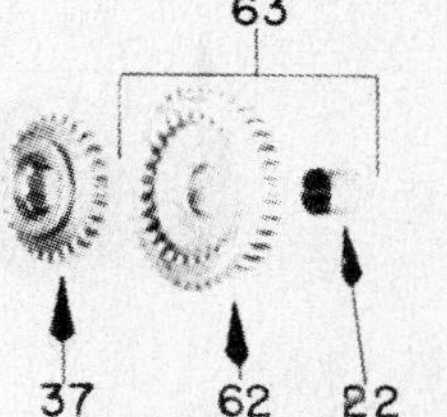

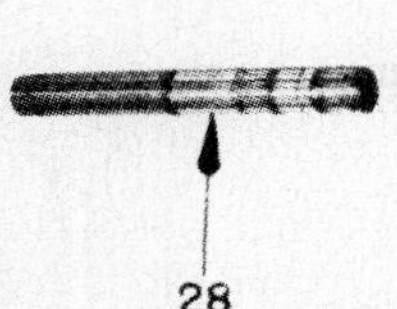

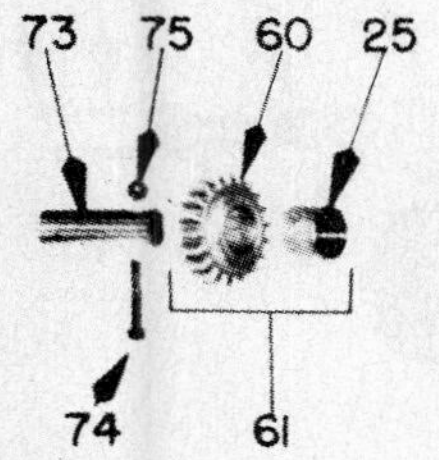

Fig. O378A—Series 66 transmission shafts and gears—exploded view.

3. "O" ring
9. Gasket
10. Bevel pinion & shaft
14. Bearing cage
15. Shield
16. Bearing cage cover
19. Countershaft gear (607)
20. Mainshaft gear (608)
21. Countershaft gear (609)
22. Bushing
23. Countershaft gear (611)
24. Countershaft gear (612)
25. Bushing
26. Countershaft
28. Main (Input) shaft
29. Spacer
30. Bearing retainer
33. Oil seal
34. Felt seal
35. Bevel pinion shaft gear (621)
36. Gasket
37. Bevel pinion shaft gear (624)
42. Shim
43. Washer
58. BP unit drive gear (1004)
59. Gasket
60. Reverse idler gear (1021)
62. Bevel pinion shaft gear (1023)
64. Countershaft gear (1024)
65. Bearing retainer
69. Snap ring
70. Snap ring
71. Snap ring
72. Snap ring
73. Reverse idler shaft
74. Cap screw
76. "O" ring
79. Mechanical lift drive gear (1542)
80. Bearing cup & cone
81. Bearing cup & cone
82. Bearing cup & cone
83. Bearing

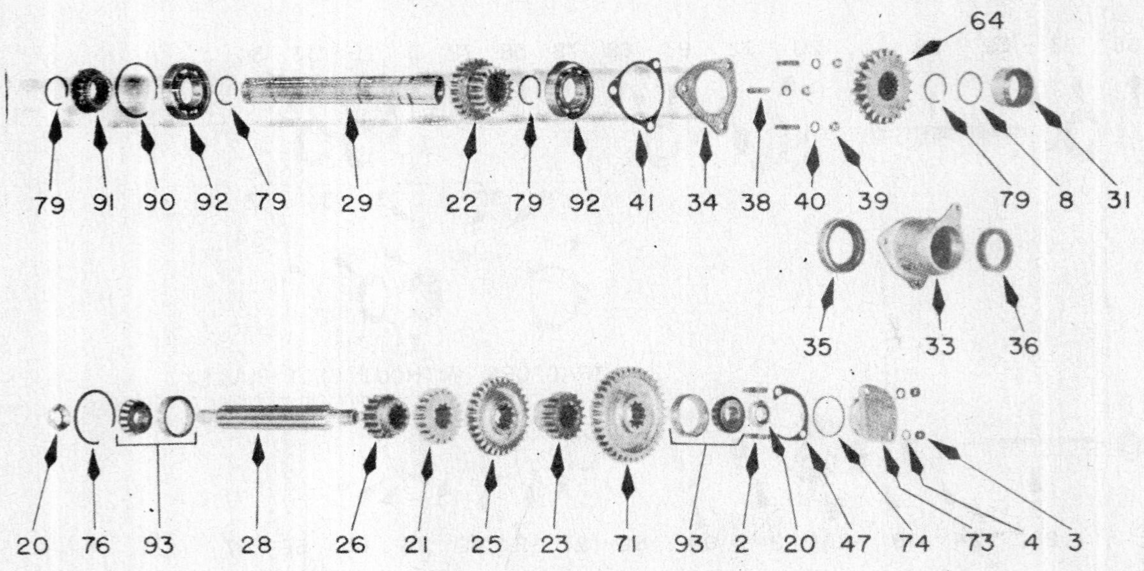

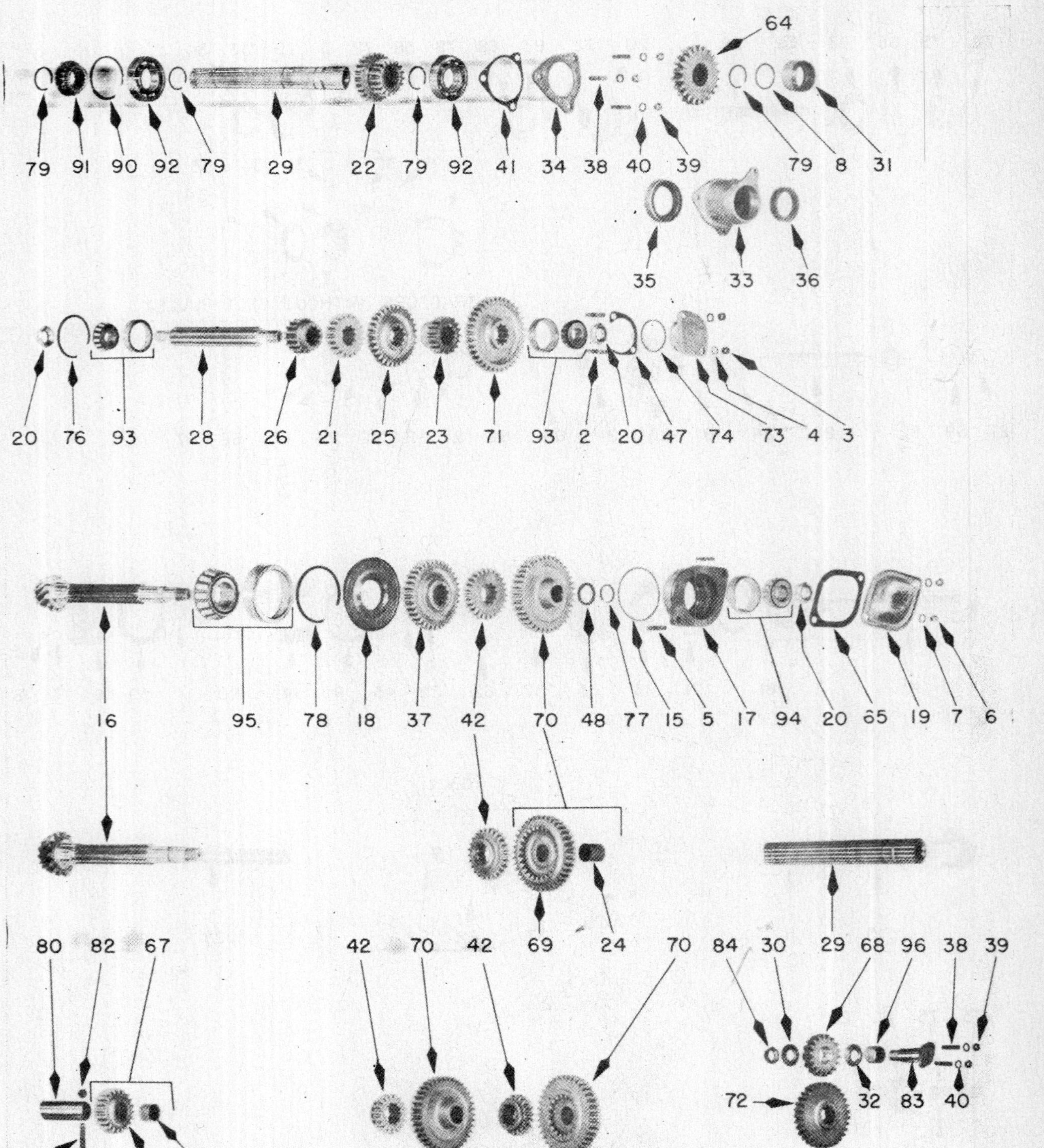

Fig. O378B—Series 77 transmission shafts and gears—exploded view.

8. "O" ring
15. "O" ring
16. Bevel pinion & shaft
17. Bearing cage
18. Shield
19. Bearing cage cover
21. Countershaft gear (607)
22. Mainshaft gear (608)
23. Countershaft gear (609)
24. Bushing
25. Countershaft gear (611)
26. Countershaft gear (612)
27. Bushing
28. Countershaft
29. Main (Input) shaft
30. Washer
31. Spacer
32. Collar
34. Bearing retainer
35. Oil seal
36. Felt seal
37. Bevel pinion shaft gear (621)
41. Gasket
42. Bevel pinion shaft gear (624-A)
47. Shims
48. Washer
64. BP unit drive gear (1004)
65. Gasket
66. Reverse idler gear (1021)
68. Multiple reverse idler (1021-A)
69. Bevel pinion shaft idler gear (1023)
70. Bevel pinion shaft gear (1023-A)
71. Countershaft gear (1024)
72. Countershaft gear (1024-A)
73. Bearing retainer
74. "O" ring
76. Snap ring
77. Snap ring
78. Snap ring
79. Snap ring
80. Reverse idler shaft
81. Cap screw
83. Multiple reverse idler shaft
90. Snap ring
91. Mechanical lift drive gear (1542)
92. Bearing
93. Bearing cup & cone
94. Bearing cup & cone
95. Bearing cup & cone
96. Needle bearing

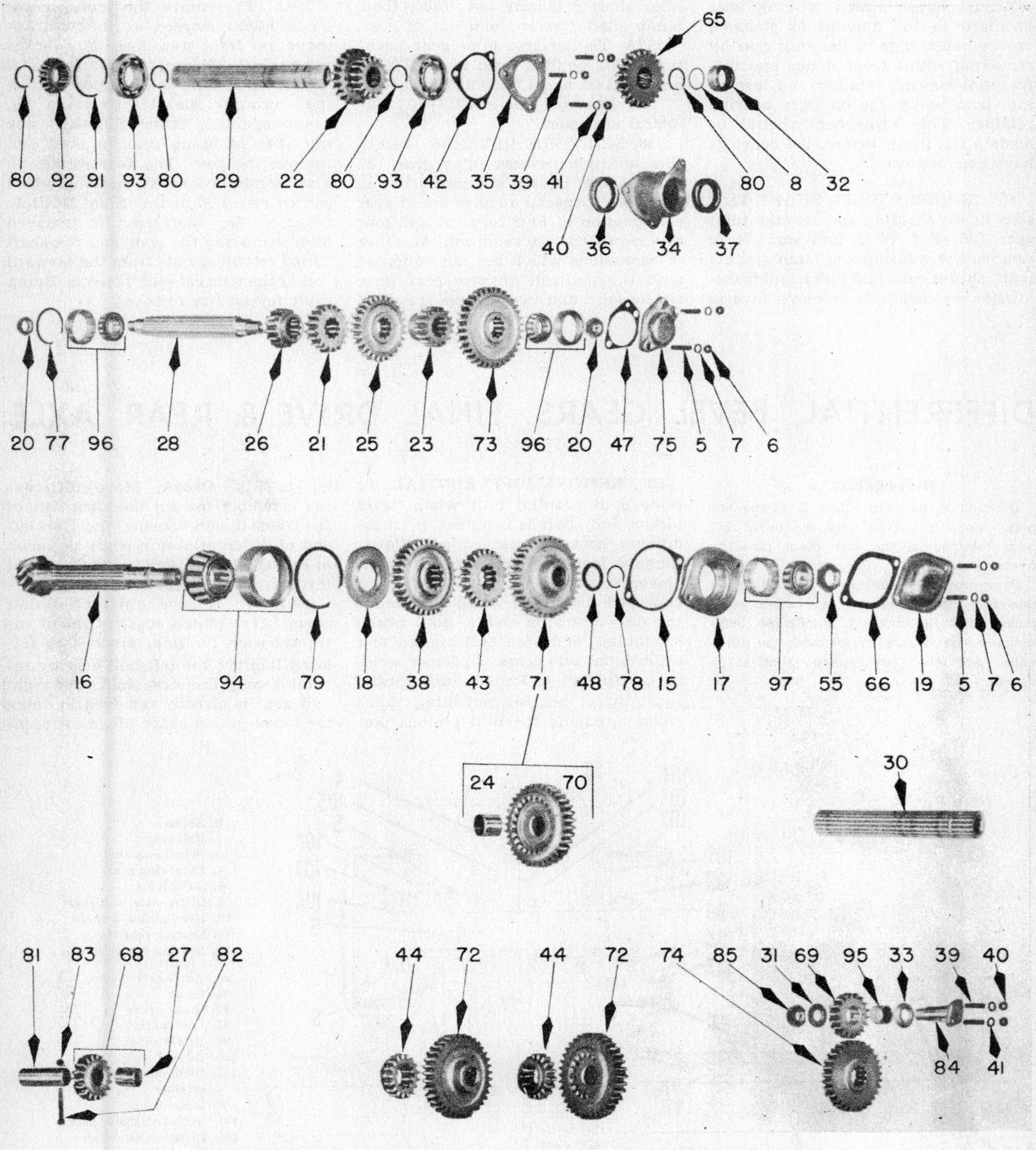

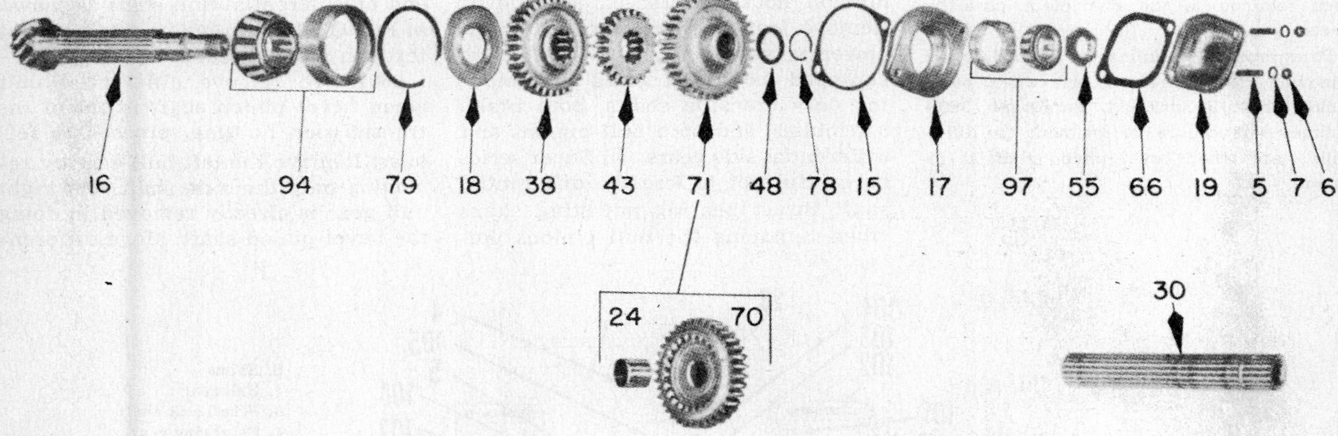

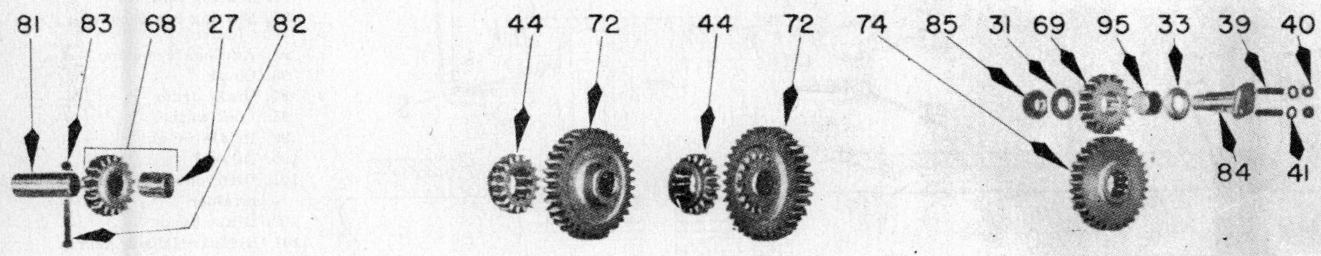

Fig. O378C—Series 88 transmission shafts and gears—exploded view.

8. "O" ring
15. Gasket
16. Bevel pinion & shaft
17. Bearing cage
18. Shield
19. Bearing cage cover
21. Countershaft gear (607)
22. Mainshaft gear (608)
23. Countershaft gear (609)
24. Bushing
25. Countershaft gear (611)
26. Countershaft gear (612)
27. Bushing
28. Countershaft
29. Main (Input) shaft
31. Washer
32. Spacer
33. Collar
34. Bearing retainer
35. Bearing retainer
36. Oil seal
37. Felt seal
38. Bevel pinion shaft gear (621)
42. Gasket
43. Bevel pinion shaft gear (624)
44. Bevel pinion shaft gear (624A)
47. Shims
48. Washer
65. BP unit drive gear (1004)
66. Gasket
68. Reverse idler gear (1021)
69. Multiple reverse idler gear (1021-A)
70. Bevel pinion shaft gear (1023)
71. Bevel pinion shaft gear (1023-A)
72. Bevel pinion shaft gear 37 external teeth
73. Countershaft gear (1024)
74. Countershaft gear (1024-A)
75. Bearing retainer
77. Snap ring
78. Snap ring
79. Snap ring
80. Snap ring
81. Reverse idler shaft
82. Cap screw
84. Multiple reverse idler shaft
91. Snap ring
92. Mechanical lift drive gear (1542)
93. Bearing
94. Bearing cup & cone
95. Bushing
96. Bearing cup & cone
97. Bearing cup & cone

Adjust bevel pinion to ring gear backlash to the amount as stamped on the outer edge of the ring gear by removing shims from under one differential bearing retainer and installing them under the opposite bearing retainer. This adjustment should be made **after,** never **before,** the bearings have been adjusted.

557. REVERSE IDLER SHAFT. First step in overhauling the reverse idler gear (26—**Fig. O376**) and shaft is to remove the transmission main (input) shaft, shifter rails and forks, and transmission countershaft. Remove reverse idler shaft retaining bolt (88); then, bump shaft forward and out of gear.

557A. The reverse idler gear bushing should be final sized to 1.000-1.001 on model 66, and 1.252-1.253 on models 77 and 88 to provide a .0015-.003 diametral clearance.

558. MULTIPLE REVERSE IDLER. The multiple reverse idler gear (81 —Fig. O374) is used on models 77 and 88 where a special reverse speed gear combination of four forward and four reverse speeds are required. All other transmissions which are not equipped with the multiple reverse gear have six forward and two reverse speeds.

558A. To remove the multiple reverse idler, proceed as follows: Remove top front transmission cover, or mechanical lift unit or hydraulic lift unit from models so equipped. Working through the transmission top cover opening, remove Elastic stop nut (84) retaining gear to shaft and remove the gear. The gear rotates on a self-contained needle bearing which can be renewed at this time. Multiple reverse idler shaft can be removed after removing the gear and the shaft flange retaining nuts from the forward end of the transmission housing. Bump shaft forward to remove.

DIFFERENTIAL, BEVEL GEARS, FINAL DRIVE & REAR AXLE

DIFFERENTIAL

Differential unit, Fig. O383, is of the two pinion open case type, with the bevel ring gear retained to the one piece case by rivets.

Paragraph 560 outlines a procedure for removing the differential unit when bevel pinion shaft is installed. Paragraph 560A outlines a procedure for removing the differential unit when bevel pinion shaft is removed.

560. **REMOVE DIFFERENTIAL.** To remove differential unit when bevel pinion and shaft is installed in transmission housing, proceed as follows: Remove transmission rear top cover, power take-off unit drive shaft if so equipped, both bull gears by removing only the axle shafts, both brake assemblies, and both bull pinions and differential side gears. On Super series be careful not to lose the differential shaft thrust bearing adjusting shims when removing the bull pinions. Refer to Fig. O386A. Move differential assembly toward the right side of the transmission housing; then, tip left end of differential unit shaft as shown in Fig. O381, and remove assembly out through rear top cover opening.

560A. To remove differential unit when bevel pinion shaft is out of the transmission housing, proceed as follows: Remove the left bull gear by removing only the axle shaft. The right bull gear is already removed in doing the bevel pinion shaft. Move differen-

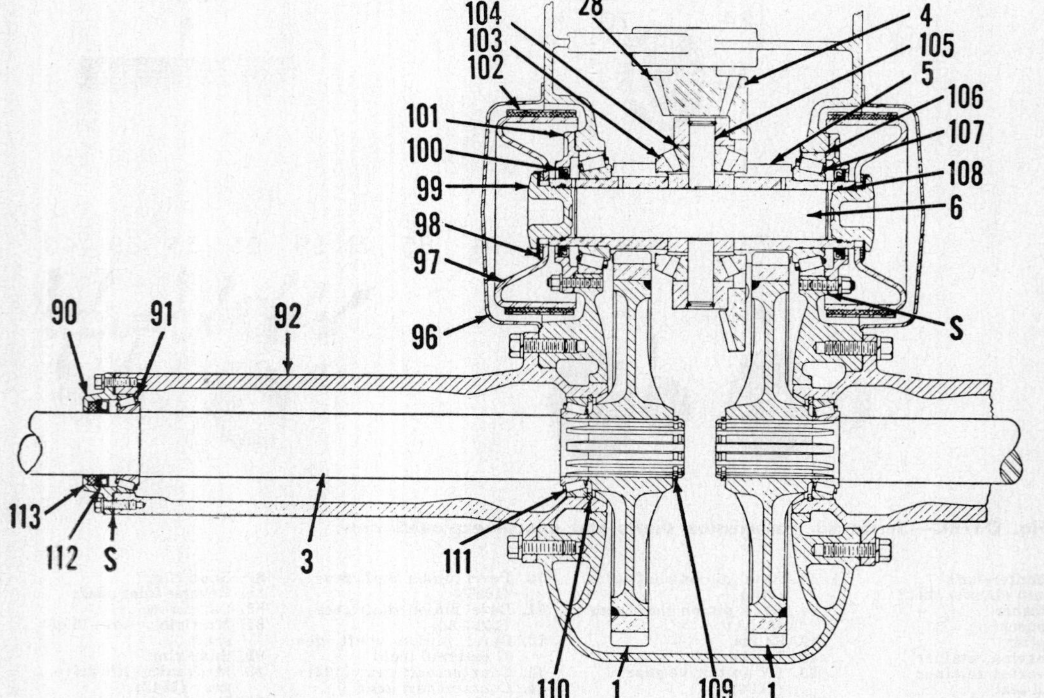

S. Shims
1. Bull gear
3. Wheel axle shaft
4. Bevel ring gear
5. Bull pinion
6. Differential unit shaft
28. Bevel pinion & shaft
90. Bearing retainer
91. Bearing cup & cone (outer)
92. Axle shaft housing
96. Cover
97. Brake drum
98. Lock washer
99. Retaining nut
100. Oil seal
101. Differential bearing retainer
102. Brake band
103. Differential side gear
104. Differential pinion
105. Bevel pinion shaft
106. Bearing cup
107. Bearing cone
108. "O" ring
109. Snap ring
110. Snap ring
111. Bearing cup & cone (inner)
112. Oil seal
113. Felt seal

Fig. O380—Main drive bevel gears, differential unit, bevel pinions, brakes, bull gears, and wheel axle shafts installation. Shown are the band type brakes as installed on early production non-Super models. Later production tractors are equipped with disc type brakes.

tial assembly toward the right side of the transmission housing; then, tip left end of differential unit shaft as shown in Fig. O381, and remove the assembly.

561. **REINSTALL DIFFERENTIAL.** Reinstall bevel ring gear and differential unit in reverse order of removal, and adjust the bevel ring gear to bevel pinion backlash and differential shaft bearings.

Refer to paragraph 562A for method of adjusting differential shaft thrust bearings on all Super models.

561B. BEARINGS ADJUSTMENT. Differential shaft bearings are adjusted with shims (S—Fig. O380) inserted under differential carrier bearing retainers (101) to provide the shaft with .000-.003 end play.

On some non-Super 88 tractors equipped with disc brakes, a condition may exist where removal of all shims (10—Fig. O382) will not decrease the differential unit shaft end play. If this condition is encountered, it will be necessary to make-up and install shims (X) between the bearing cup and bearing retainer on both sides. Shims (X) have an inside diameter of 4 3/16 inches, and an outside diameter of 4 31/32 inches. After installing shims (X), proceed to adjust the bearings with shims (10) to provide the differential unit shaft with an end play of .000-.003.

After adjusting the differential carrier bearings, adjust bevel pinion to ring gear backlash, as indicated on the bevel ring gear by removing shims from under one retainer and installing them under the opposite bearing retainer. To increase the backlash, remove shims from ring gear side (right) and install the same number on the opposite side. This adjustment should be made **after,** never **before** the bearings have been adjusted.

Fig. O381—Differential and bevel ring gear removal. Tip left end of differential shaft up; then, remove unit out through top of housing.

Refer to 770 & 880 Supplement, paragraph 31.

562. **OVERHAUL.** To disassemble the two pinion, open case type differential unit, proceed as follows: Remove pinion shaft retaining snap rings (116—Fig. O383), and using a suitable puller which is threaded into the pinion shaft (105), remove shafts and pinions.

On some differential and bevel ring gear assemblies it may be necessary to remove the bevel ring gear from differential case before the pinion shafts (105—Fig. O383 or 21—Fig. O384A) can be removed.

562A. On super series tractors make sure that the gap (X—Fig. O384A) between inner end face of each bull pinion (2) and outer end of face of each differential side gear (14) is within the limits of .015-.018 when side gears are pushed in toward differential pinions (13) as far as they will go. This is actually a measurement of the backlash between teeth of side gears and pinions. If gap is not within the stated limits add or remove shims (31) at ends of differential shaft until stated gap has been obtained. Refer also to Figs. O384B and O390B.

562B. The bevel ring gear, available separately or in a matched set with the bevel pinion, is riveted to the one-piece differential case. The preferred method of removing the rivets is by drilling. When re-riveting, pull bevel gear into tight contact with differential case using temporary bolts. Rivet ring gear to case, being careful not to distort either piece in the process. Check trueness of ring gear back face by mounting the unit in its normal

Fig. O383—Main drive bevel ring gear and differential unit. Bevel pinion and ring gear are available in a matched set, or separately. Note bevel ring gear backlash is etched on ring gear.

4. Bevel ring gear
6. Differential unit shaft
105. Pinion shaft
115. Differential case
116. Snap ring

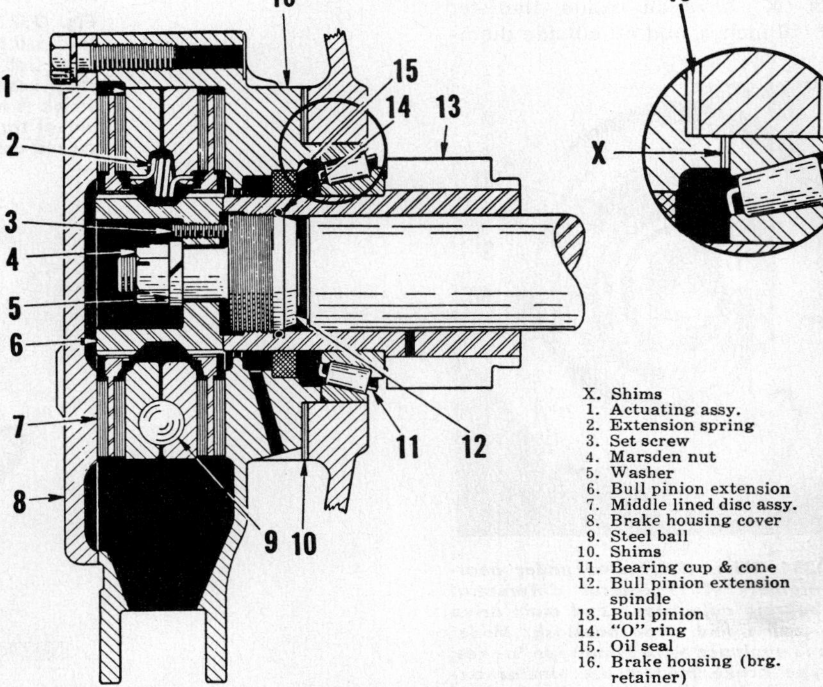

X. Shims
1. Actuating assy.
2. Extension spring
3. Set screw
4. Marsden nut
5. Washer
6. Bull pinion extension
7. Middle lined disc assy.
8. Brake housing cover
9. Steel ball
10. Shims
11. Bearing cup & cone
12. Bull pinion extension spindle
13. Bull pinion
14. "O" ring
15. Oil seal
16. Brake housing (brg. retainer)

Fig. O382—Bull pinion and double disc type brake installation on a non-Super model 88. Shims (X) are used on some model 88 tractors to permit the adjustment of the differential carrier bearings and bevel ring gear backlash with shims (10).

operating position in the differential compartment or between centers in a lathe. Total run-out should not exceed .004.

MAIN DRIVE BEVEL GEARS

The bevel ring gear and pinion are available separately or in matched pairs.

563. **BEARINGS AND BACKLASH ADJUSTMENT.** The mesh or fore and aft position of the main drive bevel pinion is fixed and non-adjustable. The bevel pinion shaft bearings are adjusted by means of nut (18—Fig. O378), located on forward end of pinion shaft, to provide a bearing preload of 2-3 lbs. spring scale pull required to rotate the shaft on the 66, 3 to 5 lbs., on the 77, and 5-6 lbs. on the 88. Spring scale pull is measured on bevel pinion shaft's largest gear which is stamped "1023".

563A. To adjust the differential carrier bearings, first remove both brake assemblies, and transmission rear top cover; then, adjust differential unit shaft to .000-.003 end play by varying shims (S—Fig. O384) located under differential bearing retainers (101).

On some non-Super 88 models equipped with disc brakes, a condition may exist where removal of all shims (10—Fig. O382) will not decrease the differential unit shaft end play. If this condition is encountered, it will be necessary to make-up and install shims (X) between the bearing cup and bearing retainer on both sides. Shims (X) have an inside diameter of 4 3/16 inches, and an outside diameter of 4 31/32 inches. After installing shims (X), proceed to adjust the bearings with shims (10) to provide the differential unit shaft with an end play of .000-.003.

563B. Backlash adjustment of bevel gears should be made after adjusting the differential carrier bearings. Recommended backlash, as indicated on the bevel ring gear, is obtained by removing shim or shims from under one retainer, and inserting same thickness of shims under the other carrier. To increase the backlash, remove shims from ring gear side (right) and install the same number on the opposite side. This adjustment should be made **after,** never **before** the bearings have been adjusted.

564. **R & R AND OVERHAUL.** Data on R & R and overhaul of the bevel ring gear and bevel pinion are outlined in the following paragraphs. The bevel ring gear and/or bevel pinion is available separately or in a matched set. Bearing adjustment and gear backlash adjustment are performed as outlined in the preceding paragraphs, 563A and 563B.

564A. BEVEL PINION. For removal and overhaul of main drive bevel pinion, refer to BEVEL PINION AND SHAFT, paragraph 556.

564B. BEVEL RING GEAR. For removal of bevel ring gear and differential unit, refer to DIFFERENTIAL, REMOVE & REINSTALL, paragraphs 560 and 561.

564C. BEVEL PINION AND RING GEAR. To renew the bevel pinion and ring gear at the same time proceed as follows: Renew bevel pinion as outlined in paragraph 556 then renew the ring gear (pinion out of tractor) as per paragraph 560A through 562.

Check trueness of ring gear back face by mounting the unit in its normal

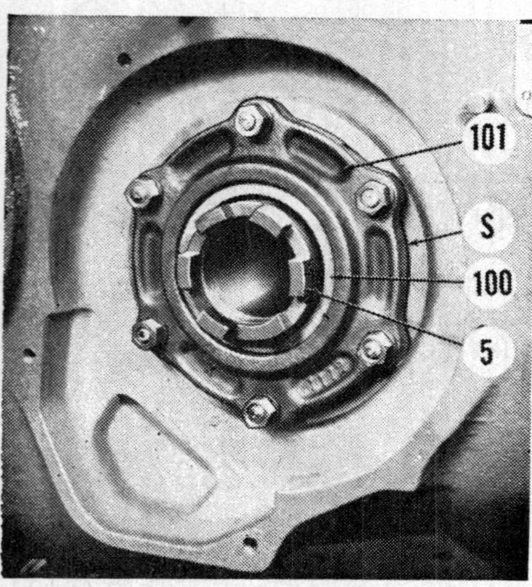

Fig. O384—Shims (S) located under bearing retainers (101) control differential shaft bearing adjustment, and main drive bevel pinion and gear backlash. Model shown is equipped with band type brakes. Disc type brake models are similar except for design changes in the bearing retainer (101).

S. Shims
5. Bull pinion
100. Oil seal
101. Bearing retainer

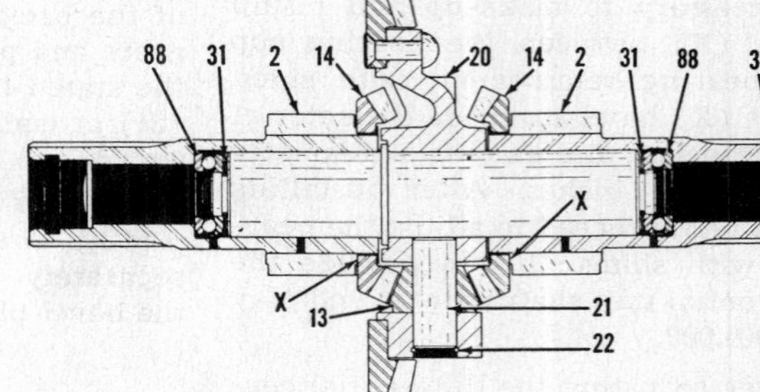

2. Bull pinion
3. Expansion plug
4. Bevel drive gear
13. Differential pinion
14. Differential side gear
20. Shaft and spider
21. Pinion pin
22. Pin snap ring
31. Thrust bearing shim
88. Thrust bearing

Fig. O384A—Sectional view of bevel ring gear, differential, and bull gears assembly used on Super series tractors. Note ball type thrust bearings (88) at ends of differential shaft used to adjust tooth backlash of gears (13 and 14) which is measured by amount of gap (X). On some individual tractors the differential pinions (13) cannot be removed until ring gear (4) is de-riveted from carrier (20).

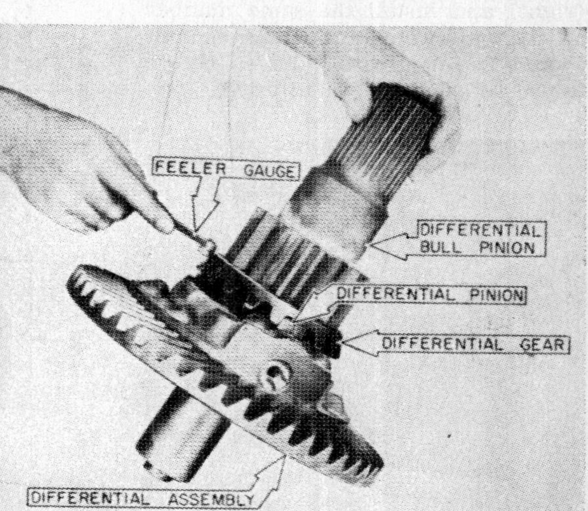

Fig. O384B—On Super series tractors the backlash of differential pinions and gears is determined by measuring gap between bull pinions and side gears as outlined in paragraph 562A.

operating position in the differential compartment or between centers in a lathe. Total run-out should not exceed .004.

FINAL DRIVE GEARS

The final drive gears consist of two bull pinions (5—Fig. O380) which rotate on the differential unit shaft, and bull gears (1) which are splined to the inner ends of the wheel axle shafts.

565. **R&R BULL PINION.** To remove a bull pinion, it will be necessary to remove transmission rear top cover, and brake assembly. Remove differential bearing retainer (101—Fig. O384). Support differential unit. On models equipped with band type brakes use the drum retaining nut (99) as an anchor and pull the pinion and bearing cup as shown in Fig. O385. On disc type brake models, use a suitable puller attached to the extension spindle stud (models equipped with disc brake field installation package) or install a puller as shown in Fig. O386 and pull the bull pinion and bearing cup out of the transmission housing.

565A. Install oil seal (100—Fig. O380) in bearing retainer with lip of seal facing differential unit. Adjust differential carrier bearings to provide the shaft with .000-.003 end play with shims (S) inserted under the bearing retainers. These same shims also, control backlash of bevel ring gear and pinion.

On Super series tractors make sure that the gap (X—Fig. O384A) between inner end face of each bull pinion (2) and outer end face of each differential side gear (14) is within the limits of .015-.018 when side gears are pushed in towards differential pinions (13) as far as they will go. This is actually a measurement of the backlash between teeth of side gears and pinions and if not within the stated limits add or remove shims (31) at ends of differential shaft until stated gap has been obtained. Refer also to Fig. O384B.

566. **R&R BULL GEARS.** To remove any one bull gear, it will be necessary to remove power take-off unit drive shaft, and transmission top rear cover. Support rear portion of tractor and remove rear wheel. Working through transmission top rear cover opening, remove bull gear retaining snap ring. From outer end of axle sleeve, remove bearing retainer (90—Fig. O387) and pull the axle shaft out of the bull gear and lift gear out of transmission housing. Remove the other bull gear by removing the other axle shaft.

566A. Reinstall bull gears so that the machined concave side of the gear hub faces toward the outside.

WHEEL AXLE SHAFTS & HOUSINGS

567. **BEARING ADJUSTMENT.** To adjust wheel axle shaft bearings, proceed as follows: Support rear portion of tractor and remove wheel and hub assembly. Shims (S—Fig. O387) inserted between the bearing retainer (90) and outer end of axle sleeve (92) control bearing adjustment. Add or remove shims (S) to slightly preload the bearings.

568. **R & R WHEEL AXLE & HOUSING.** To remove the wheel axle shaft and housing as an assembly, first drain transmission housing rear portion. Re-

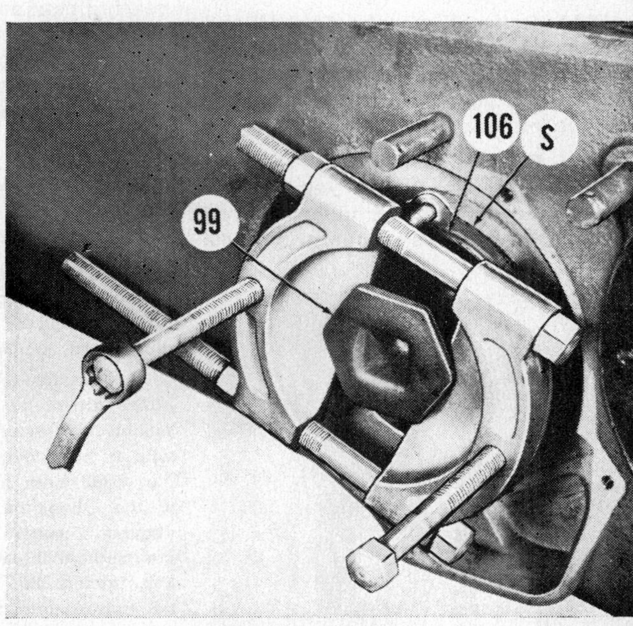

Fig. O385—Puller arrangement used in removing the bull pinion on models equipped with band type brakes.

S. Shims
99. Brake drum retaining nut
106. Bearing cup

Fig. O386—Puller arrangement used in pulling the bull pinion on models equipped with factory installed disc type brakes. As shown, the puller is attached to an annular groove located in the bore of the bull pinion. On some models, the bores of the pinions are threaded.

Fig. O386A—Bull pinion can be removed without removing differential. Only later series tractors are equipped with the differential thrust bearings shown here.

move power take-off unit drive shaft. Remove transmission top rear cover, wheel, hub and tire unit, and fender. Remove bull gear retaining snap ring from inner end of axle shaft, and axle housing to transmission housing retaining bolts. Remove assembly by withdrawing same outward and away from the transmission housing and bull gear.

568A. Reinstall wheel axle shaft and housing assembly to transmission housing so that the machined concave side of bull gear hub faces toward the outside or rear wheel.

569. **RENEW SHAFTS AND/OR BEARINGS.** To renew either the wheel axle shaft and/or inner or outer bearing, proceed as in paragraph 566.

Fig. O387—Location of wheel axle shaft bearing adjusting shims.

S. Shims
90. Bearing retainer
91. Bearing cup & cone
92. Axle shaft housing

Fig. O388—Full-wrap type brake installation as used on early production non-Super models 77 and 88.

A. Band to drum clearance adjusting screw
C. Brake band actuator
98. Retaining nut lock
99. Drum retaining nut
102. Band & lining

Then remove snap ring at inner bearing cup and bump same out of housing. Adjust wheel axle shaft bearings with shims (S—Fig. O387) to slightly preload the bearings.

570. **RENEW SHAFT OIL SEAL.** Rear axle shaft oil seal is a combination felt and a self contained neoprene seal. Seal assembly is located in bearing retainer (90), and can be renewed after removing the rear wheel and hub unit, and bearing retainer. Install neoprene seal with lip facing inward or toward the bull gear.

BRAKES

BAND TYPE

The brakes are of the mechanically actuated, contracting band type with the brake drums located on the outer ends of the bull pinion gear shafts.

On early non-Super 77 and 88 tractors, the bands are of the full wrapping (in forward motion of tractor) type and are provided with a band-to-drum clearance screw (A—Fig. O388) near the anchor. On this version of the Oliver brakes, the foot pedals are attached directly to the band cams without intermediate linkage. On some later production tractors the bands were anchored at the approximate center of their length by a welded lug (L—Fig. O390) and a slot in the transmission housing. On this half-wrap brake, the band actuator is connected to the foot pedal by a lever and linkage. On some of the later non-Super 77 and 88 models, the welded lug type of anchor is omitted, and the band is anchored and centered by an anchor plate (X—Fig. O391). The non-Super 66 remains as in Fig. O390.

First production linings were of the woven type riveted to the bands. Later tractors had bonded linings, while the linings of very latest models equipped with band type brakes are a combination of molded and woven pieces riveted to each band.

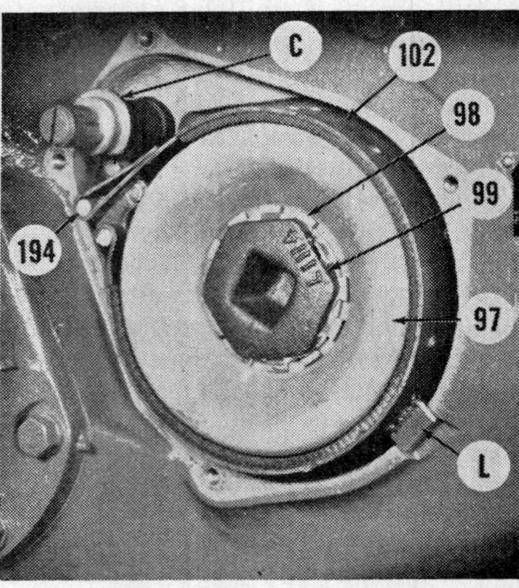

Fig. O390—Half-wrap brake installation as used on non-Super model 66 tractor.

C. Brake band actuator
L. Brake band anchor
97. Drum
98. Retaining nut lock
99. Drum retaining nut
102. Band & lining
194. Brake band pin

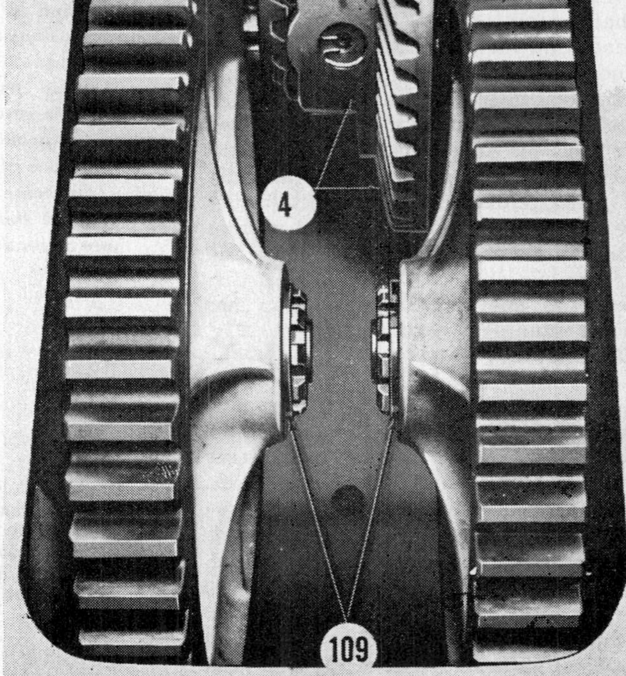

4. Main drive bevel ring gear and differential unit
109. Bull gear retaining snap ring

Fig. O389—Bull gear installation. Install bull gears so that machined concave side of gear hub faces toward axle shaft housing.

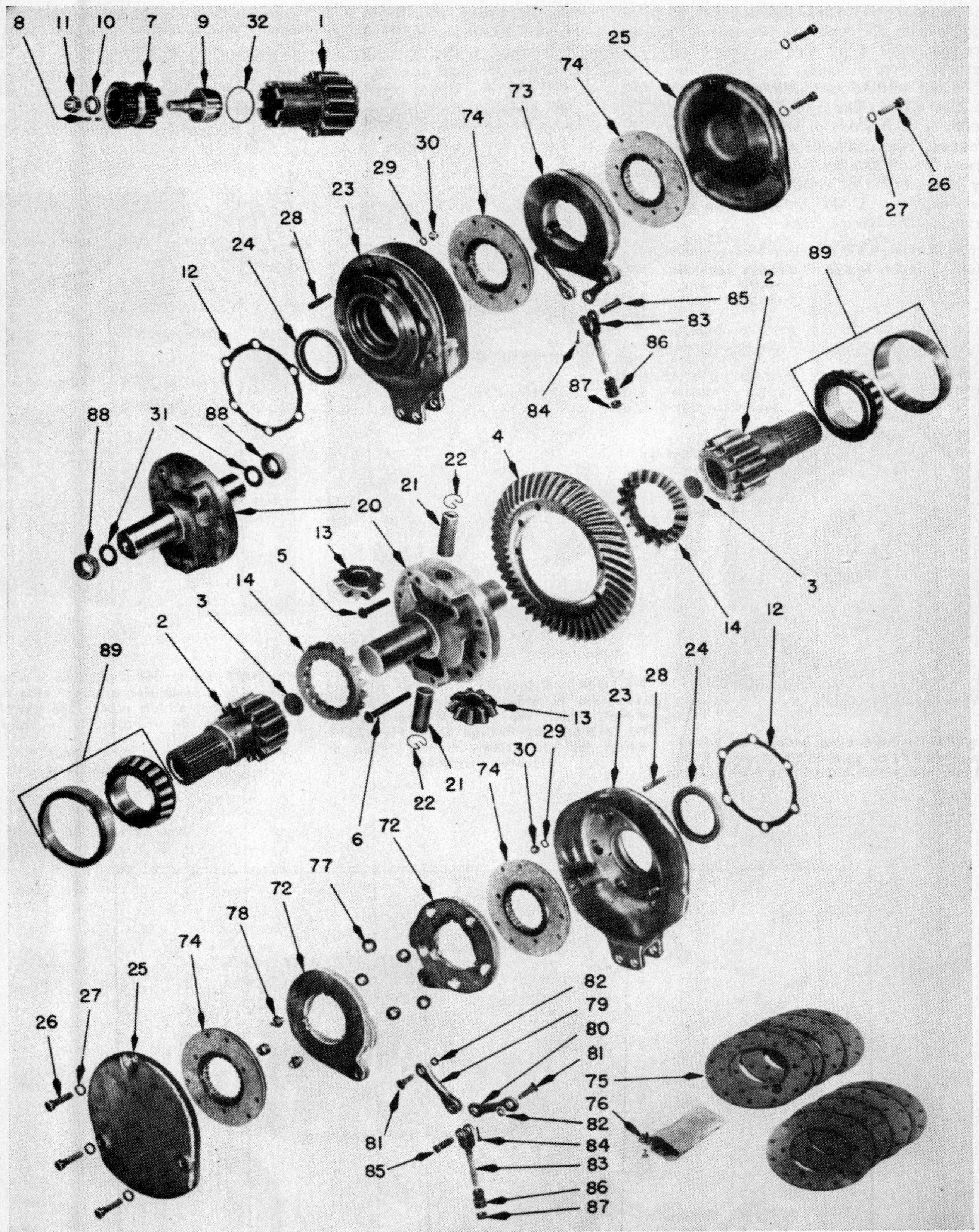

Fig. O390B—Components of differential and disc brakes used on Super series tractors. Shown is Super 77, which is basically similar to Super 66 and Super 88.

3. Expansion plug
1. Bull pinion
2. Bull pinion
4. Bevel drive gear
7. Bull pinion extension
9. Extension spindle
10. Washer
11. Marsden nut
12. Differential bearing cover shim
13. Differential pinion
14. Differential side gear
20. Shaft and spider
21. Pinion pin
22. Snap ring
23. Brake housing
24. Oil seal
25. Brake cover
31. Thrust bearing shim
72. Actuating disc
74. Middle disc
75. Lining
77. Ball ⅞"
78. Extension spring
79. Link yoke
83. Actuating rod
86. Adjusting nut
88. Differential thrust bearing
89. Differential roller bearing

575. **ADJUSTMENT.** Adjustment to compensate for wear of the lining is accomplished by removing the band actuator lever or pedal and relocating it on the actuator serrations, as shown in Fig. O392. On models where the pedal is connected to the actuator by linkage, Fig. O393, adjustment of the brake pedal position is accomplished by lengthening or shortening the link between the brake pedal, and the lever on the band actuator.

576. **R & R AND OVERHAUL.** Brake bands can be removed after removing the pedals, or brake rod lever and brake rod, and brake assembly cover.

576A. To remove brake drum for renewal of drum, or bull pinion shaft oil seal, first remove brake cover and brake band. Using a piece of one inch square steel bar stock and a wrench, remove brake drum retaining nut (99—Fig. O391). Check the condition of the neoprene ring seal in the nut. The bull pinion shaft oil seal can be renewed after removing the differential bearing retainer. Install seal with lip facing inward toward the differential.

Fig. O391—Brake band anchor and centering plate (X) as used on non-Super 77 and 88 late production band brake installations.

Fig. O392—Relocating pedal on actuator serrations to adjust for lining wear. On models where the pedal is connected to the actuator by linkage as in Fig. O393, adjust by relocating actuator lever on actuator serrations.

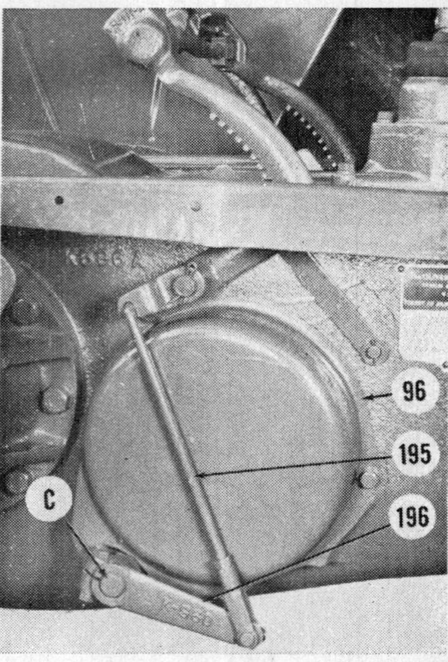

Fig. O393—Brake pedal position is accomplished by lengthening or shortening linkage (195) on models where the pedal is connected to the actuator (C) by linkage.

C. Brake band actuator
96. Brake cover
195. Linkage
196. Actuator lever

Non-Super Model 77 Row Crop

Fig. O394—Adjusting the double disc type brakes on a model 66 tractor.

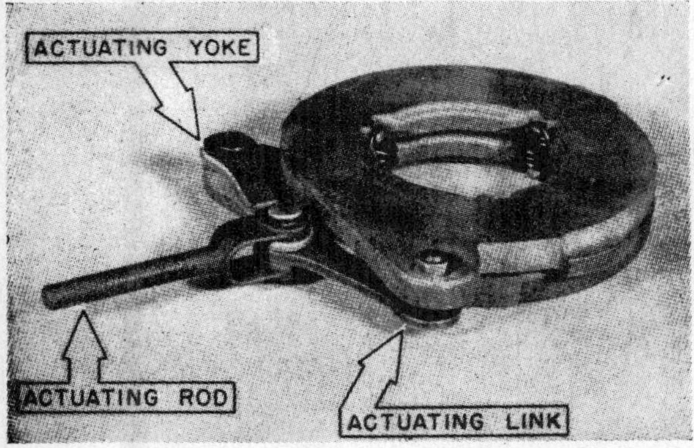

Fig. O395—Assembled view of the double disc type brakes actuating assembly used on 77 and 88. Series 66 is similar.

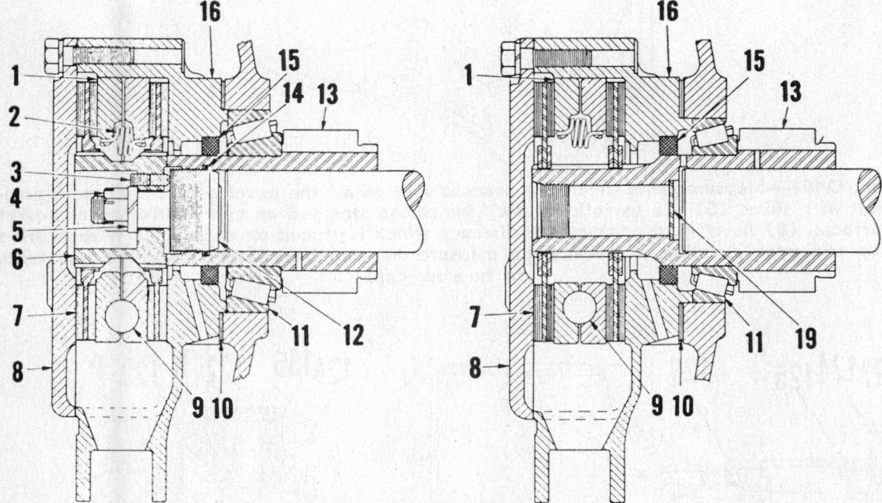

Fig. O396—Double disc type brakes used on later production non-Super models and all later models are mounted on the outer ends of the bull pinion. Left view: Showing construction details of the first type as used on non-Super series in production and field conversion units. Right view: Showing construction details of a later production version for all series. Bull pinions on some of this group are internally threaded (as shown) for the use of a puller; others have an annular groove.

1. Actuating assy.
2. Extension spring
3. Set screw
4. Washer
5. Marsden nut
6. Bull pinion extension
7. Middle lined disc assy.
8. Brake housing cover
9. Steel ball
10. Shims
11. Bearing cup & cone
12. Bull pinion extension spindle
13. Bull pinion
14. "O" ring
15. Oil seal
16. Brake housing (brg. retainer)
19. Expansion plug

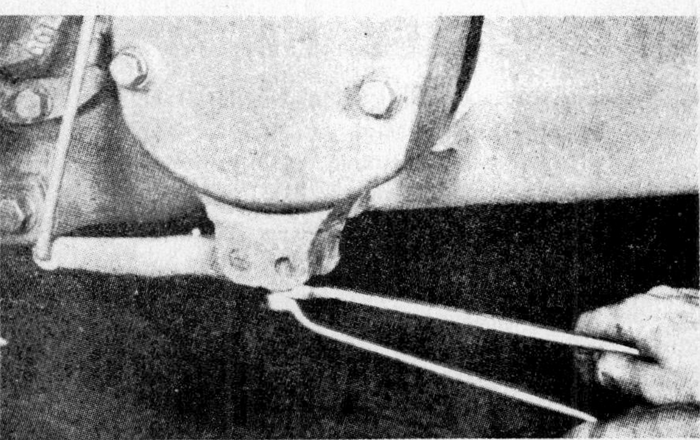

Fig. O397—Adjusting the double disc type brakes on models 77 and 88 tractors.

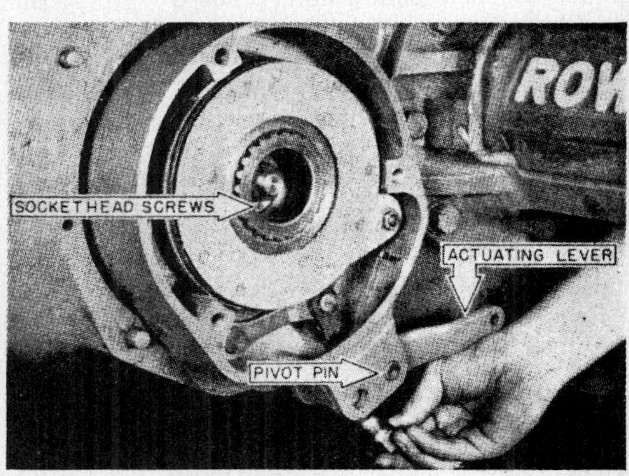

Fig. O398—An installation view of the double disc type brakes on series 77 and 88. Series 66 is similar.

DOUBLE DISC TYPE

Refer to 770 & 880 Supplement, paragraph 32.

The double disc brakes are mounted on the outer ends of the bull pinions as shown in Fig. O396. These brakes were first installed on production tractor non-Super models 66 Row Crop after 430673, 66 Standard after 3500001 and 66 Standard with a prefix of "3" or "35", 77 Row Crop after 347904, 77 Standard after 273375, 88 Row Crop after 136899, and 88 Standard after 826416.

Disc type brakes are also available in kit form for installation on models not so equipped.

All Super series tractors are equipped with double disc brakes.

578. **ADJUSTMENT.** To adjust brakes, first jack-up and support rear of tractor. Turn adjusting nut as shown in Fig. O394 or O397, in a clockwise direction until a slight drag is obtained when rotating the wheel; then back off the adjusting nut ¾ of a turn or until the brake latch will just enter the third notch from the lower end when the pedal is depressed by hand. Lock the adjustment with the jam nut. Do the same to the other brake. Equalize the brakes by backing off the adjusting nut on the tight brake.

BELT PULLEY UNIT

The belt pulley unit, Fig. O399, which is mounted on the transmisssion housing forward wall, is a self contained drive unit that is driven by a spur gear (44) located on the forward end of the transmission main (input) shaft.

Pulley shaft no-load speed at engine high idle speed is: Non-Super 66, 1079 rpm @ 1750 engine rpm; Super 66, 1357 rpm @ 2200 engine rpm; 1085 at 1750 engine rpm on all 77 and 88 models and 1193 at 1925 rpm on 770 and 880.

585. **REMOVE & REINSTALL.** To remove the belt pulley unit, proceed as follows: Disconnect belt pulley engaging mechanism. Remove pulley, clutch compartment panels, and the bolts from the coupling which joins the clutch shaft to the transmission main (input) shaft. Slide coupling forward on clutch shaft. Remove power take-off unit drive shaft or hydraulic lift pump drive shaft on models so equipped. On models 77 and 88 only, unbolt the instrument panel by removing hood, battery, fuel tank; disconnecting head light wires at fuse block; disconnecting steering shaft at front universal joint and removing instrument panel to frame retaining cap screws. Lay the panel assembly with remaining wires still attached, on top of engine. On all models, remove cap screws retaining belt pulley unit to transmission housing and remove the unit.

586. **OVERHAUL.** With belt pulley unit removed from tractor, proceed to disassemble the unit as follows: Remove pulley shifter arm and cover assembly (135—Fig. O399), drive shaft bearing cage (120), shims (124), drive shaft bearing cover (133) and shims (132). Move pulley drive shaft (122) and bevel drive gear (123) out oι mesh with bevel driven gear (130). Using a suitable puller, remove bearing cup (134). Place the drive shaft sliding spur gear (131) against rear bearing cone and block same in this position. Bump pulley drive shaft out of spur gear and rear bearing cone. Do not lose the sliding spur gear detent ball and spring. Unstake the pulley shaft bevel ring gear retaining nut, and with a 1½ inch socket on the 66, or a 2 inch socket on the 77 and 88 remove the pulley shaft bevel gear retaining nut. Press shaft toward pulley end of unit and out of pulley shaft bevel gear (130).

Pulley shaft oil seal (128) can be renewed at this time after removing pulley shaft outer bearing cone (129).

586A. GEAR MESH ADJUSTMENT. If the belt pulley unit housing or bevel gears are renewed, it will be necessary to adjust bevel gear mesh position and backlash as follows: Assemble the inner bearing cup and cone (125—Fig. O401) to pulley shaft gear (130) but do not install the assembly in the housing as yet. Now accurately measure and record the distance (C) from farthest face of gear teeth to bearing cup. Refer to the pulley shaft gear and note and record the cone center dimension (B—Fig. O402) as 1.112 etched thereon. This dimension should be recorded as B. Now refer to the gasket surface of the pulley housing

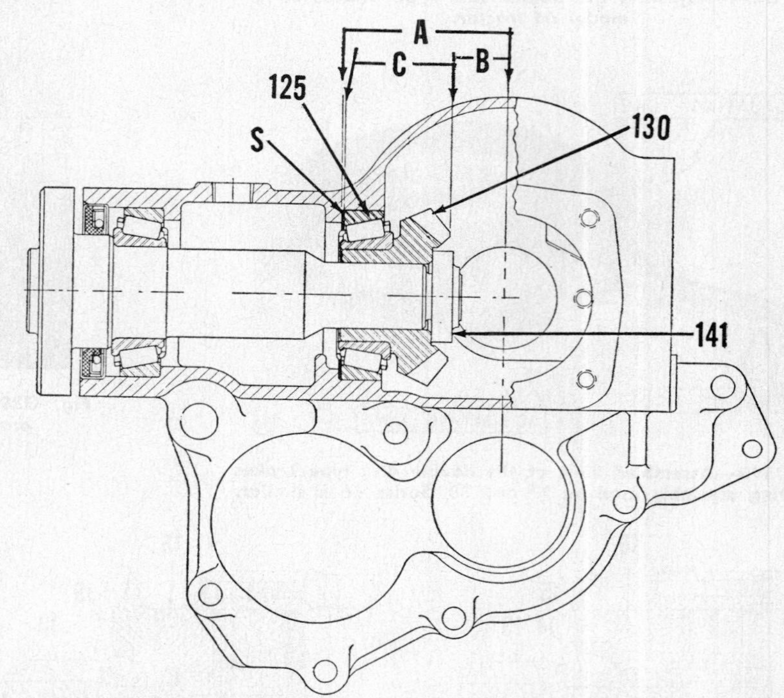

Fig. O401—Measurements which are necessary to adjust the bevel gear (130) mesh position with shims (S) are as follows: (A) Dimension stamped on unit control cover gasket surface. (B) Bevel gear cone center distance which is etched on gear. (C) With bearing cup and cone installed on bevel gear, measure distance from farthest face of gear teeth to bearing cup.

C. Transmission main (input) shaft
P. PTO unit drive shaft
14. Felt seal
44. Belt pulley unit drive gear (1004)
120. Bearing cage
121. Bearing cone
122. Pulley drive shaft
123. Drive shaft bevel gear
124. Shims
125. Bearing cup & cone
126. Pulley shaft
127. Pulley
128. Oil seal
129. Bearing cup & cone
130. Pulley shaft bevel gear
131. Sliding spur gear
132. Shims
133. Bearing retainer
134. Bearing cup & cone
135. Unit control cover
136. Bearing cup
137. Oil seal
138. Detent ball & spring

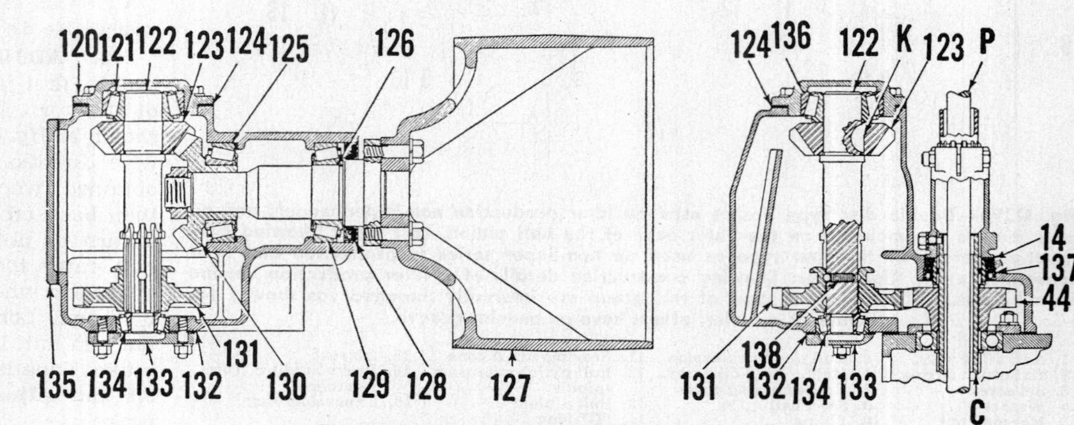

Fig. O399—Belt pulley unit used on all models.

shifter cover, and note the number stamped thereon as 3.409 in Fig. O403. This should be recorded as dimension A.

The pulley shaft bevel gear mesh position is controlled with shims (S—Fig. O401) inserted between bevel gear bearing cup (125) and bearing cup bore in housing. With the above dimensions A, B, and C recorded, subtract the total of dimension B plus C from dimension A. The result minus 0.003 will be the amount or thickness of shims to be inserted at point (S).

586B. Assemble pulley shaft to housing, but tighten nut (141) only finger tight. Mount the pulley housing in a vise so that the pulley shaft flange is between the vise jaws. Place a spring scale in lowest pulley housing mounting cap screw hole, and note amount of torque (pull) required to rotate the pulley housing about the pulley shaft. This torque represents the oil seal drag. Now adjust the bearings to a slight preload by tightening the nut (141) until the drag registered on the spring scale is ½ to 1½ pounds greater than the amount shown when nut (141) was only finger tight; then stake the adjusting nut to lock the adjustment. With this work completed, the pulley gear has been correctly positioned (mesh), and the pulley shaft bearings are correctly adjusted.

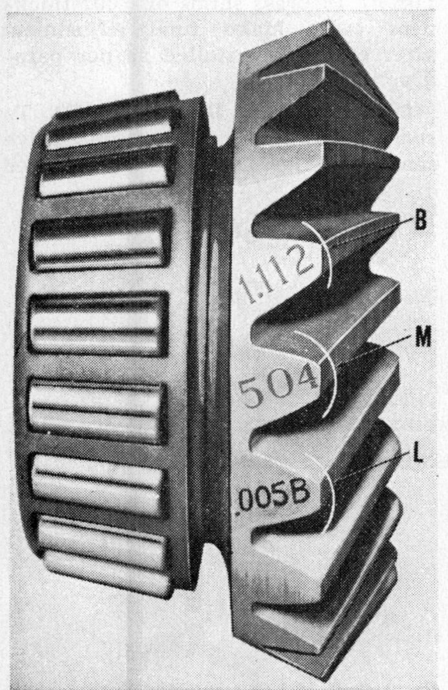

Fig. O402—Pulley shaft bevel gear is available only as a matched set with the bevel drive gear. The following numbers are etched on the gear: (B) Gear cone center distance. (L) Gear backlash. (M) Bevel gears matched set number.

586C. Before setting the backlash, adjust the pulley drive shaft bearings with shims (124 & 132) to provide zero end play to the shaft, and yet permit same to rotate freely. The backlash should now be adjusted to the value (L—Fig. O402) as etched on the pulley shaft bevel gear by removing shims (124 & 132—Fig. O399) from under one drive shaft bearing retainer or cage, and inserting the same amount under the other retainer or cage.

POWER TAKE-OFF

Refer to 770 & 880 Supplement, paragraphs 33 & 34 for Super 77 and Super 88 tractors; to 660 Supplement for Super 66 tractors.

The power take-off unit is mounted on the rear portion of the transmission housing and receives its drive directly from the engine, via shaft (P—Fig. O404) which is splined into a hub mounted on the engine flywheel as shown in Fig. O406, to provide independent operation of the unit. Power to the pto external drive shaft is controlled by use of an over-center type clutch, shown in Fig. O405.

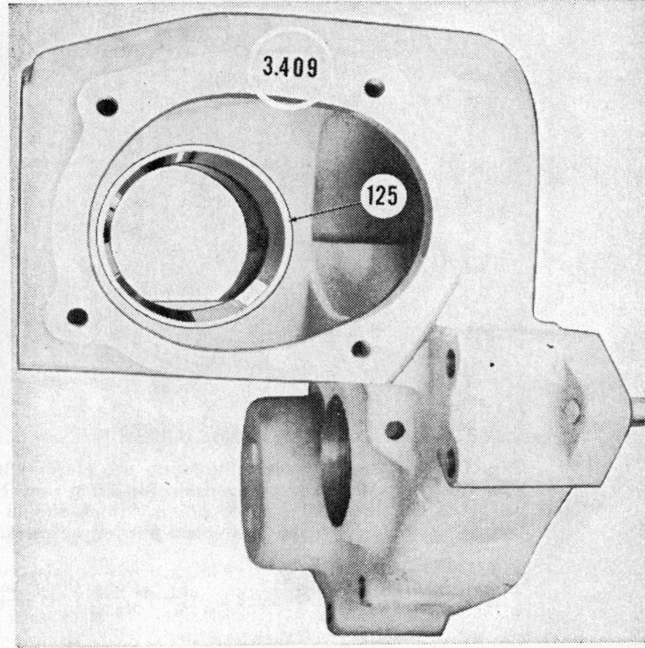

Fig. O403 — Location of dimension (A) on belt pulley unit control cover gasket surface. Mesh position is controlled with shims inserted under bearing cup (125).

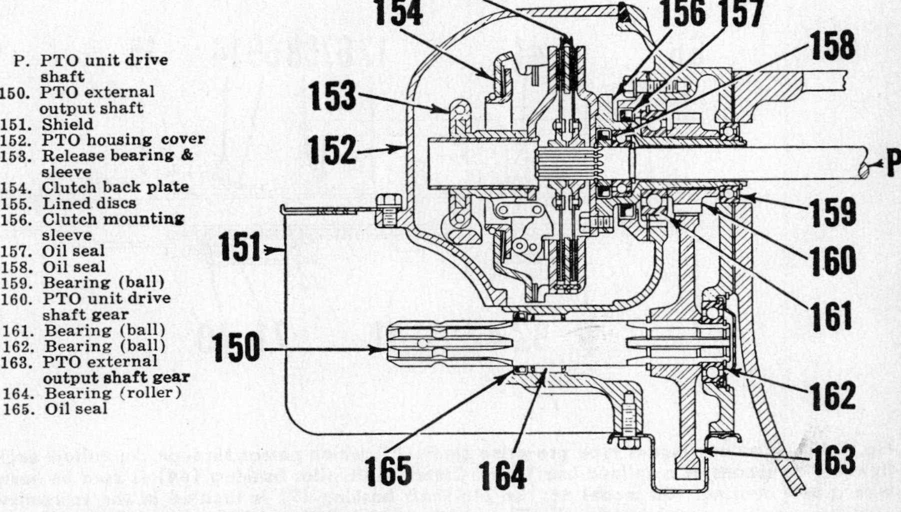

P. PTO unit drive shaft
150. PTO external output shaft
151. Shield
152. PTO housing cover
153. Release bearing & sleeve
154. Clutch back plate
155. Lined discs
156. Clutch mounting sleeve
157. Oil seal
158. Oil seal
159. Bearing (ball)
160. PTO unit drive shaft gear
161. Bearing (ball)
162. Bearing (ball)
163. PTO external output shaft gear
164. Bearing (roller)
165. Oil seal

Fig. O404—Double plate power take-off clutch used on 6-cylinder tractors. The 66 series unit is a single plate type. All later clutches are provided with an overload spring washer in series with the plates to permit temporary slippage under high initial starting loads. The washer is not shown in this illustration.

On models 66, 77, and 88 the pto shaft rotates at 582 rpm at an engine high idle speed of 1750 rpm. On Super 66 the pto shaft rotates at 733 rpm at an engine high idle speed of 2200 rpm. On 770 and 880 pto rpm is 597 at 1925 engine rpm.

590. **ADJUST PTO CLUTCH.** The Rockford CLA 2075 single plate over-center type clutch used on series 66 and the Rockford CLA 2022A double plate over-center type used on the six cylinder tractors have a shim type adjustment for wear. Clutches should engage fully, (go over-center) when a pull of 32-34 pounds registers on a spring scale attached to control lever at a point 22 inches above the center of the lever pivot point.

Fig. O405—View of over-center type pto clutch with pto cover removed. Adjustment to compensate for lining wear is by removal of shims "S" located at three points. Shims should be removed only from the laminated portion of pack.

S. Shims
150. PTO shaft
153. Release bearing & sleeve
154. Back plate
155. Lined disc
166. Clutch housing
167. PTO housing
168. Pressure plate

590A. Adjustment procedure is as follows: Remove the pto cover (152—Fig. O404) and loosen or remove the capscrews retaining the three shim packs shown at (S) in Fig. O405. Peel two or three shims from each laminated pack (do not remove any of the steel shims) then reassemble and recheck until specified engaging pull of 32-34 pounds at 22 inches is obtained. Long lobed ends of shims must all face counter-clockwise.

591. **CLUTCH OVERHAUL.** The pto clutch can be disassembled and overhauled without removing the pto drive shaft after removing the housing cover (152—Fig. O404). Disassembly of the clutch unit is self-evident.

591A. The large pressure washer used on early production tractors exerted a pressure of 1675 pounds but has no paint markings, a later washer with white paint on the outer edge exerted 1300 pounds. The latest wash-Oliver IK668B replaces both former washers and can be identified by yellow paint on the outer edge.

On series 66 reinstall the lined disc with hub forward; on 6 cylinder models with hub of front disc forward and rear disc rearward. When new or relined plates are installed discard the old shims and install kit of three shim packs each containing four loose steel shims (one of 0.005 and three of 0.010 thickness) and a laminate containing 18 brass shims of 0.007 thickness each. Make final adjustment after clutch is installed as per paragraph 590A.

592. **R&R PTO DRIVE SHAFT.** To remove the power take-off unit drive shaft (P—Figs. O404 & O406), proceed

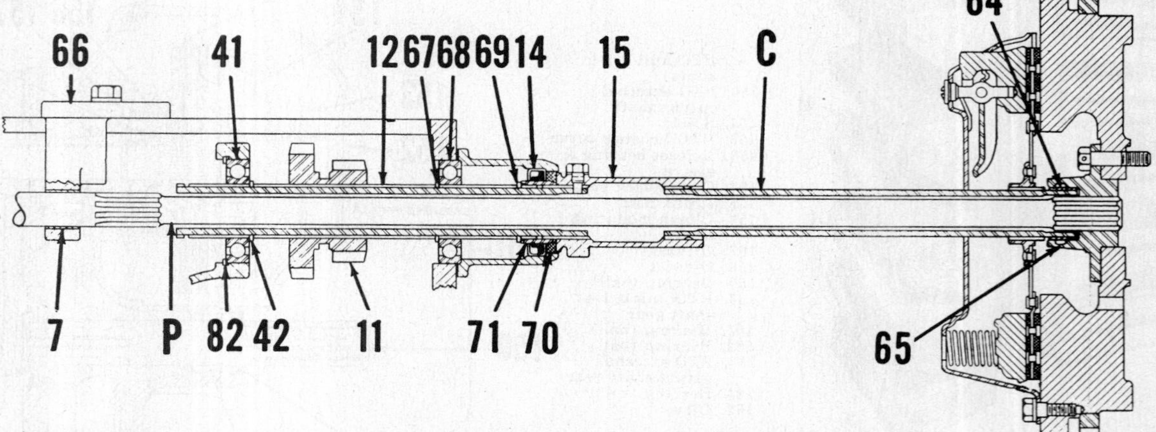

Fig. O406—The continuous type pto drive shaft (P), which passes through the hollow engine clutch shaft (C), is connected to the engine flywheel by means of a splined hub (65). Clutch shaft pilot bushing (64) is used on early production models; later models are equipped with a ball bearing. On model 66, the pto shaft bushing (7) is located in the transmission housing dividing wall instead of a bracket as shown for the 77 and 88. Models 770 and 880 without Power Booster are similar to 77 and 88.

C. Engine clutch shaft
P. PTO unit drive shaft
7. Bushing
11. Mainshaft gear (608)
12. Transmission mainshaft
14. Oil seal
15. Coupling
41. Bearing
42. Snap ring
64. Bushing (early production)
65. Drive shaft hub
66. PTO shaft support (models 77 & 88)
67. Snap ring
68. Bearing
69. Snap ring
70. Felt seal
71. Sleeve
82. Snap ring

as follows: Remove pto shield and cover. Remove cap screws (X—Fig. O405) which retain unit to pto housing (167) and withdraw pto clutch and drive shaft assembly rearward and out of the transmission housing as shown in Fig. O407.

593. **OVERHAUL PTO DRIVE SHAFT.** The pto drive shaft is supported midway between the flywheel and the clutch by a steady bearing (7—Fig. O406). The steady bearing which is located in the transmission compartment can be renewed after removing the drive shaft, and the transmission top front cover or power lift unit. The bushing (7), which is located in the transmission housing dividing wall on the 66 or in a bracket (66) on 77 and 88, requires final sizing to provide a diametral clearance of .003-.004.

Renewal of pto or hydraulic power lift pump drive shaft hub (65), which is mounted on the engine flywheel, requires removal of the engine clutch as outlined in the CLUTCH section.

594. **R & R COMPLETE PTO UNIT.** To remove the complete unit, drain the unit by removing the lower pan. Remove four capscrews (Z—Fig. 405) retaining the pto housing to the transmission housing, and remove the pto drive shaft, clutch, external output shaft, gear and housing as an assembly.

595. **R & R AND OVERHAUL PTO EXTERNAL OUTPUT SHAFT.** To renew the pto external output shaft (150—Fig. O404), oil seal (165), bearing (162 or 164), and/or gear (163), it will be necessary to remove the complete pto unit from the tractor as outlined in paragraph 594.

595A. Remove large snap ring and bearing stamped cover from front of housing. Drill snap ring and break same if necessary. Remove small snap ring retaining ball bearing (162) to the external output shaft; then bump shaft on ball bearing end and out of bearing and gear. Remove gear out through oil pan opening in the pto housing. Using a suitable puller, remove the ball bearing from the housing bore and remove the output shaft. The oil seal (165) and/or bearing (164) can be renewed at this time.

595B. Install the new oil seal so that the lip of same faces the driven gear (163). The needle roller type bearing (164) should be installed as shown in Fig. O407A. When installing the bearing, press against the stamp marked end of the bearing.

Fig. O407—Removing the pto drive shaft and clutch unit as an assembly.

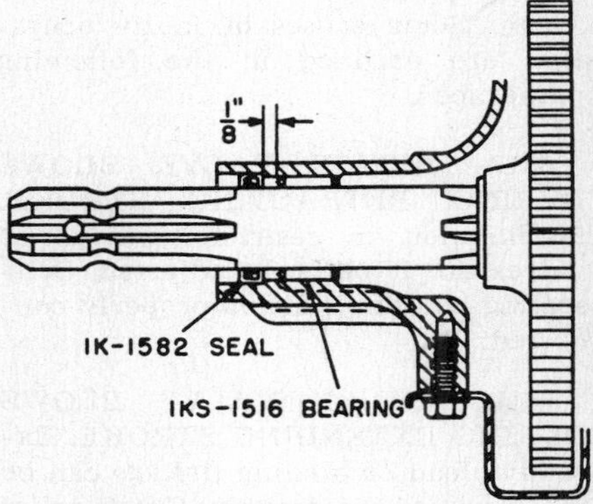

Fig. O407A—Install the pto external output shaft IKS-1516 bearing as shown.

HYDRAULIC LIFT SYSTEM

Refer to 770 & 880 Supplement, paragraph 35 for Super 77 and Super 88 tractors; to 660 Supplement for Super 66 tractors.

598. The hydraulic power lift system can be equipped either with 6 or 12 volt electric controls for actuating both the selector control valve and work cylinder follow-up stop collar, or manual controls for actuating the selector control valves. The manual and electric systems are similar in construction with regards to the pump, reservoir, mounting base, and selector control valves.

Since the introduction of the first Hydra-Lectric Control system there have been 6 variations of the basic hydraulic unit. These are the No. 1 Hydra-Lectric, No. 1 Hydraulic Control, No. 2 Hydra-Lectric, No. 2 Hydraulic Control, No. 3 Hydra-Lectric and the No. 3 Hydraulic Control. The No. 3 units went into production on the Super series tractors with the "L" model for the Super 66 and the "K" model for Super 77 and 88. The No. 3 basic units are not interchangeable with the older basic models but repair parts are available for all past models.

Most of the troubles encountered with the hydraulic system on modern tractors are caused by dirt or gum deposits. The dirt may enter from the outside, or it may show up as the result of wear or partial failure of some part of the system.

CONTROL SYSTEM

599. The Hydra-Lectric lift system used on some models 66, 77 and 88 comprises the usual hydraulic assemblies but utilizes straight electrical control of the selector valves and work cylinders. Gear type pump, fluid reservoir, control valves and the electrical solenoids which actuate the valves are all contained in a single housing which forms the top cover of the transmission. The pump is driven by the drive shaft of the independent power take-off when so equipped, or a similar but shorter shaft which passes through the hollow clutch shaft on models not equipped with a power take-off. The continuous electrical controls operate from the 6 or 12 volt battery on the tractor.

Work cylinders which are of the double-acting type are equipped with either a solenoid controlled cam lever follow-up collar stop (early production non-Super models), or a magnetic controlled type follow-up collar stop as shown in Figs O421 and O422.

Tractors with two work cylinders are equipped with two selector control valves, two hand switches, two thermal relief valves, two restrictor valves and two manually controlled actuating shafts. Tractors with only one work cylinder are equipped with one each of the preceding mentioned items for tractor equipped with two cylinders.

TROUBLE-SHOOTING

600. **TESTING.** Trouble with the "Hydra-Lectric" unit may be divided into two categories: Electrical and mechanical. If there is doubt as to whether the trouble is electrical or mechanical after checking the fuse in the electrical circuit make the check outlined in paragraph 600A, which involves operating the lift system manually and without the use of the electrical circuit. Refer also to paragraphs 601 through 601G.

600A. The selector control valves located inside the reservoir can be operated manually by turning the stub shafts (H—Fig. O408) or the hydraulic control levers on tractors so equipped. If the lift system will not function when it is manually operated, the trouble is mechanical and there is no need to check the electrical system.

601. Some causes of faulty operation are outlined in the following paragraphs.

601A. RELIEF VALVE BLOWS DURING RETRACTING STROKE. Malfunction in restrictor valves or valves are improperly adjusted. Self-sealing couplings are improperly connected.

601B. RELIEF VALVE BLOWS DURING EXTENDING STROKE. Excessive load or binding linkage can be the cause of the trouble. Stuck relief valve or restricted relief valve passage are other possibilities.

601C. RELIEF VALVE BLOWS WHEN CYLINDER IS RETRACTED OR EXTENDED. Selector valve sticking, binding solenoid linkage, weak, or damaged selector valve centering spring. Friction stop collar on work cylinder too close to piston rod yoke.

601D. UNIT WILL NOT LIFT. Insufficient oil in reservoir. Worn pump. Relief valve not seating. Leaking "O" ring or packing on work cylinder piston. Selector control valve solenoid linkage improperly adjusted. Selector control valve leaking. Interlock valve seals leaking. Selector valve body gasket leaking. Pump to reservoir seals leaking. Thermal relief valve leaking.

601E. OIL FOAMS OUT OF RESERVOIR BREATHER. Wrong type of hydraulic operating fluid. Relief valve not properly adjusted. Oil level too high. Air leak at pump inlet. External leaks.

601F. SLUGGISH LIFTING ACTION. Worn pump or obstructed pump inlet passage. Air in system.

601G. UNIT WILL NOT HOLD LOAD. Leaking "O" ring or packing on work cylinder piston. Damaged selector control valve.

602. Possible electrical faults are: Relay switch remaining closed on No. 1 or No. 2 systems. Look for short in the Visual Up or Down circuit.

SYSTEM MAINTENANCE

604. **HYDRAULIC FLUID.** Recommended operating fluid for the hydraulic lift system is SAE 10W engine lubricating oil. Fill reservoir and hydraulic system to full level mark indicated on the bayonet type gage which is attached to the reservoir air breather.

After draining and refilling the system, it will be necessary to bleed the system as outlined in paragraph 605.

604A. **AIR BREATHER.** The air breather screws into the top of No. 1 units and is a press fit on the No. 2 and No. 3 units. It should be cleaned daily.

605. **BLEEDING LIFT SYSTEM.** The presence of air in the system will be evidenced by a retarded and spongy operating condition. The hydraulic power lift system will also require bleeding whenever the system is drained and refilled.

To bleed the system, loosen one hose connection at each cylinder. Start the engine and operate the hand control lever through one cycle of cylinder operation. Then tighten the hose connection. Repeat this procedure on the other hose for the same cylinder. Continue this procedure until there is a continuous flow of oil without air bubbles at the loosened connection. Replenish the oil supply in the reservoir. Refer to Fig. O408A.

606. **FLUSHING THE SYSTEM.** Once each year or every 1000 hours of operation the system should be drained and flushed with engine flushing oil. Operate the system for about 5 minutes on flushing oil then completely drain the reservoir, hoses and cylinders. Refill with new SAE 10W engine oil and bleed as outlined in paragraph 605.

RESERVOIR, PUMP AND VALVES

607. **UNIT R&R.** To remove the hydraulic pump, valves and reservoir unit, first drain fluid from unit by removing pipe plug (44—Fig. O408) at left rear side of the reservoir base. Disconnect hoses from the unit and remove the wire harness plug from the right side of the unit. Remove pto drive shaft as per paragraph 592 from models so equipped.

607A. On models not equipped with a pto unit, remove transmission housing platform and bull gear cover. Working through bull gear cover opening, remove snap ring from pump drive gear sleeve. Thread a 3/8 inch-16 cap screw into the end of pump drive shaft; then attach a suitable puller, as shown in Fig. O410, and remove the pump drive shaft.

H. Hand control stub shaft
P. Pipe plug
R. Restrictor valve
1. Breather and filler cap
2. Cover
4. Reservoir
11. Mounting base
12. Relief valve
41. Wiring harness connection
42. Transmission filler cap
43. Pump unit
44. Reservoir drain

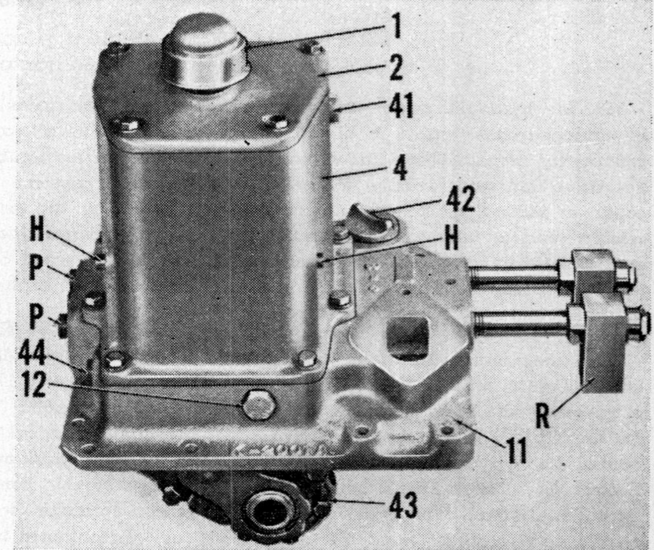

Fig. O408—Hydra-Lectric hydraulic lift unit as used on non-Super tractors equipped with one work cylinder. Units for tractors equipped with two work cylinders are similar.

607B. Remove transmission shift lever and socket being careful not to drop the two lever socket springs in the transmission housing. Refer to paragraph 554 and to Fig. O373A, for information concerning the use of a special tool to prevent the loss of the lever socket springs. Remove the cap screws retaining the unit to the transmission housing, and lift unit from tractor.

607C. To reinstall the hydraulic unit, reverse the removal procedure and fill the unit reservoir with SAE 10W engine lubricating oil until the level is within the work range as indicated on the bayonet gauge. After refilling, it will be necessary to bleed the system of air as outlined in paragraph 605.

608. **PUMP OVERHAUL.** The gear type hydraulic pump can be removed from the lower side of the hydraulic unit mounting base after removing the complete pump and valves unit as per paragraphs 607, 607A and 607B.

608A. When disassembling the pump, mount same in a vise, and remove the pump rear plate attaching cap screws. Use a soft hammer to separate the pump housing. Do not pry the end plates off of the pump body (center plate). Clean all parts in solvent. Refer to Fig. O409.

608B. Inspect needle bearings and shafts for signs of failure but do so without removing them from front and rear plates. Bearings can seldom be removed without damage to the roller (needle) cage.

When installing needle bearings, press against the end of the bearing which is stamped with the manufacturer's name.

Install pump drive and driven shaft bearings in bearing bores so that the end of the bearing is 3/32 inch below the machined surfaces of both the front and rear pump plates (51 & 54).

608C. Check pump gear and shaft assemblies, pump body (center plate), and pump end plates for wear and signs of failure. Any derangement of gears, pump body or pump end plates should be corrected by renewal of the entire pump unit.

Diameter of drive gear sleeve is 1.3743. Idler gear shaft diameter is 0.9995. Total gear side clearance is 0.003.

608D. New oil seals (49) should be installed with the lips of both inner seals facing the pump gears, and the outer seals with the lips facing outward. Apply Nos. 210 or 215 Lubriplate between the seal lips to insure proper lubrication at this point.

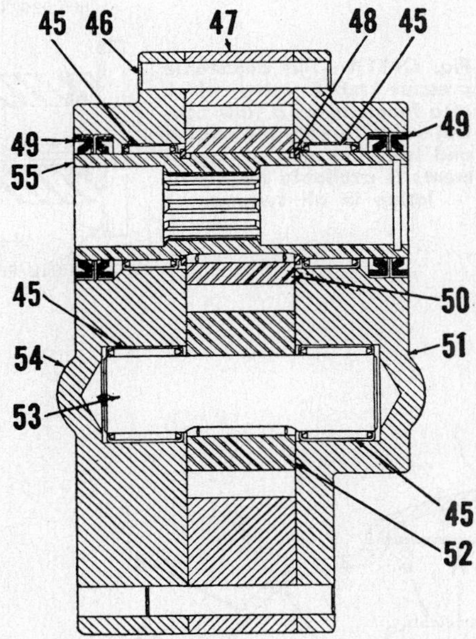

Fig. O409—Hydraulic lift unit pump used on both electric and manual type of control systems. Pump unit is mounted on the lower side of the reservoir mounting base.

45. Bearing
46. Dowel pin
47. Center plate
48. Snap ring
49. Oil seals
50. Drive gear
51. Rear plate
52. Driven gear
53. Driven gear shaft
54. Front plate
55. Drive gear sleeve

608E. When reassembling the pump, torque tighten the pump body cap screws to 33-34 ft. lbs. torque and lock same with a piece of 16 gauge annealed steel wire threaded through the heads of the screws.

The 1⅛ and 1⅜-inch pumps at 1750 engine rpm should show a gauge reading of 700 psi minimum when oil temperature is 150°F. and outlet test orifice of 0.1405 is installed. The ¾-inch pump used on series 66 should show 500 psi minimum under same test conditions.

608F. After reinstalling the complete pump and valves unit on the tractor, check the relief valve unloading pressure as per paragraph 609.

Relief Valve Adjust

609. Pump and valves unit must be installed on the tractor when checking the relief valve unloading pressure. Disconnect either of the two hose connections located on the right side of the unit and install a pressure gauge of 1500 psi capacity. Start the tractor engine and allow the pump to operate until the hydraulic fluid reaches operating temperature. Move control lever to "Visual Up" or "Visual Down" (depending on whether gauge

Fig. O410—Removing hydraulic lift unit pump drive shaft on models not equipped with a continuous PTO by using a suitable puller attached to a ⅜-inch—16 cap screw which is threaded into the end of the shaft.

H. Hand control stub shaft
R. Restrictor valve
X. ⅜ inch-16 cap screw
11. Mounting base
12. Relief valve
41. Wiring harness connection

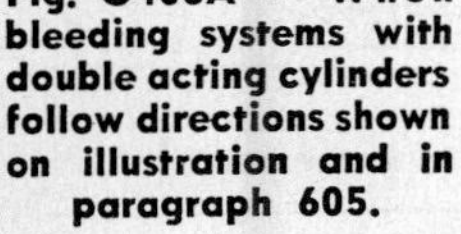

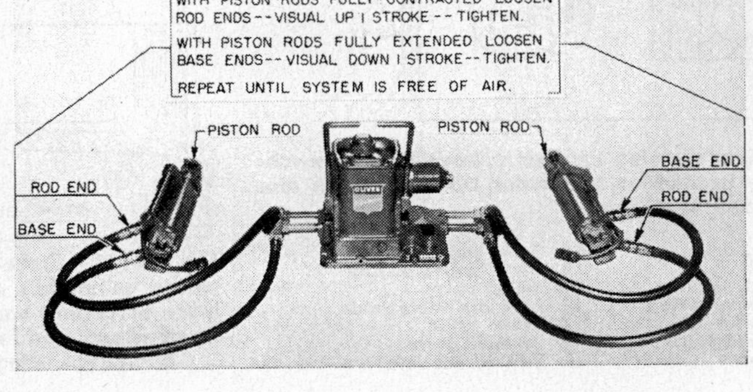

Fig. O408A — When bleeding systems with double acting cylinders follow directions shown on illustration and in paragraph 605.

is in front or rear reservoir port) and note the gauge reading.

The regular type relief valve (14—Fig. O411) should unload when the pressure is within the range of 1000-1100 psi. If the gauge reading is lower than 1000 psi either the pump is at fault or there is a derangement of the relief valve. If the reading is higher than 1100 psi, the trouble is in the relief valve. The unloading pressure can be varied by adding or removing shims located between relief valve spring (13) and relief valve guide (12).

609A. An adjustable type pressure relief valve, as shown in Fig. O411A, can be installed as optional equipment in place of the regular type pressure relief valve. This adjustable relief valve permits the operator to adjust and control the raising or lowering rate of an implement. When the adjustable relief valve restrictor (1K-2105—A) is turned in so as to seal off the by-pass ports (A) in plunger (1K-2019-A), the valve functions as a regular relief valve and the unloading pressure can be adjusted with shims like the regular relief valve. Malfunctioning of the valve is corrected by renewing the individual parts.

Valves Overhaul

610. Procedures for the removal, disassembly and overhaul of the selector control valve and related parts, interlock valve, relief valve, thermal relief valve, restrictor valve, momentary restrictor valve, and free flow valve are outlined in the following paragraphs 611 through 615.

611. **SELECTOR CONTROL VALVE & INTERLOCK.** Selector valve controls the direction of oil flow to and from the work cylinder. It also contains an interlock lock system (3—Fig. O411) for locking the oil in the cylinder. Four different selector valve bodies have been used. Refer to Figs. O414, O416 and O431A. The solenoid operated spool type selector control valve, body and solenoid assembly can be removed from the hydraulic lift unit as follows: Drain hydraulic fluid from the unit by removing the pipe plug located at the rear or left rear side of the reservoir mounting base. Remove reservoir cover and disconnect the wires from the solenoid unit. Remove reservoir housing from mounting base. The selector control valve, body and solenoid can be removed after removing the attaching cap screws.

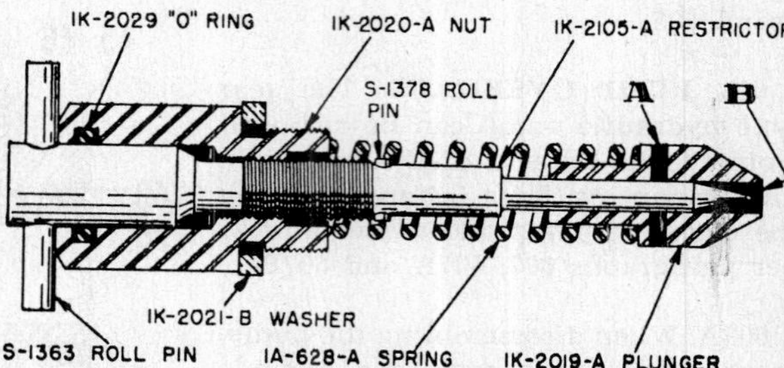

Fig. O411A—This adjustable pressure relief valve which also functions as a flow control valve to vary the raising and lowering rate of implements is available for installation in all systems.

In the event of solenoid failure, the selector valves can be manually operated.

611A. Disassembly of the selector control valve and body assembly varies. First production non-Super units, as shown in Fig. O411, were equipped with two centering springs. Later production units have only one control valve centering spring located on the forward end of the valve, as shown in Fig. O412. On No. 3 systems the valve is sealed by "O" rings, Fig. O431A.

611B. Thoroughly inspect the selector control valve, and the bore in the valve body for wear, scratches, and signs of failure. Any derangement should be corrected by renewal of the control valve and body.

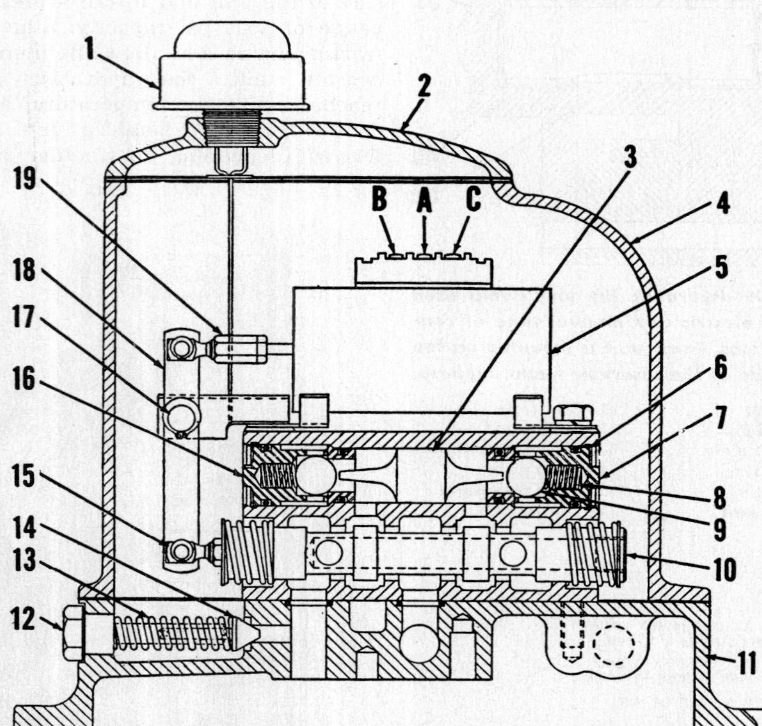

Fig. O411—Hydra-Lectric control, No. 1 version, showing lift unit reservoir, valve mechanism, and solenoid assembly as used on early production non-Super tractors. Refer also to Figs. O412 and O431A.

A. Center terminal (blue wire)
B. End terminal (red wire)
C. End terminal (green wire)
1. Breather & filler cap
2. Cover
3. Interlock valves plunger
4. Reservoir
5. Solenoid
6. Snap ring
7. Interlock valve guide
8. Interlock valve spring
9. Interlock valve ball
10. Selector control valve
11. Mounting base
12. Relief valve guide
13. Relief valve spring
14. Relief valve
16. Interlock valve guide
17. Lever pivot
18. Solenoid lever
19. Ball socket joint

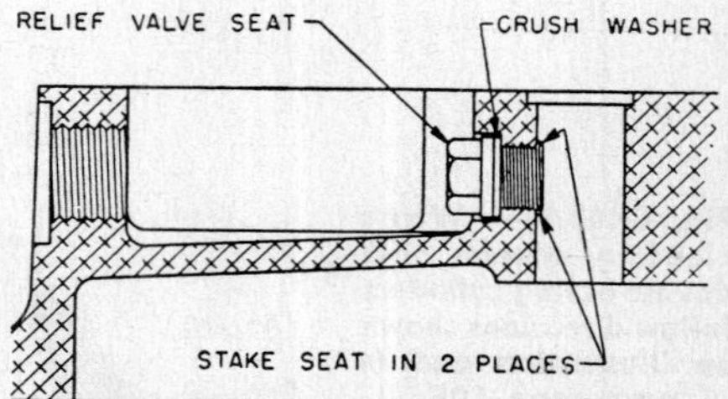

Fig. O411B—Stake the relief valve seat as shown on models equipped with an aluminum mounting base. A relief valve seat insert is not used in a cast iron type mounting base.

611C. INTERLOCK VALVE. Interlock valves (9—Figs. O411 & O412) or (43—Fig. O431A) which are located in selector control valve body can be removed after removing selector control valve, body and solenoid assembly as outlined in paragraph 611.

611D. Remove either the interlock valve retaining snap ring or steel end plate from each end of the body. Using pliers, remove the two interlock valve guides (7 & 16—Figs. O411 & O412) and springs. Remove the ¾ inch steel balls. Remove the two interlock valve seats by bumping on the end of the interlock plunger (3) with a 3/8 inch brass rod and hammer. Remove the interlock plunger.

611E. Renew "O" rings and check the condition of the valve seats and ball valves. Renew valve seats and/or ball valves if they show signs of wear, grooving and/or pitting. Renew the interlock plunger if the ends are chipped or upset.

611F. Reassemble the interlock valve assembly by reversing the disassembly procedure. An interlock valve can be reseated by placing the ¾ inch ball on the seat and tapping it lightly with a hammer and soft drift.

613. RELIEF VALVE. The poppet type relief valve (14—Figs. O411 & O412) is located in hydraulic unit mounting base. Relief valves as used in early production models, seat directly in the mounting base. Later production units are equipped with a renewable steel seat (S—Fig. O412).

613A. The relief valve can be removed for inspection after removing the relief valve guide (12—Fig. O411 & O412), spring (13) and shims. Check the line contact of both the valve and seat, and if same are grooved, it will be necessary to renew the valve and mounting base, or valve and seat. Valve seat renewal on models so equipped will require removal of the hydraulic lift unit mounting base.

An adjustable type pressure relief valve, as shown in Fig. O411A, can be installed as optional equipment in place of the regular type pressure relief valve. Refer to paragraph 609A.

614. THERMAL RELIEF VALVE. Each selector control valve body is equipped with a poppet type safety or thermal relief valve assembly (57—Fig. O415 or O416) to protect the hoses against expansion of the oil arising from temperature increases when the oil is locked in the cylinder and hoses. The non-adjustable thermal relief valve is designed to open when the pressure within the cylinders exceeds 3500 psi.

Early production non-Super models were equipped with a disc type thermal relief valve using a brass disc which would rupture to relieve the pressure. This necessitated renewing the ruptured disc whenever the valve unloaded. Where this valve is encountered in servicing the hydraulic system, it should be changed to the later poppet valve type.

614A. The thermal relief valve can be removed after draining the reservoir and removing the reservoir cover and housing.

615. RESTRICTOR VALVE. A restrictor valve, Fig. O415A, is often inserted between the reservoir and hose connection to restrict the flow of fluid from the work cylinder on the retracting stroke. The adjustable type restrictor valve consists of a valve body, ball check valve, and a needle valve or only a spring loaded plunger.

To retard the lowering speed, loosen the adjusting screw Palnut then rotate the guide in a clockwise direction.

Method of removal and disassembly of the restrictor valve is self-evident.

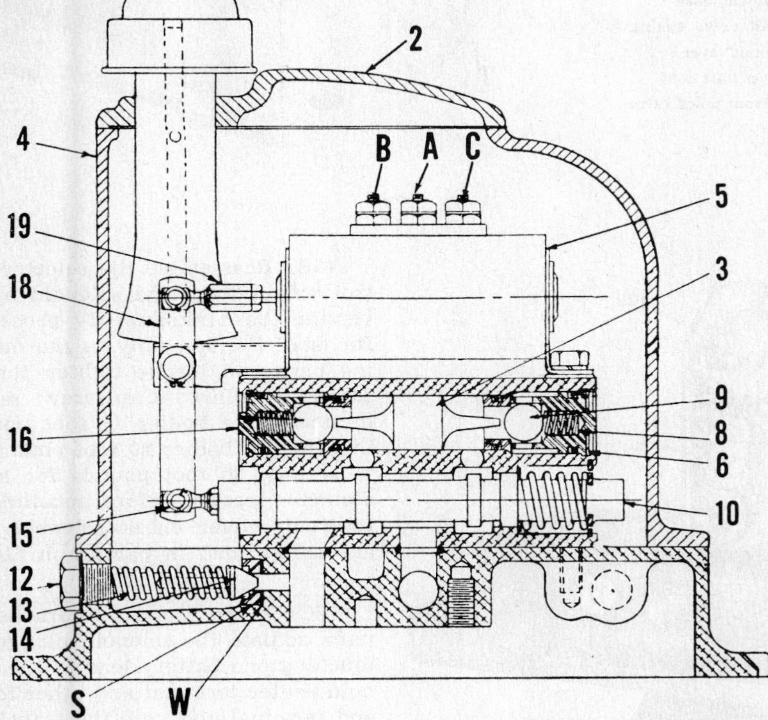

Fig. O412—Hydra-Lectric (electric control), No. 2 type, hydraulic lift unit reservoir, valve mechanism and solenoid assembly used on later production non-Super tractors. Refer also to Figs. O411 and O431A.

A. Center terminal (blue wire)
B. End terminal (red wire)
C. End terminal (green wire)
S. Relief valve seat
W. Washer
2. Cover
3. Interlock valves plunger
4. Reservoir
5. Solenoid
6. Snap ring
8. Interlock valve spring
9. Interlock valve ball
10. Selector control valve
12. Relief valve guide
13. Relief valve spring
14. Relief valve
16. Interlock valve guide
18. Solenoid lever
19. Ball socket joint

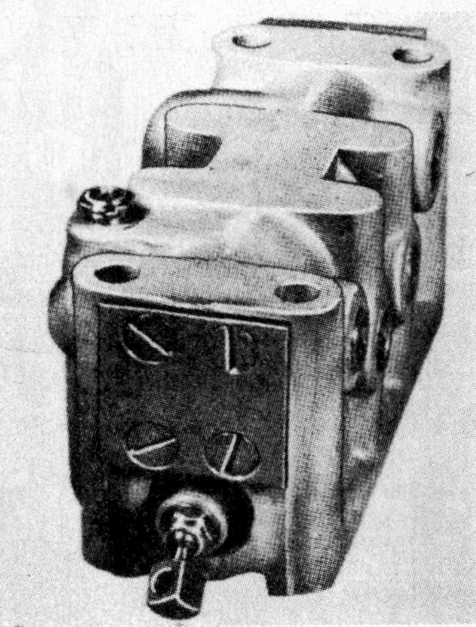

Fig. O414—On later production units, the interlock valve assemblies are retained in the selector control valve body with an end plate as shown.

ELECTRICAL UNITS

616. The 6 or 12-volt system of the Hydra-Lectric Control for a model equipped with two work cylinders includes the following: Two double-acting solenoids located on the selector control valve body for actuating the selector control valves; two hand control switches having two toggle and two self-return positions for raising or lowering and selecting the operating depth; work cylinder follow-up stop collar which is electrically controlled by either solenoid cam operated levers (early production) or a magnetic type of control; two relays used with the magnetic type of follow-up; one 30 ampere fuse and wiring harness.

Systems which are equipped with only one work cylinder have one each of the mentioned items.

616A. **SOLENOIDS.** Hydra-Lectric Control units No. 1 and No. 2 are equipped with a separate solenoid for moving each of the selector valve spools. These solenoids are integral units and cannot be disassembled for repairs. Use Oliver kit IKS2032A solenoid assembly for repairs.

The solenoid used on the No. 3 Hydra-Lectric Control is a part of the solenoid and switch assembly. Some parts of this assembly are furnished separately for repairs. The electrical contacts can be cleaned with a contact point file.

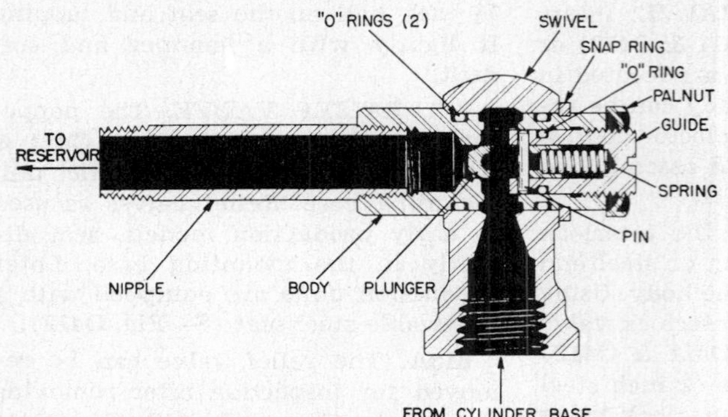

Fig. O415A—Adjustable type restrictor can be used to vary the lowering speed of an implement.

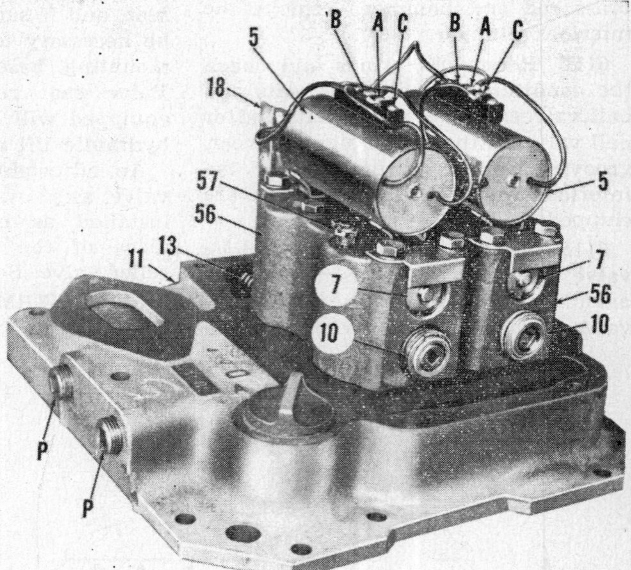

Fig. O416—Right side and front view of the Hydra-Lectric (electric control) type hydraulic lift unit showing the valve units and solenoids. An early production non-Super unit is illustrated.

A. Center terminal (blue wire)
B. End terminal (red wire)
C. End terminal (green wire)
5. Solenoid
7. Interlock valve guide
10. Selector control valve
11. Mounting base
13. Relief valve spring
18. Solenoid lever
56. Valves unit body
57. Thermal relief valve

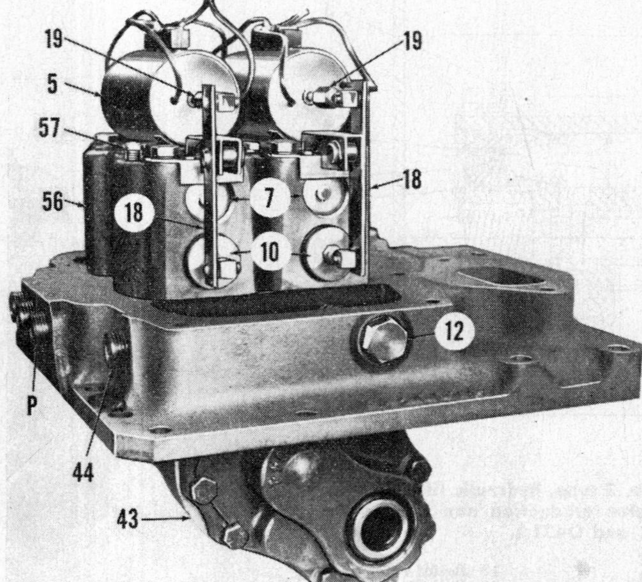

5. Solenoid
12. Relief valve guide
18. Solenoid lever
19. Ball socket joint
43. Pump
56. Valves unit body
57. Thermal relief valve

Fig. O415—Left side and rear view of the Hydra-Lectric (electric control) type hydraulic lift unit showing the valve units and solenoids. An early production non-Super unit is illustrated.

616B. Reassemble the selector control valve, body and solenoid by reversing the disassembly procedure. Reinstall the assembly to the mounting base and torque tighten the cap screws as follows: Cap screws in aluminum valve bodies 18 foot pounds; in cast iron bodies 20 foot pounds for $\frac{5}{16}$ screws, 30 foot pounds for larger diameter screws. Before installing the reservoir cover check solenoid linkage as outlined in paragraph 516B.

616C. On No. 1 and 2 Hydra-Lectric units actuate the solenoid plunger by touching one battery lead to the solenoid center terminal and other to one end terminal and note the travel distance of selector control valve. Note the valve travel in the opposite direction by connecting the center terminal and the other end terminal. Rotate the solenoid linkage ball socket joint (19) Fig. O412 **in** or **out** until the selector control valve travels an equal distance in both directions.

On No. 3 Hydra-Lectric systems loosen the lock nut on the ball swivel (72—Fig. O431A) and turn swivel **in** or **out** until the distance between end of solenoid housing and roll pin in each end of rail is equal at each end.

616D. Reinstall the unit on the tractor and refill the reservoir as per paragraph 604. Bleed the system as per paragraph 605.

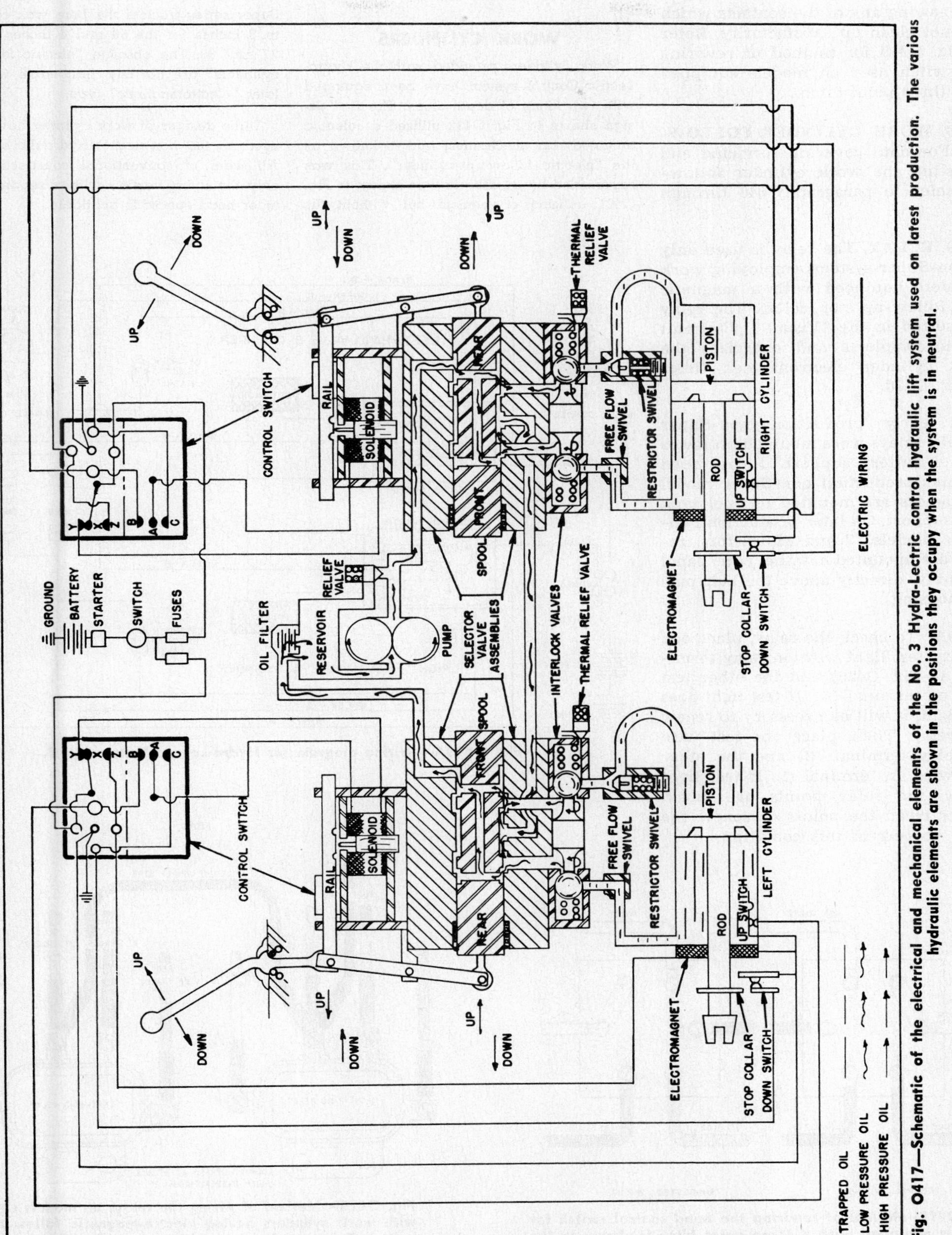

Fig. O417—Schematic of the electrical and mechanical elements of the No. 3 Hydra-Lectric control hydraulic lift system used on latest production. The various hydraulic elements are shown in the positions they occupy when the system is in neutral.

617. **HAND CONTROL SWITCH.** Removal and disassembly of the four position hand control switch is self-evident after an examination of the unit. Normal servicing includes cleaning the switch contacts with a file, or renewing any of the contacts which will not clean up satisfactorily. Refer to Fig. O419 for method of rewiring the switch used on models equipped with three point hitch.

618. **WORK CYLINDER FOLLOW-UP.** For data covering servicing and adjusting the work cylinder follow-up, refer to paragraphs 640 through 641B.

619. **RELAY.** The relay is used only in non-Super systems employing work cylinders equipped with a magnetic type follow-up stop collar. The relay is inserted in the "Visual Up" circuit which completes and energizes the work cylinder electromagnet. Refer to Fig. O420.

On early production non-Super models, relays were mounted on lower side of battery support. Relays used on later production non-Super model 66 tractors are mounted on fuel tank rear support. On later production non-Super models 77 and 88 tractors, relays are mounted on the rear panel assembly, directly above the belt pulley housing.

619A. To check the relay, place one battery test light wire on relay terminal (4—Fig. O420) and the other test wire on terminal (3). If test light does not light, it will be necessary to renew the relay. Then, place one test wire on relay terminal (2) and the other test wire on terminal (1). If test light lights, the relay points are stuck. Either clean the points or renew the relay to correct this condition.

620. **FUSE.** The electrical circuit of the hydraulic lift system contains a fuse. The fuse is inserted in the main circuit connected to the combination switch which is mounted on the tractor instrument panel.

WORK CYLINDERS

Work cylinders provided with the Hydra-Lectric Control system have been equipped with two kinds of depth stops. The first design shown in Fig. O422 utilized a solenoid actuated cam mechanism and is known as the "electric follow-up cylinder". This was superseded by an electric type shown in Fig. O421 utilizing a solenoid but without the cam mechanism as shown in Fig. O421. This currently used model is known as the "electromagnetic follow-up cylinder". The first electromagnetic units were on cylinders of 2½-inch bore for the 66 and 3-inch bore for the 77 and 88. After introduction of the Super series tractors the bore was changed to 3 inches for the 66 and 4 inches for the 77 and 88. The obsolete "electric follow-up cylinders" are not interchangeable with the later "electromagnetic" type.

Three designs of work cylinder have been used on the non-electric hydraulic systems. All were of conventional construction but only the latest version with a separate cylinder head spacer is available.

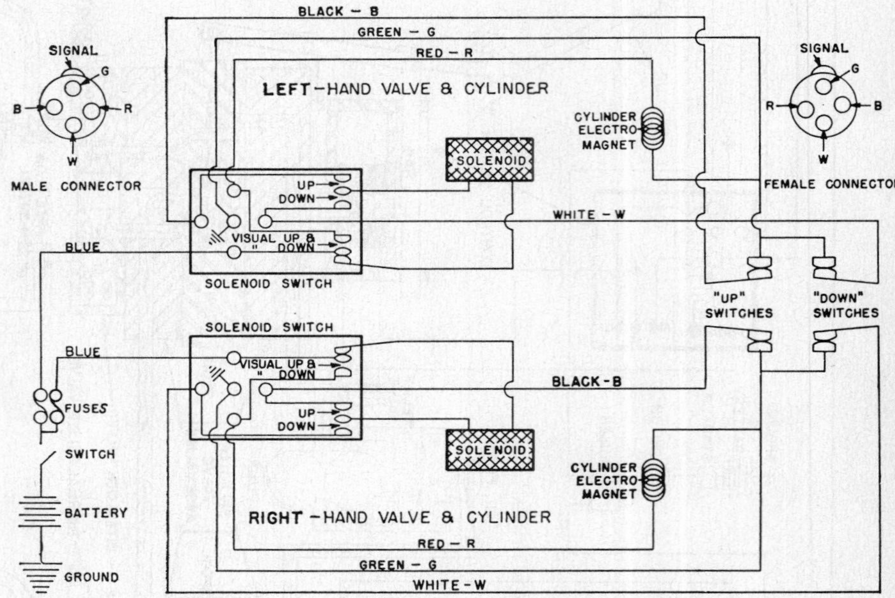

Fig. O419A—Wiring diagram for Hydra-Lectric System No. 3.

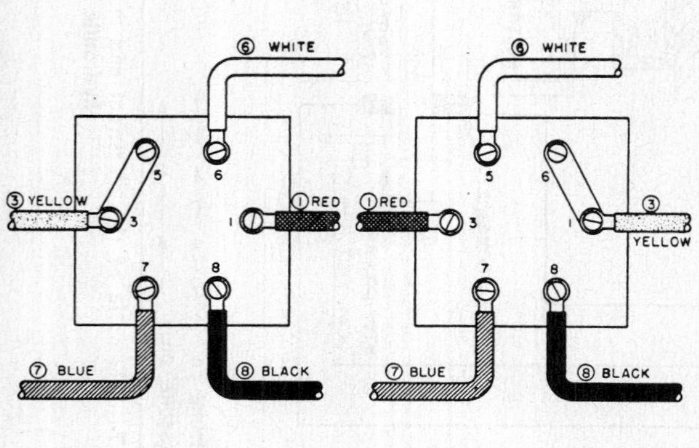

Fig. O419—Method of rewiring the hand control switch for models equipped with a three point hitch is shown in the right view.

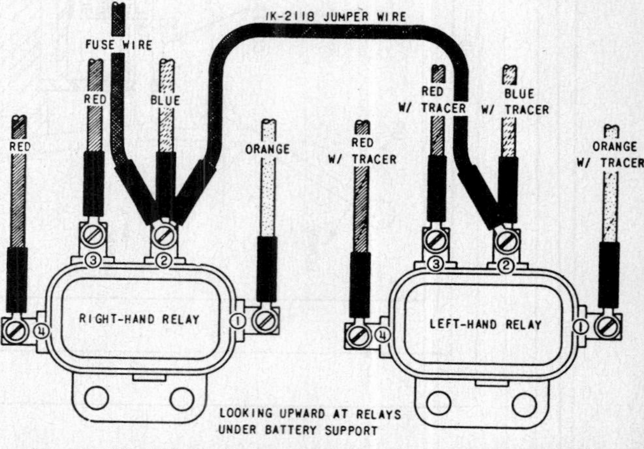

Fig. O420—Method of wiring the relays on models equipped with work cylinders having electro-magnetic follow-up stop collar. Connect all wires with tracer identification to one relay.

640. **CAM LEVER ELECTRIC FOLLOW-UP.** Refer to Fig. O422. It is recommended by the Oliver Corp. that this type of work cylinder be reworked to the "electromagnetic" follow-up type. Kits which include the necessary wiring, relays, and work cylinder head are available for this changeover.

641. **ELECTRO MAGNETIC FOLLOW UP.** Work cylinders equipped with a magnetic type follow-up stop collar are used on later Hydra-Lectric. Refer to Fig. O421.

Procedure for removal of work cylinder is self-evident. To disassemble the cylinder, first clean it thoroughly and remove external pipe nipple and elbow. Bump the cylinder head until it is driven into the cylinder as far as it will go. Insert a punch through the small hole in the cylinder to move snap ring (78) out of its groove; then using a screw driver, inserted through the open end of the cylinder, remove the snap ring. Withdraw piston and rod assembly in a series of short jerks to dislodge the head from the cylinder. Remove solenoid assembly. Further disassembly is self-evident.

Wash all parts and renew the seals and "O" rings. Renew piston rod and/or cylinder if rubbing surfaces of same are scratched, flaked or pitted.

If cylinder head is not equipped with a Nylon locking pin to hold the Up switch terminal stud in adjustment it can be re-worked at overhaul to include this feature. The re-work diagram is shown in Figs. O422A and O423. The pin is available under Oliver No. IK2247 and the required cup point socket set screw as No. S-1184.

Solenoid wires should be connected as follows: Using the small self tapping screw as a terminal, ground the green (No. 10) wire from the cylinder cable, short non-insulated wire from solenoid coil and short white jumper wire from the micro switch. Black (No. 8) wire from cylinder cable connects to the Up switch stud and should be fitted with an insulating sleeve. The orange or red (No. O) wire from cylinder cable should be connected by screw and nut to the insulated wire from the solenoid coil. The white (No. 6) wire from the cylinder cable and the white jumper wire from the ground connection should be connected to the terminals of the micro switch.

641A. Before reassembling the cylinder, check and adjust the cylinder "Up" switch. The "Up" switch should open when the piston is ⅛ inch away from the cylinder head as shown in

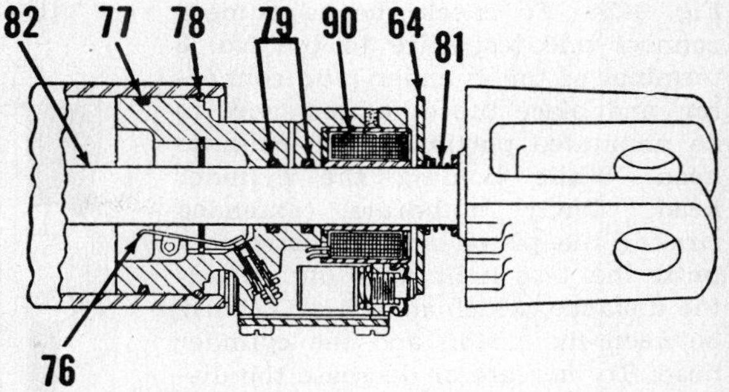

64. Friction collar stop
76. Follow-up contact arm
77. "O" ring
78. Snap ring
79. "O" ring
81. Spring
82. Rod assembly
90. Electromagnet

Fig. O421—Work cylinders equipped with a magnetic type follow-up stop collar without cam levers are known as "electromagnetic follow-up" cylinders.

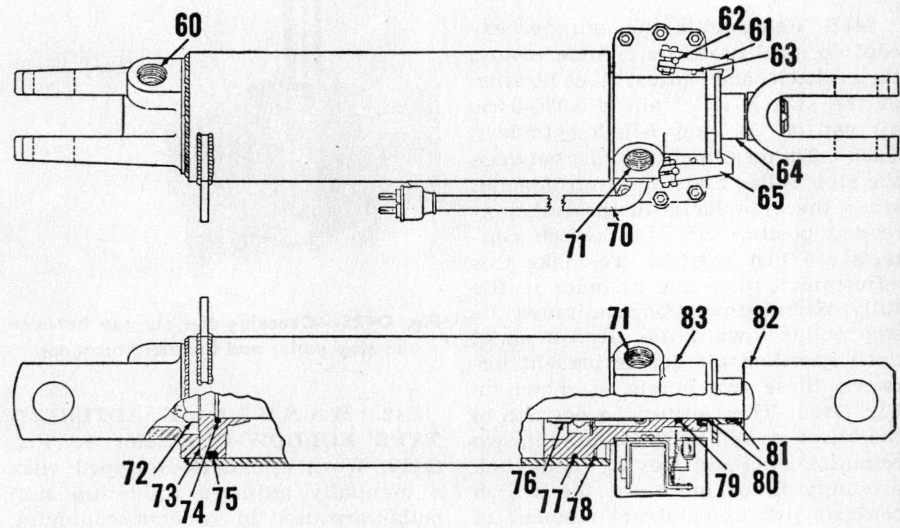

Fig. O422—Work cylinders equipped with a solenoid operated cam lever type follow-up stop collar are known as "electric follow-up" cylinders.

61. Cam lever (right)
62. Follow-up solenoid cam
63. Groov pin
64. Friction stop collar
65. Cam lever (left)
73. Piston
74. "O" ring
75. "O" ring
76. Follow-up contact arm
77. "O" ring
78. Snap ring
79. "O" ring
80. Seal
81. Spring
82. Rod assembly
83. Head

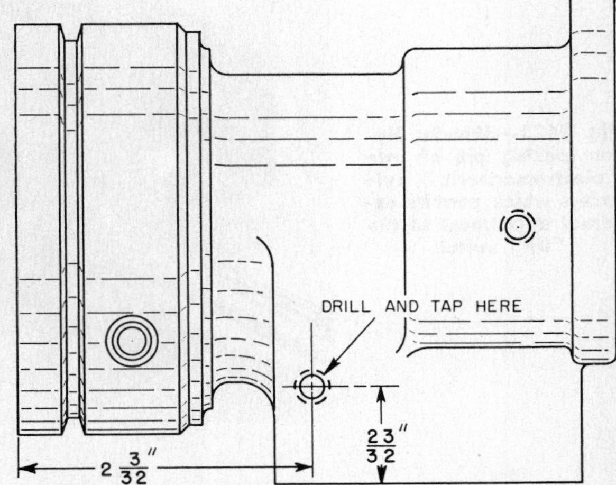

Fig. O422A — On later production "electromagnetic" type work cylinders the "UP" switch can be adjusted externally. This feature can be obtained with the earlier cylinders by re-working the cylinder head as shown here and in Fig. O423.

Fig. O424. To check the adjustment connect one test wire to the No. 8 terminal of the cylinder plug connector, and place the other test wire to an unpainted portion of the cylinder head. While holding the cylinder head, slowly withdraw (extended stroke) the piston and rod assembly until the test light goes out. Check the distance, which should be ⅛ inch, between the piston and the cylinder head. To increase or decrease the distance on cylinders **not** equipped with the Nylon locking pin it will be necessary to bend the "Up" switch contact arms as shown in Fig. O424. If cylinder has the Nylon locking pin adjust by turning the terminal stud assembly in or out as required.

641B. Reassemble the cylinder except do not install the cylinder cover, then check and adjust the position of the switch to obtain a 0.030-0.050 air gap on 2½ and 3-inch cylinders; 0.030-0.070 on 4-inch models, between the stop collar and the electromagnet when the cylinder is in the fully extended position and the switch contacts are just opening. To make this adjustment, place the cylinder in the fully extended position, and move the stop collar toward the electromagnet until specified air gap is present between these two points as shown in Fig. O425. Then adjust the position of the Micro switch by loosening the two retaining nuts and moving the switch assembly **in** or **out** until the switch contacts just open. Exact moment of switch contact opening can be determined with a test light by placing one test wire on the No. 9 terminal and the other wire on the No. 6 terminal of the cylinder wiring harness connector.

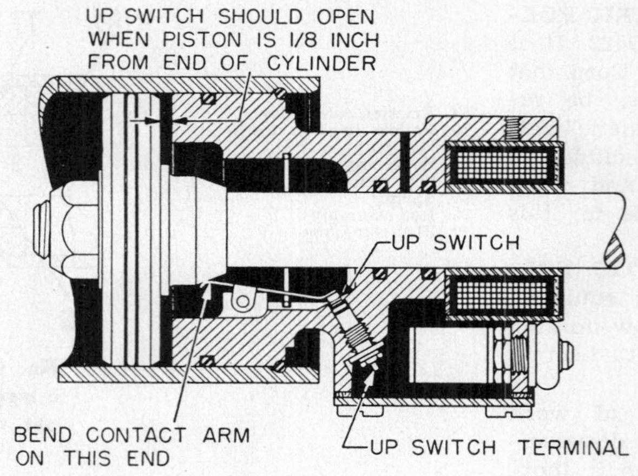

Fig. O424—Cross section of "electromagnetic" type follow-up stop collar work cylinder showing the "Up" switch adjustment.

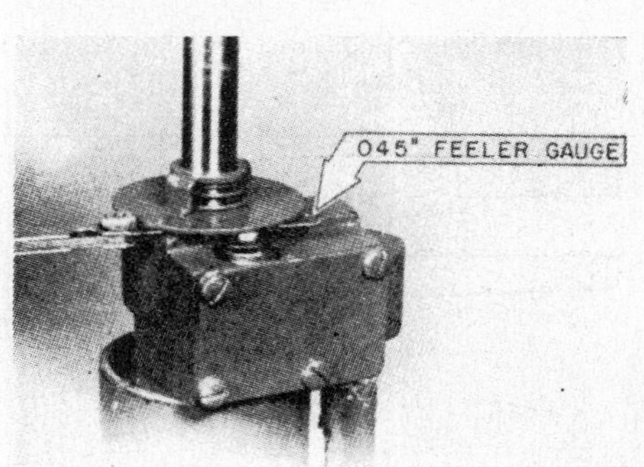

Fig. O425—Checking the air gap between the stop collar and the electromagnet.

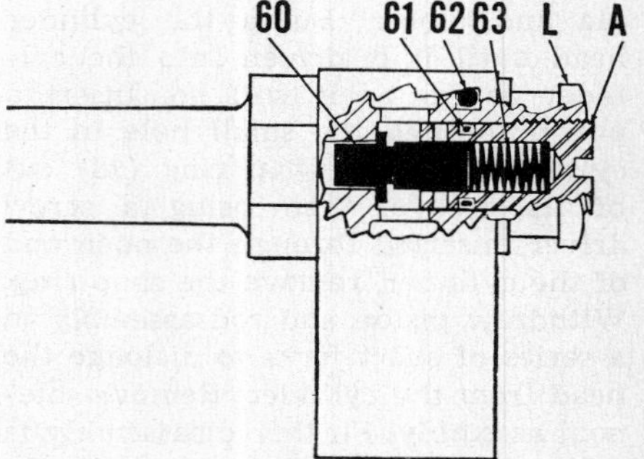

Fig. O426—Momentary restrictor can be made by converting a standard swivel fitting with an Oliver K2114 plunger and K2115 spring.

A. Adjusting screw
L. Lock nut
60. Plunger
61. "O" ring
62. "O" ring
63. Spring

642. **MANUALLY ADJUSTED TYPE FOLLOW-UP.** Refer to Fig. O429. Work cylinders equipped with a manually adjusted follow-up stop collar are used in systems employing non-electric control. Procedure for removal of the work cylinder is self-evident. To disassemble the work cylinder, remove the four bolts (9) attaching the base assembly (10), to the head assembly (7). The cylinder poppet valve (12), is located in the base assembly as shown in Fig. O427. Further disassembly is self-evident.

Wash all parts and renew the seals and "O" rings. Renew piston rod and/or cylinder sleeve if rubbing surfaces of same are scratched, flaked or pitted.

Fig. O423—Showing Nylon locking pin of late "electromagnetic" cylinders which permits external adjustment of the "UP" switch.

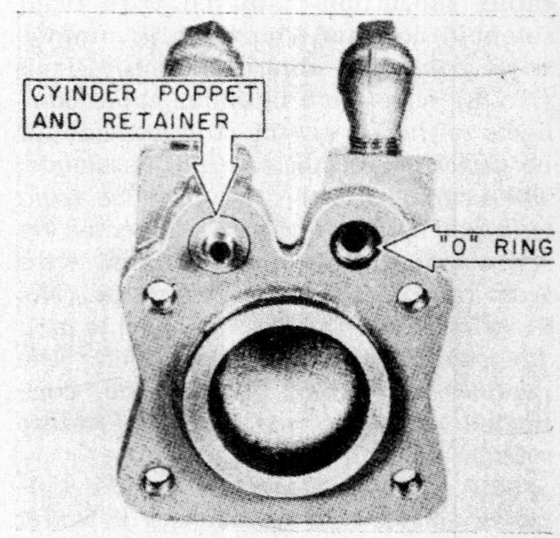

Fig. O427—View of the cylinder base showing the installation of the cylinder poppet valve and retainer.

642A. When reassembling the work cylinder, insert a lock washer and an "O" ring in the tube assembly bore of the cylinder head as shown in Fig. O431. On model 1KS-2061-C cylinders, insert a lockwasher in the other tube bore of the cylinder head as shown. Continue to reassemble the work cylinder and after tightening the barrel bolts (9—Fig. O429); rotate the cylinder tube (8), which is threaded at the head end, in a counter-clockwise direction until it is tight against the poppet valve retainer (15).

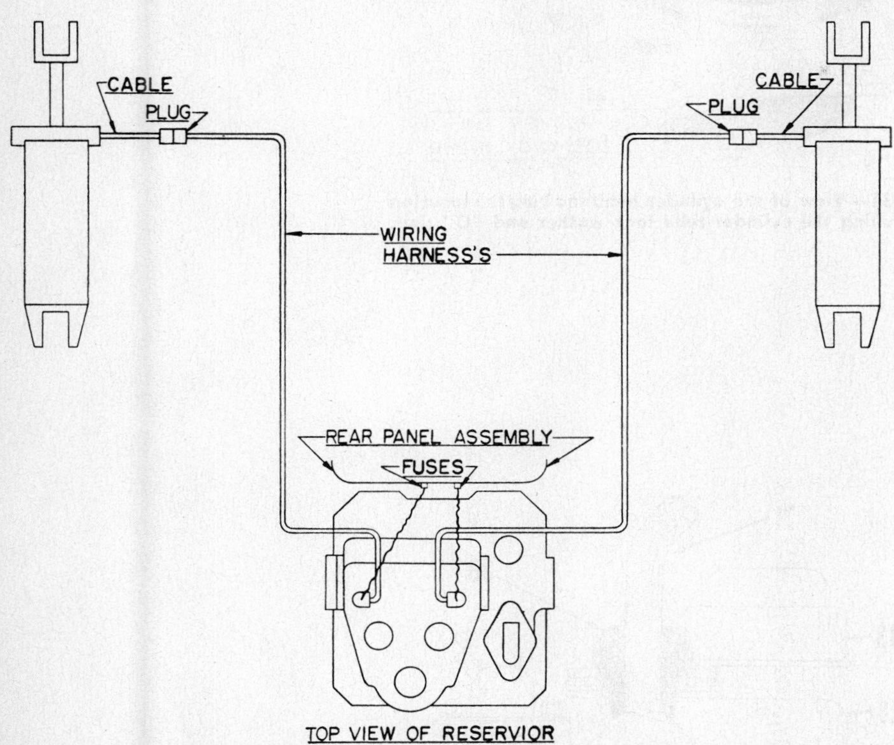

Fig. O428—Schematic of wiring for work cylinders of the No. 3 Hydra-Lectric hydraulic system.

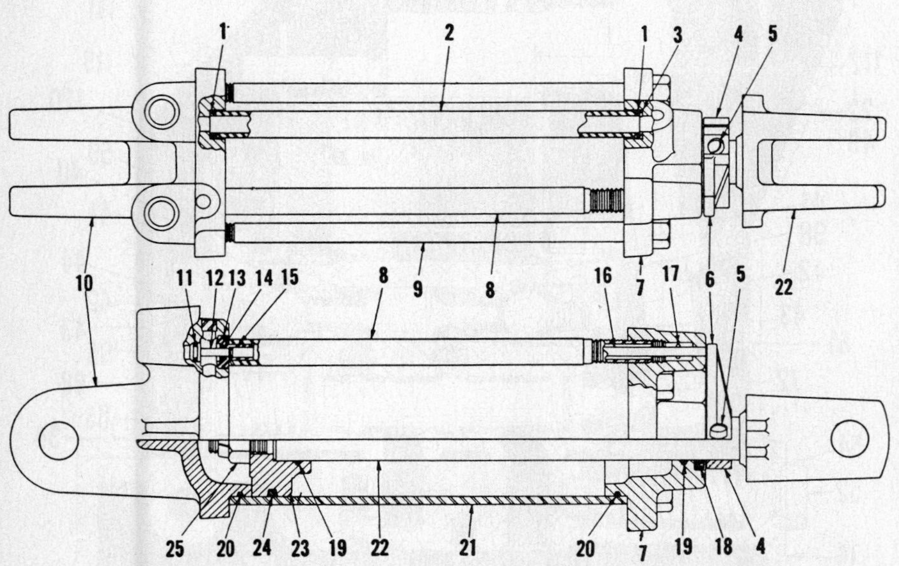

Fig. O429—Work cylinder as used on models equipped with a non-electrical type hydraulic lift unit. Note the use of lock washers (3) and an "O" ring (1) on cylinder tube (2).

1. "O" ring
2. Tube assembly
3. Lock washer
4. Stop cap
6. Collar stop
7. Head assembly
8. Tube
9. Barrel bolt (4)
10. Base assembly
11. "O" ring
12. Poppet valve
13. "O" ring
14. "O" ring
15. Poppet valve retainer
16. Stop rod
17. Roll pin (used on ¼" dia. rods only)
18. Seal
19. & 20. "O" ring
21. Barrel
22. Piston rod assembly
23. Piston
24. "O" ring
25. Lock nut

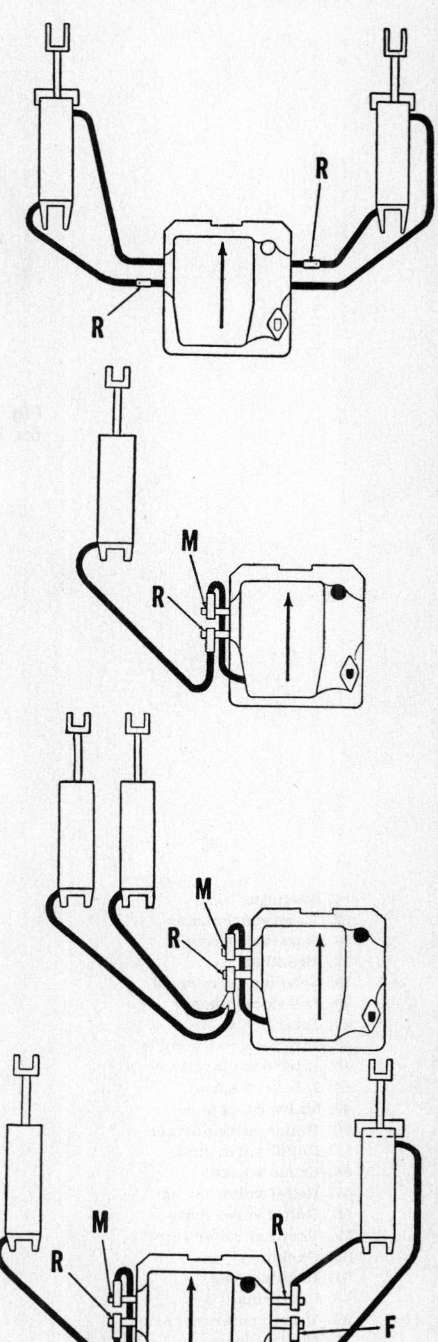

Fig. O430—Hydraulic lift unit may be connected to one or two work cylinders as shown in accompanying sketches.

Top. Two double-acting cylinders and regular restrictor valves installation.

2nd. Single-acting cylinder using a momentary and regular restrictor valve.

3rd. Single-acting cylinders using a momentary and regular restrictor valve.

4th. Single and double-acting using two regular restrictors and a momentary restrictor. Plain swivel fitting is shown at (F).

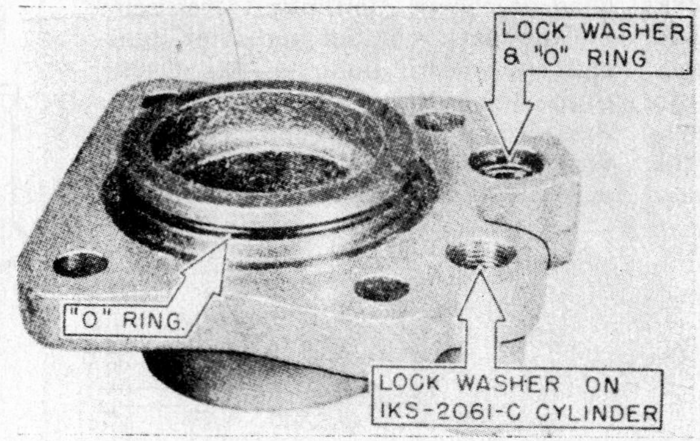

Fig. O431—View of the cylinder head showing the location for installing the cylinder tube lock washer and "O" ring.

14. Mounting base
22. Reservoir housing
28. Reservoir cover
33. Breather
39. Selector valve spool
40. Interlock plunger
41. Interlock valve seat
42. Interlock valve guide
43. Interlock valve ball
44. Interlock spring
49. Valve spool lever
50. Relief valve plunger
52. Relief valve guide
54. Crush washer
57. Relief valve spring
58. Relief valve seat
59. Thermal relief valve
60. Packing ring
61. Packing ring
63. Snap ring
68. Valve centering spring
71. Control lever
72. Ball socket joint
85. Packing ring
88. Lever boot
95. End plate, front
98. End plate, rear
111. Solenoid rail
112. Roll pin
115. Solenoid and switch
119. Solenoid coil
120. Solenoid plunger

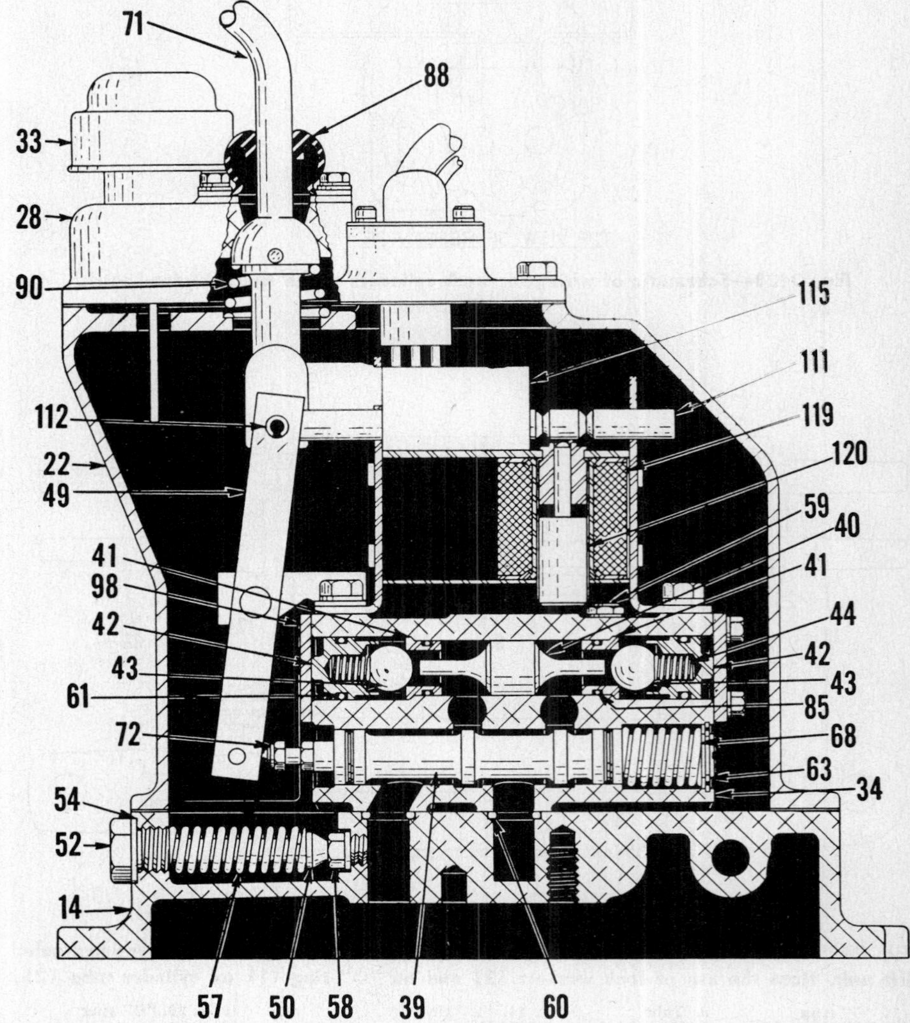

Fig. O431A—Sectionalized view of latest No. 3 Hydra-Lectric lift system reservoir housing which contains control and relief valving and valve operating solenoids. Shown is Super 88, Super 66, 77, 770 and 880 are similar.

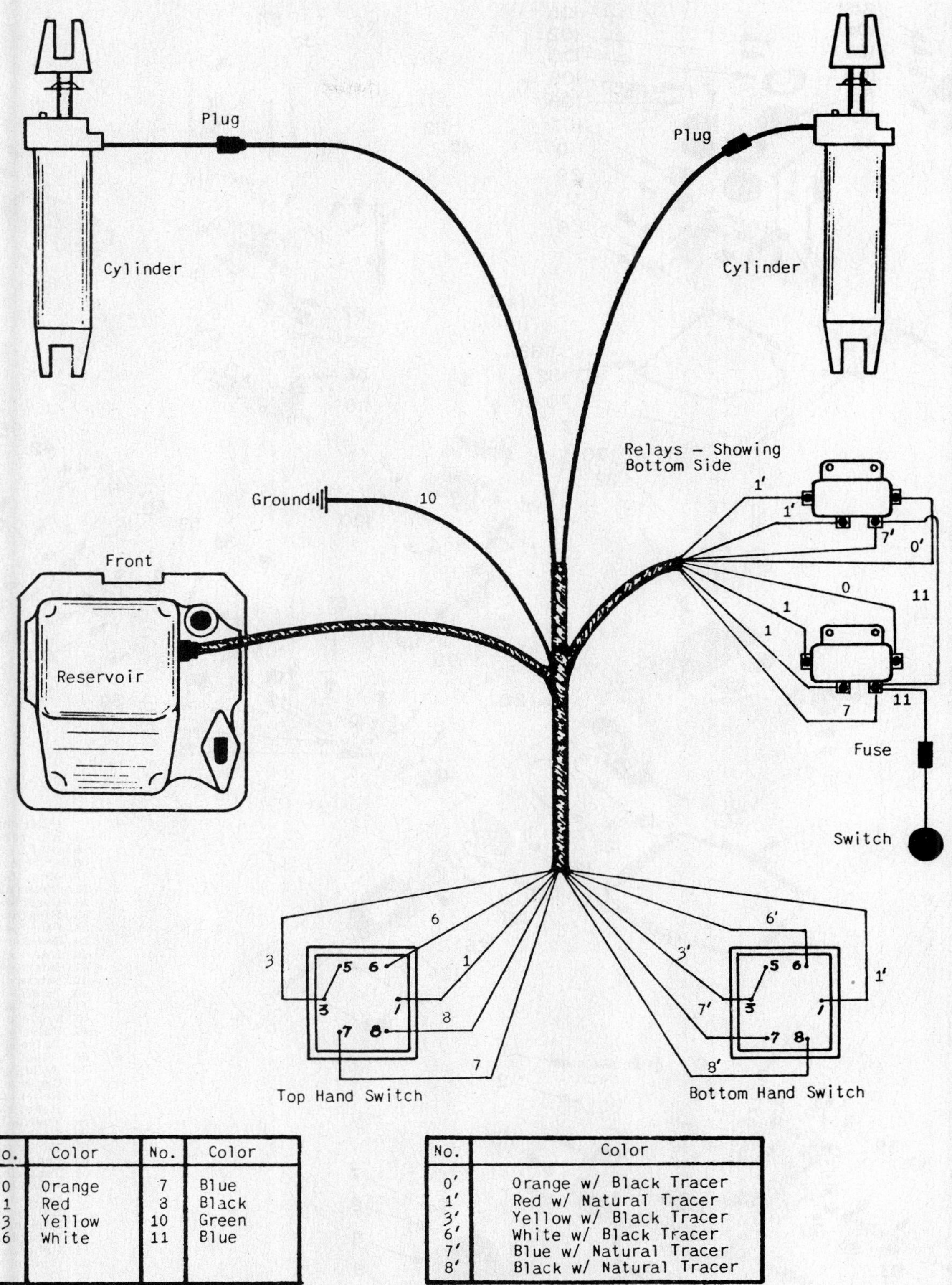

No.	Color	No.	Color
0	Orange	7	Blue
1	Red	8	Black
3	Yellow	10	Green
6	White	11	Blue

No.	Color
0'	Orange w/ Black Tracer
1'	Red w/ Natural Tracer
3'	Yellow w/ Black Tracer
6'	White w/ Black Tracer
7'	Blue w/ Natural Tracer
8'	Black w/ Natural Tracer

Fig. O431B—Wiring of the No. 2 Hydra-Lectric hydraulic system.

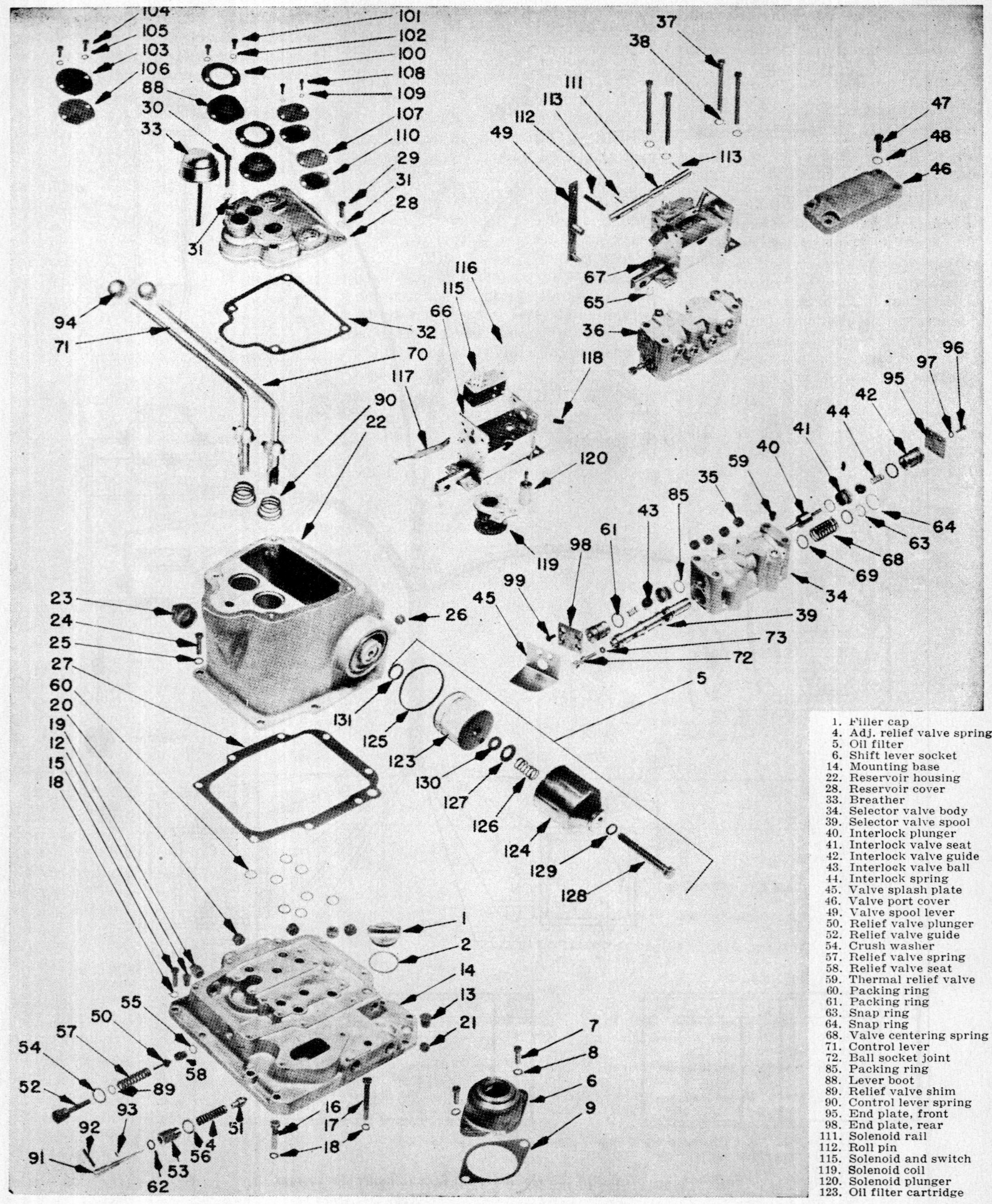

Fig. O431C—Exploded view of Hydra-Lectric lift system reservoir housing solenoids and valves assembly for Super 88 tractors shown sectionalized in Fig. O431A.

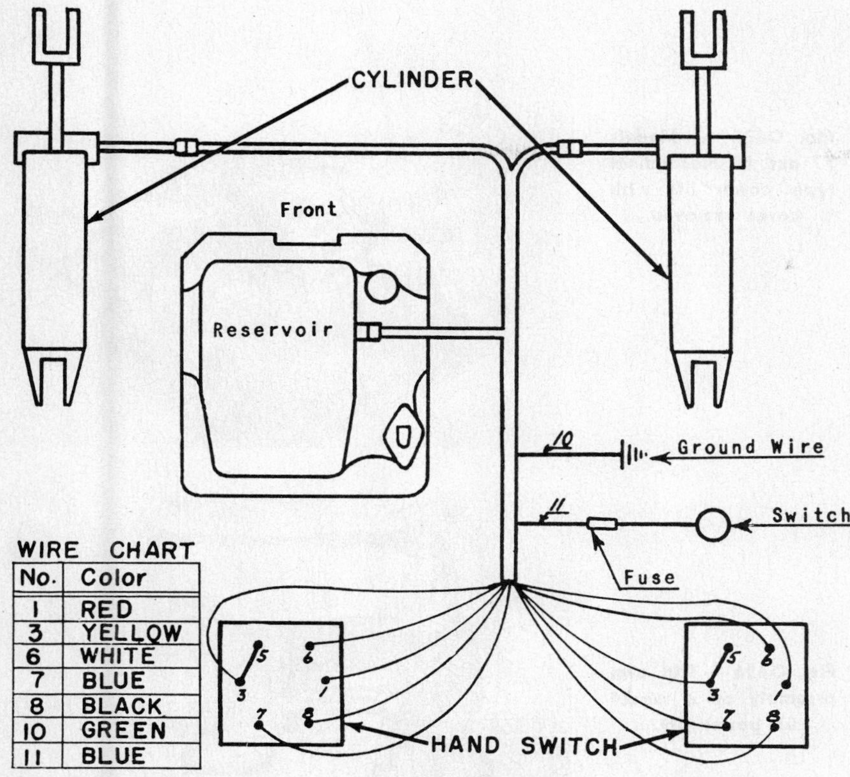

WIRE CHART

No.	Color
1	RED
3	YELLOW
6	WHITE
7	BLUE
8	BLACK
10	GREEN
11	BLUE

Fig. O431D—Wiring of the No. 1 Hydra-Lectric hydraulic system.

MECHANICAL LIFT SYSTEM

The mechanical type lift unit is of the worm and gear type. Refer to Figs. O431F and O432. The unit, which is mounted on top of the transmission housing in place of the transmission top front cover, receives its drive from a spur gear splined to the end of the transmission main (input) shaft.

650. **R & R LIFT UNIT.** To remove the power lift unit, remove the brake pedals equalizer bar and cap screws retaining the lift unit housing to the transmission housing. Lift up and remove the power lift unit.

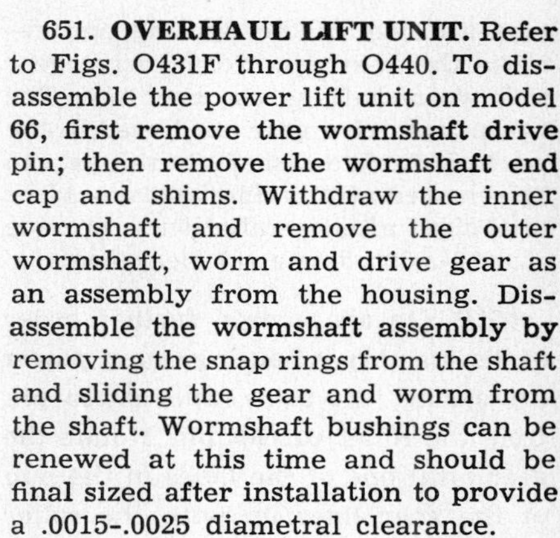

651. **OVERHAUL LIFT UNIT.** Refer to Figs. O431F through O440. To disassemble the power lift unit on model 66, first remove the wormshaft drive pin; then remove the wormshaft end cap and shims. Withdraw the inner wormshaft and remove the outer wormshaft, worm and drive gear as an assembly from the housing. Disassemble the wormshaft assembly by removing the snap rings from the shaft and sliding the gear and worm from the shaft. Wormshaft bushings can be renewed at this time and should be final sized after installation to provide a .0015-.0025 diametral clearance.

651A. On models 77 and 88, to disassemble the power lift unit, first remove front and rear wormshaft bearing retaining cap screws. Bump the front bearing rearward until the dowel

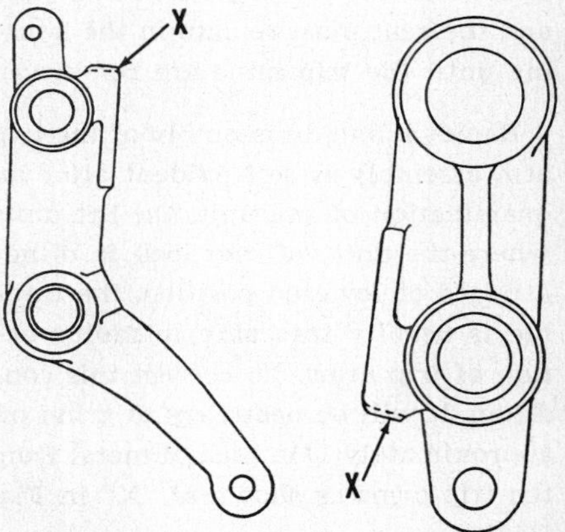

Fig. O433—To correct a condition where the lift unit will not lock in either a raised or lowered position, remove approximately 1/16 inch of metal (X) from trip arms.

Fig. O431F—Mechanical power lift as used on model 66, lower side view.

Fig. O432—Mechanical power lift as used on models 77 and 88, lower side view.

pins are free of the housing and remove the bearings, worm, drive gear and wormshaft as an assembly. The worm and drive gear are keyed to the wormshaft. Wormshaft bushings can be renewed at this time and should be final sized after installation to provide a .0015-.0025 diametral clearance.

651B. On all models, further order of disassembly is housing cover, power lift spring, lift arms, and cotter pin from the roller carrier pin. Rotate the gear until one of the holes in the hub of the gear lines up with the roller carrier pin and remove the pin. Be careful not to lose the roller carrier spring. Remove the lift shaft bearing and withdraw the lift shaft; then remove the gear. (On some model 66 units, there is not enough clearance to remove the worm gear at this time and the gear must remain in the housing until the trip arms are removed).

Removal and disassembly of the trip arm assembly is self-evident after an examination of the unit. On lift units where the unit will not lock in either a raised or lowered position, the trouble is usually traceable to faulty action of trip arms. To correct this condition, it will be necessary to grind off approximately 1/16 inch of metal from the trip arms as shown at "X" in Fig. O433.

651C. Reassembly is the reverse of disassembly. On model 66, when reinstalling the worm shaft bearing end cap, place a sufficient amount of shims under the bearing flange to provide the wormshaft with a .005-.010 end play.

Fig. O435 — Models 77 and 88 mechanical type power lift with cover removed.

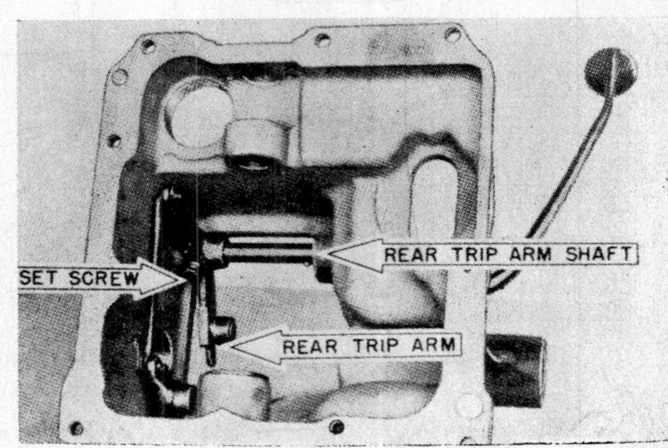

Fig. O436 — Trip arm assembly on a model 66 power lift.

Fig. O437 — Trip arm on 77 and 88 mechanical power lift.

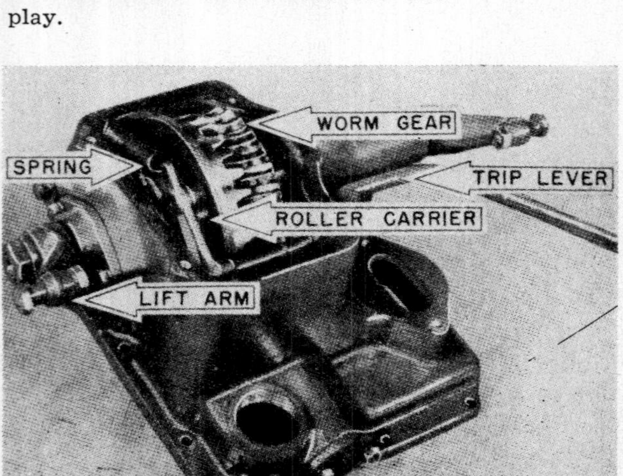

Fig. O434—Model 66 power lift wirh housing cover removed.

Fig. O438—Location of shims for adjusting the wormshaft end play on a model 66 power lift unit.

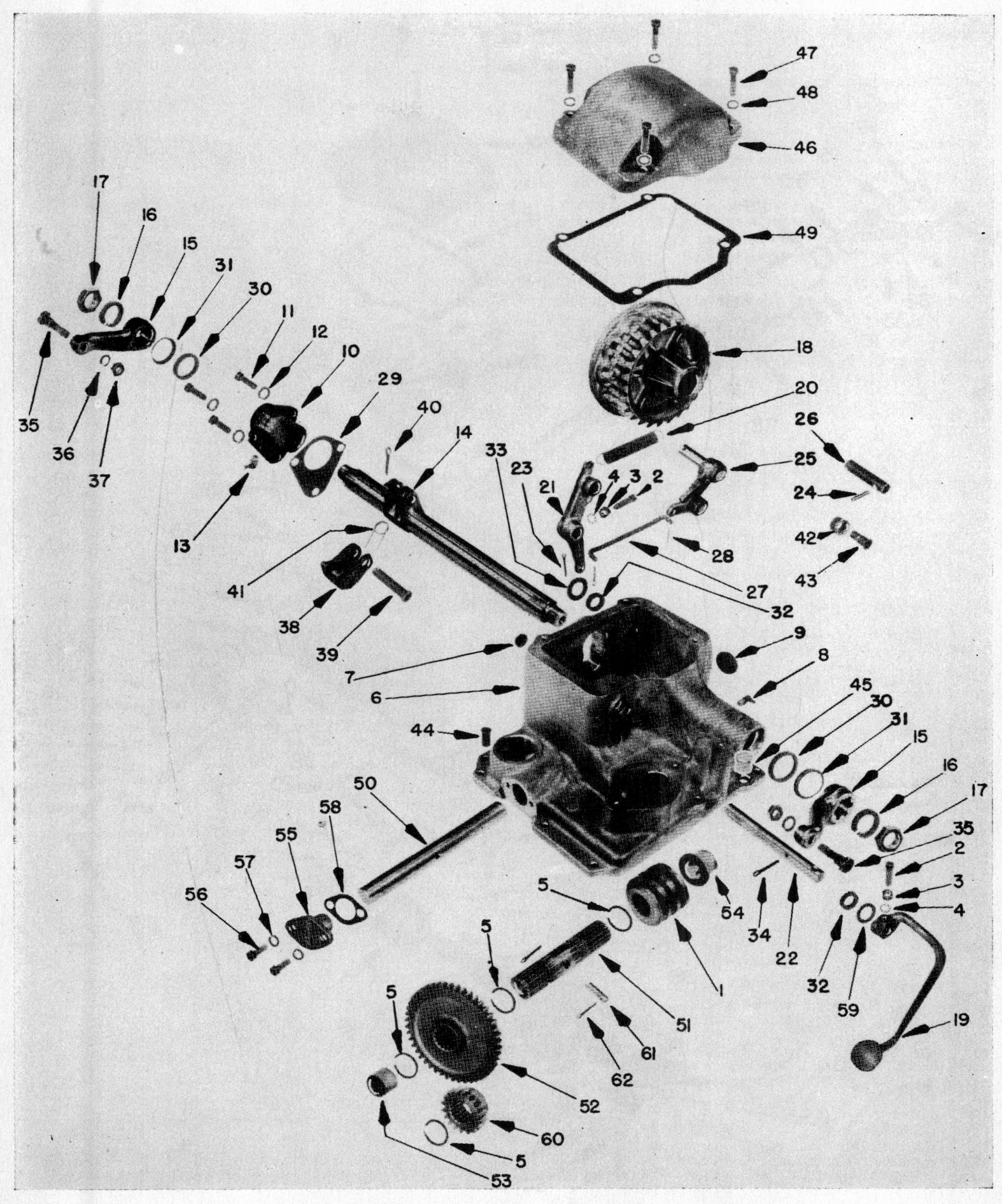

Fig. O439—Mechanical type power lift as used on the model 66 tractors.

1. Power lift worm
5. Snap ring
6. Lift housing
7. Plug
9. Plug
10. Lift shaft bearing
14. Lift shaft
15. Lift shaft lever
16. Lock washer
17. Retaining nut
18. Worm gear
19. Trip pedal
20. Spring
21. Trip arm (rear)
22. Trip arm shaft
23. Cotter pin
24. Trip arm roll
25. Trip arm (front)
26. Trip arm shaft
27. Trip arm link
28. Cotter pin
29. Gasket
30. Oil seal
31. Seal retainer
32. Oil seal
33. Washer
34. Cotter pin
38. Carrier
39. Carrier pin
40. Cotter pin
41. Roller carrier spring
42. Trip arm roller
43. Roller pin
44. Dowel pin
45. Dowel pin
46. Cover
49. Gasket
50. Wormshaft (inner)
51. Wormshaft (outer)
52. Drive gear (1536)
53. Bushing
54. Bushing
55. End cap
58. Shim
59. Washer
60. Drive pinion (1542)
61. Drive shaft pin
62. Cotter pin

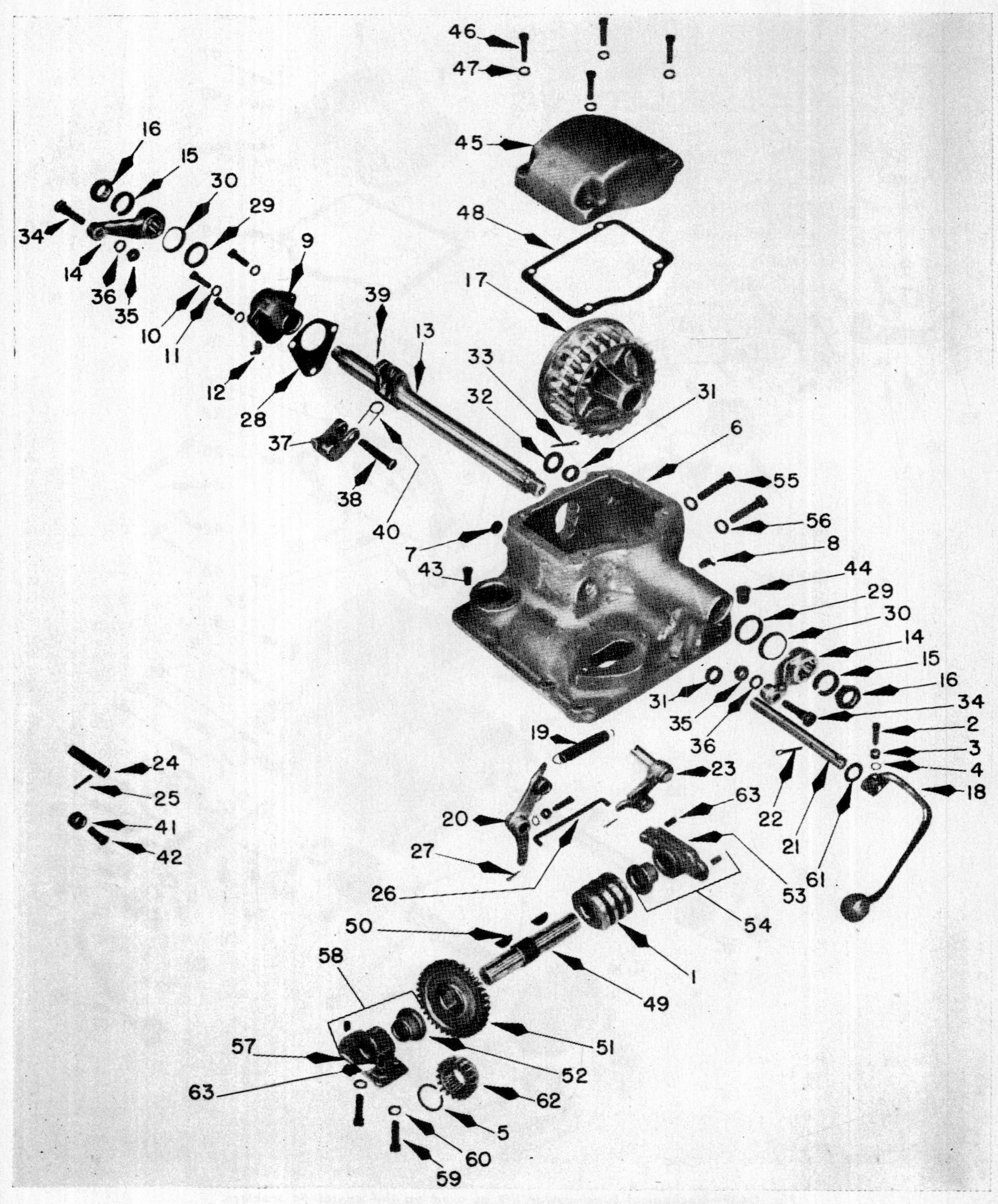

Fig. O440—Mechanical type power lift as used on models 77 and 88 tractors.

1. Worm
5. Snap ring
6. Power lift housing
7. Plug
9. Bearing
13. Lift shaft
14. Shaft lever
15. Lock washer
16. Lift shaft nut
17. Worm gear
18. Trip pedal
19. Spring
20. Trip arm rear
21. Trip arm shaft
22. Cotter pin
23. Trip arm (front)
26. Trip arm link
27. Cotter pin
28. Gasket
29. Oil seal
30. Seal retainer
31. Oil seal
32. Washer
33. Cotter pin
37. Roller and pin carrier
38. Carrier pin
39. Cotter pin
40. Spring
43. Dowel pin
44. Dowel pin
45. Cover
48. Gasket
49. Wormshaft
50. Woodruff key
51. Drive gear
52. Bushing
53. Wormshaft bearing
54. Wormshaft bearing assembly
57. Wormshaft bearing
58. Wormshaft bearing assembly
61. Washer
62. Drive pinion (1542)
63. Dowel pin

OLIVER

SHOP MANUAL SUPPLEMENT

Series ■ 660 Supplement

This supplement provides methods for sevicing the 660 series tractors which, except for the differences covered in this supplement, are serviced in the same manner as the Super 66 series covered in the previous section of this book (covering Series Super 66) or in the I & T SHOP SERVICE MANUAL No. 0-10.

CYLINDER BORE

As stated in manual No. O-10, the Super 66 tractors had a cylinder bore of 3½ inches and a piston displacement of 144 cu. in. The larger 660 engine has a bore of 3⅝ inches and a piston displacement of 155 cu. in. The compression ratio on 660 HC engines is 7.75:1.

POWER STEERING

On models equipped with power steering, refer to paragraphs 425A through 426J in Manual O-10 for service information. The Gemmer unit used on model 660 is the same as that used on models 770 and 880.

ROCKER COVER

Model 660 tractor engines are fitted with cast iron rocker covers. When gaskets which have shellac on one side and graphite on the opposite side are used, install same with shellac side toward rocker cover.

VALVES

A ½-degree interference angle is being used on the intake and exhaust valves on both HC and diesel engines of the 660 tractors. Valve seat angles are ground to 45 degrees and valve face angles are ground to 44½ degrees.

PISTON PIN BUSHINGS

The connecting rods on 660 tractors are fitted with steel-backed piston pin bushings. On Super 66 tractors, the steel-backed bushings are interchangeable with the early bronze bushings providing they are installed in pairs. Please make this notation in paragraph 444 on page 30 of manual No. O-10.

CARBURETOR

The carburetor applications on model 660 tractors are a Marvel-Schebler TSX775 or Zenith 12427. Refer to the following identification and calibration data.

Marvel-Schebler TSX775

Idle jet 49-101-L
Main adjusting needle.........43-277
Idle adjusting needle...........43-33
Main nozzle 47-A24
Venturi 46-A125
Float needle and seat........233-543
Float 30-739
Gasket kit 16-654
Repair kit 286-1252
Float setting ....¼-inch

Zenith 12427

Idle jet C55-22-10
Main adjusting needle........C71-54
Idle adjusting needle.........C46-25
Main nozzle C66-109-55
Venturi B38-73-23
Float needle and seat.......C81-1-40
Float C85-115
Gasket kit C181-325
Repair kit K12427
Float setting 1 5/32-inch

INJECTION PUMP

The first 8 digits in the model designation of the injection pump used on 660 tractors are PSB 4A 70V. These pumps differ from the Super 66 pumps primarily in the pump drive hub and pump camshaft. Instead of being splined, the late type hub and camshaft are tapered and fitted with a Woodruff key.

WATER PUMP SEAL

When renewing pump seal, be sure to install the type which has an exposed spring. This also applies to the Super 66 and a notation should be made in paragraph 522A on page 50 in manual O-10.

STARTING MOTOR

Delco-Remy starting motor 1108632 is used on 660 tractors. Test specifications for this unit are: Brush spring tension 24 ounces. No load test is 75 amperes at 11.25 volts at 6000 rpm. Lock test is 29 ft.-lbs. at 615 amperes at 5.85 volts.

BRAKES

On 660 series tractors, increased ratio brake actuating assemblies (73—Fig. O390B) have been installed These brake actuating assemblies are identified by a "5" stamped on the actuating lug. When servicing the Super 66 tractors, do not intermix the early and late type assemblies. When installing the late type actuating assemblies in early tractors, it will be necessary to grind out the inside of existing brake housings (23) to provide operating clearance for the longer actuating lug, or install new type brake housings.

HYDRAULIC LIFT SYSTEM

The hydraulic lift system control levers on 660 series tractors are made of steel tubing. These levers are interchangeable with the old type used on the Super 66; however, they are slightly longer and the Oliver Corporation recommends that they be renewed in pairs.

Beginning with Hydra-Lectric unit number 66 471, the solenoid rails were changed in that the angular grooves were cut at a 30 degree angle rather than the previous 45 degree angle. Only the late type rails will be furnished for service.

OLIVER

SHOP MANUAL SUPPLEMENT

Series ■ 770 & 880 Supplement

This supplement informs you of changes that were made to later production Oliver 770 and 880 tractors. Use the information contained in this supplement in conjunction with the information contained in the previous section of this book (covering Series 770 and 880) or in I & T SHOP SERVICE MANUAL No. 0-10.

VALVES

1. A ½-degree interference angle is being used on the intake and exhaust valves on both HC and diesel engines of the 770 and 880 series tractors. Valve seat angles are ground to 45 degrees and valve face angles are ground to 44½ degrees. Please make a note of this in the Condensed Service Data table on page 5; also in paragraph 432 on page 25.

PISTON PIN BUSHINGS

2. The connecting rods on 770 and 880 series tractors are now fitted with two steel-backed piston pin bushings. These steel-backed bushings are interchangeable with the early bronze bushings provided they are installed in pairs. Note this in paragraph 444 on page 30.

FINAL FUEL FILTER HOSE

3. In some cases, uneven engine operation and power loss have been found to be caused by pressure impulses from the injection pump being transmitted back through the copper tube to the final stage fuel filter. When this complaint is encountered, replace the copper tube with a rubber tube (Oliver part No. 102 317-A) and flared male elbow (Oliver part No. 939 161).

WATER PUMP

4. When renewing water pump seal on engines prior to serial number 1082091, be sure to install the type which has the exposed spring. Engines after serial number 1082090 have water pump fitted with this type seal. Note this in paragraph 522A on page 50.

OIL ADDITIVE

5. All 770 and 880 transmissions, as well as the later Super 77 and 88 transmissions, are filled with a special oil. Use SAE 5W-10 oil for temperatures below 0° F, or SAE 10W-30 oil for temperatures above 0° F and add Oliver additive (part No. 102 082-A) in the ratio of 16 parts oil to 1 part additive. This is one quart of additive to each four gallons of oil. Do not use SAE 90 or SAE 140 transmission oil in tractors equipped with Power Booster.

This blend of oil is also used in the Power Booster, belt pulley, power take-off, final drives, and if so desired, the power steering. Additive need not be added to those units using Type "A" Automatic Transmission fluid.

POWER BOOSTER

6. The Power Booster unit has had several modifications made to it, since its inception, to improve its service and operation. Please make appropriate notations on pages 54, 55 and 56.

7. A repair package is available which replaces the split bronze, externally lubricated throw-out bearing with a prepacked ball bearing throw-out assembly. This package also includes a wider throw-out fork which is needed to accommodate the prepacked ball bearing throw-out assembly.

8. A new gear set, comprised of input gear and output gear, is available. These gears differ from earlier gears in that the input gear has a shoulder machined on rear side and the snap ring and oil slinger are no longer used on the output gear. The shoulder of the input gear rides in the inside diameter of the output gear and in addition to maintaining gears in alignment, acts as a retainer for the overrunning clutch. When installing these gears on older tractors, pay close attention to the input shaft and if same is damaged, it will be necessary to install new input shaft. On tractors prior to serial number 72385, rework housing cover by grinding a ⅛-inch wide by 3/16-inch deep slot between countershaft retaining hole and oil seal bore when using these gears. On tractors prior to serial number 72100, a new grooved input gear ball bearing must also be installed.

9. Tension washers are now used under heads of link pins as well as improved pin retaining rings. Pins are installed with heads pointing in direction of engine rotation and concave part of tension washer facing away from pin head.

10. The straight type clutch pressure spring is no longer used and at tractor serial number 76247, three hook type springs were substituted to dampen center plate vibration and wear. At tractor serial number 93691, three additional hook type springs were added to the clutch pressure plate. These six hook type springs are installed through holes in center plate and pressure plate and hooked onto clutch housing in the direction opposite to engine rotation. Note: If holes in clutch pressure plate lugs are not bored completely through, complete the drilling using a No. 20 (.161) drill.

11. A felt dust washer has been added to front of input shaft to prevent dust from entering input shaft. On models with counter-bored input shaft and clutch shaft, a sleeve is provided as a means of holding felt washer in place. On models without counterbored shafts, glue felt washer to end of input shaft.

REVERSE-O-TORC DRIVE

Oliver industrial 770 and 880 tractors may be equipped with "Reverse-O-Torc" drive and torque convertor. The following paragraphs can be used to service this type of drive.

When disassembling any part of the "Reverse-O-Torc" drive it is a good practice generally to not remove any parts which can be thoroughly inspected while they are installed. All parts, when removed, should be handled with the same care as would be accorded the parts of a diesel injection pump or nozzle unit and should be soaked or cleaned with an approved solvent to remove gum deposits. Unless you practice good housekeeping (cleanliness) in your shop, do not undertake the repair of "Reverse-O-Torc" equipped tractors.

REVERSE-O-TORC DRIVE LUBRICATION

12. The "Reverse-O-Torc" operating oil capacity is approximately 10 quarts of Automatic Transmission Fluid Type A. The operating oil and filter element should be changed at least every 500 hours of operation or sooner under severe dust conditions. On those models with no filter, change operating oil every 200 hours. Oil should be to low mark on dip stick when unit is cold and at full mark when warm. NOTE: Make certain unit is not overfilled. All joints of the control system should be lubricated with heavy oil or gun grease.

REVERSE-O-TORC DRIVE TROUBLE SHOOTING

13. The following paragraphs list some of the troubles which may be encountered with "Reverse-O-Torc" drive equipped industrial tractors. The procedure for correcting most of the troubles is evident; for those not readily remedied, refer to the appropriate subsequent paragraphs.

A. OIL TEMPERATURE TOO HIGH. Could be caused by:

1. Indicator on instrument panel not operating properly.
2. Oil level too high or too low.
3. Operating the tractor too long in too high a work range (gear). Shift to lower gear.
4. Restricted oil tube to or from radiator.
5. Dirty or plugged oil radiator and/or water radiator.
6. Engine overheated due to lack of water, incorrect engine timing, etc.

B. JERKING STARTS. Could be caused by:

1. Engine idle speed too high.
2. Bent, binding, worn or maladjusted "Reverse-O-Torc" control and governor control linkage. Check and adjust as in paragraph 14.
3. Leaking or sticking regulator valve or valves. Refer to paragraph 16

C. SLUGGISHNESS. Could be caused by:

1. Engine not operating correctly.
2. Slipping "Reverse-O-Torc" unit. Check as in paragraph 14.

D. TRACTOR WILL NOT PULL LOAD. Could be caused by:

1. Low oil level.
2. Restricted oil lines or passages.
3. Incorrect regulator valve pressures. Refer to paragraph 16.
4. Incorrect control or linkage adjustment. Check and adjust as in paragraph 15.
5. Faulty control valve. Refer to paragraph 18.
6. Damaged clutch assembly. Refer to paragraph 21.
7. Faulty "Reverse-O-Torc" pump. Refer to paragraph 21.

E. CLUTCH FAILS TO RELEASE PROPERLY. Could be caused by:

1. Oil level too high.
2. Dirty oil.
3. Incorrectly adjusted linkage. Refer to paragraph 15.
4. Leaking control valve. Refer to paragraph 18.
5. Damaged "Reverse-O-Torc" internal parts. Refer to paragraph 21.

F. NOISY CONVERTOR UNIT. Could by caused by:

1. Low oil level.
2. Worn or damaged convertor internal parts. Refer to paragraph 21.

G. NOISY "REVERSE - O - TORC" UNIT. Could be caused by:

1. Low oil level.
2. Damaged "Reverse-O-Torc" internal parts. Refer to paragraph 21.

REVERSE-O-TORC DRIVE TESTS AND ADJUSTMENTS

14. **PERFORMANCE TEST.** First check the "Reverse-O-Torc" oil level, shift the transmission to the neutral position and then place control handle in the forward (upper) position and depress the foot accelerator pedal. At this time the engine speed should be approximately 1925 rpm for both 770 and 880 tractors. Check in a similar manner in the reverse position.

If the speeds are below specified rpm, check the engine governed speeds and the "Reverse-O-Torc" control and linkage adjustments as in paragraph 15.

Shift the transmission to sixth gear, lock the brakes, place control handle in the forward (upper) position and depress the accelerator pedal. Engine speed more than 1500 rpm for 770 tractors, or 1700 rpm for 880 tractors indicates a slipping "Reverse-O-Torc" front clutch or damaged torque convertor.

A similar check should be made with the control handle in the reverse (lower) position. If the engine rpm is not as specified it indicates a damaged torque convertor or slipping rear clutch.

If engine speeds were above specified rpm, check and adjust the pressure regulator valves as in paragraph 16. If the pressures are correct or if correcting the pressure fails to prevent slipping, the following are possible causes: Restricted oil lines or passages; incorrect control or linkage adjustment (refer to paragraph 15); damaged "Reverse-O-Torc" clutch assembly or pump (refer to paragraph 21).

15. **CONTROL AND LINKAGE ADJUSTMENTS.** To adjust the "Reverse-O-Torc" control valve linkage, proceed as follows: Determine the amount of unrestricted valve movement when the bell crank is moved from neutral position to either forward or reverse positions. There should be 5/16 ± 1/32 inch of unrestricted valve movement. This movement may be regulated by loosening the control stop plate cap screws and moving the control stop plate rearward until the shifter rocker will not strike the throttle lever when the hand lever is moved.

The control valve spool should protrude ¾ ± 1/32-inch from the end of the body when the controls are in the neutral position. Adjustment is accomplished by turning jam nuts (17—Fig. OS20).

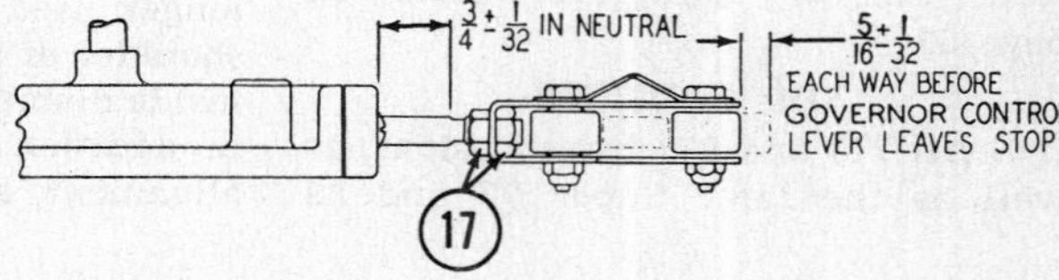

Fig. OS20—View showing control linkage adjustments. Refer to paragraph 15.

If the recommended engine low idle speed or the high idle no-load speed cannot be obtained, it may be necessary to vary the length of the governor control rod.

16. **PRESSURE TEST AND ADJUSTMENT.** Before checking the pressure regulator valves, make certain oil level is correct.

Install a pressure gage of at least 200 psi capacity in place of the pipe plug shown at (3—Fig. OS21), start engine and run at approximately 1800 rpm. The pressure should be 160-170 psi. NOTE: Engaging either clutch should drop the pressure approximately 100 psi momentarily, but pressure should return to within 5 psi of gage reading with clutches disengaged. To check the condition of the lower valve (1—Fig. OS25), remove the upper plug (2) and withdraw the spring (17). NOTE: Make certain shims (18) are not lost or damaged. Reinstall the plug (2) leaving out the spring (17) and run engine at approximately 1800 rpm. Gage should register 60 psi for units built prior to serial number 1051 and approximately 80 psi for units after serial number 1050. If this pressure is satisfactory, but the first pressure check (160-170 psi) was not, the upper valve (2) is faulty.

If the pressures are not correct with the clutches disengaged, remove the valves and clean; if cleaning doesn't restore the pressure, adjusting washers can be added. If the correct pressures cannot be obtained by the addition of washers, check for leaking or damaged "Reverse-O-Torc" internal parts.

If when a clutch is engaged the pressure will not return to within 5 psi of the pressure when clutches are disengaged, it indicates leaking internal seals.

REVERSE-O-TORC CONTROL VALVE AND LINKAGE

18. **R&R AND OVERHAUL.** Removal and disassembly procedures for the "Reverse-O-Torc" control valve and linkage will be evident after an examination of same and reference to Fig. OS21.

Renew any excessively worn parts and all seals. The valve spool and body are available only with a complete control valve assembly.

Reassemble and install using Figs. OS20 and OS21 as a guide.

With the controls installed, the control valve and linkage should be adjusted as in paragraph 15. Center the control valve by turning the jam nuts (17—Fig. OS20) until the control valve spool extends ¾ ± 1/32-inch from the body when the controls are in the neutral position. Recheck the control and linkage adjustment

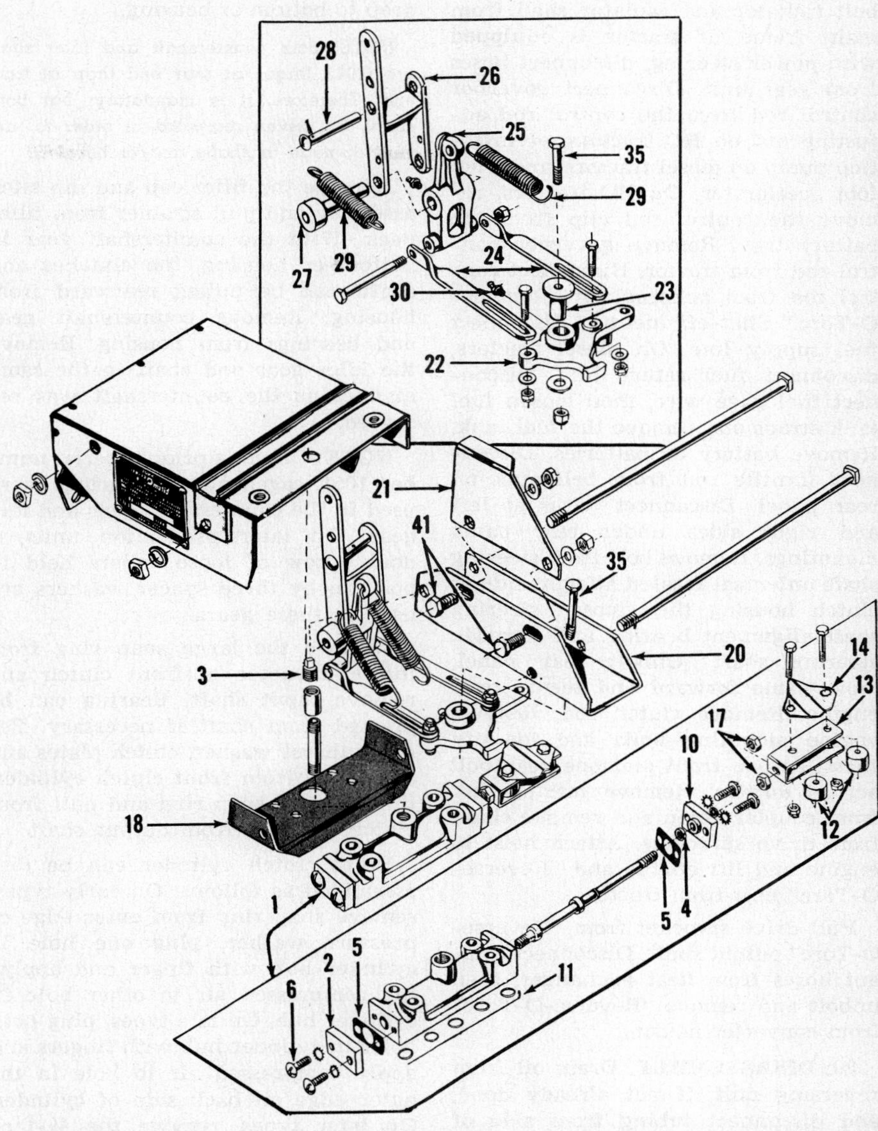

Fig. OS21—Exploded view showing control linkage and control valve of "Reverse-O-Torc." When testing, install pressure gage in place of plug (3).

1. Control valve
2. End cover
3. Pipe plug
4. Oil seal
5. Gaskets
10. End cover
11. "O" rings
12. Rollers
13. Roller cage
14. Roller cage spring
17. Adjusting nuts
18. Base
20. Controls bracket
21. Stop bracket
22. Bellcrank
23. Bushing
24. Connecting links
25. Throttle rocker
26. Governor control lever
27. Bushing
28. Pin
29. Springs
30. Grease fitting
35. Cap screw
41. Stop adjusting bolts

"REVERSE-O-TORC" DRIVE

19. **DESCRIPTION.** The "Reverse-O-Torc" drive consists of a pair of hydraulically actuated, multiple disc clutches. When the front clutch is engaged, the output shaft turns the same direction as the engine and this produces forward motion of the tractor. The rear clutch is driven by a simple gear train and it rotates in the opposite direction of the front clutch. When the rear clutch is engaged, the output shaft turns and produces rearward motion of the tractor.

19A. **REMOVE AND REINSTALL.** To remove the "Reverse-O-Torc" unit, first drain the oil from "Reverse-O-Torc," then remove the engine with the reversing unit attached as follows: Drain cooling system, then remove hood side panels and hood. Remove radiator grille and screen, then unbolt radiator and radiator shell from main frame. If tractor is equipped with power steering, disconnect hoses from gear unit. Disconnect governor control rod from the control rod adjusting nut on HC tractors, or injection pump on diesel tractors, and from foot accelerator. On HC tractors, remove the control rod clip from the battery tray. Remove governor control rod from tractor. Disconnect control rod from bellcrank of "Reverse-O-Torc." Shut-off fuel and disconnect fuel supply line. On diesel tractors, disconnect fuel return line. Disconnect fuel gage wire, then loosen fuel tank straps and remove the fuel tank. Remove battery or batteries. Disconnect throttle rod from bellcrank on rear panel. Disconnect wires at left and right sides under rear panel mountings. Remove bolt from steering shaft universal located at right side of clutch housing then loosen steering shaft alignment bearing and separate steering shaft. Unbolt rear panel, move same forward and secure over engine. Remove clutch rod. Remove engine mounting bolts and identify them, as one front and one rear bolt act as dowels. Remove master link from coupler chain and remove chain from drive sprockets. Attach hoist to engine and lift engine and "Reverse-O-Torc" unit from tractor.

Pull drive sprocket from "Reverse-O-Torc" output shaft. Disconnect coolant hoses from heat exchanger, then unbolt and remove "Reverse-O-Torc" from converter housing.

20. **DISASSEMBLY.** Drain oil from reversing unit, if not already done, and disconnect tubing from side of case. Remove the heat exchanger from air scoop, and if so equipped, the oil filter. Disconnect the two springs from rocker and throttle lever, then remove the tie bolts and remove the control assembly from air scoop. Remove nipple from control valve, unbolt and remove air scoop, then remove control valve assembly. Remove the high pressure regulator valve (upper) and converter charge regulator valve (lower) from left front side of housing. Note: Some valves may not have the valve pins and spacer washers shown; however, if same are included, note their locations so they can be reinstalled in the same positions. Remove the converter charging pump and gasket from front of housing. Complete removal of rear cover and gasket. Remove the corks from bores at front of countershaft and idler shaft, then using a brass drift, drive countershaft out rear of case and allow countershaft gear to drop to bottom of housing.

NOTE: Both countershaft and idler shaft are 0.002 larger at rear end than at front end. Therefore, it is mandatory that both shafts be driven rearward in order to prevent damage to shafts and/or housing.

Remove the filler cap and dip stick assembly and pull strainer from filler neck. With the countershaft gear in bottom of housing, the clutches and shafts can be pulled rearward from housing. Remove countershaft gear and bearings from housing. Remove the idler gear and shaft in the same manner as the countershaft was removed.

NOTE: On units prior to serial number 1051, caged needle bearings were used in the countershaft gear and idle gear. On later production units, a double row of loose rollers held in position by three spacer washers are used in these gears.

Remove the large snap ring from inside diameter of front clutch and remove input shaft. Bearing can be pressed from shaft if necessary. Remove thrust washer, clutch plates and drive hub from front clutch cylinder, then remove snap ring and pull front clutch cylinder from output shaft.

Front clutch cylinder can be disassembled as follows: On early types, remove snap ring from outer edge of pressure washer, plug one hole in cylinder hub with finger and applying compressed air to other hole in cylinder hub. On late types, plug both holes in cylinder hub with fingers and apply compressed air to hole in the outer edge on back side of cylinder. On later types, remove the ¼-inch steel ball from clutch cylinder. Rear clutch and cylinder assembly is disassembled in a similar manner.

21. **OVERHAUL.** Clean all parts in a suitable solvent and inspect for wear or other damage. The oil pump except for oil seal, is available only as a complete unit. Use all new seals and gaskets when reassembling. A complete seal and gasket kit is available under Oliver part number 103 515-AS.

The "Reverse-O-Torc" contains nine cast iron sealing rings which should not be removed unless necessary. Check condition of cast iron rings as follows: Use a 0.015 feeler gage blade and measure between side of ring and groove and if the 0.015 feeler blade will enter between ring and groove, renew ring. Also check for fractures or loss of tension. When renewing the cast iron sealing rings, use only fingers as the rings are easily broken.

The external toothed clutch drive plates are conical and have a 0.015-0.018 slope between outside and inside diameters. Check same by placing drive plate concave side down on a flat surface and measuring under inside diameter.

Check regulator valves, check valve and control lever springs against the following values: High pressure regulator valve and converter charge regulator valve springs, 2¼ inches free lengths; check valve spring, 1¾ inches free length; throttle lever spring (inner), 4¼ inches free length; rocker spring (outer) 3¾ inches free length.

NOTE: On early units in serial number range 139-999 the high pressure regulator valve spring had a free length of 2 inches and the converter check valve spring had a free length of 1½ inches.

Disassembly and overhaul of control valve is obvious after removing end caps. Be sure valve spool moves smoothly in its bore. Light scratches or scoring may be cleaned up with crocus cloth. Install seals in open cap so that inner seal lip faces inward and outer seal lip faces outward.

Any further overhaul required is obvious.

22. **REASSEMBLY.** When reassembling the "Reverse-O-Torc" unit, lubricate all clutch plates, seals and surfaces which contact seals with Automatic Transmission Fluid Type A. Renew all seals and gaskets.

To reassemble early production type clutches, refer to Fig. OS23. The inner edge of spring (53) should contact the back-up ring (54). Pressure plate (52) should be installed with smooth side toward clutch plates. Internally splined clutch plates (50) and externally splined plates (51) should be installed alternately, starting with an internally splined plate. The missing teeth of the externally splined plates should be aligned with the drain holes in the hub and BE SURE to keep the plates sloping in the same direction. Take care not to damage the sealing rings (40) when installing the clutch cylinders (drum) (48).

To reassemble later production type "Reverse-O-Torc" clutches, refer to Fig. OS24. The steel ball (48A) should be positioned in clutch cylinders (drums) before the pistons (55) are installed. Internally splined clutch plates (50) and externally splined plates (51) should be installed alternately starting with an internally splined plate. The missing teeth of the externally splined plates should be aligned with the drain holes in the

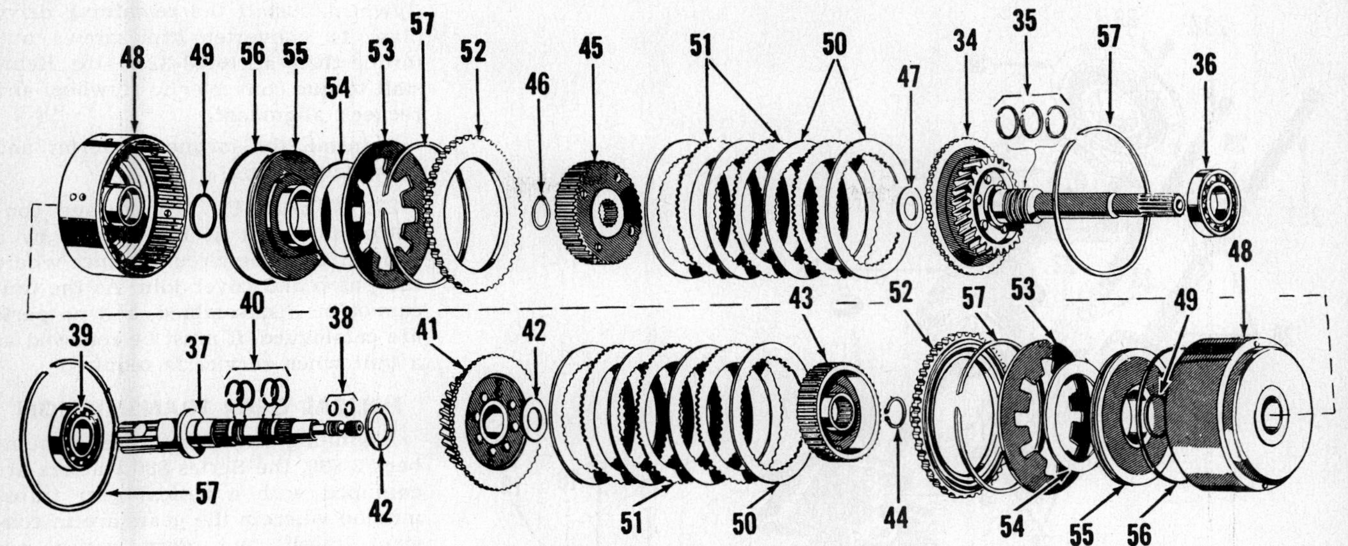

Fig. OS23—Exploded view of early production "Reverse-O-Torc" clutches assembly. Refer to Fig. OS24 for the later production type.

34. Input shaft
35. Seal rings (1⅜ O.D.)
36. Ball bearing
37. Output shaft
38. Seal rings (⅝ O.D.)
39. Ball bearing
40. Seal rings (1 1/16 O.D.)
41. Output gear
42. Thrust washer
43. Output clutch hub
44. Snap ring
45. Input clutch hub
46. Snap ring
47. Thrust washer
48. Clutch cylinders
49. "O" rings
50. Internal spline clutch plates
51. External spline clutch plates
52. Pressure plates
53. Springs
54. Back-up rings
55. Clutch pistons
56. "O" rings
57. Snap rings

hub and BE SURE to keep the plates sloping in the same direction. Take care not to damage the sealing rings (40) when installing the clutch cylinders (drums) (48).

On both type "Reverse-O-Torc" units, install the idler gear (21—Fig. OS25), shaft (29Z), spacer (24), washers (23), rollers (22) and thrust washers in the case. Position the counter shaft gear (25), bearings (26), spacer (27) and thrust washers (28) in the case; then, thread a length of wire through these parts instead of shaft (29Y). With the wire threaded through the counter shaft assembly, drop counter shaft to the bottom of the case. Enter the clutches assembly in the case. With the clutches assembly correctly positioned and with the rear end of the "Reverse-O-Torc" unit facing up, use two large screw drivers and the wire to work the thrust washers (28) and the counter shaft assembly into correct alignment with the holes in the case. Install the shaft (29Y) making certain that it goes through the thrust washers (28).

The remainder of the reassembly procedure is the reverse of the disassembly procedure. Vary the 0.015 thickness shims under the rear cover to provide an output shaft end play of 0.015-0.030. Adjust the control valve and linkage as in paragraph 15 and the pressures as in paragraph 16.

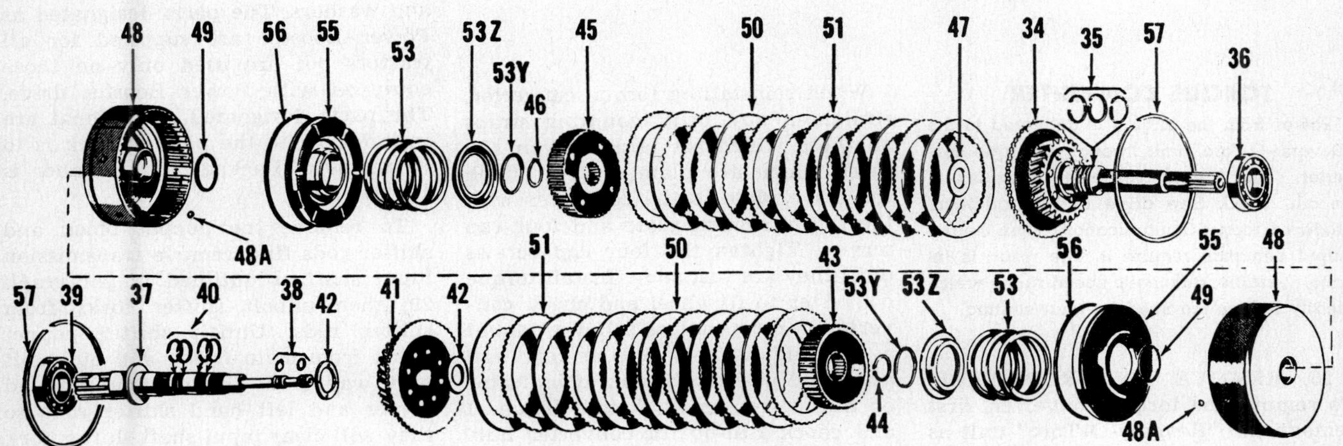

Fig. OS24—Exploded view of later production "Reverse-O-Torc" clutches assembly. Refer to Fig. OS23 for early production type.

34. Input shaft
35. Seal rings (1⅜ O.D.)
36. Ball bearing
37. Output shaft
38. Seal rings (⅝ O.D.)
39. Ball bearing
40. Seal rings (1 1/16 In.)
41. Output gear
42. Thrust washer
43. Output clutch hub
44. Snap ring
45. Input clutch hub
46. Snap ring
47. Thrust washer
48. Clutch cylinders
48A. Steel balls (¼ in.)
49. "O" rings
50. Internal spline clutch plates
51. External spline clutch plates
53. Springs
53Y. Snap rings
53Z. Spring retainers
55. Clutch pistons
56. "O" rings
57. Snap rings

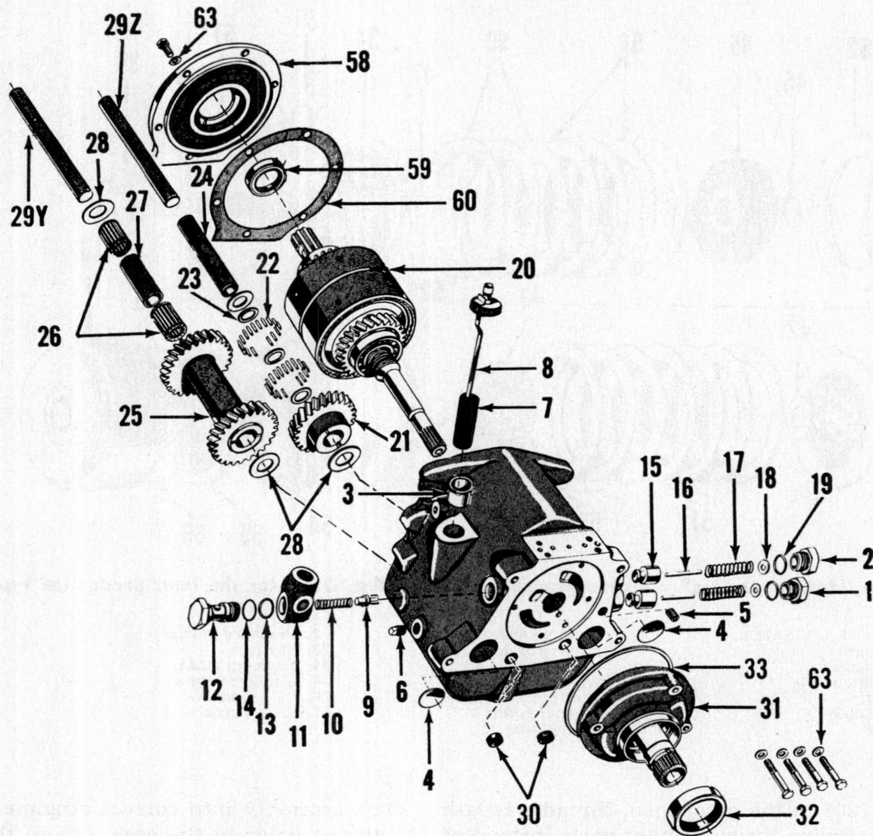

Fig. OS25—Partially exploded view of the "Reverse-O-Torc" unit showing component parts and their relative positions. Refer to Figs. OS23 and OS24 for exploded views of the clutch assemblies. Bearings (26) are loose rollers on latest production units.

1. Plug
2. Plug
3. Oil filler neck
4. Expansion plugs
5. Pipe plug
6. Drain plug
7. Screen
8. Dip stick
9. Check valve
10. Check valve spring
11. Valve block
12. Cap
13. "O" ring
14. "O" ring
15. Regulating valves
16. Valve pins
17. Regulating valve springs
18. Adjusting washers
19. "O" rings
20. Clutch aassemblies
21. Idler gear
22. Rollers
23. Bearing spacer washers
24. Idler gear spacer
25. Countershaft gear
26. Countershaft gear bearings
27. Bearing spacer
28. Thrust washers

29Y. Countershaft gear shaft

29Z. Idler gear shaft

30. Corks
31. **Oil pump**
32. Oil seal
33. **Gasket**
58. Rear cover
59. **Oil seal**
60. Gasket
63. Sealing washers

TORQUE CONVERTER

Power from the engine is delivered to the "Reverse-O-Torc" unit through a torque converter. The torque converter provides a smooth, shock free drive and in addition, produces torque multiplication when heavy output demands require it. The result is infinite, variable, automatic speed ratios which greatly reduce the need for gear shifting.

26. **REMOVE AND REINSTALL.** To remove the torque converter, first remove the "Reverse-O-Torc" unit as outlined in paragraph 19A. Remove starting motor, timing hole cover and bottom front dust cover, then unbolt and remove the flywheel and torque converter housing from engine. Unbolt and remove torque converter from engine flywheel.

When reinstalling torque converter, straighten any bent mounting straps on those which are strap driven. On those which are plate driven, attach the drive plate to the converter with the reinforcement plate and four cap screws. Tighten the four cap screws until they are just snug. Install torque converter to flywheel and check converter hub run-out as follows: Mount a dial indicator to engine rear end plate and position button of indicator on hub of converter. Rotate flywheel and check run-out of converter hub. The run-out must not exceed 0.015. If run-out exceeds 0.015, correct it on strap driven types by prying on converter and bending straps; on plate driven types correct it by shifting converter on drive plate. On plate driven types, remove converter from flywheel, install the remaining drive plate to converter cap screws and torque them all to 28-32 ft.-lbs. Reinstall torque converter to flywheel and recheck alignment.

Reinstall the torque converter and flywheel housing.

27. **OVERHAUL.** The torque converter is sealed at the factory by a weld around the circumference where the pump and cover join. As the unit cannot be disassembled, and no parts are catalogued, it must be renewed as a unit when service is required.

HELICAL GEAR TRANSMISSION

Beginning with tractor serial number 73 639, the Series 880 tractors are equipped with a helical gear transmission wherein the gears are in constant mesh and gear ratios are changed by sliding couplings.

While differences in construction exist, the removal and installation of shafts, EXCEPT for the input shaft, remain similar to those outlined in paragraphs 551 through 558 in I&T Shop Manual O-10.

HELICAL GEAR TRANSMISSION SHIFTER RAILS AND FORKS

28. All model 880 helical gear transmissions built after tractor serial number 82 239 have an improved shifter mechanism. Parts required to make the change on prior models are as follows: Interlock yoke and support assembly, shifter rods, poppet block gasket, oil trough (Power Booster), oil trough spacer (Power Booster), shifter poppet block (optional), shifter rod end cover (optional), shifter rod end cover gasket optional) and the required cap screws and washers. The parts designated as Power Booster are supplied for all tractors but are used only on those equipped with Power Booster drive. The parts designated as optional are included when the poppet block is to be renewed. Package part number is 104 041-AS.

To remove the poppet block and shifter rods first remove transmission input shaft as outlined in paragraph 29, then unbolt shifter forks from shifter rods. Unbolt shifter poppet block from main frame and pull unit part way from main frame. Spread center and left hand shifter rods so they will clear input shaft shifter fork as shown in Fig. OS27, then complete removal of poppet block and shifter rods. Remove pipe plugs from poppet block, turn shifter rods 90 degrees and pull same from poppet block after making provision to catch any parts that might fly due to releasing the spring pressure. Remove the poppet balls, springs and interlock balls.

Note: If original poppet block is being reinstalled, be sure to discard the three 3/8-inch steel balls.

Install the improved interlock mechanism as follows: Install the poppet springs in the shifter rod holes, place the 7/16-inch balls on springs and depress same with a small rod or punch. Insert the shift rods in the side of the poppet block which has the two oil drain holes.

Note: As shifter rods are being installed, do not forget to install the interlock balls.

The two upper shift rods are correctly installed when the detents face downward and the shifter lever slots face inward. The center (lower) shifter rod is correctly installed when the lug on same is between the two upper shifter rods. Install plugs in poppet spring holes. Refer to Fig. OS28.

Use new gasket and insert new shifter mechanism part way through front mounting hole. Spread center and left hand shifter rods and slide input shaft shifter fork into position. Position the bevel pinion shaft shifter fork (rear) on the shifter coupling and shifter rod. Position front shifter fork on bevel pinion shaft, then insert aft ends of shifter rods in holes of main frame. Secure poppet block in place and secure shifter forks to shifter rods. If a new poppet block is being installed, install shifter rod end cover before securing poppet block. Install the interlock yoke support over the top of the shifter rods so that the tab of the cradle aligns with the slots in the shifter rods and on models without Power Booster, secure with cap screws. On models with Power Booster, install the oil trough, with spacer underneath, at the front cap screw hole of the interlock yoke support. Shift the center shifter rod into the forward neutral position and install input shaft. Adjust input shaft to 0.001-0.003 end play by varying the shims under input shaft bearing retainer. Shims are available in thicknesses of 0.004, 0.007 and 0.0149.

Complete reassembly of tractor.

Fig. OS27—Spread shifter rails as shown when removing poppet block and rails.

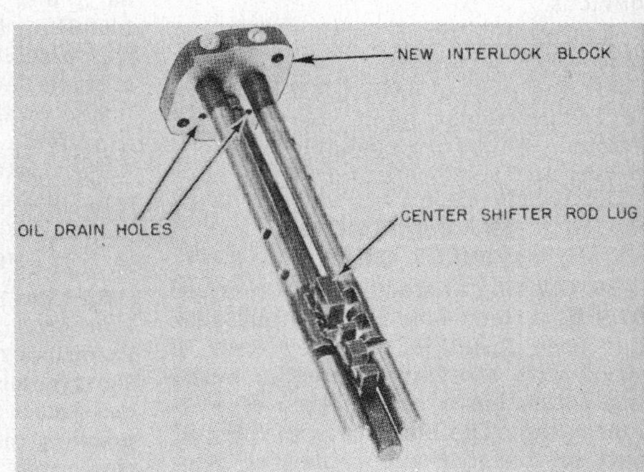

Fig. OS28 — Assembled poppet block and shifter rails. Note position of center shifter rod lug.

Fig. OS29—When removing input shaft use cap screws as shown to hold shifter coupling in place.

HELICAL GEAR TRANSMISSION INPUT SHAFT

29. To remove the helical gear transmission input shaft, proceed as follows: Remove components necessary to gain access to the input shaft. In cases where tractor is equipped with "Power Booster" or "Reverse-O-Torc," this will include removal of the engine. Remove the input shaft front bearing retainer and on tractors not equipped with a hydraulic unit, remove the power take-off drive shaft support. Remove the input gear oil trough and if so equipped, the Power Booster drive oil trough. Shift the center shifter rod to move the input shaft shifter collar to the rear position, then on tractors which are so equipped, rotate input shaft so input gear oil scoops will not strike the countershaft gear when shaft is moved forward. Place two 3/8 x 2½-inch cap screws and nuts, or their equivalents, between input shaft shifter coupling and the front of rear main frame. Thread nuts toward shifter coupling until cap screws and coupling are held firmly in place. See Fig. OS29. Use a shaft driver, or a suitable pusher and move input shaft forward until input gear almost con-

tacts the front countershaft gear, then slide input shaft drive collar rearward and expose snap ring. Lift snap ring from groove and slide same rearward on shaft.

Note: It may be necessary to rework a pair of snap ring pliers so that jaws are thin enough to lift snap ring.

Continue to move input shaft forward until input shaft front bearing is free from the rear main frame and the input shaft is free of the rear bearing. Now remove the input shaft, rear bearing, gear mounting sleeve and gears from the rear main frame. Be careful not to drop the input shaft rear snap ring into the transmission. Further disassembly of input shaft is obvious.

Reassemble by reversing the disassembly procedure. Adjust input shaft end play to 0.001-0.003 by varying shims under input shaft bearing retainer. Shims are available in thicknesses of 0.004, 0.007 and 0.0149.

TRANSMISSION INPUT GEAR

30. Starting at tractor serial number 87 135, a new low range input gear has been installed. The new gear is fitted with a nylatron (plastic) bushing which has a high degree of self-lubrication. The bushing is an integral part of the gear and cannot be purchased separately. When using the new type gear, the oil scoops are not required, however, oil scoops will continue to be furnished to service the early type gear.

Low range input gear can be renewed after removing input shaft as outlined in paragraph 29.

DIFFERENTIAL

31. Riveted differential assemblies will no longer be furnished for repair. Bolts with self-locking nuts have been released for field repair for model 880 tractors.

Should loose rivets be encountered, drill out same and install bolts and nuts. Torque nuts to 110-115 ft.-lbs. Please make a note of this in paragraph 562 on page 65 in manual O-10.

BRAKES

32. Beginning with 770 tractor serial number 71 319, and 880 tractor serial number 73 639, increased ratio brake assemblies have been installed. These brake actuating assemblies are identified by a "5" stamped on the actuating lug. When servicing brakes on these tractors, do not intermix the early and late type assemblies. When installing the late type brake actuating assemblies in early tractors, it will be necessary to grind out the inside of the existing brake housings to provide operating clearance for the longer actuating lug, or install new type brake housings.

POWER TAKE-OFF

33. A 1000 rpm external power take-off unit is available for all Fleetline 77, 88; Super 77, 88; 770 or 880 tractors. One unit is geared for the 1600 rpm rated engine speed tractors and one is geared for the 1750 rpm rated engine speed tractors. This unit is used in conjunction with the 540 rpm unit already on tractor. Install unit as follows:

34. Remove pto lower housing and catch oil, then scrape off gasket and clean mounting surface. Remove pto shield. Unbolt and remove pto clutch cover, control handle and rubber gasket. Remove cap screws from pto shaft bearing retainer and pull pto clutch and shaft from tractor. To provide ample working room, remove the drawbar vertical adjustment bolts and lower drawbar to floor.

Install two 0.004, one 0.007 and one 0.0149 shims on the 1000 rpm unit mounting surface and install the unit on the lower side of the 540 rpm pto housing. Tighten cap screws to 25-30 ft.-lbs. Check the backlash between the large gear in 540 rpm pto and the idler gear in the 1000 rpm unit as follows: Remove filler hole plug from 1000 rpm unit, place a small brass rod through filler hole and lock (wedge) idler gear in place. Mount a dial indicator to the 540 rpm pto housing and position contact button of indicator against side of tooth on 540 rpm pto large gear. Check amount of backlash by turning external shaft of 540 rpm pto. Backlash should be 0.009-0.015. If backlash is not as specified, remove 1000 rpm unit and add shims to increase, or remove shims to decrease the backlash.

Install the auxiliary pto shield in the master pto shield 5½ inches below top of master shield and ¾-inch forward of rear side.

Reassemble pto and drawbar and fill to level of filler plug of 1000 rpm unit housing with SAE 10W-30 oil mixed with Oliver additive in the ratio of 16 parts oil to 1 part additive.

Note: When tractor is used frequently at temperatures below 0° F use SAE 5W-20 oil, plus additive.

HYDRAULIC LIFT SYSTEM

35. Beginning with Hydra-Lectric unit number 66 471, the solenoid rails were changed in that the angular grooves are cut at a 30 degree angle instead of the 45 degree previously used. Only the late type rails will be furnished for service.

OLIVER

Models ■ 99 (6 Cyl.) 4 Speed
■ Super 99 (6 Cyl.) 6 Speed
■ Super 99GM (3 Cyl.) 6 Speed

Previously contained in I & T Shop Service Manual No. 0-7

SHOP MANUAL

OLIVER

MODELS 99 (6 Cyl.), SUPER 99 (D & HC), SUPER 99 GM

IDENTIFICATION

On 6 cylinder models tractor serial number is on plate aft of starting motor. Engine serial number stamped on right rear flange. On Super 99 GM tractor serial number is on plate at left side of clutch housing cover. Engine serial number is on plate right side of rocker arms cover and stamped on block upper left side.

The non Super models are equipped with a four speed transmission, all Super series have six speed transmissions.

INDEX (By Starting Paragraph)

CONDENSED SERVICE DATA

GENERAL

	99	Super 99		
	HC & D	HC	D	GM
Engine Make	Own	Own	Own	G.M.
Engine Model	....	....	....	3023
Cylinders, Number of	6	6	6	3
Cylinder Bore—Inches	4.0	4.0	4.0	4¼
Stroke—Inches	4.0	4.0	4.0	5.0
Displacement—Cubic Inches	302	302	302	213
Compression Ratio—Non-Diesel	6.2	6.2	....	....
Compression Ratio—Diesel	15.5	....	15.5	17.0
Pistons Removed From	Above	Above	Above	Above
Main Bearings, Number of	4	4	4	4
Main & Rod bearings, Adjustable?	No	No	No	No
Cylinder Sleeves	Dry	Dry	Dry	Dry
Forward Speeds	4	6	6	6

TUNE-UP

	99 HC & D	Super 99 HC	Super 99 D	Super 99 GM
Firing Order	1, 5, 3, 6, 2, 4	1, 5, 3, 6, 2, 4	1, 5, 3, 6, 2, 4	1, 3, 2
Valve Tappet Gap—Inlet	0.009 Cold	0.009 Cold	0.009 Cold	None
Valve Tappet Gap—Exhaust	0.016 Cold	0.016 Cold	0.016 Cold	0.009H
Valve Face and Seat Angle	45	45	45	30
Generator, Distributor and Starter, Make	Delco-Remy	Delco-Remy	Delco-Remy	Delco-Remy
Ignition Distributor Model	1111731	1111731	None	None
Ignition Magneto Make, Optional	Wico	Wico	Wico	None
Ignition Magneto Model, Wico	XVD2291	XVD2291	None	None
Distributor Breaker Gap	0.022	0.022	None	None
Magneto Breaker Gap	0.015	0.015	None	None
Injector Timing	Refer to paragraph 36			par. 90
Distributor Timing High Idle	28°B	28°B	None	None
Flywheel Mark Distributor High Idle Timing	IGN	IGN	None	None
Flywheel Mark Distributor Retard Timing	TC	TC	None	None
Flywheel Mark Magneto Running Timing	MAG	MAG	None	None
Distributor Governor Advance Curve	Refer to paragraph 52			None
Spark Plug Make	A-C, Champion or Auto-Lite			
Plug Model for Normal Duty	A-C 88S Comm. or equivalent			
Injection Pump, Make and Type	Bosch PSB	Bosch PSB	Bosch PSB	G.M. 70
Nozzle Opening Pressure	1750	No	1750	350-850
Plug Electrode Gap	0.025	0.025	None	None
Carburetor Make	Marvel-Schebler		None	None
Carburetor Model	TSX477 or	TSX581	None	None
Carburetor Calibration and Parts	Refer to paragraph 29A			
Engine Low Idle RPM—Battery Ignition	350	350	....	....
Engine Low Idle RPM—Diesels	600	....	600	500
Engine High Idle RPM—Non-Diesels	1925	1925	....	....
Engine High Idle RPM—Diesels	1840	....	1840	1840
Engine Governed RPM—Loaded	1675	1675	1675	1675
Belt Pulley RPM—Loaded	1000	980	980	980

SIZES—CAPACITIES—CLEARANCES

(Clearances in Thousandths)

	99 HC & D	Super 99 HC	Super 99 D	Super 99 GM
Crankshaft Journal Diameter	2.625	2.625	2.625	3.500
Crankpin Diameter Prior Engine Serial 964432	2.250	2.250	2.625	....
Crankpin Diameter Engine Serial 964432 and up	No	2.625	2.625	2.750
Connecting Rod Center to Center Length—Inches	6¾	6¾	6¾	10⅛
Camshaft Journal Diameter—Nominal	1¾	1¾	1¾	1½
Piston Pin Diameter	1.250	1.250	1.250	1.500
Valve Stem Diameter—Nominal	⅜	⅜	⅜	11/32
Compression Ring Width	⅛	⅛	⅛	⅛
Oil Ring Width	3/16	3/16	3/16	3/16
Main Bearings Clearance, New—Diesel	Refer to paragraph 24D			1.5-3
Main Bearings Clearance, New—Non-Diesel	Refer to paragraph 24D			....
Main Bearings Clearance, Maximum	4.5	4.5	4.5	6
Rod Bearings, Clearance, New—Diesel	Refer to paragraph 23A			1.5-3
Rod Bearings Clearance, New—Non-Diesel	Refer to paragraph 23A			....
Rod Bearings Clearance, Maximum	2.5	2.5	2.5	6
Piston Skirt Clearance	5-10 pounds checked with 0.003 feeler			4-9
Camshaft End Play	Spring Loaded			4-18
Crankshaft End Play	5-12	5-12	5-12	4-18
Camshaft Bearings Running Clearance	Refer to paragraphs 19A and 67B			
Cooling Systems—Gallons	5¼	5½	5½	5½
Crankcase Oil, Refill—Quarts	6	6	6	11
Crankcase Oil, if Filter is Changed	7 or 8	7	8	13
Transmission and Differential—Quarts	48	32	32	32
Hydraulic System, Maximum—Quarts	9	9	9	9
Add for Belt Pulley and PTO—Quarts	....	4	4	4

TIGHTENING TORQUES

(In Foot-Pounds)

	99 HC & D	Super 99 HC	Super 99 D	Super 99 GM
Oil Line Cylinder Head Bolts	96-100	No	96-100	....
Cylinder Head, Non-Diesel—Foot-Pounds	91-100	91-100	No	No
Cylinder Head, Diesel—Foot-Pounds	112-117	No	112-117	165-175
Main Bearings, 6 Cylinder—Foot Pounds	108-112	108-112	108-112	No
Main Bearings, Bolts—Foot-Pounds	No	No	No	180-190
Connecting Rods—⅜ Bolts	44-46	44-46	44-46	No
Connecting Rods—7/16 Bolts on 6 Cylinder	87-92	87-92	87-92	65-75
Manifold Nuts and Rocker Shaft Brackets	25-27	25-27	25-27	....
Flywheel	67-69	67-69	67-69	150-160

STEERING GEAR

1. All models are equipped with a Saginaw recirculating ball type gear basically similar to the unit used on the 4 cylinder, series 99 tractors as shown in Fig. 0450. As will be seen from the illustration, the teeth cut on the sector shaft (70) are tapered. The sector operating nut (71) rides on steel balls interposed between the nut and the wormshaft. Power steering of the linkage booster type is available. Construction of the gear unit and the method of adjusting the unit are the same with and without power steering.

1A. **ADJUST.** On tractors having 6 speed transmission, it is necessary to remove the steering gear or the transmission shifter tower when making a complete adjustment of the steering gear. On tractors having 4 speed transmission, it is not necessary to remove the gear or disturb the shifter tower.

2. WORM END PLAY. This adjustment is controlled by the threaded plug (B). To adjust worm end play, loosen the locknut (A) and rotate the adjuster nut (B) clockwise until there is no perceptible end play and no binding. Recheck the adjustment after tightening the lock nut.

2A. MESH (BACKLASH) ADJUSTMENT. This adjustment is controlled by the lash adjuster screw (C), shown in Fig. 0450 which, on 6 speed tractors, cannot be manipulated without first removing the gear housing from the transmission shifting tower or the tower from the gear housing. Gear unit can be removed from tower by removing the three attaching cap screws after disconnecting throttle rod and removing drop arm from gear. The mesh (backlash) adjustment should be made **after** the wormshaft end play has been adjusted. Place steering gear in mid-position (exactly half way between full left and full right turn position) and loosen the lock nut (D). Rotate lash adjuster (C) clockwise until a very slight drag is felt only when the steering wheel passes through the mid or central position. Wheel should revolve freely at all other points in its rotation. Gear is correctly adjusted when 16 to 20 inch pounds of torque at steering wheel rim is required to pass through the mid-position, wheels off floor, linkage disconnected.

Note: Backlash adjusting screw (C) should have from zero to 0.002 end play in sector shaft. If end play exceeds 0.002, it will prevent correct adjustment; in which case, the sector cover (76) should be removed and shim washers (S) added at head of adjuster screw to remove the excess backlash.

Caution: Do not turn steering wheel hard against stops when drag link is disconnected as ball guides may be damaged in doing so.

Fig. 0449 — Steering backlash adjustment screw (C) is not accessible until gear unit is separated from gear shifter tower (T).

3. **R & R AND OVERHAUL.** To remove the steering gear unit, disconnect throttle rod at gear, remove drop arm from sector shaft and the three cap screws holding the unit to the shifter tower. Lift gear unit out from above.

Procedure for bench overhaul of the gear unit is as follows: Remove sector shaft arm and cap screws from sector cover (76) and withdraw sector and cover as a unit from housing.

Remove worm (screw) shaft cover (E) and withdraw worm shaft through

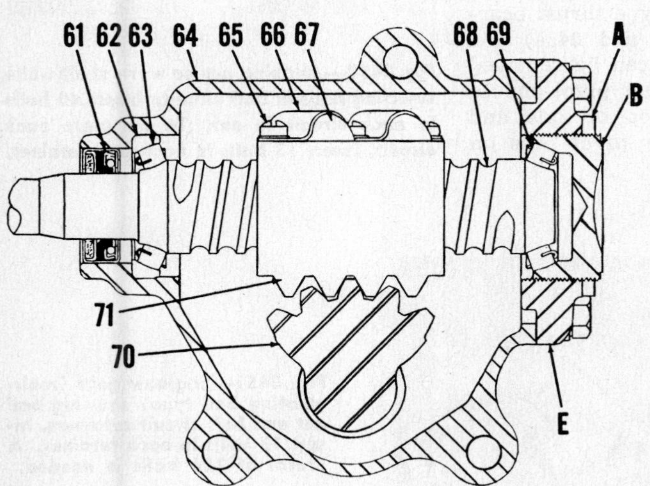

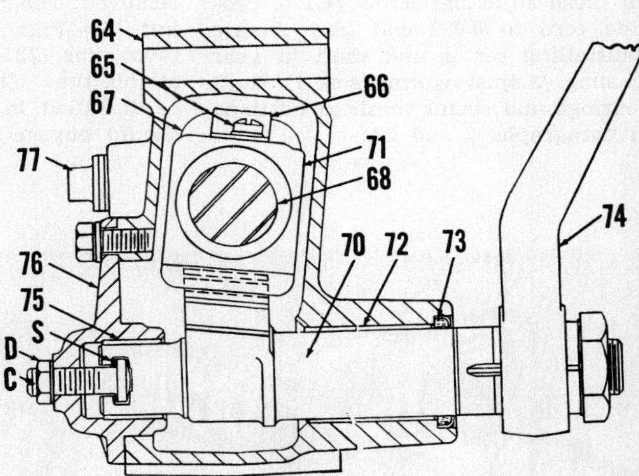

Fig. 0450—Saginaw steering gear (recirculating ball type) as used on later production series 90 and 99 tractors.

A. Lock nut
B. Worm shaft bearing adjuster nut
C. Lash adjuster
D. Lash adjuster lock nut
S. Shim washers
61. Seal
62. Bearing cup
63. Bearing cone
64. Gear housing
65. Clamp retaining screw
66. Retainer clamp
67. Retainer
68. Worm and shaft
69. Bearing cone
70. Sector and shaft
71. Ball nut
72. Bushing (large)
73. Seal
74. Sector shaft arm
75. Bushing (small)
76. Sector cover
77. Lubricant filler plug

this opening. Worm shaft bearing cup (62) and/or oil seal (61) can be renewed at this time.

Ball nut (71) should move along grooves in wormshaft smoothly without lumpiness and with minimum end play. If worm shows signs of wear, scoring or other derangement, it will be advisable to renew the worm and nut as a unit. The nut can be disassembled for inspection by removing the ball retainer clamp (66) and the retainers (67).

To reassemble ball nut, place nut over middle section of wormshaft as in Fig. 0452. Drop bearing balls into one retainer hole in nut and rotate worm shaft slowly to carry balls away from the hole. Continue inserting balls into hole until 40 balls are in each circuit and circuit is full to bottom of both holes. If end of worm is reached before 40 balls are in circuit, hold balls in position with a clean blunt rod as shown while shaft is turned in opposite direction, and then drop remaining balls in hole. Make certain that no balls are outside regular ball circuits. If balls remain in groove between two circuits or at ends, they cannot circulate and will cause gear failure.

Next, fill each ball retainer, Fig. 0451, with 13 balls and plug retainer ends with heavy grease; then, insert retainer units in worm nut and install retainer clamp. Exactly 106 balls are used in the circuit.

Sector shaft large bushing (72—Fig. 0450) has an inside diameter of 1.375 inches. Other sector shaft bushing (75) has an inside diameter of 1.0625 inches. I&T suggested clearance of shaft in bushings is 0.0015-0.003.

Select and insert shims (S) on sector mesh adjusting screw (C) to provide zero to 0.002 end play, before reinstalling sector and shaft in gear housing. Adjust worm (screw) shaft bearings and sector mesh as outlined in paragraphs 2 and 2A.

FRONT AXLE SYSTEM

SPINDLES

(Applies to tractors serial 5193000 and up. Refer to manual 0-2 for earlier tractors having conventional spindles.)

4. Front wheels are carried on "live" (rotating) spindles which revolve in tapered roller bearings mounted in the spindle carriers (16—Figs. 0454 and 0456A). The latter are supported by stationary pivots (23 and 24) anchored in the axle yokes and provided with needle type roller bearings (57) and separate thrust bearings.

5. Procedure for adjustment or renewal of front wheel bearings is conventional except that knuckle arms must be removed to obtain access to spindle nuts. Cups and cones may be renewed without removing the spindle carriers from the axle.

SPINDLE CARRIERS

(Applies to tractors serial 519300 and up.)

6. To remove or renew the spindle carrier pivot pins, proceed as follows: Remove front wheels and drive out the Rollpins (49—Fig. 0455) which retain the pivot pin in the axle yoke. Disconnect knuckle arm (17) from tie rod or remove from carrier. Remove grease fitting from pivot pins and extract same using a screw, nut and pipe as a puller as shown in Fig. 0456. Pivot pins may also be removed by using a slide hammer (inertia) type puller attached to a screw or rod on which has been cut a ⅛ inch pipe thread, threaded into the hole from which the grease fitting was removed. The roller type thrust bearing (58—Figs. 0456A and 0454) and pivot pins (23 & 24) can be renewed at this time. The longer pivot should be installed in the top of axle and should engage shorter pivot boss on the carrier. After carrier is installed, insert ½ pint of engine oil into same through pipe plug opening.

To remove the needle type pivot bearings, refer to paragraphs 6B and 6C.

6A. **SEALS, WEAR CUPS & WHEEL BEARINGS.** These parts can be renewed without removing spindle carriers from axle, after removing wheel spindles from carriers. Larger diameters of wheel bearing cups should all face in manner shown in **Fig.** 0454. Apply a thin coat of shellac or gasket cement to outer edge of metal portion of each grease seal. Do the same to the inner surface of the seal wearing cups (32). It is important that wearing surface of wearing cup be smooth and square with seal. Use

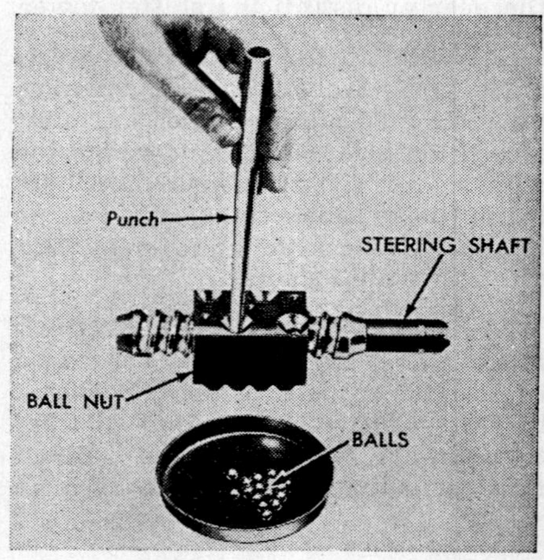

Fig. 0452—Aligning nut on worm shaft while inserting balls in ball circuit. Insert 40 balls in each circuit in nut. To complete each circuit, insert 13 balls in each ball retainer.

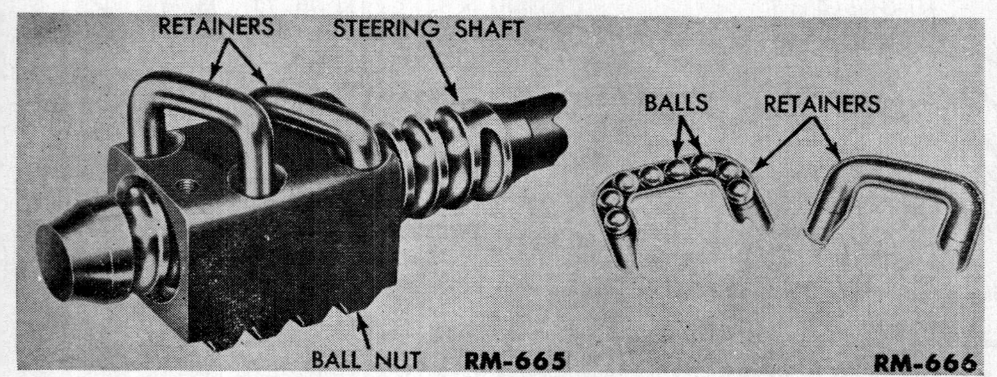

Fig. 0451—Saginaw gear (recirculating ball type) showing ball nut and ball circuit retainers. Insert 13 balls in each retainer. A total of 106 balls is needed.

crocus cloth or very fine sandpaper to remove slight nicks or foreign matter from wearing cup. The wearing cup and seal can be installed squarely without special tools if care is exercised, but Oliver Wear Cup Driver ST-124 and Seal Driver ST-123 are recommended.

6B. **PIVOT NEEDLE BEARINGS.** These bearings are of the closed end type and the Oliver Corporation states that a driver engaging only the top or the bottom of the bearing will damage it seriously. Recommended tool is the Owatonna Driving Mandrel No. 815 on which is placed the OTC No. 0-6 Driving Collar. The specified driver and collar combination applies pressure to upper edge of bearing to prevent distortion of same.

6C. To renew the needle roller type pivot bearings when spindle carrier is removed from axle, proceed as follows: Remove spindle (15) from carrier being careful not to damage the oil seal. Use one of the pivot pins (23) or a drift to drive the bearings out of the carrier. **Bearings will be damaged during removal, thus should not be removed unless new ones are available.** Using the OTC 815 driving mandrel with the 0-6 collar mounted on same, press or drift new bearings into carrier until top of each bearing is flush with carrier.

OTHER PARTS OF AXLE SYSTEM

7. Data in paragraphs 7A and 7B covers the important additional assemblies in the front axle steering system. Servicing of any other parts is conventional and self-evident after viewing exploded view, Fig. 0456A.

7A. **BOLSTER.** Front axle support (bolster) shown at (1) in Fig. 0456A, can be removed by unscrewing the retaining cap screw (7), bumping pivot (6) rearward and removing three bolts which attach bolster to front frame. Pivot is not bushed.

7B. **RADIUS ROD.** This item shown at (34) in Fig. 0456A can be removed without disturbing the front axle after removing two nuts from axle end of rod and unbolting stay rod ball socket (38) from lower side of front frame.

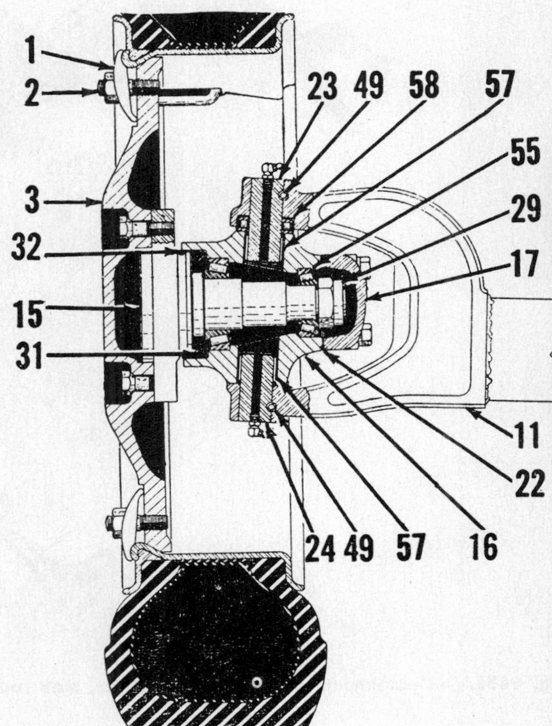

Fig. 0454 — Sectional view of "live" type front axle knuckle spindle.

1. Rim clamp
3. Front wheel
11. Front axle
15. Knuckle spindle
16. Spindle carrier
17. Knuckle arm
22. Gasket
23. Upper pivot
24. Lower pivot
29. Spindle nut
31. Seal
32. Seal wear cup
49. Roll pin
55. Wheel bearing inner
57. Needle bearing
58. Thrust bearing

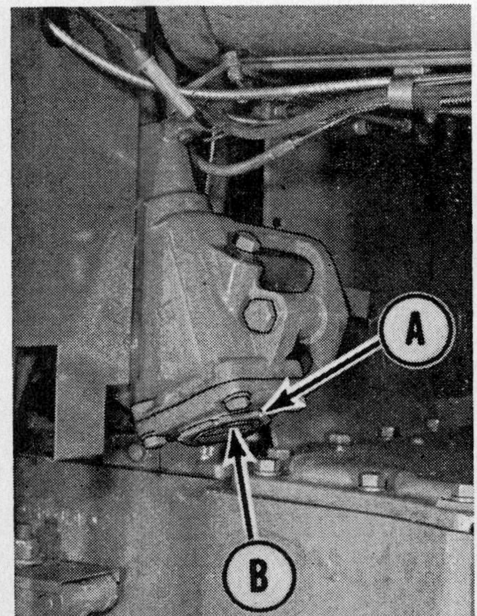

Fig. 0453—End play of steering gear wormshaft is controlled by the threaded adjusting plug (B) and lock nut (A). Both of these parts are also shown in left view in Fig. 0450.

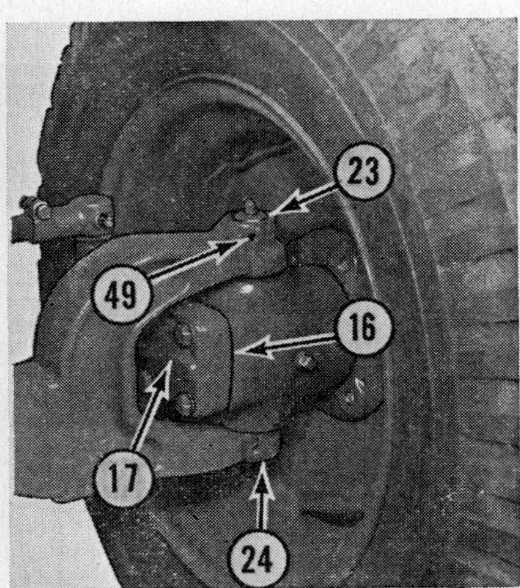

Fig. 0455—Center point easy steering is obtained by this "live" type wheel spindle and carrier. Refer to Fig. 0454 for details.

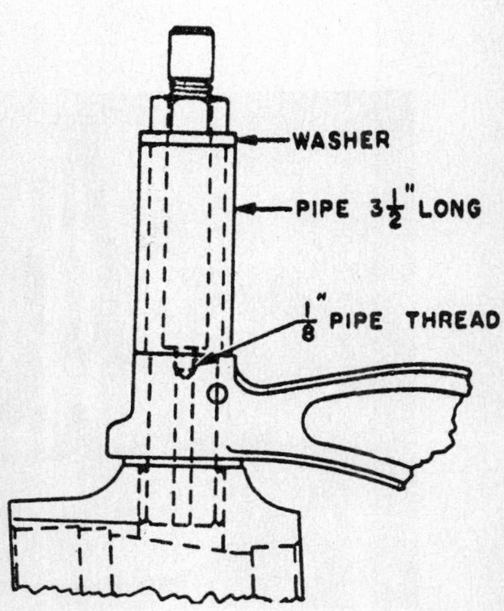

Fig. 0456—The ⅛ pipe thread in each knuckle pivot pin can be the attaching point for a pulling rig in removing pivot pins from axle.

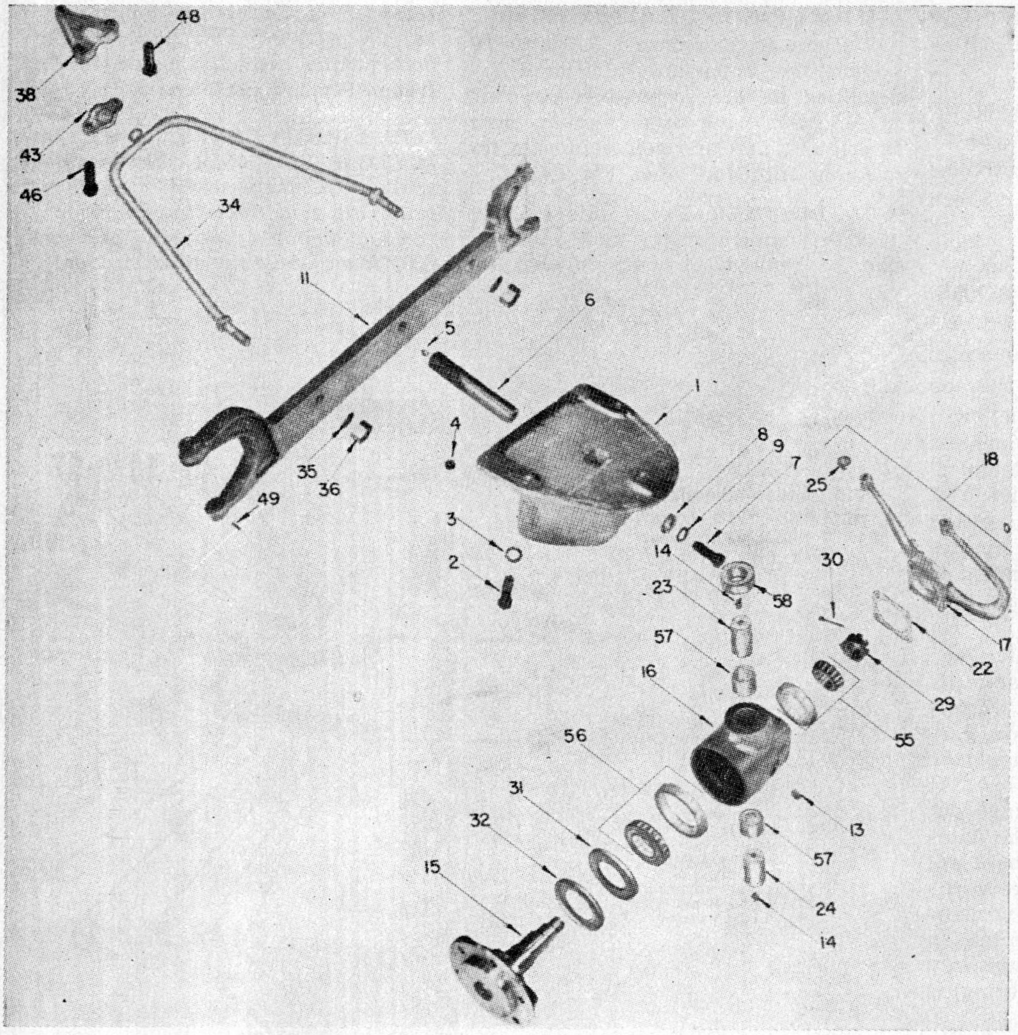

1. Front bolster
6. Front axle pivot
11. Front axle
15. Steering spindle
16. Spindle carrier
18. Knuckle arm
22. Gasket
23. Upper pivot
24. Lower pivot
25. Knuckle arm ball
26. Tie rod
28. Front wheel
31. Seal
32. Wearing cup
34. Stay rod
38. Ball socket
43. Ball cap
45. Tie rod end
49. Roll pin
55. Wheel bearing
56. Wheel bearing
57. Needle bearing
58. Roller thrust bearing

Fig. 0456A—Components of front axle, bolster, stay rod and one wheel spindle assembly. Spindle (15) is "live" type and rotates in spindle carrier (16).

Fig. 0456B—Right view shows live front wheel spindle being removed from carrier. Spindle wearing cup shown at left must be square with grease seal and free of nicks or scoring

ENGINE ASSEMBLY (6 CYLINDER)

(Three cylinder type begins with paragraph 60.)

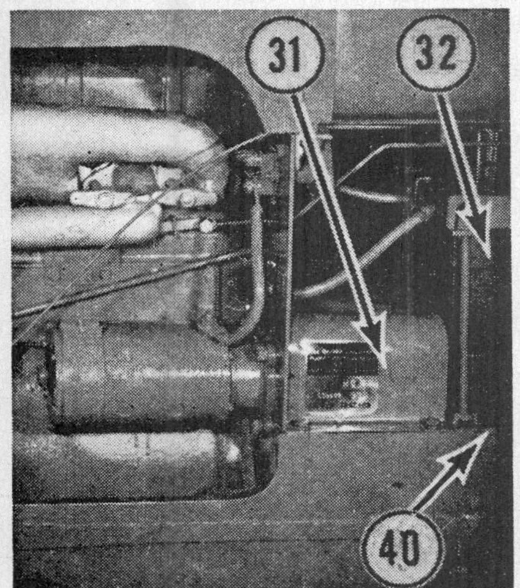

Fig. 0457—On tractors with 6 speed transmission access to clutch shaft coupling is by removal of battery, front cover (31) and flat dust cover (40).

REMOVE AND REINSTALL

10. To remove 6 cylinder engine assembly from tractor, proceed as outlined in paragraphs 10A through 10F.

10A. HOOD. Remove hood as follows: Loosen front grille center strap bolts, and remove two cap screws attaching front grilles to support and lift off grilles. Detach both headlights (each held by two screws) and disconnect wires. Remove hood straps, precleaner, muffler, and wiring harness clips from right side of hood. Lift off hood.

10B. PTO DRIVE SHAFT. If tractor is equipped with pto, remove the drive shaft of same as outlined for continuous type in paragraph 145J.

10C. FUEL TANK. Remove battery or batteries. Disconnect upper and lower radiator hoses and hour meter cable. Remove coolant temperature sending unit and air cleaner. On Diesel, remove leak-off line connecting the injectors to the tank. Disconnect fuel line at tank. Remove the four nuts attaching tank to bracket, three nuts attaching throttle and wiring harness to bottom of tank. Lift tank from tractor.

10D. Remove injection pump stop cable from fuel tank front bracket. Disconnect wiring harness at generator regulator and right headlight. Pull harness out of fuel tank front bracket. Remove throttle rod after disconnecting at rear of vertical rod crank and at injection pump.

Fig. 0457A—On tractors with 4 speed transmission access to clutch shaft coupling can be obtained without removing the belt pulley carrier (BC).

Remove the clutch housing front cover (31—Fig. 0457 and 0457A) after removing the cap screws attaching it to front frame. If tractor has 6 speed transmission, remove the flat dust cover (40—Fig. 0457) by removing the screws attaching it to frame and to the clutch release bearing carrier. On 4 speed tractors, remove screws retaining release bearing carrier to the carrier support.

10E. CLUTCH SHAFT AND COUPLING. Refer to Fig. 0458. Remove chain (22) after extracting master link. Slide clutch shaft coupling (21) forward on shaft after loosening coupling clamping bolt. Disconnect outer end of the clutch shifter shaft, and loosen the set screw which retains shifter fork (26) to shifter shaft then bump shifter shaft out of fork toward left side of tractor. Withdraw clutch shaft (23) from clutch.

10F. LIFTING. Remove the two front and two rear engine mounting bolts which may be provided with aligning shims. Be careful not to mix these shims. Remove fan blades to prevent damaging the radiator core. Attach a lifting bracket to the two special hoisting studs located on top of engine cylinder head between exhaust manifold and rocker cover. Hoist engine out of front frame.

10G. **REINSTALL AND ALIGN.** Install felt strips at locations where engine rear adapter plate contacts main frame. Engine to transmission alignment should be checked whenever a new front frame or engine is installed or premature main clutch failures have occurred. Dial indicator rig shown in Fig. 0460 can be made up from rod and bar stock. Make first check with bottom of indicator in contact with bore wall of clutch pilot bushing. If total indicated runout exceeds 0.015, vary the shims at front and rear engine mounts to bring the reading within the 0.015 limit. Shims are available in thicknesses of 0.005 and 0.008.

After engine alignment has been corrected, check face runout of flywheel with gauge hooked up as shown in Fig. 0460. If total indicator reading exceeds 0.015, check for a bent crankshaft mounting flange.

CYLINDER HEAD

Six Cylinder

11. Herewith is procedure for R & R of head: Remove hood assembly as outlined in paragraph 10A. Remove air cleaner pipe, air cleaner and upper radiator hose. Remove two cap screws attaching pump by-pass to thermostat housing. Remove rocker arms cover, rocker oil feed line, rocker arms and shaft assembly and long pushrods. On Diesels, remove all pump to injectors pipes, leak-off pipe and the fuel return line connected to No. 6 injector. Disconnect the ether primer line at inlet manifold. On all models, disconnect temperature gauge sending unit and hour meter cable. Remove cap screws retaining head to block.

Retorque head screws after engine has operated a few hours. Screws can be retorqued without removing rocker shaft assembly by using ¾ inch crowfoot. Note that the drilled cap screw used for oil supply requires less tightening torque. Head screws on HC gasoline engines can be torqued to Diesel values if washers 1C-111 used on Diesel head retaining screws are installed.

Drilled (Oil Supply)
Head Screw ....95-100 Ft.-Lbs.
Other Head Screws—
Non-Diesel 91-100
Other Head Screws—
Diesel 112-117
Inlet Tappet Gap.....0.009 Cold
Exhaust Tappet Gap..0.016 Cold

Fig. 0458—Clutch to transmission connections. Dust cover (40-Fig. 0457) is anchored to front frame (F) and to release bearing carrier support (29). The pto drive shaft on six cylinder models with continuous type pto passes through hollow clutch shaft (23) and is splined into flywheel. Bolt (28B) should be adjusted to provide ½" maximum release travel.

Fig. 0458A—Manifold side of 6 cylinder Super 99 gasoline engine.

VALVES, SEATS AND GUIDES

Six Cylinder

12. Intake valves are provided with seals to prevent oil flow into combustion chambers via the valve stems. Install new seals each time the valves are reseated. Exhaust valves seat on renewable inserts in the cylinder head and are provided with rotators on non-Diesel models only. Shoulder type valve guides should stand with respect to top machined face of cylinder head as indicated in table below. Guides may require one pass with reamer after installation to remove burrs or high spots. Inlet and exhaust guides are not interchangeable. Neither the valves nor the guides for non-Diesels are interchangeable with those used on Diesels. Refer also to information in table below.

Stem Diameter,
Nominal 3/8
Intake Clearance
in Guides 0.0015-0.0035
Exhaust Clearance
in Guides 0.0025-0.0045
Renew If Clearance Exceeds:
Inlet 0.0055
Exhaust 0.0075
Guide Height Above Machined
Top Surface Cylinder Head:
Non-Diesel, Inlet . . 3/4 inch
Non-Diesel, Exhaust 27/32 inch
Diesel (below top
surface) . . . . . . 1/64 inch

VALVE ROTATORS

Six Cylinder Non-Diesel

13. Exhaust valves are fitted with Roto-cap positive rotators which require no maintenance. Observe when engine is running, to make certain that each exhaust valve rotates slightly. Renew the rotator of any exhaust valve which fails to rotate, as individual parts are not available for repairs. Exhaust valve springs used with rotators are of a different length than those used on inlet valves of same engine.

VALVE SPRINGS

Six Cylinder

14. On Diesel engines, the inlet and exhaust springs are the same, whereas on the non-Diesel, the exhaust springs are shorter, due to the use of the valve rotators. Renew any springs which are rusted or pitted or which do not conform with these specifications: On Diesel engines, exhaust and intake spring (part No. 1K-198, which supersedes and is interchangeable with part No. 1KA-198) should require 65-77 pounds to compress spring to length of 1 19/32 inches. On non-Diesel engines, equipped with valve rotators, intake springs (part No. 7AA-198) should show 114-134 pounds at 1 15/16 inches length; exhausts (part No. 7A-198B) 122-132 pounds at 1 53/64 inches. Free length of Diesel springs is 2 25/32 inches; of non-Diesel inlet springs, 2 49/64 inches; exhaust, 2 11/16 inches.

VALVE TAPPETS OR LIFTERS (Cam Followers)

Six Cylinder

15. Valve lifters are of the mushroom type operating directly in the unbushed bores in the cylinder block. It is necessary to remove the camshaft as per paragraph 19, before the lifters can be removed. Lifters are furnished in standard size only.

Lifter Diameter 0.6243
Running Clearance
in Bores 0.0005-0.002
Renew Block and/or
Lifter When Clearance Exceeds 0.007

ROCKET ARMS AND SHAFT

Six Cylinder

16. Rocker arms are provided with non-renewable, steel backed, babbitt lined bushings. Rockers are of the wickless type; those shown in Fig. 0463 were used on tractors serial 518300 through 519144 and the type shown in Fig. 0462 on tractors begin-

Fig. 0458B—Injector nozzle side of 6 cylinder Super 99 Diesel engine.

Fig. 0458C—Spark plug side of 6 cylinder Super 99 gasoline engine. Opposite side is shown in Fig. 0458A.

ning with serial 519145. The later design is most efficient in preventing over-oiling of valve stems and consequent excessive oil consumption. Later type is interchangeable on same rocker shaft as early type, but must be changed in complete sets, not individually. Parts numbers for late rockers are IKS206D left hand; IKS-206E right hand.

Rocker Shaft Diameter	0.742-0.743
Reject Shaft If Less Than	0.740
Running Clearance	0.0015-0.0035
Renew If Clearance Exceeds	0.005
Contact Button Radius	⅜ inch
Tappets Gap, Inlet Cold	0.009
Tappets Gap, Exhaust Cold	0.016
Brackets Tightening Torque	25-27 Ft.-Lbs.

TIMING GEAR COVER AND SEAL

Six Cylinder

17. To remove timing gear cover, it is necessary to first remove the hood and the radiator as follows: Loosen front grille center strip bolts and remove cap screws attaching grilles to support. Detach both headlights and disconnect wires from same. Remove hood straps, precleaner, muffler, and two wiring harness clips from right side of hood. Lift off hood.

17A. To remove radiator after hood is off, remove the two cap screws from each side which attach the radiator shell to the front frame. Also, remove the radiator to front frame screws and baffle to radiator shell screws.

17B. To remove timing gear cover when radiator is off, remove the hydraulic pump or blank plate from right side of gear cover. Remove generator, fan belts, crankshaft pulley and pulley felt seal. On Diesels, remove the bolts which attach the injection pump to the timing gear cover. The inner one of these bolts is provided with a copper sealing washer. Remove lubricating oil drain line from bottom of injection pump and from cylinder block. Disconnect the fuel line connecting the secondary filter to the pump.

Remove cap screws attaching timing gear cover to engine and the three which attach it to the front of oil pan. Remove water pump. A copper sealing washer is used under the head of the lower left pump attaching screw. Disconnect water by-pass from thermostat housing and lift off the timing gear cover.

17C. The treated cork oil seal (33—Fig. 0461) can be renewed at this time. Soak new seal in engine oil before installing.

While cover is off, check condition of thrust washer and the spring loaded camshaft thrust button in end of camshaft. On Diesel engines, an additional thrust button and washer set up is provided for the idler gear as shown at (49) in Fig. 0464.

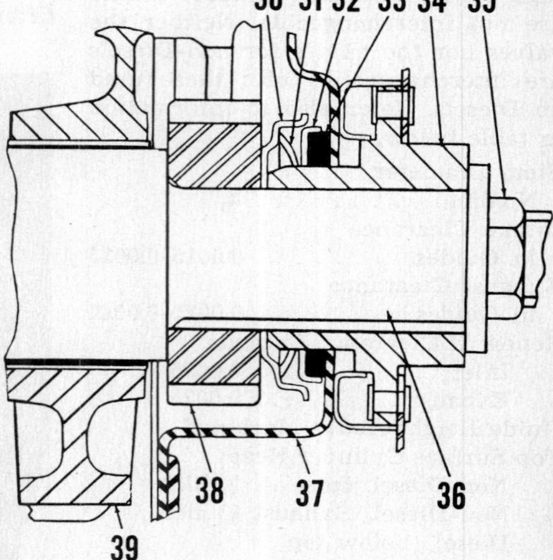

Fig. 0461—Crankshaft front oil seal of treated cork (33) can be renewed after removing the timing gear cover. Parts (30), (31), (32) and (33) are available only as an assembly.

30. Seal housing
31. Spring
32. Seal retainer
33. Cork seal
34. Crankshaft pulley hub
35. Pulley retaining nut
36. Crankshaft
37. Timing gear cover
38. Crankshaft gear
39. No. 1 main bearing cap

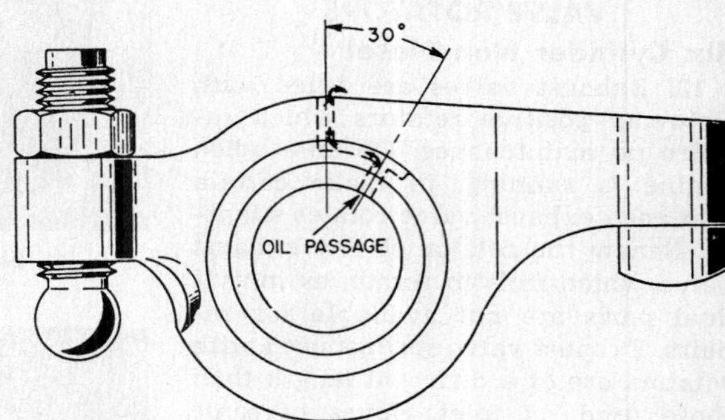

Fig. 0462—Latest rocker arm (IKS206E and F) has better oil control than type shown in Fig. 0463.

Fig. 0460—Checking alignment of flywheel face with transmission input shaft. Misalignment greater than .015 vertical, is corrected by varying the engine rear mounting shims.

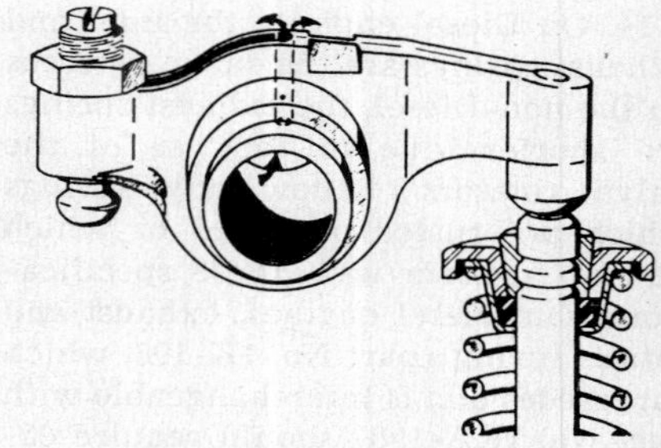

Fig. 0463—Beginning with engine 519145 this rocker arm is superseded by one shown in Fig. 0462.

TIMING GEARS

Six Cylinder

18. Timing drive on Diesel engines consists of three helical gears and the injection pump gear; on non-Diesels, two helical gears are used. The cylinder block for all models, however, is designed for the installation of an idler gear as shown in Fig. 0464. For all non-Diesel models, check to make certain that a plug (42—Fig. 0465) is installed in what would normally be the idler shaft bushing bore on the Diesel engines.

18A. Remove camshaft gear from shaft as shown in Fig. 0466, by using a puller attached to two ⅜-16 cap screws which can be threaded into gear. Crankshaft gear can be removed in a similar manner. Avoid pulling the gears with pullers which clamp or pull on the gear teeth.

The camshaft and crankshaft gears are stamped either "S" (standard), "O" (oversize), or "U" (undersize), and amount of oversize or undersize. Recommended backlash of camshaft to crankshaft gear is 0.001-0.003. Renew gears when backlash exceeds 0.007.

When reinstalling the cam gear, remove the oil pan and buck-up the camshaft at one of the lobes near the front end of the shaft with a heavy bar. The crankshaft gear should be heated in oil to facilitate installation.

Reinstall gears with the "C" mark on crankshaft gear meshed with a similar mark on camshaft gear as shown in Fig. 0465.

18B. The idler gear (48—Fig. 0464) is installed only on Diesel engines and can be removed after removing timing gear cover. Running clearance of gear spindle in bushings should not exceed 0.004. When gear is in contact with front face of sleeve, idler gear face should be flush with crank gear face.

If clearance is greater than 0.004, or if flush condition is not obtained, renew the sleeve and bushings assembly as neither part is available separately.

In rare cases, it may be necessary to reface the thrust face of a new sleeve assembly in order to obtain the flush setting. Bushings in new sleeve may require final sizing to provide the desired 0.0015-0.002 running clearance.

CAMSHAFT

Six Cylinder

The camshaft front journal on all models rotates in a steel-backed, babbitt lined bushing. The three remaining journals rotate in machined bores in the cylinder block.

19. To remove camshaft, first remove timing gear cover as outlined in paragraph 17. Remove ignition distributor from non-Diesel engines, rocker arms and shaft assembly, engine oil pan, and oil pump. On Diesel engines, remove primary fuel pump from right side of engine. Remove long push rods and block-up or support tappets (cam followers). Thread shaft and gear forward out of block bores.

19A. Camshaft lobe lift for all models is 0.247. Shaft journal sizes are: Front, 1.7495-1.750; others, 1.7485-1.7495. Recommended running clearance of number one journal is 0.0015-0.003 with a maximum of 0.005; all others, 0.0025-0.0045 with a maximum permissible clearance of 0.007.

When the running clearance in unbushed bores exceeds 0.007, correction can be made by renewing camshaft and/or cylinder block, or by reboring the bores in cylinder block to a diameter of 1.8745-1.8755 inches and installing bushings. Presized service

Fig. 0466—Removing camshaft gear on non-Diesel engines by using a puller attached to two ⅜-16 bolts which can be threaded into gear. Diesel engine camshaft gear removal is similar.

Fig. 0464—Timing gear installation on Diesel engines. Mesh camshaft gear mark "C" with an identical mark on crankshaft gear. Note call-out "12" indicates Diesel injection pump outlet for No. 1 cylinder. Shown is early PSB pump and Oliver 77 engine but 6 cylinder Super 99 is similar.

1. Nozzle return line
10. Timing port
38. Crankshaft gear
44. Camshaft gear
45. Thrust button
47. Injection pump drive gear
48. Idler gear
49. Thrust button
50. Crankshaft engaging ratchet
51. Location of non-Diesel engine governor drive gear bushing and sleeve

Fig. 0465—Crankshaft and camshaft gear installation on non-Diesel engines. Mesh camshaft gear mark "C" with an identical mark on crankshaft gear. Diesel engines are similar. Plug "42" is installed on all non-Diesel engines.

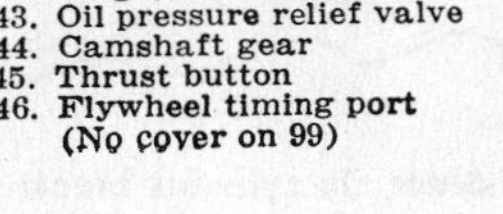

38. Crankshaft gear
40. Governor shaft sleeve
41. Governor shaft bushing
42. Plug (non-Diesel)
43. Oil pressure relief valve
44. Camshaft gear
45. Thrust button
46. Flywheel timing port (No cover on 99)

bushings as supplied for number one journal can be used and should be installed using a close fitting piloted driver.

Camshaft end play is controlled by a spring loaded thrust button. Thrust button spring has a free length of 1 3/16 inches and should have a pressure of 15.5-18.5 lbs. when compressed to a height of 25/32 inch.

If for any reason, the cam gear was removed from the shaft, the gear should be pressed on the shaft before installing the shaft in the engine.

ROD AND PISTON UNITS
Six Cylinder

20. On early production non-Diesels, the connecting rod lower end bearing split line is at a right angle to length of rod; on all Diesels and later production non-Diesels, the lower end bearing split line is diagonal and the bearing inside diameter is larger than on early non-Diesels.

Piston and connecting rod units are removed from above after removing the cylinder head as per paragraph 11 and the oil pan as in paragraph 20A.

20A. Oil pan is removed in the conventional manner, but difficulty arises in removing the two attaching cap screws at rear end of pan. A vertically mounted, stamped steel, dust shield interferes with the removal of these two rear cap screws. Rather than cut or distort the dust shield which is held to the engine adapter plate by two cap screws, remove the shield. This is accomplished by removing the clutch housing front cover (31—Fig. 0457) and the two screws which attach the dust shield to the engine adapter plate. Remove nuts from rod bolts and push piston and rod assemblies upward out of the cylinder block.

Torque the 7/16 inch bolts of diagonally split connecting rods to 87-92 foot pounds. Straight split type, with 3/8 inch bolts, should be torqued to 44-46 foot pounds.

PISTONS, SLEEVES AND RINGS
Six Cylinder

21. **PISTONS & SLEEVES.** Cast iron pistons are supplied only in standard size and are available for repairs as a set for one or more cylinders. Each set contains a piston, sleeve, piston pin and pin retainers.

Sleeves should be renewed when either of the following conditions exist: Taper, 0.008; out-of-round, 0.002; wear, 0.010.

Desired piston skirt clearance is checked with a spring scale pull of 5-10 lbs., using a 0.003 x ½ inch feeler gage. Wear limit of pistons and sleeves is when a 0.006 x ½ inch feeler gage requires less than a 5-10 lb. pull on a spring scale to withdraw it.

Before installing the dry type cast iron (alloy No. 210) sleeves, clean cylinder block surfaces, and apply a coat of oil-mixed aluminum powder (Thermo-Lock) to outside surfaces of sleeves and bores in cylinder block.

21A. Effective with engine serial No. 952651, square flanged sleeves (7A-178A—Fig. 0467) instead of beveled flange type 7A178 are factory fitted in all cylinder blocks.

Both types will be available for service. The later square flange type can be installed in cylinder blocks originally fitted with the older beveled type, if the cylinder bores are counterbored at their upper ends to the dimensions shown in Fig. 0469.

21B. **RINGS.** There are three compression rings and one oil control ring per piston. The lower compression ring is a scraper type. Factory supplied service rings are stamped "Top" and should be installed in this manner.

Recommended end gap for all compression rings is 0.015-0.025 with a reject value of 0.045; oil control ring, 0.010-0.020 with a reject value of 0.045. Recommended side clearance, compression rings, 0.0015-0.002 with a reject value of 0.006; oil control ring, 0.0015-0.003 with a reject value of 0.006.

PISTON PINS AND BUSHINGS
Six Cylinder

22. The 1.2495-1.2498 diameter full floating type piston pins are retained in piston bosses by snap rings and are available in oversizes of 0.005 and 0.010.

The split type graphite bronze piston pin bushings should be installed in the rod so that bushing outer edge is flush with outer edge of rod bore and the split side is at the top of the rod. Two bushings are installed in each rod and those for Diesels are longer than non-Diesel bushings. Bushings should be sized after installation to provide a 0.0005-0.001 clearance on pin, and the piston bosses sized to provide a 0.0002-0.0004 clearance in piston.

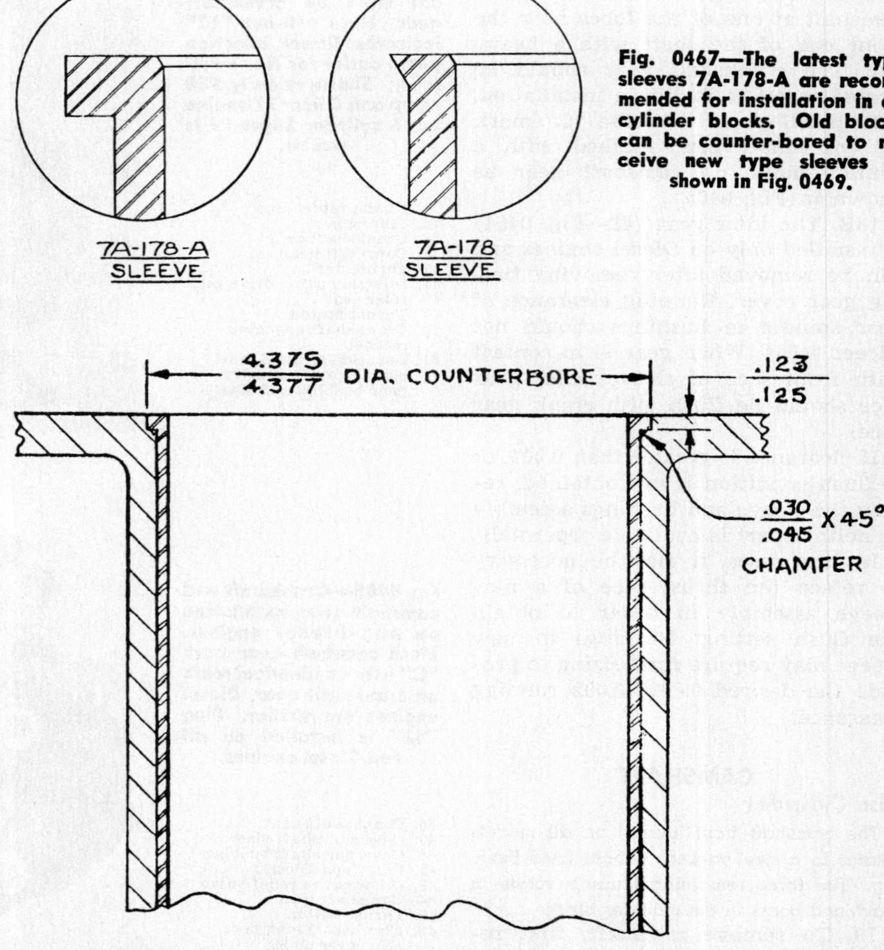

Fig. 0467—The latest type sleeves 7A-178-A are recommended for installation in all cylinder blocks. Old blocks can be counter-bored to receive new type sleeves as shown in Fig. 0469.

Fig. 0469—When older six cylinder blocks are counter-bored as shown, the new 7A-178-A sleeves can be installed.

CONNECTING RODS AND BEARINGS

Six Cylinder

23. Diesel connecting rods are split diagonally at the lower bearing parting line and are fitted with non-adjustable, copper-lead lined, precision bearing shells which ride on 2⅝ inch diameter crankshaft crankpins. Similar specifications apply to non-Diesel HC engines beginning with serial 964432.

Non-Diesel engines prior to serial 964432 have babbitt-lined precision type rod bearings which ride on 2¼ inch diameter crankpins and the lower bearing parting line is at a right angle to the length of the rod. The rod and bearing set up specified for later production non-Diesel engines can be installed in early production non-Diesel units if the later crankshaft, main bearings and pistons are also installed.

Refer to CRANKSHAFT AND BEARINGS for data on re-working older style Diesel connecting rods when they are to be used with new style Diesel crankshaft.

23A. Bearings are renewable from below without removing the rods from engine. Bearing shells for 2⅝ inch crankpins are furnished in one undersize of 0.003. For non-Diesels prior to engine serial number 964432 which have 2¼ inch crankpins, the available undersizes are 0.003, 0.030 and 0.033.

Crankpin Diameter HC prior 964432 . . . . . . 2.250
Crankpin, Diesel & HC after 964431 . . . . 2.625
Bearing Running Clearance, Prior 964432 . . . 0.001-0.0015
Bearing Running Clearance, HC after 964431 . . . . .0.0005-0.0015
Renew If Clearance Exceeds 0.003
Rod Bolt Torque ⅜ Inch Bolt . . . . . . 44-46 Ft.-Lbs.
Rod Bolt Torque 7/16 Inch Bolt. . . . . 87-92 Ft.-Lbs.

CRANKSHAFT AND BEARINGS

Six Cylinder

24. Tocco hardened crankshafts part number 7A-119 having four main journals and 2⅝ inch diameter crankpins are used in early production Diesel engines. In Diesels beginning with engine serial number 952769, except engines 952840 through 952887, and in non-Diesels (HC) beginning with engine serial number 964432, crankshaft number 7A-119A having offset crankpins but otherwise the same as 7A-119 is used. In non-Diesel (HC) engines prior to engine serial 964432, the crankshaft part number IK-119B, is not Tocco hardened and the crankpins are of 2¼ inch diameter.

24A. Latest crankshaft, number 7A-119A can be installed in early production Diesels by using later pistons number 7A-228A shown in Fig. 0471 or, with the older 7A-228 pistons if the upper ends of the connecting rods are re-worked as shown in Fig. 0470. In HC non-Diesels, latest 7A-119A shaft can be used in early production engines if the later connecting rods, main bearings, and pistons are also installed.

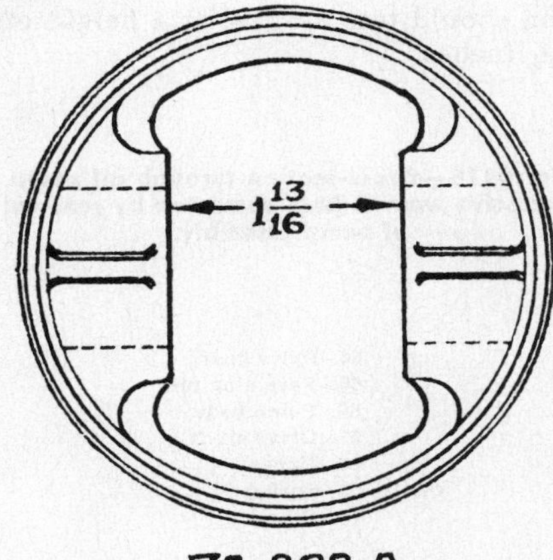

Fig. 0471—Latest piston for use with latest 7A-119-A crankshaft. Refer also to Fig. 0470.

24B. Rod and main bearings used with crankshafts having 2¼ inch crankpins are precision type renewable shell inserts lined with babbitt. Bearings used with shafts having 2⅝ inch diameter crankpins are of similar type but are lined with copper-lead. All bearings are of the non-adjustable type.

24C. All main bearings can be renewed from below without removing the crankshaft. Main bearings of 0.003, 0.030 and 0.033 undersize are supplied for shafts with 2¼ inch diameter crankpins. Bearings for the later Tocco hardened shafts with 2⅝ inch crankpins are available in only one undersize of 0.003.

24D. Check shaft and main bearings for wear, scoring and out-of-round.

Main Journal Dia. 2⅝
Crankpin Dia.-Early . . 2¼
Crankpin Dia.-Late . . . 2⅝
End Play Control . . . . No. 3 Brg.
Permissible Out-of-Round, Taper 0.0005
Running Clearance, Prior 964432 . . . . 0.002-0.0035
Running Clearance, After 964431 . . . . .0.0005-0.003
Renew If Clearance Exceeds 0.005
End Play 0.005-0.012
Mains, Tightening Torque—Ft.-Lbs. 108-112

CRANKSHAFT REAR OIL SEAL

Six Cylinder

25. Procedure for renewal of the one piece, treated cork oil seal shown in Figs. 0473 and 0476A is outlined in paragraphs 25A through 25D.

25A. On models with continuous (live) type pto the shaft is splined

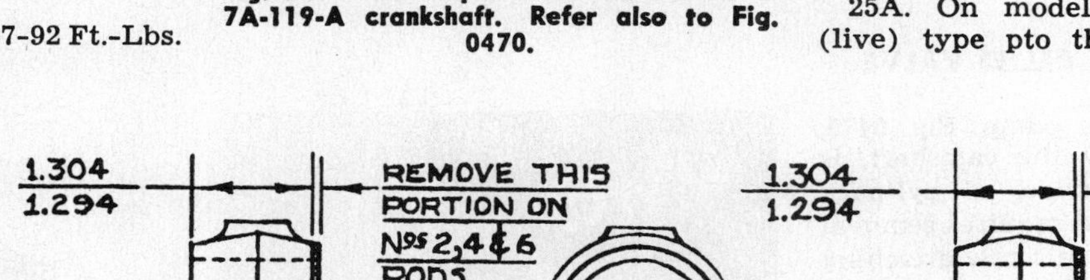

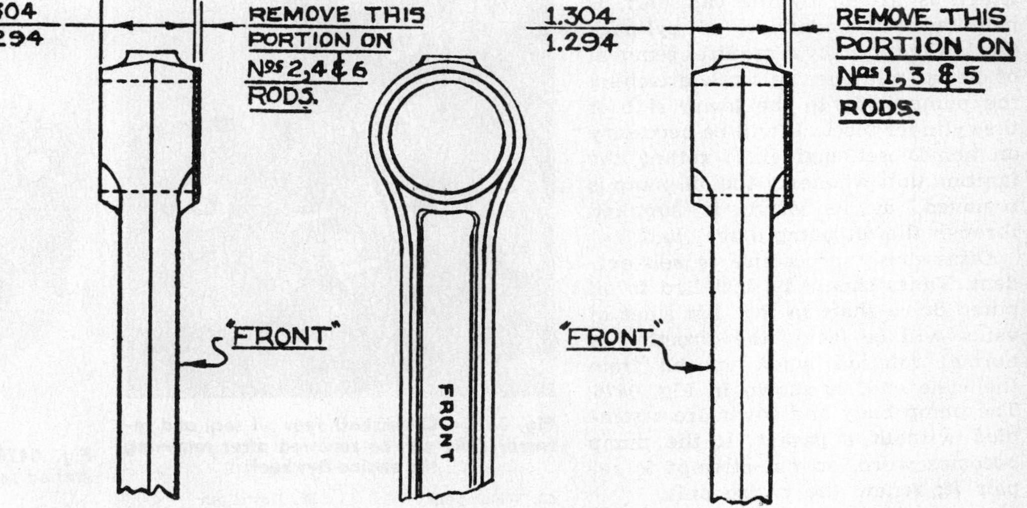

Fig. 0470—Old style Diesel connecting rods and pistons can be used with latest crankshaft if rods are reworked at upper ends as shown.

into the flywheel and should be removed as per paragraph 145J.

25B. Remove the clutch housing front cover (31—Fig. 0457) after removing 4 cap screws attaching it to front frame. Remove storage battery and the flat dust cover (40) which is attached to front frame by two screws and on models with spring loaded clutches, to clutch release bearing carrier by two additional screws. It is not necessary to remove belt pulley carrier on 4 speed tractors. Refering to Fig. 0458, remove master link from coupling chain (22). Remove chain and slide coupling (21) forward after loosening the coupling clamp bolt. Disconnect outer end of clutch shifter shaft and loosen the set screw which locates the shifter fork (26) on shaft. Bump shifter shaft out of shifter fork toward left side of tractor. Lift clutch shaft (23) rearward and out.

25C. Correlation mark clutch cover and flywheel then unbolt clutch from flywheel and flywheel from crankshaft and lift out.

25D. Remove the vertically mounted dust shield by removing two screws attaching it to the engine adapter plate. Unbolt seal retainer from adapter plate and from rear end of oil pan. The seal assembly, Fig. 0476A, can be renewed at this time. Soak new seal in oil before installing same.

FLYWHEEL AND RING GEAR
Six Cylinder

26. Remove flywheel as outlined in paragraphs 25B through 25C.

Flywheel run-out checked at rear outer face of wheel should not exceed 0.015. To install a new ring gear, heat same to 450 degrees F. and install with beveled end of teeth facing the engine.

OIL PUMP AND RELIEF VALVE
Six Cylinder

27. The vane type pump, Fig. 0476, which is driven by the camshaft, is mounted on underside of cylinder block. Pump removal requires removal of oil pan and one cap screw attaching the pump body to the lower side of the cylinder block. It will be necessary on non-Diesel engines to retime the ignition unit whenever the oil pump is removed, as its drive is supplied through the oil pump drive shaft.

Disassembly procedure is self-evident. Vanes should be installed to oil pump drive shaft so that flat sides of vanes will be facing the direction of normal rotation when viewed from the vane end, as shown in Fig. 0476. The pump body and cover are assembled without a gasket. If the pump becomes worn, do not attempt to repair it; renew the entire unit.

27A. Before installing the oil pump to non-Diesel engines, rotate the engine crankshaft until the number one piston is on compression stroke and the flywheel mark "TDC" is indexed at the inspection port. Install the oil pump so that the narrow side of the pump shaft, as divided by the slot (ignition unit drive slot) is on the crankshaft side and parallel to the crankshaft. Refer to Fig. 0474 which shows correct position of the ignition unit drive slot when viewed from above through the ignition unit shaft hole in the cylinder block. If the drive slot is not in the position as shown, remove the oil pump and remesh the pump drive gear. Reinstall the ignition unit and check the timing.

28. The non-adjustable piston type oil relief valve is located externally on the right side of the engine in the vicinity of timing gear cover gasket surface. Correct pressure is 26-34 psi at a crankshaft speed of 1675 rpm.

Relief valve spring (Oliver's part 7K-303) has a free length of 2 inches and should test 5-6 lbs. at a height of $1\frac{1}{16}$ inches.

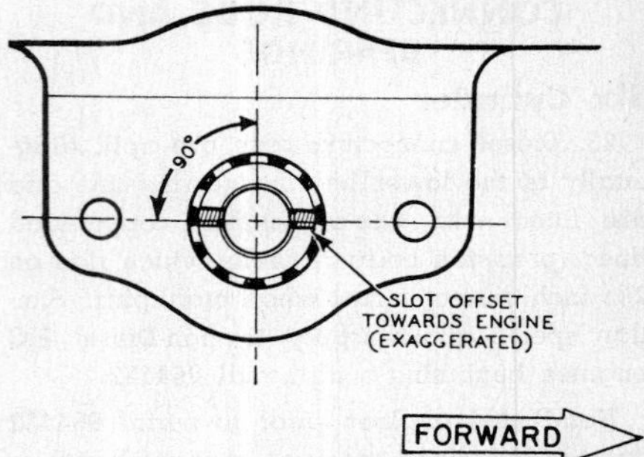

Fig. 0474—Correct position of the ignition unit drive slot when viewed from the ignition unit mounting pad surface.

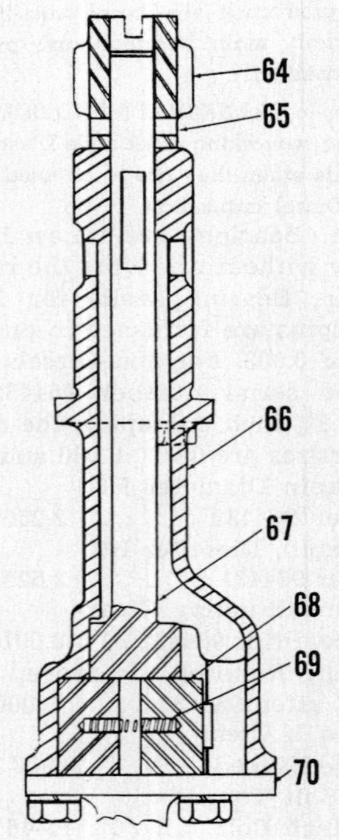

Fig. 0475—Cross-section through oil pump. Excessive wear is best corrected by renewal of pump assembly.

64. Drive gear
65. Retaining pin
66. Pump body
67. Drive shaft
68. Vanes
69. Spring
70. Cover

Fig. 0473—Crankshaft rear oil seal and retainer (58) can be renewed after removing the engine flywheel.

62. Welch plug 63. Dowel pin

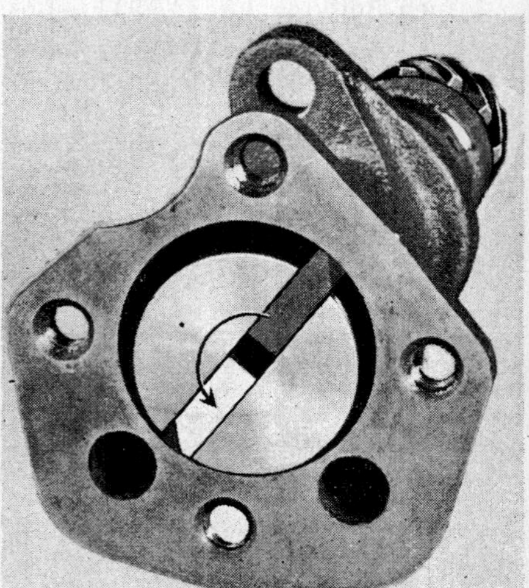

Fig. 0476—Oil pump vanes should be installed so that the flat sides will face in the direction of rotation.

CARBURETOR

29. Carburetor is a Marvel-Schebler TSX model. Early production engines used the TSX-477; latest production engines are fitted with TSX-581 units. Clockwise rotation of idle adjusting needle richens the idle mixture; whereas similar rotation of the high speed needle leans the high speed (power) mixture. Bowl fuel height is indicated on outside of carburetor body.

29A. Calibration and repair parts are as indicated herewith. Parts numbers shown are Marvel-Schebler, not Oliver.

	TSX 477	TSX 581
Idle Jet	49-165	49-285
High Speed Needle	43-631	
Idle Adjusting Needle	43-58	
Main Nozzle	47-221	47-365
Venturi $1\frac{1}{16}$"	46-490	
Economizer Jet	49-262	49-354
Float Needle & Seat	233-543	
Throttle Bushings	60-260	
Float	30-621	
Gasket Kit	16-594	
Repair Kit	286-1000	

DIESEL SYSTEM - 6 Cylinder

(System for Three Cylinder Begins with Paragraph 85)

Six Cylinder Only

The extreme pressure of the injector nozzle spray is dangerous and can cause the fuel to penetrate the human flesh. Avoid this source of danger when checking the nozzles by directing the spray away from your person.

GENERAL TROUBLE SHOOTING
Six Cylinder

32. The following data, supplied through the courtesy of American Bosch Company, should be helpful in shooting trouble on 4 cycle type Diesel tractor engines.

32A. **SYMPTOM.** Engine does not idle well; erratic fluctuations.

CAUSE. Could be caused by faulty nozzle or nozzles, also by pump overflow valve remaining in open position. The overflow valve if pump is so equipped, should be removed and washed in cleaning solvent.

32B. **SYMPTOM.** Intermittent or continuous puffs of black smoke from exhaust.

CAUSE. Faulty nozzle or nozzles, also improper engine operating temperature can be the cause of the trouble.

32C. **SYMPTOM.** Fuel oil builds up (dilution) in the engine crankcase.

CAUSE. The trouble could be caused by a leaking gasket under the delivery valve, or badly worn plunger. The remedy for any of these conditions would be renewal of the complete hydraulic head as a unit as outlined in paragraph 35.

32D. **SYMPTOM.** Sudden heavy black smoke under all loads.

CAUSE. This calls for removal of the entire injection pump assembly for handling by competent personnel. The difficulty possibly is caused by a stuck displacer piston. Other possible causes are improperly adjusted smoke cam or dilution of the fuel by engine oil being by-passed by a damaged hydraulic distributor head lubricating oil filter.

32E. **SYMPTOM.** Poor fuel economy.

CAUSE. Water temperature too low. Check thermostat for proper functional control. Check for fuel leakage.

32F. **SYMPTOM.** Engine low in power.

CAUSE. Filter between supply pump and injection pump may be clogged; or, a faulty supply pump. Due to type of fuel used, it may be necessary to advance the timing. Under no circumstances should the timing be advanced more than 4 degrees. Refer to paragraph 36.

32G. **SYMPTOM.** Engine rpm too low at full throttle position.

CAUSE. Could be caused by improper setting of the throttle linkage. Remove pump control lever cover and check if full travel is obtained at full load position of throttle control lever.

INJECTION PUMP
Six Cylinder

33. The injection pump as used on the Super 99 Diesel engine is an American Bosch PSB6A. The Oliver Super 77 and Super 88 Diesel engines use the same pump but with a different delivery calibration.

The single plunger, constant stroke, sleeve control type pump is driven at crankshaft speed by an idler gear which meshes with the crankshaft gear as shown in Fig. 0464.

Early production engines were equipped with PSB6A70Y 2920E; later production engines are equipped with PSB6A70Y 3826-A. Later pumps are fitted with Bosch HD9012-2A hydraulic head having rounded edges instead of the square type of early tractors.

A flyweight type governor, which is used to control the fuel delivery as a function of speed control, is an integral part of the injection pump.

The automotive, diaphragm type primary fuel supply pump, which furnishes fuel to the inlet side of the injection pump, is mounted on the right side of the engine and is driven by the engine camshaft. Refer to paragraph 43 for service data.

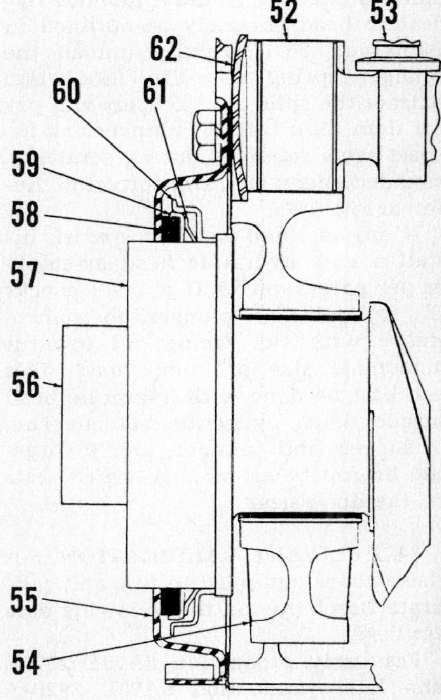

Fig. 0476A—Cross-sectional view showing the crankshaft rear oil seal which is made of specially treated cork.

52. Camshaft
53. Cam follower
54. Rear main bearing cap
55. Oil retainer
56. Crankshaft
58. Cork seal
59. Seal retainer
60. Seal assembly retainer
61. Spring
62. Expansion plug

33A. **PUMP TROUBLESHOOTING.** If the engine fails to accelerate or to respond to the throttle the trouble may be caused by excess friction in the control sleeve unit. With engine running, remove timing window from side of pump and observe action of control arm to which the governor rod is linked. If arm is sticking or moving erratically, stop the engine and disconnect governor rod. When arm is manually moved to either extreme of its movement, it should drop to neutral by its own weight. If movement is sticky, remove the control arm unit after unscrewing the two retaining screws. Wash unit in approved solvent and if this treatment does not give free movement, disassemble the unit and clean the bearing surfaces by lapping with mutton tallow.

If control arm swings freely when disconnected from governor rod, check governor rod for full travel and smooth operation. If binding in movement of governor rod is encountered, remove governor housing from end of pump and clean or renew parts as needed to remove wear or eliminate sticking. Refer to Fig. 0484A.

33B. HARD STARTING, POOR PERFORMANCE. Before condemning the pump assembly as the cause of these troubles, check condition of pump plunger as follows: Remove hydraulic head assembly as outlined in paragraph 35. Carefully unload the plunger spring (48—Fig. 0486) and extract the split cone keepers and pry off item (50). Lift out plunger and inspect same for scuff marks, scratches, rounded edges, rust and corrosion. Refer to Fig. 0479.

If any of these conditions exist, install a new hydraulic head assembly as per paragraph 35. If plunger checks O. K., reinstall plunger to control sleeve with slot facing out towards nameplate side of pump head. This can best be done with hydraulic head upside down by gentle maneuvering of sleeve and plunger. Don't forget the bronze thrust washer which seats on the drive gear.

34. **DELIVERY CALIBRATION.** For those shops equipped to test and calibrate Bosch pumps, the following data applies:

For early production 6A70Y 2920E and later production 6A70Y 3826-A versions, pump delivery for the Super 99 should be:

27.5-28.5 cc for 500 strokes @ 1674 pump rpm.

28.0-29.5 cc for 500 strokes @ 1100 pump rpm.

7.0-8.0 cc for 500 strokes @ 500 idle rpm.

It should be remembered, however, that for replacement purposes, one pump serves for the Super Series of the 77, 88 and 99. The universal replacement pump as received, is calibrated either for the 88 or the Super 88 engines, and must, therefore, be recalibrated when installed on the Super 99 engine. Recalibration is accomplished by turning the full load adjusting nut (99—Fig. 0477) located under the governor housing end plug (103) in or out as required. It is recommended that such recalibration be performed by the official Bosch station or some establishment equipped with the necessary test stand equipment. In an emergency where such facilities are not available, proceed as in paragraph 34A.

34A. Load the engine until the speed drops to 1000-1200 rpm when the throttle hand control is in the wide open position. Using a ⅜ inch deep socket, rotate the control rod nut (99) until the exhaust shows a slight indication of black smoke. Rotating the nut in a counter-clockwise direction lessens the smoke or leans the mixture by decreasing the amount of fuel delivered. Generally speaking, a pump calibrated for the 88 tractor engine will be approximately calibrated for the Super 99 by turning the nut inward (richer) 8 to 10 flats or 1⅓-1⅔ turns. Setting should be such that no exhaust smoke is present at light loads.

35. **HYDRAULIC HEAD UNIT.** Installation of a new or exchange hydraulic head unit is sometimes the indicated remedy when injection pump trouble is encountered. To R&R or renew the hydraulic head, refer to Figs. 0479 and 0482 then, remove the pump timing window cover and rotate the engine until the marked tooth of gear (51) is approximately in register with the "O" mark on early pumps or the arrow head on late pumps. Remove control arm unit from surface (CU) and engine oil small filter (38) from pump. The fuel lines and head nuts can now be removed and the head lifted off as shown in Fig. 0479. Do not attempt to lift off the head without first bringing the marks into register, as the sheet metal plate on top of the quill shaft gear prevents unmeshing of plunger drive gear (51) except in specified position. Use new "O" rings when reinstalling the hydraulic head.

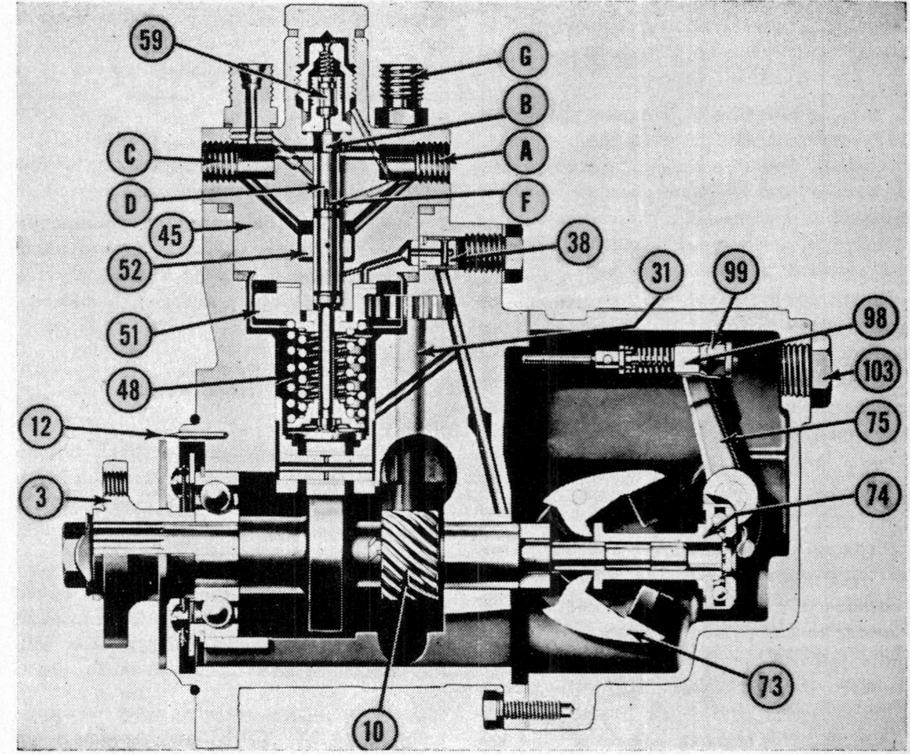

Fig. 0477—Phantom view of latest version Bosch PSB type injection pump is identified by the round cornered (instead of square) hydraulic head (45). Refer to Fig. 0486 for parts legend.

Courtesy—Diesel Publications, Inc.

36. **TIMING TO ENGINE.** Pump is correctly timed to the engine when the plunger port for No. 1 pump outlet closes either 28 degrees or 24½ degrees before TDC. These two settings are mentioned because the Oliver Corporation recommends the 28 degree setting for standard operation medium load conditions where the engine speed is not allowed to drop below the governed 1675 rpm. For severe load conditions where the engine is lugged down to 1400-1500 rpm, the recommended setting for maximum power is 24½ degrees.

36A. FLYWHEEL TIMING MARKS. Factory affixed timing mark on the flywheel is "FP". Beginning with engine serial 949518, all Super 99 Diesel tractors have the "FP" mark located 28 degrees before TDC; some earlier engines have the mark at 24½ degrees before TDC. Distance on flywheel from TDC to a 28 degree "FP" mark is $3\frac{9}{16}$ inches, to a 24½ "FP" mark 3⅛ inches.

When it is known that the pump internal timing is O. K., the pump can be timed to the engine as follows:

36B. Crank the engine until the flywheel mark "FP" is visible through the timing opening in the flywheel housing. Using a pair of dividers or other means, measure the distance on the flywheel from the "FP" mark to the "TDC" mark. If distance is $3\frac{9}{16}$ inches, the mark is at 28 degrees; if 3⅛ inches, the mark is at 24½ degrees. If actual mark is 3⅛ inches away from TDC, scribe a new "FP" mark at a point $3\frac{9}{16}$ distant from the "TDC" or $\frac{7}{16}$ distant from the other "FP" mark. You now have a 24½ degree and a 28 degree timing mark.

Remove inspection port plug (11—Fig. 0481) located in the timing gear cover directly above the pump drive gear. Remove the timing window cover (22) from side of pump.

Rotate engine crankshaft until air is felt escaping from the energy cell opening or until both valves of No. 1 cylinder are closed; then slowly, until the flywheel mark "FP" is aligned with the flywheel housing inspection port notch.

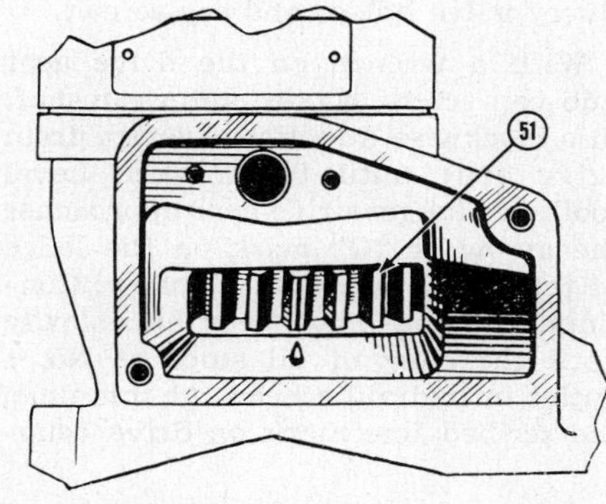

Fig. 0480—Pump is phased to fire number one cylinder when the marked tooth of plunger drive gear (51) is approximately in register with the "O" mark or arrow on timing window ledge as shown.

Note: Compression stroke of No. 1 cylinder can be determined either by removing the pipe plug and cap from the No. 1 energy cell or by removing valve rocker cover and observing the closing of No. 1 inlet valve.

At this time, the line mark on the injection pump coupling hub should be in register within $\frac{1}{32}$ inch with the pointer extending from the front face of the pump as shown in left circle in Fig. 0481. The mark and pointer are viewed through the inspection port plug opening (11). At the same time, the line marked tooth on plunger drive gear (51) will be in register within ¼ inch with the arrow or the "O" mark on window ledge. If the coupling mark is not in register with pointer, as stated, it will be necessary to remesh the pump gear as per paragraph 36C.

36C. Remove pump gear cover located on the front face of the engine timing gear cover.

Working through the pump gear cover opening, remove cap screws retaining the pump drive gear (PG—Fig.

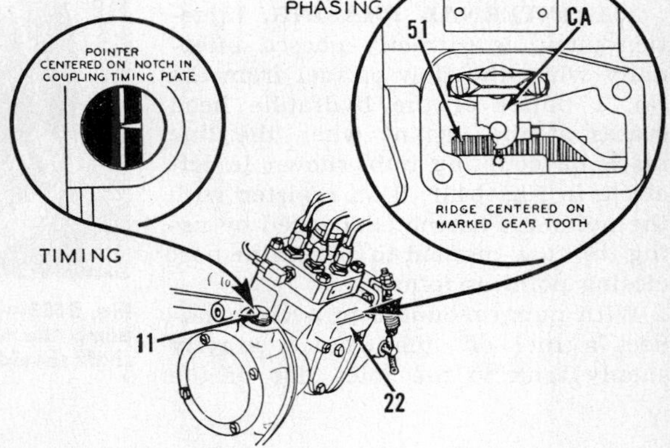

Fig. 0481 — Injection pump showing coupling hub timing marks in small circle and phasing position marks in large circle. Timing marks are visible through port when plug (11) is removed.

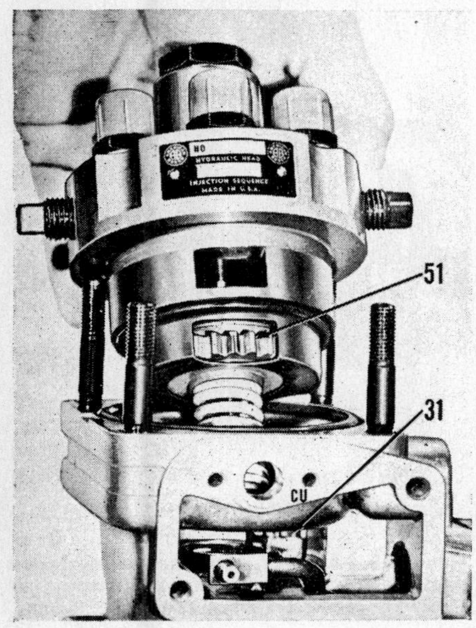

Fig. 0479 — To remove hydraulic head from injection pump, remove timing window and control arm unit from surface "CU" and oil filter (38—Fig. 0482) from pump body. Rotate engine crankshaft until marked tooth of plunger gear (51) is in register with "O" mark or arrow on window ledge then remove head nuts and lift off as shown.

Courtesy—Diesel Publications, Inc.

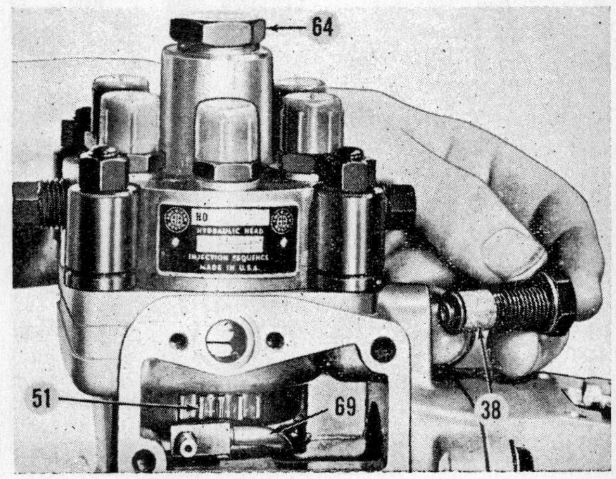

Fig. 0482 — Injection pump with control arm unit removed from rod (69) and from pump. Oil filter (38) is being removed preparatory to removing the hydraulic head unit.

0482A). Using a socket wrench, rotate the pump camshaft until the pump coupling hub line mark exactly registers with the pointer; then reinstall pump drive gear retaining cap screws. Several trials may be necessary before finding the holes which will admit the cap screws without throwing the marks out of register. Lock the cap screws with wire.

36D. **INTERNAL TIMING.** If pump has been disassembled, it should be internally timed at reassembly as follows:

Insert pump camshaft into body with the wide groove at splined end in register with the "CLW" mark on bearing retainer plate as shown in Fig. 0483. Install the quill gearshaft (31—Fig. 0486) through the bottom of the pump housing so that when the spiral gear (33) at bottom end is meshed with camshaft spiral gear, the open tooth of the spur upper gear on gearshaft (31) will be in register with the drill mark on the counterbore of the pump housing as shown in Fig. 0484. Now, install the hydraulic head assembly with the line marked tooth of its plunger gear in register with the arrow (an "O" on early production pumps) on the timing window recess as shown in Fig. 0480. Refer also to Fig. 0487.

36E. **INTERNAL PHASING.** Injection pump is correctly phased internally when the flow of fuel from the No. 1 outlet of the hydraulic head ceases at the instant when the line mark on coupling hub, shown in left circle in Fig. 0481, is in register with the pointer. Phasing is checked by using the flow method to determine port closing point as follows:

With pump mounted in a vise, connect a fuel oil line from a gravity supply tank to the inlet side of the hydraulic head. Remove constant bleed (overflow) line fitting from hydraulic head and replace with a ¼ inch pipe plug. Bleed pump of all air by loosening this pipe plug and rotating pump drive camshaft. Next, place operating lever in the full load position. Remove delivery valve cap screw (64) shown in Fig. 0482. With a $\frac{7}{16}$ inch socket wrench, unscrew the delivery valve holder and lift out the delivery valve spring and valve (Fig. 0485). The delivery valve body must remain in position. Reinstall delivery valve holder and cap screw.

With a wrench on the drive gear hub cap screw, rotate pump camshaft in a clockwise direction (viewed from drive end) until the marked bevel tooth of plunger drive gear approaches the arrow or "O" mark on the ledge of the pump inspection window. Continue rotating drive gear hub slowly until the flow of oil stops at No. 1 outlet in hydraulic head. At this time, the scribed line mark on drive coup-

Fig. 0483—When reassembling the injection pump the wide spline at drive end of camshaft should be registered with mark "CLW" as shown.

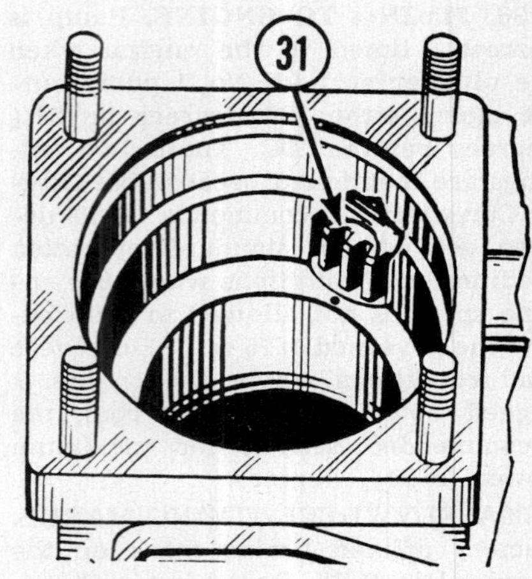

Fig. 0484—When assembling injection pump, the center open tooth of gear at top of quill shaft (31) should be registered with mark on body as shown, when pump camshaft is positioned as shown in Fig. 0483.

Fig. 0484A—When pin "P" is removed from control arm the control arm shaft should rotate freely in shaft bearings. This is important.

Courtesy—Diesel Publications, Inc.

Fig. 0485—A stuck, scratched or corroded delivery valve may cause hard starting or non-starting of engine.

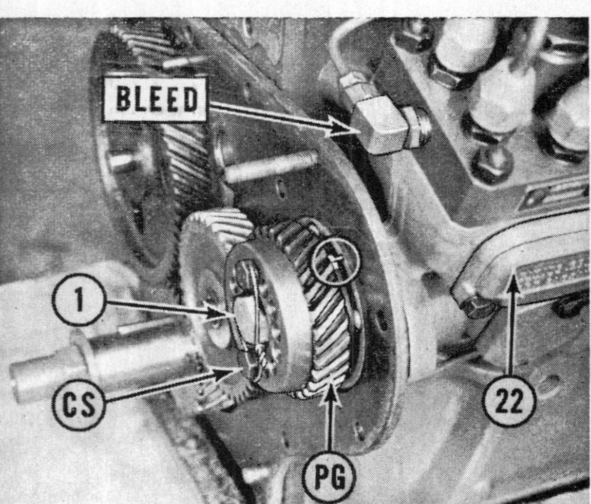

Fig. 0482A—View of injection pump and drive with timing gear cover removed. Pump timing pointer is shown in circle. Note bleed line.

ling hub should be in register with body pointer within $\frac{1}{32}$ inch and marked plunger gear tooth in timing window will be about one tooth to the right of the arrow or "O" mark on inspection window ledge.

If pointer and mark on pump drive hub are out of register more than ⅜ inch, the pump is incorrectly assembled. If out-of-register is ⅛ or less, remove old timing mark and affix a new line mark on drive gear hub. Repeat this procedure until constant results are obtained. Reinstall delivery valve, spring, and overflow valve, which were previously removed.

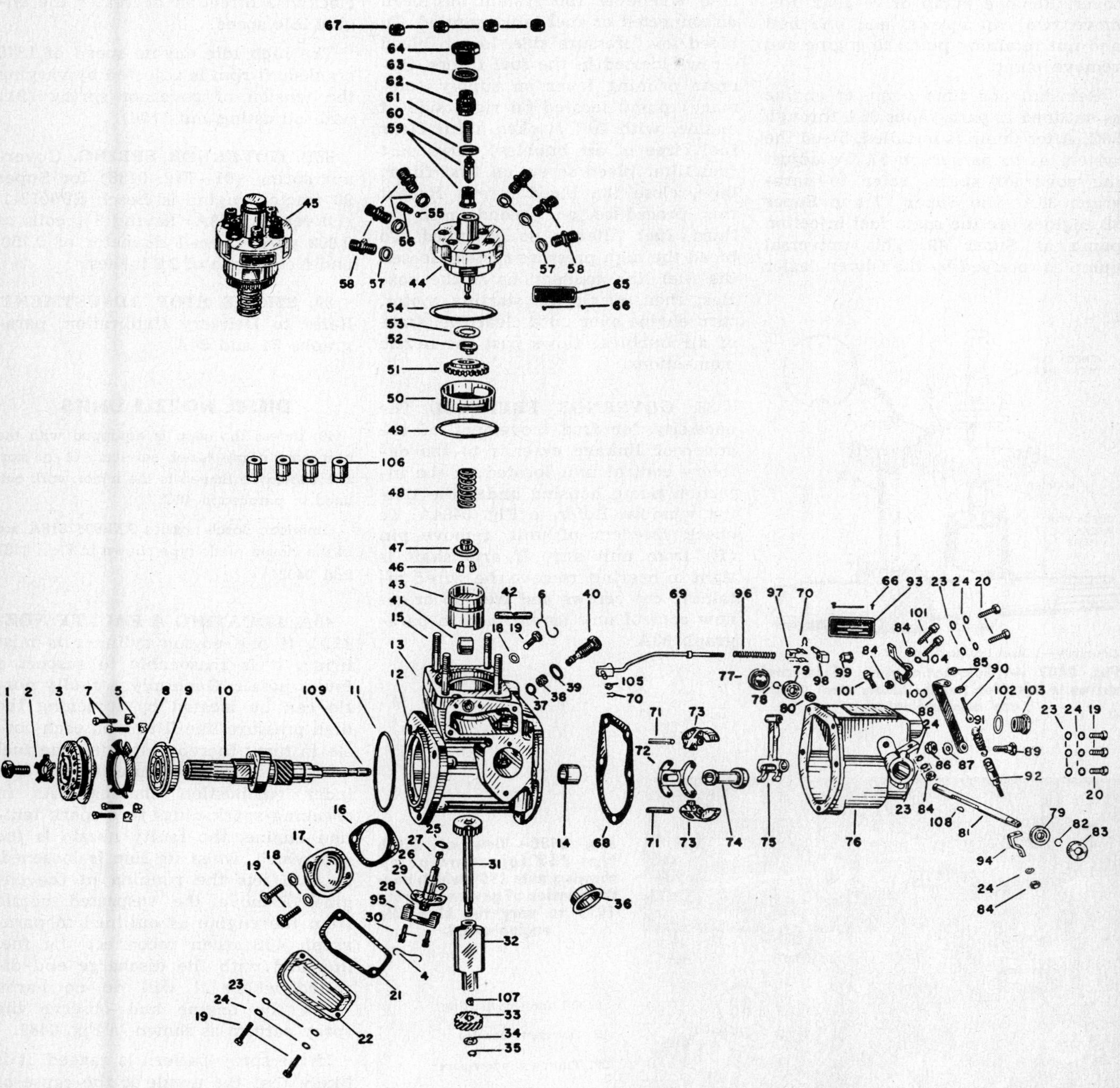

Fig. 0486—Disassembled view of Bosch PSB pump 6A70Y3826-A used on Super series tractors. Note round cornered hydraulic head assembly (45). Refer also to Fig. 0477.

A. Fuel inlet
B. Space above plunger
D. Plunger
F. Plunger slot
G. To nozzles
3. Drive hub
8. Retainer for item 9
9. Camshaft bearing
10. Camshaft
11. Rubber ring
12. Timing pointer
13. Pump housing
14. Camshaft bushing
22. Window cover
26. Control assembly
31. Quill shaft and gear
32. Quill shaft bushing
33. Camshaft driven gear
36. Closing plug
38. Filter screen
41. Tappet roller
43. Tappet guide
45. Hydraulic head assembly
48. Spring, inner
51. Plunger drive gear
52. Plunger guide
55. Screw sealing ball
59. Delivery valve assembly
61. Spring for item 59
62. Holder for valve 59
69. Control rod assembly
72. Weight spider
73. Governor weight
74. Weight sleeve
75. Fulcrum lever
76. Governor housing
79. Bearing for 81
81. Fulcrum lever shaft
83. Seal for 81
88. Operating lever
91. Extension spring
94. Shut-off plate
96. Control rod spring
98. Control rod pin
99. Load adjusting nut
100. Stop plate
104. Shut-off spacer
106. Stud sleeve

36F. **REMOVE & INSTALL.** To remove injection pump from engine, proceed as follows: Shut off fuel supply at tank, disconnect all lines from hydraulic head, and disconnect hand controls Remove injection pump gear cover from front face of timing gear cover. Remove pump drive gear. Remove two cap screws, and one bolt and nut retaining pump to engine and remove pump.

Reinstall and time pump to engine as outlined in paragraphs 36A through 36C. After pump is installed, bleed the system as in paragraph 37. To adjust the governor speed, refer to paragraph 38A. The Super 77 and Super 88 engines use the same fuel injection pump as Super 99. This universal pump as received by the Oliver dealer is calibrated (fuel delivery adjusted) for the 88 engine, and should be recalibrated by an official station if installed on the Super 99.

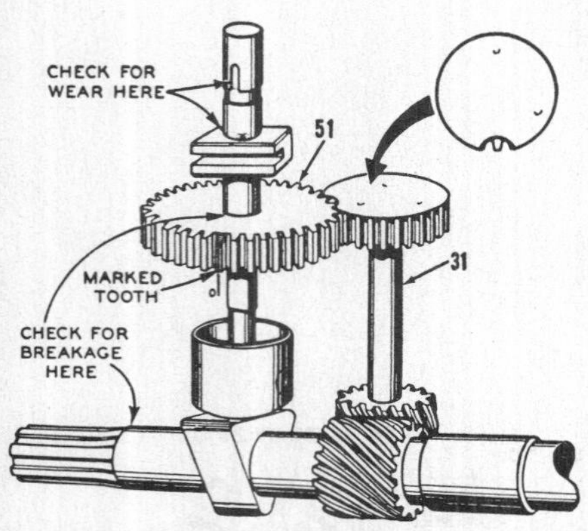

Courtesy—Diesel Publications, Inc.

Fig. 0487 — Schematic view of internal drives in injection pump. Likely wear points are also indicated.

37. **FUEL SYSTEM BLEEDING.** The Diesel fuel supply system should be bled whenever the system has been disconnected or fuel tank emptied. To bleed low pressure side, loosen bleed screws located in the fuel filters. Operate priming lever on supply (primary) pump located on right side of engine, with full strokes, until clear fuel (free of air bubbles) flows past fuel filter bleed screw on first filter; then, close the bleed screw. Repeat this procedure on second, also on third, fuel filter if so equipped. To bleed the high pressure system, loosen the fuel line connections at the nozzles; then, using the starting motor, turn engine over until clear fuel (free of air bubbles) flows past the nozzle connections.

38. **GOVERNOR FREEDOM.** The necessity for free movement of the governor linkage extends to the delivery control unit located on the injection pump housing under the timing window. Refer to Fig. 0484A. To check freedom of unit, remove pin (P) from unit arm. If arm shaft is tight in bearing, remove the wired retaining cap screws and free up or renew control unit as outlined in paragraph 33A.

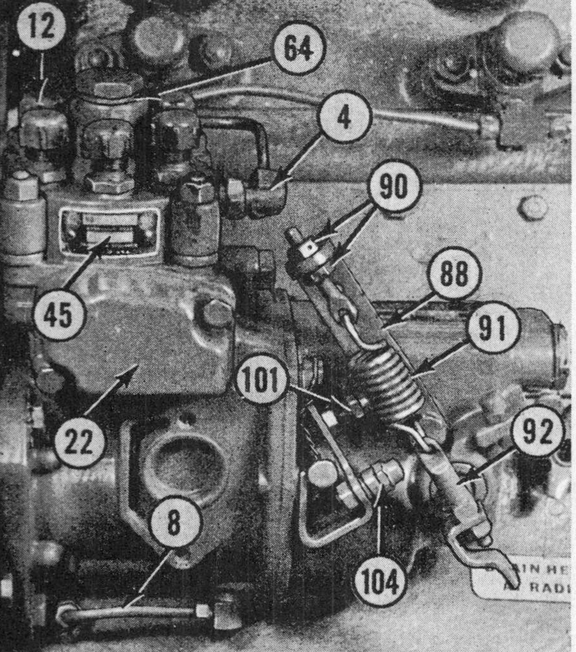

Fig. 0488—View of Bosch type PSB injection pump showing nuts (90) which control tension of governor spring (91) to vary the high idle engine speed.

8. Oil line (lubricating)
12. Outlets to nozzles
22. Timing window cover
45. Hydraulic head
64. Delivery valve cap
88. Operating lever
101. Low idle screw
104. Shut-off spacer

38A. **GOVERNOR ADJUSTMENT.** The Diesel engine governor (flyweight type) is an integral part of the injection pump. Engine low idle no-load speed of 600 crankshaft rpm is adjusted by means of screw (101—Fig. 0488). Rotating the adjusting screw in a clockwise direction decreases the engine idle speed.

The high idle engine speed of 1840 crankshaft rpm is adjusted by varying the tension of governor spring (91) with adjusting nuts (90).

38B. **GOVERNOR SPRING.** Governor spring (91—Fig. 0488) for Super 99 tractor engine is Bosch SP9013-1, (Oliver 7A5798A) having 8½ coils of 0.092 wire, overall diameter of 0.750, and free length of 2.0 inches.

39. **SMOKE STOP ADJUSTMENT.** Refer to Delivery Calibration, paragraphs 34 and 34A.

DIESEL NOZZLE UNITS

40. Unless the shop is equipped with the necessary nozzle tester, servicing of the nozzles should be limited to the minor work outlined in paragraph 40C.

American Bosch nozzles AKB6062618A are of the closed pintle type shown in Figs. 0491 and 0492.

40A. **LOCATING A FAULTY NOZZLE.** If one engine cylinder is misfiring, it is reasonable to suspect a faulty nozzle. Generally, a faulty nozzle can be located by loosening the high pressure line fitting on each nozzle in turn; thereby allowing the fuel to escape instead of entering the cylinder combustion chamber. As in checking spark plugs in a spark ignition engine, the faulty nozzle is the one which, when its line is loosened, least affects the running of the engine. Remove the suspected nozzle from the engine as outlined in paragraph 40B; then reconnect the fuel line and with the discharge end directed where it will do no harm, crank the engine and observe the spray pattern as shown in Fig. 0489.

If the spray pattern is ragged, it is likely that the nozzle is the cause of the misfiring. To prove the diagnosis, install a new or rebuilt nozzle or a nozzle from a cylinder which is firing regularly. If the cylinder fires regularly with the other nozzle, the condemned nozzle should be serviced as per paragraph 40C. If cleaning and/or renewal of tip (body) does not restore the nozzle, it should be overhauled by a shop equipped to handle such work.

40B. **R&R OF (INJECTORS) NOZZLES.** These assemblies commonly are called nozzles or injectors, although the nozzle is strictly not a complete injector. Before loosening any lines, wash all connections with fuel oil.

Procedure for R&R of injector units varies according to the cylinder location as follows:

The number 6 unit can be removed after disconnecting the leak-off line and feed line.

The numbers 2, 3, 4, and 5 units can be removed in the usual manner after removing the 4 clamps from the line harness and springing the number 6 injector line up or down slightly for removal clearance.

The number 1 unit can be removed after disconnecting all lines and lifting the line harness assembly off the engine.

After disconnecting the high pressure and leak-off lines, cover open ends of lines and pump with tape or composition caps to prevent the entrance of dirt. Remove nozzle holder stud nuts and carefully withdraw injector from cylinder head being careful not to strike end of same against any hard surface.

Thoroughly clean the nozzle recess in the cylinder head before reinstalling the nozzle holder assembly. It is important that the seating surfaces of the recess be free of even the smallest particle of carbon which could cause the nozzle to be cocked and result in blowby of hot gases. No hard or sharp tools should be used for cleaning. Bosch recommends the use of a wooden dowel or brass bar stock which can be shaped for effective cleaning. Do not reuse the copper ring gasket; install a new one. Tighten the nozzle holder stud nuts to 14-16 foot-pounds torque.

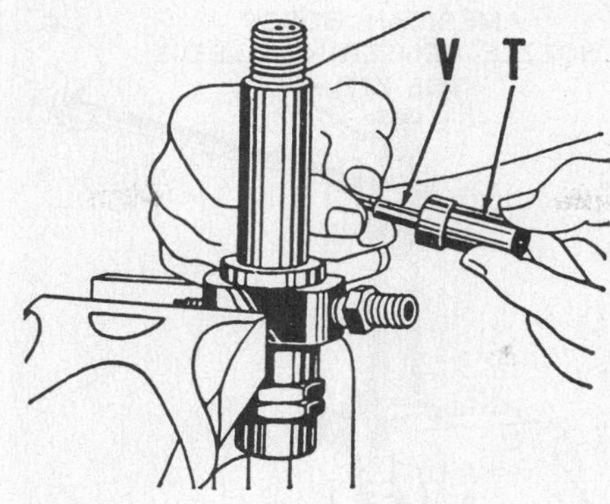

Fig. 0490—Removing Oliver Diesel engine injector nozzle tip (T) and the valve (V).

40C. **MINOR SERVICING OF NOZZLE.** Hard or sharp tools, emery cloth, crocus cloth, grinding compounds or abrasives of any kind should **NEVER** be used in the cleaning of nozzles.

Carefully clamp the nozzle holder in a vise and remove the cap nut and spray nozzle, consisting of the body or tip (T—Fig. 0490) and valve or nozzle body stem (V). Soak the nozzle in fuel oil, acetone, carbon-tetrachloride or a similar carbon solvent being careful not to permit any of the polished surfaces of the valve or tip to become nicked by contact with any hard substance.

All surfaces of the nozzle valve pintle should be bright and shiny except the contact line of the beveled seating surface. Polish the valve (pintle) with mutton tallow used on a soft cloth or felt pad. The valve may be held by its stem in a revolving chuck during this operation. A piece of soft wood well soaked in oil, or a brass wire brush will be helpful in removing carbon from the valve.

The inside of the nozzle body (tip) can be cleaned by forming a piece of soft wood to a point which will correspond to the angle of the nozzle valve (pintle) seat. The wood should be well soaked in oil. Some Bosch mechanics use an ignition distributor felt oiling wick instead of the soft wood rod, for cleaning the pintle seat in the tip. Delco-Remy part DR804076 is suitable for the purpose. Form the end of the wick, and coat the formed end with tallow for polishing.

The orifice at the end of the tip can be cleaned with a wood splinter. Outer surfaces of the nozzle body or tip should be cleaned with a brass

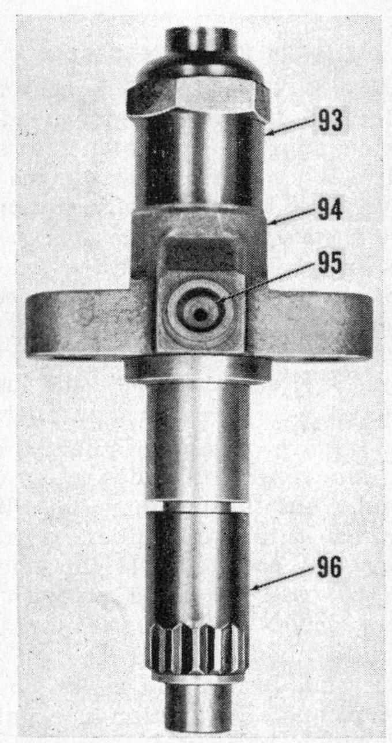

Fig. 0492—Bosch closed pintle type fuel nozzle as used on 6 cylinder Super 99 Diesel.

93. Cap
94. Nozzle holder
95. Fuel inlet
96. Cap nut

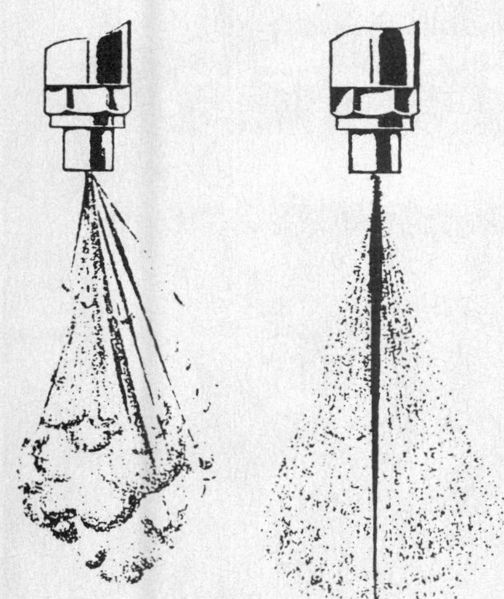

Fig. 0489 — Spray patterns of a standard pintle type nozzle. Left: A poor spray pattern. Right: Ideal spray pattern.

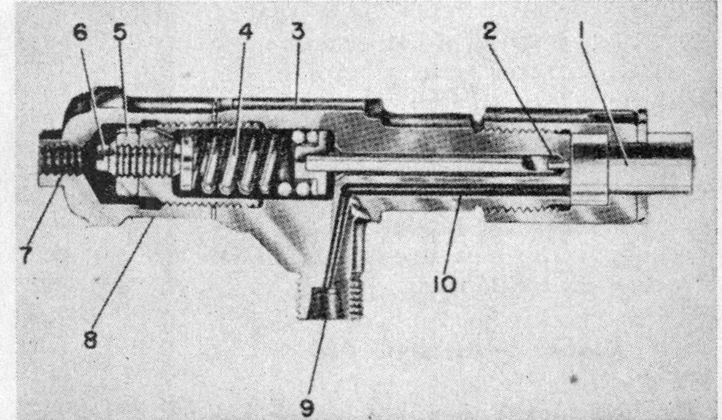

1. Nozzle tip (body)
2. Nozzle valve
3. Holder
4. Spring
5. Lock nut
6. Adjusting screw
7. Leak-off connection
8. Protection cap
9. Fuel inlet
10. Fuel passage

Fig. 0491—Sectional view of the closed pintle type nozzle used on the Oliver Diesel engines. Nozzle opening pressure of 1750 psi is controlled by adjusting screw (6).

wire brush and a soft cloth soaked in carbon solvent.

Before reassembling the nozzle to the holder, thoroughly rinse all parts in clean fuel oil and make certain that all carbon is removed from the cap nut. It is desirable that the nozzle body (tip) be perfectly centered in the cap nut. A centering sleeve, American Bosch tool TSE773, Fig. 0493, is available and should be used for this purpose. Avoid overtightening of the cap nut.

40D. Further disassembly of the nozzle holder should not be attempted except in an emergency or when a nozzle testing device is available for readjustment of the opening pressure. Recommended opening pressure for the nozzles is 1750 psi.

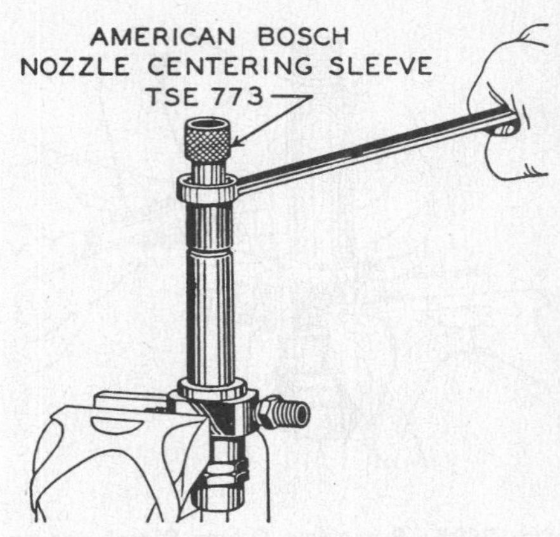

Fig. 0493—Using a Bosch tool No. TSE773 to center the nozzle tip in the cap nut.

DIESEL ENERGY CELLS

41. **R & R AND CLEAN.** Energy cells are located in the cylinder head on the side opposite to the injectors or nozzles. When the engine smokes excessively or the fuel consumption is above normal, the cause may be sometimes traced to an excessive amount of carbon deposit in the energy cells. An emergency job of cleaning the cells, Fig. 0494, can be done by removing the cell cap (99) and using a hooked wire to form a scraper.

To remove complete energy cell, first remove the threaded plug (97) and take out the retainer (98). With a pair of thin nosed pliers, remove energy cell cap (99). If the energy cell (100) will not come out with the fingers, screw a $\frac{15}{16}$—20 NF threaded bolt into the threaded end of the energy cell. A nut and collar on the bolt will make it function as a puller. If no puller is available, remove the injection nozzle and use a brass rod to drift the energy cell out of the cylinder head.

Clean all deposits from the front and rear craters of the cell body (100) and also from the small orifice using a brass carbon scraper or a shaped piece of hardwood. After cleaning the exterior surfaces with a wire brush, soak the parts in a carbon solvent. Reject any pieces that show leakage or burning anywhere on their surfaces. If parts are O. K. for leakage and burning, clean the lapped sealing surfaces by using a figure 8 motion on a tallow coated lapping plate.

DIESEL STARTING AID

Six Cylinder

42. A chevron type ether priming system utilizing ether filled cartridges is used as a starting aid. This element takes the place of the manifold preheater used on some other Oliver Diesel powered tractors. Control of the ether injector is located on a subpanel bolted to the instrument panel. Servicing procedures are conventional.

PRIMARY FUEL SUPPLY PUMP

43. Diesel engines are equipped with an AC automotive diaphragm type, primary fuel supply pump, Fig. 0494A, which is mounted on the right side of the engine. The pump, which is equipped with a hand operating priming lever for use in bleeding the low pressure side of the fuel system, is actuated by the engine camshaft. A satisfactory pump will show a 5-11 psi gage reading when checked at the outlet side.

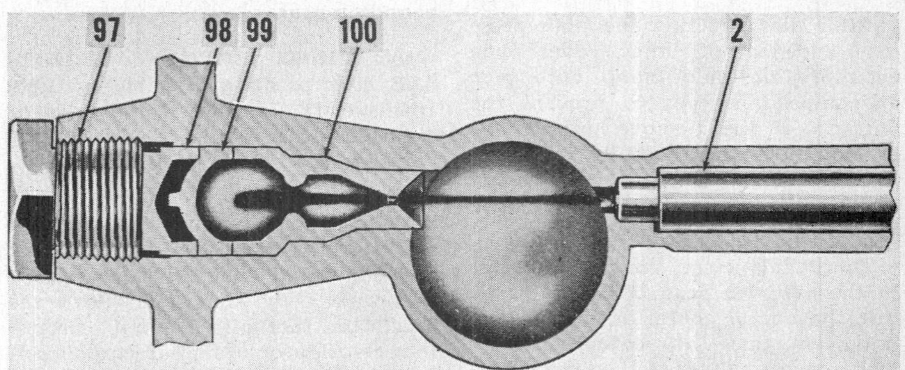

Fig. 0494—Energy cells (pre-combustion chambers) one for each cylinder, are located in cylinder head facing the injectors from opposite side of head.

2. Nozzle 97. Retaining plug 98. Retainer 99. Cell cap 100. Energy cell

Fig. 0494A—Fuel filters side of 6 cylinder Super 99 Diesel engine. Two of the 6 energy cells are indicated at arrow (EC), primary fuel pump at (FP).

NON-DIESEL GOVERNOR

The flyweight type governor, Fig. 0496 as installed on non-Diesel engines is mounted on the front face of the timing gear cover, and is driven by the camshaft gear. For adjustment data on the Diesel engine governor, refer to paragraph 38A.

ADJUSTMENT

44. To adjust the governor on non-Diesel engines, first remove air cleaner; then, place the hand control lever in the full speed position. Disconnect carburetor throttle valve control rod (133—Fig. 0495) at the governor operating arm (118). Place carburetor throttle valve in the wide-open position, and the govenor unit lever in full forward position. Adjust length of rod (133) to provide approximately $\frac{1}{16}$ inch over-travel ($\frac{1}{16}$" longer than necessary) as shown; then, reconnect the rod.

Start and warm up engine. Turn bumper spring screw (116—Fig. 0496) "out" about 4 revolutions. Adjust carburetor mixture and idle stop screw to provide a closed throttle, no-load crankshaft speed of 350 rpm for battery ignition equipped models, or 600 rpm for magneto equipped models. High speed, no-load crankshaft rpm of 1925 is obtained by lengthening or shortening the hand control rear rod (136—Fig. 0497) after disconnecting it from the front (flat) rod (134) at bolt (B).

If engine speed surges at 1925 rpm, rotate bumper spring screw just enough to eliminate the surge but not enough to increase the engine speed.

R&R AND OVERHAUL

44A. Governor can be removed without disturbing the radiator by proceeding as follows: Remove the air cleaner cup. Disconnect governor linkage and remove governor to timing gear cover retaining cap screws. **Carefully withdraw the unit so as to prevent the loss of the drive gear bronze washer (123—Fig. 0496).**

Governor drive gear (133) is a press fit and is keyed to the governor shaft (139). Gear can be removed with the use of a suitable puller. Governor weight carrier (122) is a press fit and keyed to the shaft. The weight carrier should be removed from the shaft by pressing on the gear end of the shaft. Disassembly of the weight unit is self-evident after an examination. The bushings (126 & 129 & 131) require final sizing after installation to provide running clearance of 0.0015-0.002.

Governor gear hub rotates in a steel-backed babbitt bushing (41) and sleeve assembly which is pressed into the front face of the cylinder block. To renew the assembly, it will be necessary to remove timing gear cover and use a suitable puller to remove same as shown in Fig. 0498. Renew bushing when clearance exceeds 0.005. Pre-sized service bushings have a bore diameter of 1.002, and should be installed with a piloted drift. Refer to Fig. 0499A.

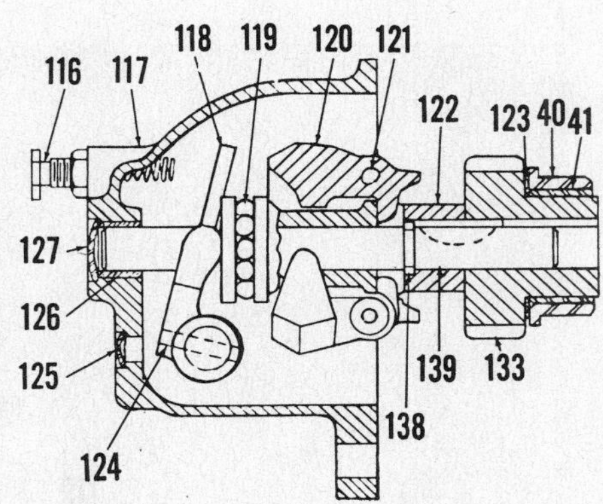

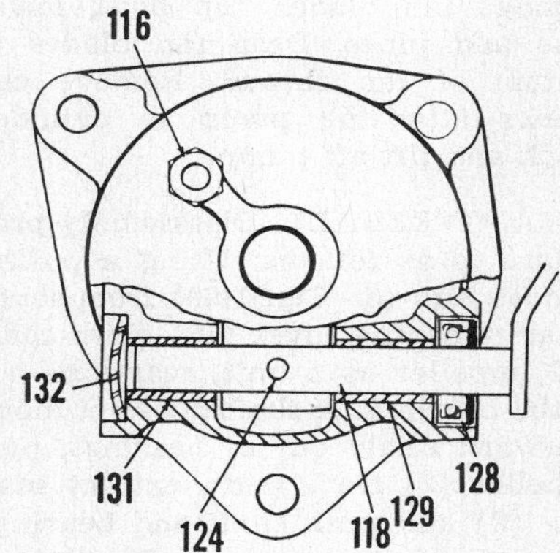

Fig. 0496—Speed governor assembly used on non-Diesel engines, cross-sectional view. Sleeve (40) and bushing (41) are located in the front face of the cylinder block.

Fig. 0495—Rod which connects carburetor throttle to governor fork lever should be adjusted to 1/16 inch over-travel, that is, it should be 1/16 inch longer than the center distance. Illustration is of 88 tractor but Super 99 is basically similar.

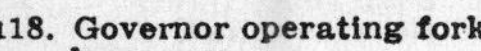
118. Governor operating fork lever arm

134. Governor control rod (front)

135. Governor spring

Fig. 0497—High idle, no-load engine crankshaft rpm of 1925 is controlled by lengthening or shortening the hand control rod (136). Shown is 88 tractor but Super 99 is basically similar with head of bolt (B) on outside.

COOLING SYSTEM (6 Cylinder)

(System for Three Cylinder Engine Begins with Paragraph 110)

RADIATOR

45. Procedure for removal of radiator which involves removal of the hood is as follows: Loosen front grille center strap bolts, remove two caps screws attaching front grilles to support and lift off grilles. Detach both headlights and disconnect the wires. Remove hood straps, precleaner, muffler, two wiring harness clips from right side of hood, and lift hood from tractor. Remove top radiator hose and disconnect lower hose. Remove radiator to front frame cap screws, two on each side, and shield to radiator shell screws, two on each side.

WATER PUMP

46. **REMOVE.** Pump can be removed without disturbing the radiator. Remove fan blades, top hose, lower hose and pipes. Drop fan blades to bottom of fan shroud. Remove cap screws attaching pump to cylinder block and lift off pump.

46A. **OVERHAUL.** Disassembly procedure is as follows: Using a puller, remove hub (8—Fig. 0499) from shaft. After removing cover (10), push shaft and impeller as a unit, rearward out of the bearings. If shaft does not move rearward easily out of bearings, pull impeller (2) from shaft, extract snap ring (9) and push shaft and bearings forward out of pump body. If bearings are renewed, install same with sealed ends facing outward and partially fill cavity in body with heavy oil or ball bearing grease. Always use a new seal unit.

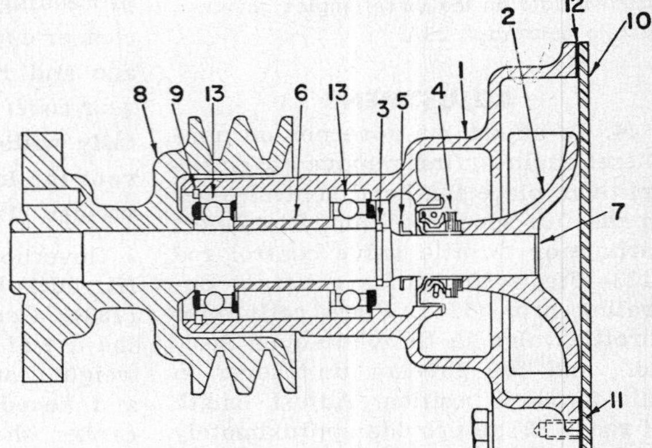

Fig. 0499 — Components of water pump. Note that impeller shaft is not integral with bearings.

1. Body
2. Impeller
3. Retaining ring
4. Seal assembly
5. Slinger
6. Bearing spacer
7. Shaft
8. Fan pulley
9. Retaining ring
10. Cover
13. Ball bearing

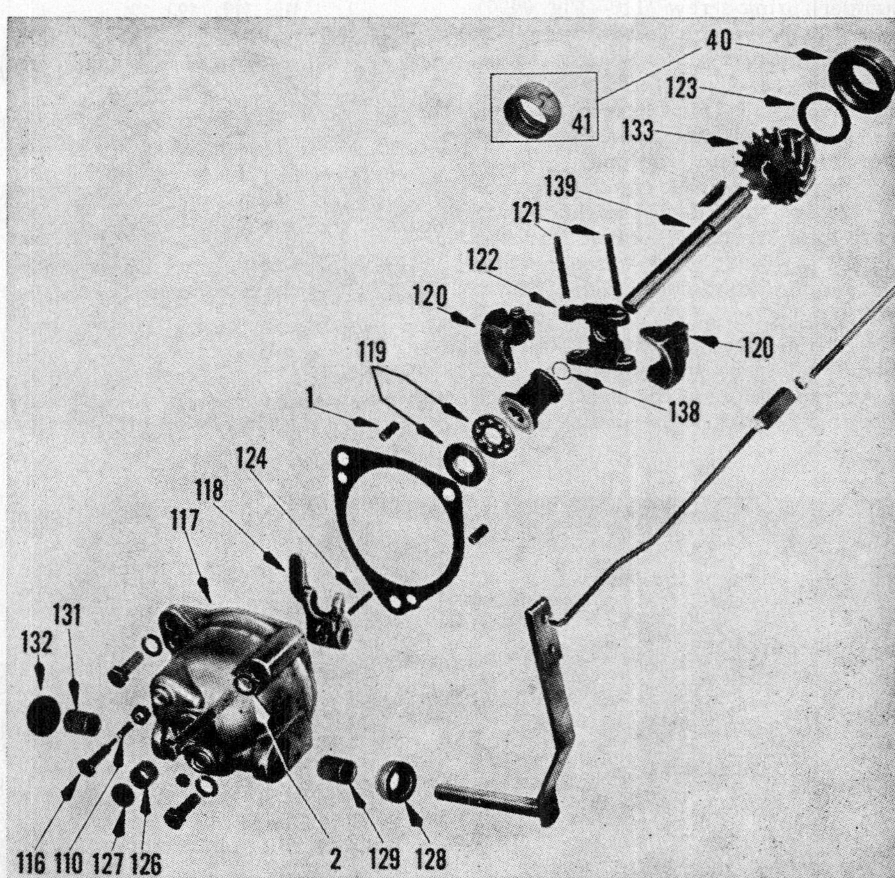

Fig. 0499A—Exploded view of speed governor. Refer also to Fig. 0496.

40. Sleeve
41. Bushing
116. Bumper spring adjusting screw
117. Housing
118. Fork
119. Thrust bearing
120. Weight
121. Weight pin
122. Weight carrier
123. Thrust washer
124. Fork retaining pin
125. Expansion plug
126. Bushing
127. Expansion plug
128. Oil seal
129. Bushing
131. Bushing
132. Expansion plug
133. Drive gear
138. Snap ring
139. Governor shaft

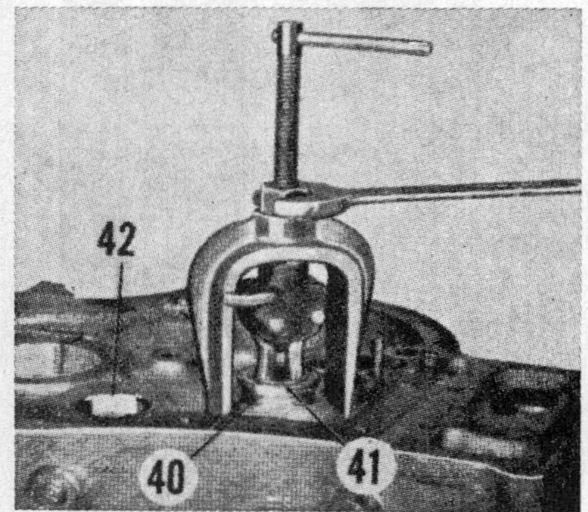

Fig. 0498—Using a puller to remove the governor gear sleeve (40) and bushing (41) assembly. Diesel engine idler gear shaft bore (42) is plugged on non-Diesel engines.

ELECTRICAL SYSTEM (6 Cylinder)

(Electrical system data for 3 cylinder engine begins with paragraph 115.)

GENERATOR AND REGULATOR

Six Cylinder

50. Non-Diesel engines are equipped with Delco-Remy 1100504, clockwise, 6 volt, third brush type generator. Hot output is 16-19 amperes @ 6.9-7.1 volts @ 2500 generator rpm. Brush spring tension is 16 ounces. Field current @ 80° F. is 2.5-2.72 amperes @ 6 volts.

Diesel engines are equipped with Delco-Remy 1100953, clockwise, 12 volt, third brush type generator. Hot output is 9-11 amperes @ 13.8-14.2 volts @ 2500 generator rpm. Brush spring tension is 16 ounces. Field current is 2.0-2.14 amperes @ 12 volts.

Generator cutout relay for 6 volt gasoline engines is Delco-Remy 1118305; for 12 volt Diesel tractors, Delco-Remy 1118306.

STARTING MOTOR

Six Cylinder

To remove starting motor from 6 cylinder tractor it will be necessary to first remove the stamped curved clutch cover.

51. Non-Diesel engines are equipped with Delco-Remy 1108951, clockwise, 6 volt motor fitted with Bendix drive. Tested at no load, current draw should be 60 amperes @ 5.7 volts @ 3300 rpm. Lock torque minimum should be 30 pounds feet @ 675 amperes @ 3300 rpm. Brush spring tension is 36-40 ounces.

Diesel engines are equipped with Delco-Remy 1109229, clockwise, 12 volt motor, fitted with over-running clutch drive. Tested at no load, current draw should be 95 amperes @ 11.7 volts @ 8000 rpm. Locked torque minimum should be 20 pounds feet @ 570 amperes @ 2.8 volts. Brush spring tension is 36-40 ounces.

IGNITION EQUIPMENT

Six Cylinder

52. Regular equipment is Delco-Remy 1118305 counter-clockwise battery distributor with centrifugal automatic advance. Optional equipment is a Wico XV6 vertical type, fixed spark, high tension magneto equipped with an impulse coupling.

Distributor specifications are as follows: Breaker point gap .022, cam angle 31°-37°, rotation counter-clockwise. Automatic advance starts @ 275 distributor rpm, should be 5-7 distributor degrees @ 400 rpm, 9-11 degrees @ 800 rpm and reach full advance of 14-16 distributor degrees @ 1300 distributor rpm. Multiply the foregoing values by 2 to obtain flywheel degrees.

IGNITION TIMING

Six Cylinder

53. **BATTERY IGNITION.** Battery distributor rotates counter-clockwise viewed from the drive end. Engine firing order is 1-5-3-6-2-4. To time distributor to engine, proceed as follows: Adjust breaker gap to 0.020-0.022.

Crank engine until No. 1 piston (timing gear end) is on compression stroke, then rotate slowly until flywheel mark "TDC" is registered with notch at inspection port. Install distributor to engine with rotor positioned to fire No. 1 cylinder. Rotate distributor housing counter-clockwise until breaker contacts have just started to open, then, temporarily tighten the distributor clamping bolt.

Run engine at 1925 rpm (high idle) at which time a timing light should show spark occurring when flywheel mark "IGN" registers with notch at inspection port. Rotate distributor body to obtain registration, then tighten clamp bolt securely.

54. **MAGNETO.** Magneto rotates counter-clockwise, viewed from the drive end. Engine firing order is 1-5-3-6-2-4. To time magneto to engine, proceed as follows: Adjust breaker contacts to 0.015 gap. Crank engine until No. 1 piston (timing gear end) is on compression stroke, then, slowly rotate flywheel until flywheel mark "IGN" is registered with notch at inspection port.

Turn magneto shaft to position where rotor is ready to fire No. 1 cylinder and temporarily install to engine. Rotate magneto body counter-clockwise until impulse coupling releases with a sharp snap. Rotate body in opposite direction until breaker contacts are just starting to open, then, temporarily tighten magneto clamping bolt.

Run engine at 1925 rpm or higher at which time a timing light should show spark occurring when "IGN" mark registers with notch at inspection port. Rotate magneto body to obtain this registration, then tighten the clamp bolt securely.

Fig. 0499B-Removing fenders, platform and bull gear cover as a single assembly. In most instances this is easiest method of removing bull gear cover from rear main frame.

ENGINE ASSEMBLY (GM Diesel 71 Series)

(Oliver Six Cylinder Type Engines Begin with Paragraph 10)

(Oliver Six Cylinder Type Engines Begin with Paragraph 10)

The model 3-71, 3023, type RB General Motors three cylinder, two-cycle Diesel engine is provided as the standard power unit for the Super 99 GM tractor. Other Super 99 tractors fitted with Oliver six cylinder Diesel and non-Diesel engines do not carry the "GM" suffix in the tractor model designation.

The two-cycle GM Diesel 71 series engines are offered in three, four and six cylinder models having the same bore and stroke, and using the same parts wherever possible. Thus, many of the engine parts are interchangeable.

Engine servicing procedures such as removal, reinstallation, overhaul and adjustment as covered herewith for the three cylinder engine, used in the Oliver Super 99 GM tractor, can also be applied to all other three, four and six cylinder 71 series engines. Exceptions to this statement apply only to the location of the various accessories and components on the engine as shown in Fig. 0499C, to the injection timing dimension which varies with the size of injector, and to the type of governor.

REMOVE AND REINSTALL

60. To remove GM engine assembly from tractor chassis, proceed as outlined in paragraphs 60A through 60G. Refer to Figs. 0500, 0501 and 0502.

60A. HOOD. Remove hood as follows: Loosen front grille center strap bolts, and remove two cap screws attaching front grilles to support and lift off grilles. Detach both headlights (each held by two bolts) and disconnect wires. Remove hood strap, precleaners, muffler, and two wiring harness clips from right side of hood. Lift off hood.

60B. PTO DRIVE SHAFT. If tractor is equipped with pto, remove the drive shaft of same as outlined in paragraph 145H.

60C. FUEL TANK. Shut off fuel supply at tank. Disconnect fuel line from strainer on fuel tank, and fuel return line at top front of tank. Remove governor control bell crank assembly and wiring harness clips from bottom of tank. Remove the fastenings attaching tank to rear bracket and lift tank from tractor.

60D. Remove batteries. Perform the following work on left side of tractor: Remove fuel tank to fuel pump line. Disconnect tachourmeter cable at drive unit (9); Chevron starting aid primer line (13) at air inlet housing on blower and remove one clip attaching line to blower housing; coolant temperature sending unit at water manifold; fuel shut-off cable (4) at governor;

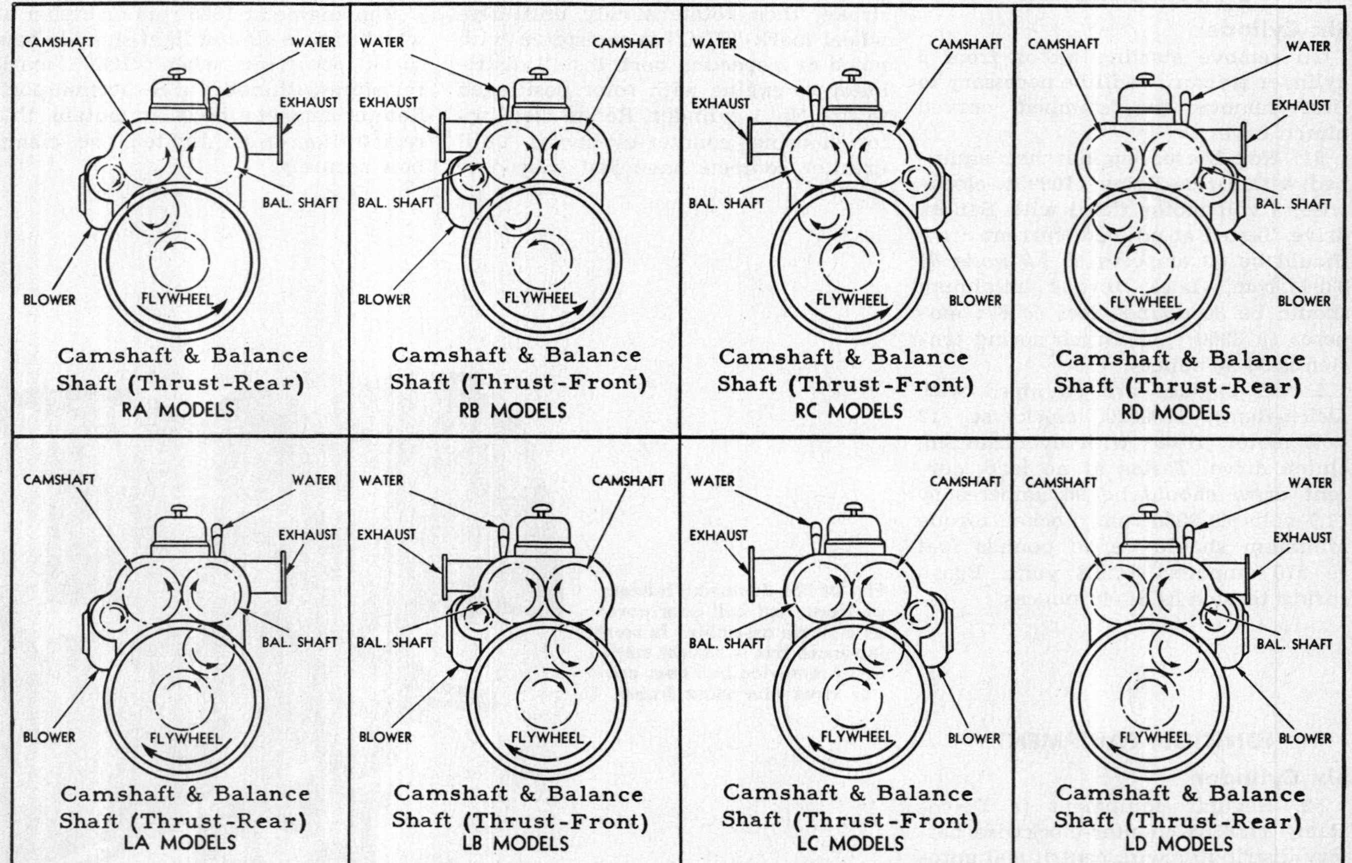

Fig. 0499C—Rotation and accessory arrangements for 3, 4 and 6 cylinder, 2 cycle, series 71 GM engines. A model RB, 3 cylinder series 71 GM engine is used in the Oliver Super 99 GM tractor. Views are from flywheel end of engine.

governor control rod (6) at governor; upper radiator hose; and oil cooler (1) to water pump hose.

Remove radiator to front frame cap screws (two on each side) and shield to radiator shell screws (two on each side). Disconnect radiator to oil cooler hose and lift radiator from tractor.

Disconnect engine oil cooler hoses at oil cooler, located in forward section of front frame.

Perform the following work on right side of tractor: Disconnect wiring harness at generator regulator; battery cables at starter; and oil pressure gage line (16) from cylinder block. Remove one clip attaching fuel shut-off cable (4) to side of cylinder head.

60E. CLUTCH SHAFT AND COUPLING. Remove clutch housing front cover after removing cap screws attaching it to front frame. Remove clutch housing flat dust cover after removing cap screws attaching it to front frame and to clutch release bearing carrier.

Refer to Fig. 0458 and remove chain (22) after extracting master link. Slide clutch shaft coupling (21) forward on shaft after removing coupling clamping bolt. Disconnect outer end of clutch shifter shaft, and loosen the set screw which retains shifter fork (26) to shifter shaft; then, bump shifter shaft out of fork toward left side of tractor. Withdraw clutch shaft (23) from clutch.

60F. LIFTING ENGINE. Remove the two front and two rear engine mounting bolts which may be provided with aligning shims. Be careful not to mix these shims. Attach a lifting bracket or sling to the two special lifting eyes on engine and hoist engine out of front frame as shown in Fig. 0502.

60G. REINSTALL AND ALIGN ENGINE. Engine to transmission alignment should be checked whenever a new front frame or engine is installed or whenever premature main clutch failures have occurred. A dial indicator rig, shown in Fig. 0460, can be used to check the alignment.

Make first check with button of indicator in contact with bore wall of clutch pilot bushing. If total indicated runout exceeds 0.015, vary the shims at front and rear engine mounts to bring reading within the 0.015 limit. Shims are available in thicknesses of 0.005 and 0.008.

After engine alignment has been corrected, check face runout of flywheel with gage hooked up as shown in Fig. 0460. If total indicator reading exceeds 0.015, check for a bent crankshaft mounting flange or dirt between the flywheel and crankshaft mounting flange.

Fig. 0500—Left side view of a GM engine installation in an Oliver Super 99 tractor.

BS. Throttle booster spring
GT. Governor control housing
GWH. Governor weight housing
1. Oil cooler
3. Thermostat housing
4. Fuel shut-off cable
5. Water manifold
6. Governor control rod
7. Fuel return line
8. Governor control rod beil crank
9. Tachourmeter drive
10. Fuel pump outlet line
11. Clutch housing flat dust cover
12. Clutch housing front cover
13. Chevron starting aid primer line
43. Control housing cover

CYLINDER HEAD

Cylinder head, a one-piece casting, can be removed from the engine as an assembly. The head assembly contains cam followers, push rods, rocker arms, exhaust valves, fuel injectors, exhaust valve seats, injector hole copper tubes and cylinder head water nozzles.

61. **REMOVE HEAD.** Herewith is a procedure for R&R of cylinder head. Remove hood assembly as outlined in

Fig. 0501—Right side view of a GM engine installation in an Oliver Super 99 tractor.

1. Oil cooler
4. Fuel shut-off cable
7. Fuel return line
14. Temperature sending unit line
15. Engine front lifting bracket
16. Oil pressure gage line

Fig. 0502—Hoisting a GM engine out of front frame by attaching a chain sling as shown.

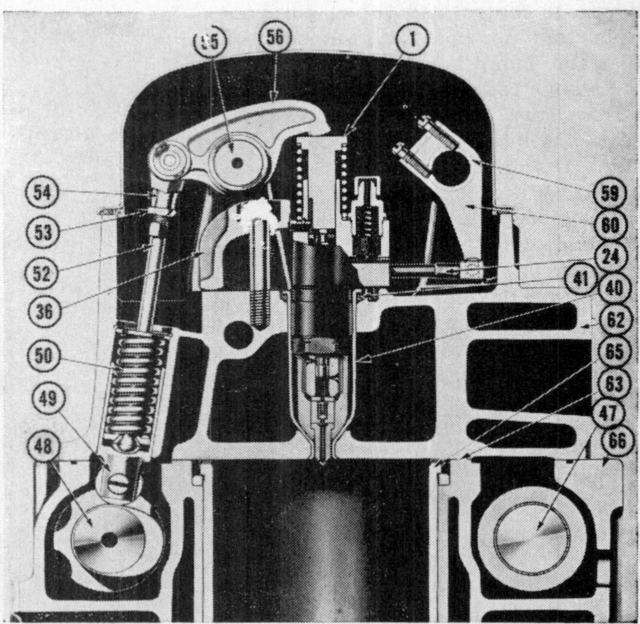

Fig. 0505A—GM engine. Showing location of injector and injector operating mechanism. A similar push rod, rocker arm and cam follower assembly is used to operate the exhaust valves.

24. Injector rack
36. Injector hold down clamp
40. Injector hole tube
41. Neoprene seal ring
47. Balance shaft
48. Camshaft
49. Cam follower
50. Follower spring
52. Push rod
54. Rocker arm clevis
56. Rocker arm
59. Injector rack control tube
60. Injector rack lever
65. Cylinder liner, dry type

paragraph 60A. Refer to Figs. 0500 and 0504.

Perform the following work on left (blower) side of engine: Drain engine coolant. Remove rocker arms cover. Disconnect upper radiator hose, and water pump to thermostat by-pass line hose. Remove thermostat housing (3) from water manifold (5). Remove coolant temperature sending unit from water manifold.

Remove governor control housing cover (43). Remove link, connecting injector rack control tube to governor differential lever. Disconnect throttle booster spring (BS) from governor. Remove two cap screws attaching governor control housing (GT) to cylinder head and four cap screws attaching governor control housing to governor weight housing (GWH). Pull upper end of control housing away from engine and at the same time push lower end of control housing in toward engine to free the dowels.

Remove cap screws attaching front lifting eye bracket (15) to balance weight cover and rear lifting bracket (40) to flywheel housing. Loosen (3 to 4 turns) the two cap screws, located directly below each lift eye bracket, which retain balance weight cover (25) and flywheel housing (22) to front and rear end plates of engine.

Perform the following work on right (starter motor) side of engine. Disconnect fuel pump to filter line (10) at the filter. Remove cap screws attaching injector rack control tube bracket to cylinder head and lift off injector rack control tube assembly.

If injector fuel lines and rocker arm shaft brackets were not removed pre-

Fig. 0504—Right side view of a GM engine showing the various components mounted on the cylinder head.

3. Thermostat housing
5. Water manifold
10. Fuel pump line
15. Engine front lifting bracket
16. Oil pressure gage line fitting
22. Flywheel housing
23. Secondary fuel filter
24. Cap screw
25. Balance weight cover
26. Air box hand hole cover
27. Oil filter
33. Engine rear end plate
40. Engine rear lifting bracket
111. Fuel inlet manifold
112. Fuel outlet manifold

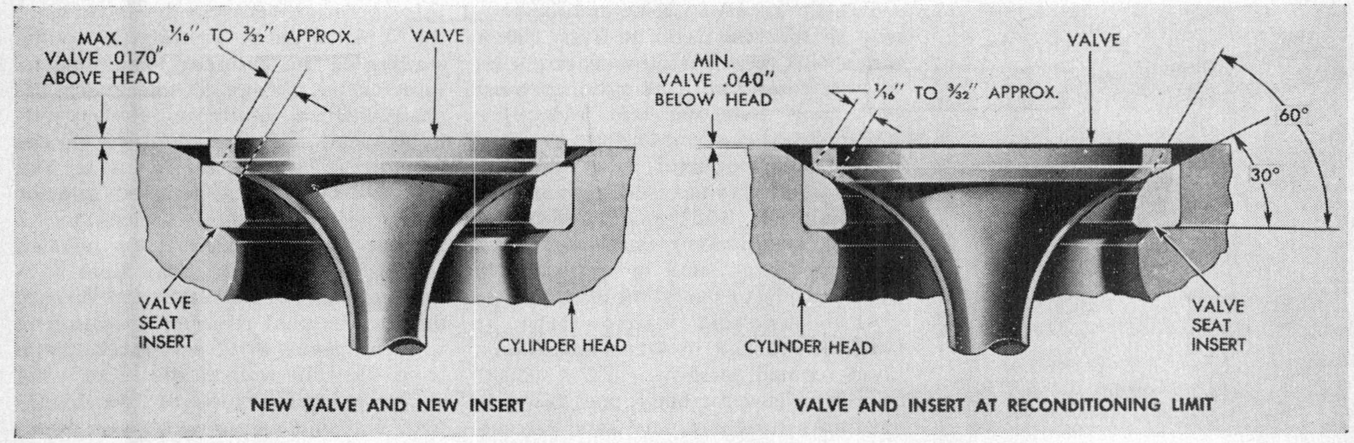

Fig. 0505B—GM engine. Showing relationship of top of valve head to machined surface of cylinder head.

viously, a thin-wall $\frac{15}{16}$ inch socket must be used to remove the head stud nuts. Remove eight cylinder head stud nuts and lift off head. Do not set cylinder head directly on the bench. Use wood blocks under head to prevent damaging tips of injectors and cam followers. Install hold-down clamps on cylinder liners to hold liners in place if crankshaft is to be rotated.

61A. **OVERHAUL.** Herewith are procedures for warpage inspections, for renewal of injector hole copper tubes and for renewal of cylinder head water nozzles.

61B. INSPECTION. Check cylinder head for warpage against the following: Maximum lengthwise warpage of 0.0055, and maximum widthwise warpage of 0.004. Reface head if above warpage limits are exceeded. Not over 0.020 total of metal should be removed from the cylinder head. When a cylinder head has been refaced, it will be necessary to check the protrusion dimensions of the valve head, Fig. 0505B, and of injector spray tip, Fig. 0506A, to prevent their striking the top of the piston. Excessive protrusion either of the valve head or injector spray tip can be corrected by reseating the valve seats, or renewing injector hole tubes.

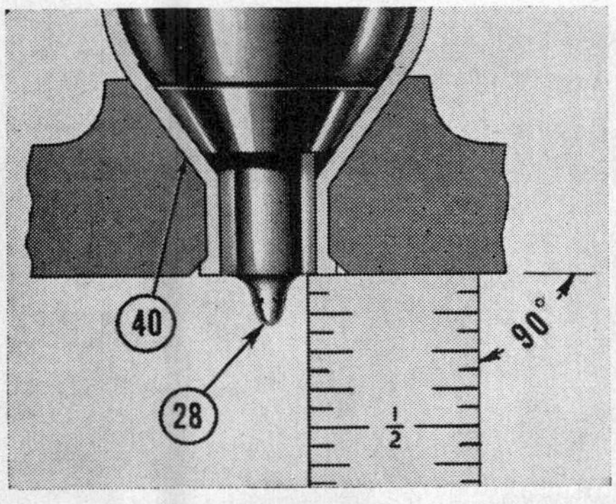

Fig. 0506A — GM engine. Showing flush condition of injector spray tip shoulder relative to machined surface of cylinder head.

61C. RENEW INJECTOR HOLE TUBES. Refer to Fig. 0505A. Each injector is inserted into a thin-walled copper tube (40) which passes through the water jacket in the cylinder head. The copper tube is flared-over at the lower end, and it is sealed at the upper with a neoprene ring (41). Effectiveness of the injector tube flared-over type seal and neoprene seal can be checked by subjecting the cylinder head water jacket to 80-100 psi air pressure and submerging the head in water heated to 180 deg. F.

Injector tubes can be removed as follows: Refer to Figs. 0507A and 0507B. With injector removed, thread tap (159), GM tool J5286-2 or equivalent, into upper end of injector tube (40) until ¾ inch of threads have been cut in tube. To prevent injector tube from turning while cutting the threads, use a tube holder tool, GM J5286-1 or equivalent, driven into lower or small end of tube. After cutting ¾ inch of threads in tube, remove tube holder tool. Then, using a small diameter rod, GM tool J5286-3 or equivalent, and inserted as shown in Fig. 0507B to contact end of thread tap (159), bump injector tube out of cylinder head.

Fig. 0507A—GM engine. First step in removing an injector hole copper tube (40) is to cut ¾ inch of threads in tube using a thread tap. GM tool J5286-2 or equivalent.

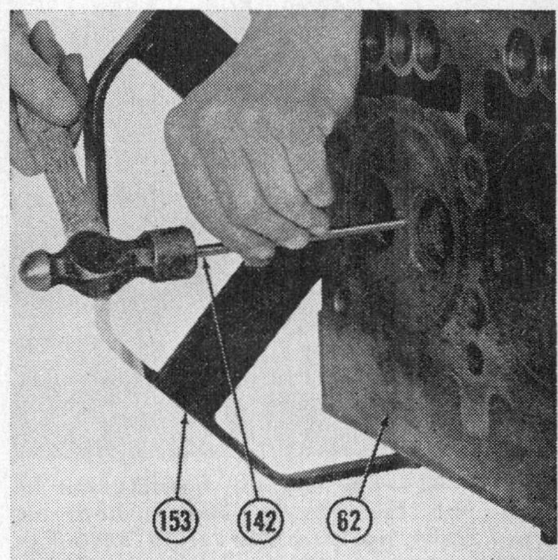

Fig. 0507B—GM engine. Second step in removing an injector hole copper tube is by bumping on end of thread tap using a small diameter rod (142).

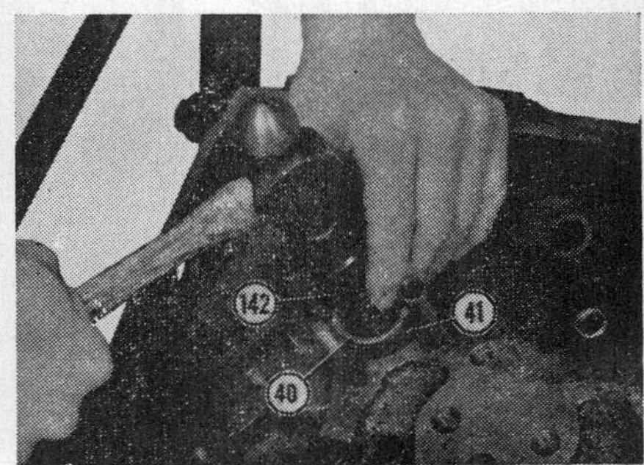

Fig. 0508A—GM engine. Showing method of installing an injector hole copper tube (40) with driver (142).

Fig. 0508B—GM engine. Upset or flare-over lower end of injector hole tube by applying 30 ft. lb. torque to upsetting die (144).

142. Driver, GM tool J 5286-4
144. Upsetting die, GM tool J 5286-6

61D. To install an injector tube proceed as follows: Refer to Figs. 0508A and 0508B. Place a new sealing ring (41) in counterbore of cylinder head and using installing tool and pilot, GM tools J5286-4 and J5286-5 or equivalent, bump injector tube in place. Using injector tube installing tool and upsetting die, GM tools J5286-5 and J5286-6 or equivalent, flare-over lower end of injector tube by applying 30 ft. lb. torque to upsetting die.

After installing injector tube, it must be finished in three operations: Hand reamed as shown in Fig. 0508C to receive injector body; spot faced at the flared-over end; and hand reamed to provide a sealing surface for the bevel seat on lower end of injector nut as shown in Fig. 0509A.

As shown in Fig. 0508C, ream injector tube, using GM tool J5286-7 or equivalent, until lower shoulder of reamer contacts injector tube.

Remove excess stock from flared-over end of injector tube using GM tool J5286-8 or equivalent, so that lower end of tube is from flush to 0.010 below finished surface of cylinder head.

The third step as shown in Fig. 0509A, reaming bevel seat in injector tube, should be performed with care as this operation controls the location of the injector spray tip relative to the surface of the cylinder head. Use an injector as a gage to check for the desired flush condition between cylinder head and the shoulder on the injector spray tip as shown in Fig. 0506A. Note: If pre-finished injector tubes are used (those which have a narrow land machined at the beveled seat as distinguished from tubes having a straight bevel seat) exercise care during the final reaming operation to prevent removal of too much metal from the thin wall of the tube.

Check effectiveness of the flared-over seal and neoprene seal on injector tube by subjecting the cylinder head water jacket to 80-100 psi air pressure and submerging the head in water heated to 180 deg. F.

61E. RENEW WATER NOZZLES. Refer to Fig. 0509B. A total of 8 water directional nozzles (5 and 6) are installed in the cylinder head. Large diameter water nozzles can be removed with a ¾ inch thread tap. Smaller nozzles can be removed in a similar manner using a small diameter thread tap.

Press new nozzles into place with nozzle openings (156) parallel to sides

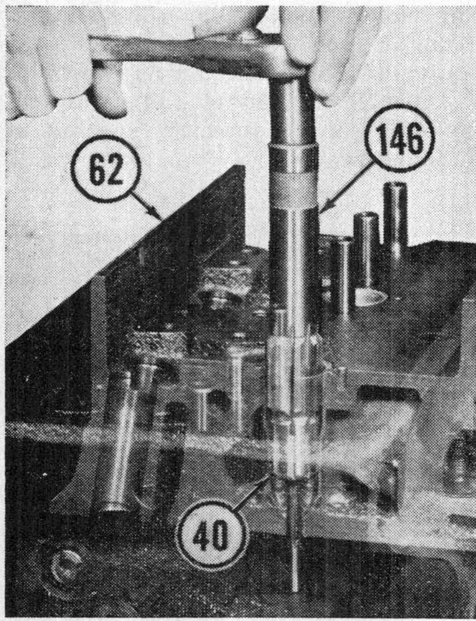

Fig. 0508C—GM engine. Resizing the injector hole tube for the injector body nut and spray tip with GM tool J 5286-7.

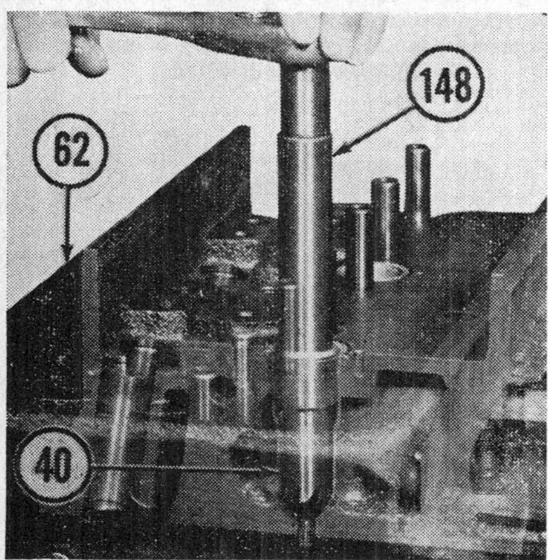

Fig. 0509A—GM engine. Reaming bevel seat in injector tube for the injector nut with GM tool J 5286-9.

5. Single jet nozzle
6. Double jet nozzle
42. Exhaust valves
43. Exhaust valve guides
47. Oil gallery
64. Neoprene seal ring

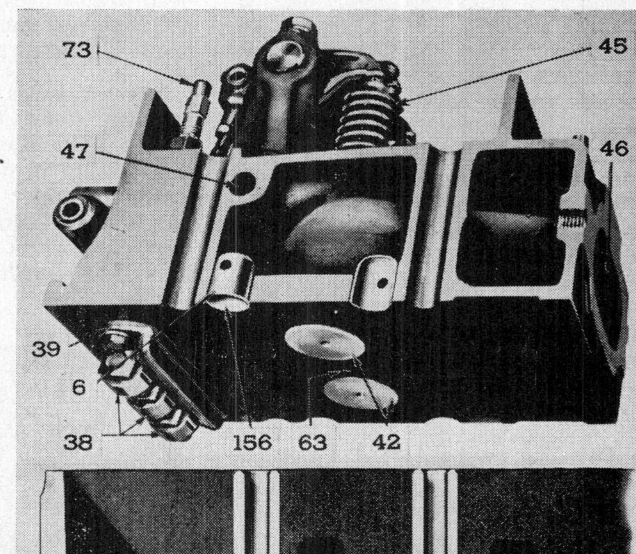

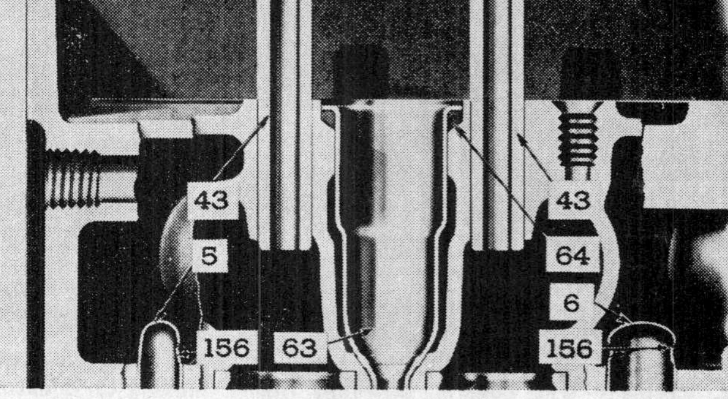

Fig. 0509B—GM engine. Showing location of water nozzles (5 and 6) in cylinder head.

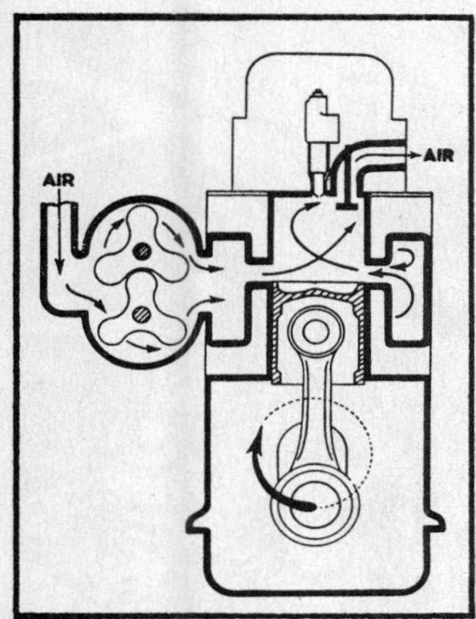

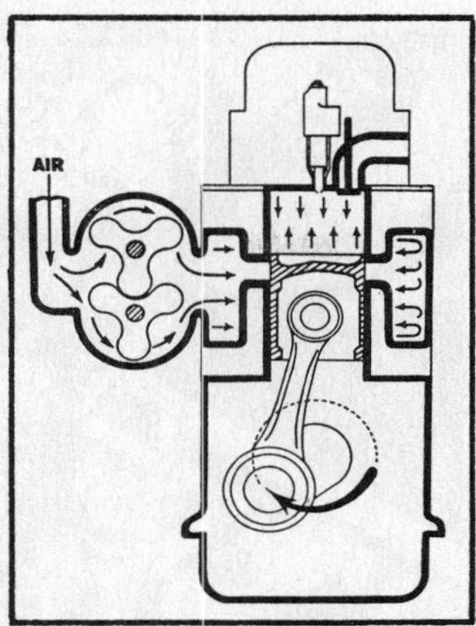

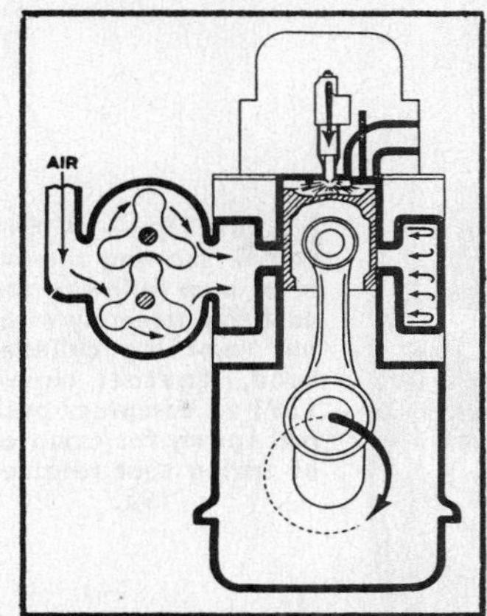

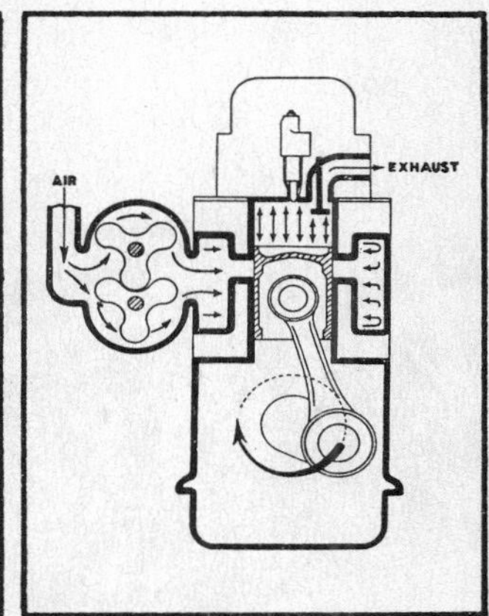

Fig. 0509C—Showing the two cycle principle of operation as employed in the GM series 71 engines. Left to right: Scavenging, compression, power, and exhaust.

of cylinder head. Nozzles should be installed from flush to $\frac{1}{32}$ inch below surface of cylinder head. If nozzle ports in cylinder head have been enlarged by corrosion, tin outside diameter of nozzles with solder so that a press fit can be obtained.

61F. REINSTALL HEAD. Before reinstalling cylinder head, check cylinder liner flange to top-of-block distance which is controlled by a liner insert (165—Fig. 0509D). On high block engines as used in the Oliver Super 99 GM, top surface of liner flange should be from 0.0465 to 0.050 below top surface of cylinder block and with not more than 0.002 variation in height between adjacent liners. On engines of the low block type, the top surface of liner flange should extend from 0.002 to 0.006 above the top surface of the block. Refer to paragraph 70D for data pertaining to this measurement.

Install new cylinder head compression gaskets (160), combined oil and water seals (163), water seals (161 and 162), and oil seal (164). Install oil seal (164) with flat surface up.

Install cylinder head and torque hold down nuts to 165-175 ft. lb. Adjust exhaust valve tappet gap to 0.012 cold and readjust to 0.009 hot.

Time injectors and position injector rack control tube levers as outlined in paragraphs 90, and 87 and 87A, respectively.

VALVES, SEATS AND GUIDES

62. Two exhaust valves, located in the cylinder head, are provided for each cylinder, Fig. 0510A. Inlet ports are in the cylinder liner. Valve head must not protrude more than 0.017 beyond the surface of cylinder head as shown in Fig. 0505B. Limiting the protrusion to a maximum of 0.017 will prevent the valve head from striking the top of the piston.

62A. Exhaust valves seat on renewable inserts in the cylinder head. Refer to Fig. 0505B for reconditioning limits of valve and insert.

62C. Shoulderless valve guides should be installed so that top or countersunk end of guide projects $1\frac{19}{32}$ inch above machined face of valve spring seat on cylinder head. Service guides, requiring a press fit of 0.0005-0.0035, are prefinished and do not require final sizing. Service valve guides are available with oversize outside diameters of 0.016. Oversize guides

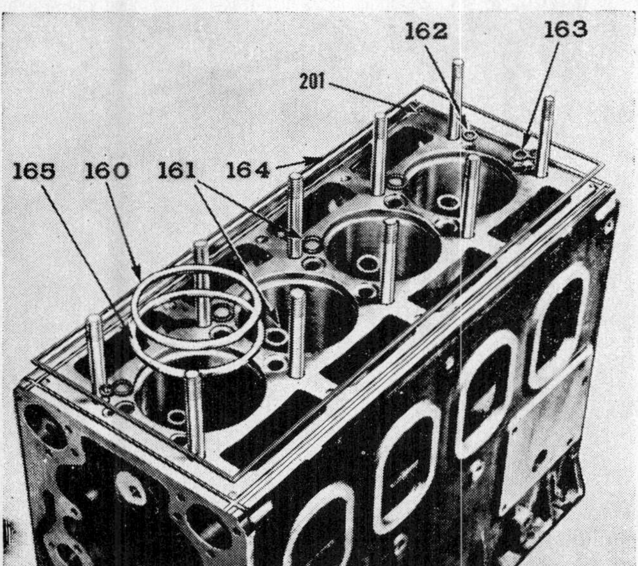

Fig. 0509D—GM engine. Location of cylinder head gaskets, oil and water seals for a high block design engine. Cylinder liner inserts (165) are installed under the liner flange.

160. Cylinder liner compression gasket
161. Water hole seal ring
162. Water hole seal ring
163. Water and oil holes seal ring
164. Cylinder head oil seal
201. Plug

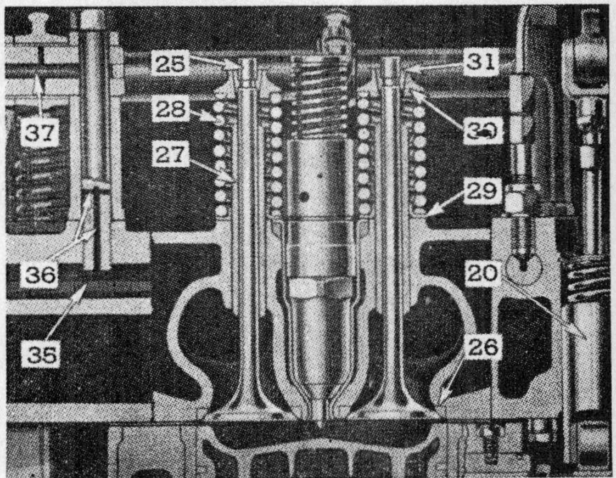

Fig. 0510A — A GM engine showing installation of exhaust valves, injector, and cam follower and push rod assembly.

20. Cam follower
26. Valve seat insert
27. Valve guide
29. Valve spring seat
30. Valve spring cap
35. Oil gallery
36. Rocker arms shaft bracket special bolt
37. Rocker arms shaft oil gallery

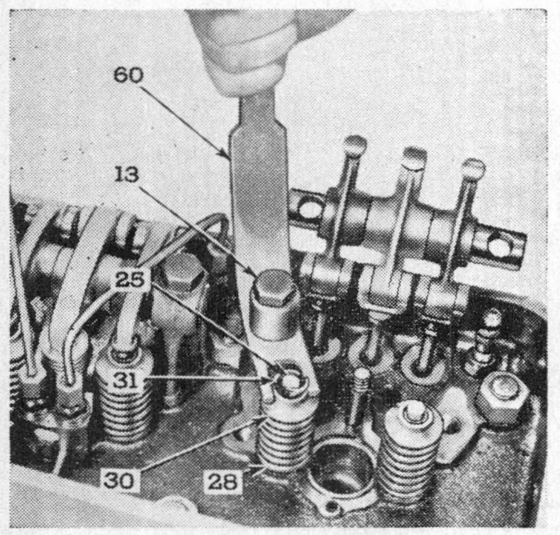

Fig. 0510B—A GM engine showing method of renewing valve springs without removing the cylinder head. Piston is at TDC.

13. Rocker arms shaft bracket bolt
60. Valve spring compressor—GM tool J 1227-A

Fig. 0511A — A GM engine showing removal of cam follower and push rod assembly without removing cylinder head. Install sleeve (39) to compress push rod spring for removal of spring seat retainer (19).

can be identified by an annular groove located on port end of guide.

Valve Face Angle..............30°
Valve Seat Angle..............30°
Valve Tappet Gap—Hot........0.009
Valve Stem Diameter..........0.342
Valve Clearance in Guide.......0.002
Renew If Clearance Exceeds..0.006
Guide Height Above Spring Seat Surface On Cylinder Head—Inches.......1 19/32
Valve Seat Counterbore Diameter 1.6265
Valve Seat Counterbore Depth 0.377
Exhaust Valve Lift—Inches.....0.394

VALVE SPRINGS

63. Any exhaust valve spring can be renewed, as shown in Fig. 0510B, without removing cylinder head. To renew a valve spring, first position the piston at TDC (TDC is indicated when injector plunger has traveled downward

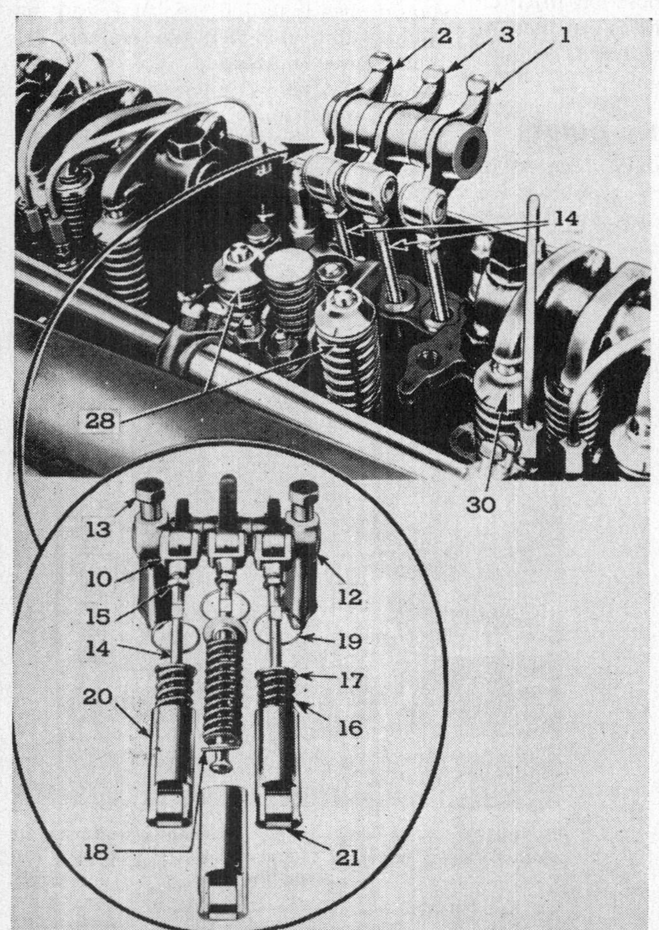

Fig. 0510C—GM engine valves and injector operating mechanism.

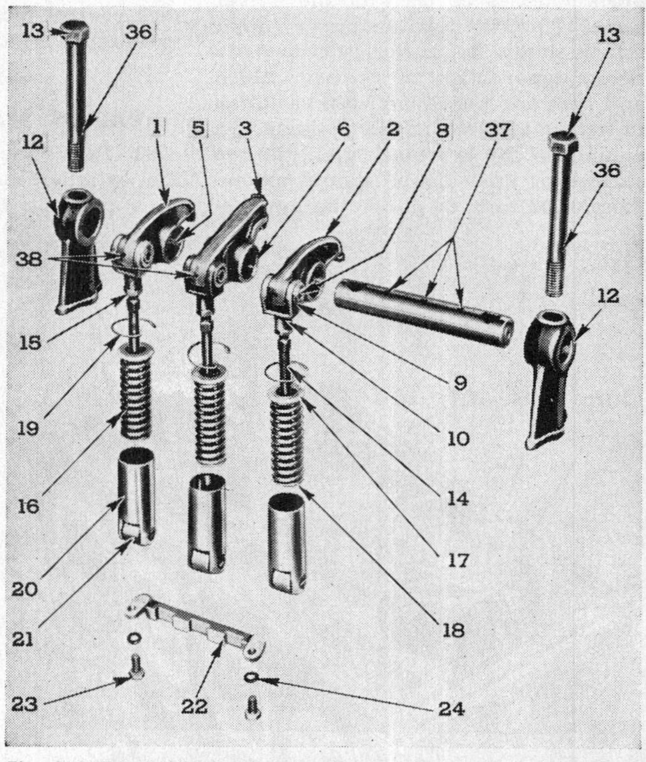

Fig. 0511B—A GM engine showing details of the valve and injector operating mechanism for one cylinder.

1. Rocker arm, left
2. Rocker arm, right
3. Injector rocker arm
5. Bushing
6. Bushing
8. Clevis pin
9. Bushing
10. Push rod clevis
12. Rocker shaft bracket
14. Push rod
15. Lock nut
16. Push rod spring
17. Push rod spring upper seat
18. Push rod spring lower seat
19. Spring seat retainer
20. Cam follower
21. Cam follower roller
22. Cam follower guide
36. Oil passage
37. Oil passage
38. Oil passage

approximately $\frac{3}{16}$ inch). Then, remove the injector fuel lines, cap screws attaching rocker arms shaft brackets to cylinder head, and injector.

Valve springs have a free length of 2⅜ inches and should require 44 plus or minus 3 lb. to compress spring to a length of $2\frac{3}{16}$ inches; and 140 plus or minus 3 lb. for a compressed length of 1 51/64 inches.

CAM FOLLOWERS AND PUSH RODS

Cam followers, operating in unbushed bores located in the cylinder head, are of the roller type.

64. Refer to Fig. 0510C. A cam follower (20), push rod (14), push rod spring (16), spring seats (17 and 18) can be removed from the cylinder head without removing the cylinder head as follows: Refer to Fig. 0511A and remove fuel lines, injectors, rocker arms shaft brackets and shaft. Unscrew rocker arm from push rod to be removed. Compress push rod spring with a tool similar to the one as shown, and remove spring retainer (19). Pull push rod assembly and cam follower out through top of cylinder head.

Push rods and cam followers can be removed from combustion chamber side of cylinder head as follows: Remove cylinder head and disconnect push rod from rocker arm. Then remove two cap screws (23—Fig. 0511B) attaching cam follower guide (22) to cylinder head and withdraw cam follower assembly.

64A. Cam followers, available in standard size only, have a running clearance in bores of 0.001-0.006. Renew the followers if clearance exceeds 0.006. Cam follower roller bushing to roller pin clearance is 0.0008-0.0015. Roller, bushing and roller pin are available as service items.

64B. Install cam followers in cylinder head with oil hole in bottom of follower pointing away from the valves.

Time injectors and position injector control racks as outlined in paragraphs 90, and 87 and 87A, respectively.

ROCKER ARMS AND SHAFTS

Three rocker arms, two for the exhaust valves and one for the injector, are provided for each cylinder. Refer to Fig. 0510C.

65. Rocker arms are provided with renewable bushings both for the rocker arm shaft and rocker arm clevis pin. Running clearance of either bushing is 0.001-0.004. Renew bushings and/or shaft and pin if clearance exceeds 0.004.

Rocker arms can be removed after removing hood, rocker arms cover, injector fuel lines, rocker arms shaft bracket cap screws and disconnecting rocker arms from push rods. Removal of the rocker arms may be facilitated by rotating engine crankshaft until the three rocker arm clevises are on the same plane.

Rocker Shaft Diameter........0.8735
Rocker Arm Bushing
Inside Diameter 0.8750
Running Clearance 0.001-0.004
Renew If Clearance Exceeds 0.004
Clevis Pin Bushing
Inside Diameter 0.5645
Running Clearance 0.001-0.004
Renew If Clearance Exceeds 0.005
Valve Tappet Gap—Hot........0.009

GEAR TRAIN COVER (FLYWHEEL HOUSING)

The gear train, Figs. 0511C and 0512A, is located at the flywheel end of the engine, and it is housed by the flywheel housing.

66. **REMOVE.** To remove gear train cover, first remove engine as outlined in paragraphs 60 through 60F. Remove engine clutch from flywheel, starting motor, and generator. Remove generator drive pulley (pulley flange is threaded for cap screw type puller). Remove five cap screws attaching generator drive pulley hub oil seal retainer to flywheel housing and remove retainer. Remove tachourmeter drive from flywheel housing. Remove four

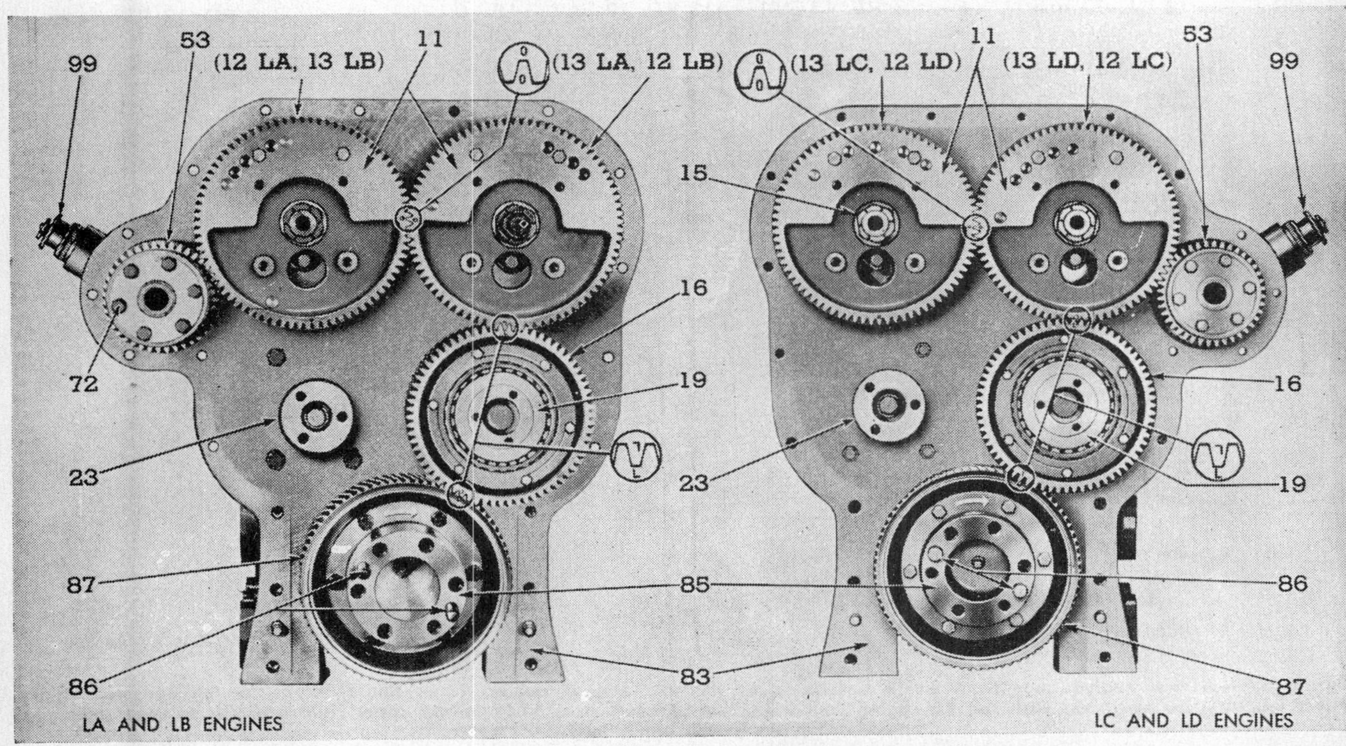

Fig. 0511C—A GM engine gear train, located at flywheel end of engine, shows timing marks for LA and LB engines. Oliver tractors are equipped with the RB engine which is shown in Fig. 0512A. Refer to Fig. 0512A for legend.

cap screws attaching oil pan to flywheel housing and loosen all others.

Support flywheel and remove flywheel attaching cap screws. Thread two $\frac{7}{16}$ x 4 inch standard thread cap screws into tapped holes in flywheel bolt flange and remove flywheel. Crankshaft rear lip type oil seal can be renewed at this time; refer to paragraph 72D.

Support flywheel housing at lifting eye bracket, and remove bolts and cap screws attaching housing to engine rear end plate. Refer to Fig. 0512B Install four guide studs as shown. Lift off housing by bumping same until it is free of the dowels.

66A. **REINSTALL.** Reinstall flywheel housing and tighten bolts and cap screws in sequence as shown in Fig. 0512C. Torque tighten flywheel attaching cap screws to 150-160 ft. lb.

TIMING GEARS

Refer to Figs. 0511C and 0512A, timing gear train, which is housed by the flywheel housing, consists of four helical gears and a blower drive gear. Normal backlash between mating gears is 0.003 to 0.008.

66B. **GEAR TIMING MARKS.** Refer to Figs. 0511C and 0512A. The four timing gears are marked for correct timing as follows: Camshaft gear mark "O", balance shaft gear marks "O" and "R", idler gear marks "R" and "R", and crankshaft gear mark "R".

With number one piston at TDC, gears are in time when "O" marks on balance shaft gear and camshaft gear are in register, and "R" mark on crankshaft gear is in register with one "R" mark on idler gear while the second "R" mark on idler gear is in register with an "R" mark on balance shaft gear.

66C. **CRANKSHAFT GEAR.** With flywheel housing removed, remove crankshaft gear (87) from shaft by using two cap screws which can be threaded into gear. Refer to Fig. 0513A. As one cap screw hole is offset, the gear can be installed in one position only.

66D. Reinstall crankshaft gear so that timing mark "R" is registered as outlined in paragraph 66B.

66E. **IDLER GEAR.** Refer to Fig. 0513B. With flywheel housing removed, remove cap screw attaching idler gear hub (19) to cylinder block.

Idler gear rotates on a double row taper roller bearing (17). Bearing preload is controlled by spacer ring (18), located between the bearing cones, and the clamping action exerted on the bearing cones by the cylinder block rear end plate (83) and a boss on the flywheel housing.

Fig. 0512B—A GM engine showing method of removing or installing the flywheel housing (gear train cover) by using aligning studs (149).

44. Dowel
89. Oil seal
148. Oil seal expander

Bearing preload is checked with idler gear assembly removed from engine. To check preload, mount idler gear assembly in a vise using two suitable metal plates which are positioned

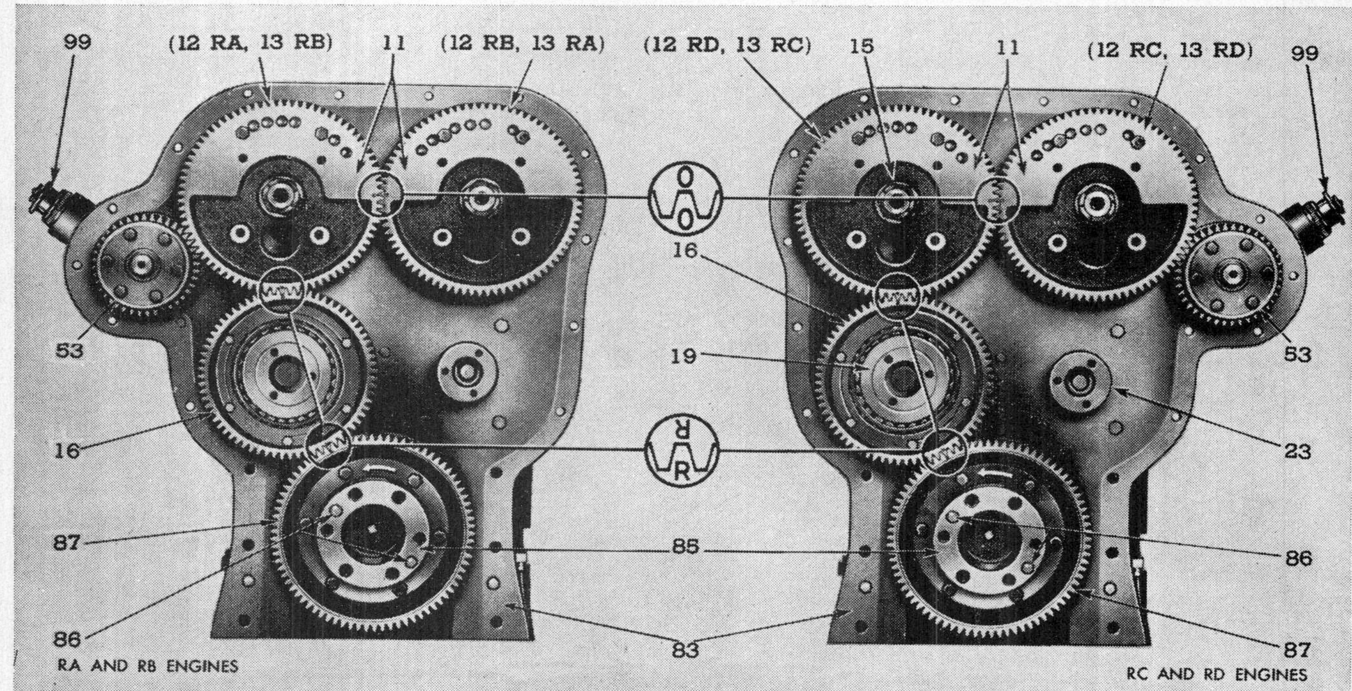

Fig. 0512A—A GM engine gear train which is located at flywheel end of engine shows timing marks for RA and RB engines. Oliver tractors are equipped with the RB engine which has the camshaft gear (12) at the upper right and the balance shaft gear (13) at the upper left.

12. Camshaft gear
13. Balance shaft gear
16. Idler gear
19. Idler gear hub
23. Spacer
53. Blower drive gear
83. Engine rear end plate
85. Crankshaft
86. Dowel
87. Crankshaft gear
99. Oil filler cap

Fig. 0512C—A GM engine flywheel housing showing bolt and cap screw tightening sequence.

Fig. 0513A—A GM engine showing use of two cap screws (102) for removing crankshaft gear (87).

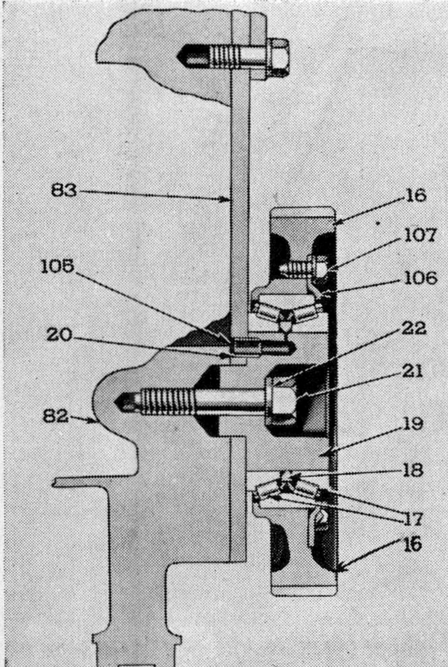

Fig. 0513B—A GM engine idler gear installation. Bearing preload is controlled by thickness of spacer ring (18) and clamping action which is exerted on bearing cones by engine rear end plate and a boss on flywheel housing.

so as to contact the ends of idler gear hub (19) and bearing cones. Using a spring scale attached to a cord wrapped around the gear, check for preload of ½ to 15 inch pounds. If preload is not within specified limits, renew the bearings, spacer (18) and idler gear hub which are available only as an assembly.

Reinstall idler gear and hub assembly so that timing marks on camshaft gear, balance shaft gear, crankshaft gear and idler gear are in register as outlined in paragraph 66B and as shown in Figs. 0511C or 0512A.

66F. **BLOWER DRIVE GEAR.** Refer to Fig. 0513C. Blower drive gear (53) can be removed after removing the flywheel housing, or blower and blower drive gear hub support assembly. If flywheel housing is removed, proceed as follows: Thread a $\frac{5}{16}$ inch—24 x 2 inch bolt into outer end of blower drive shaft (79) and remove snap ring (80) from outer end of shaft. Withdraw blower drive shaft. Remove six cap screws attaching gear and flexible coupling drive support (67) to blower drive gear hub (54) and lift off drive gear.

If blower drive gear hub (54) requires removal at this time, it will be necessary to remove the blower and drive gear hub support.

For removal and overhaul of blower drive gear hub (54), bushings (59), flexible coupling drive support (67)

20. Dowel pin, hollow
21. Hub-to-block cap screw
83. Engine rear end plate
105. Oil passage
106. Bearing retainer

47. Blower drive cover
51. Seal
52. Clamp
53. Blower drive gear
54. Gear hub
55. Lock ball
56. Thrust washer
57. Gear hub nut
58. Lock washer
59. Flanged bushings
61. Support
63. Gasket
67. Flexible drive coupling drive support
68. Flexible coupling drive cam
71. Retainer
75. Oil line
76. Oil line fitting
79. Blower drive shaft
80. Snap ring
81. Blower rotor gear hub
83. Engine rear end plate
91. Flywheel housing
92. Flywheel housing cover
93. Gasket

Fig. 0514A—A GM engine showing method used in removing either the camshaft gear or balance shaft gear.

and drive gear hub support (61), refer to paragraphs 78 through 79.

Reinstall blower drive gear so that cam lobes of flexible coupling are in line with the two oil grooves in blower drive gear hub.

66G. **BALANCE SHAFT GEAR.** Refer to Figs. 0511C or 0512A. With flywheel housing removed, remove balance shaft gear as shown in Fig. 0514A by using a puller attached either to two or four cap screws which can be threaded into gear.

When reinstalling balance shaft gear, heat same in oil to facilitate installation. Reinstall balance shaft gear so that timing marks on camshaft gear, balance shaft gear, idler gear and crankshaft gear are in register as outlined in paragraph 66B and shown in Figs. 0511C or 0512A.

66H. **CAMSHAFT GEAR.** Refer to Figs. 0511C or 0512A. With flywheel housing removed, gear can be removed from camshaft as shown in Fig.

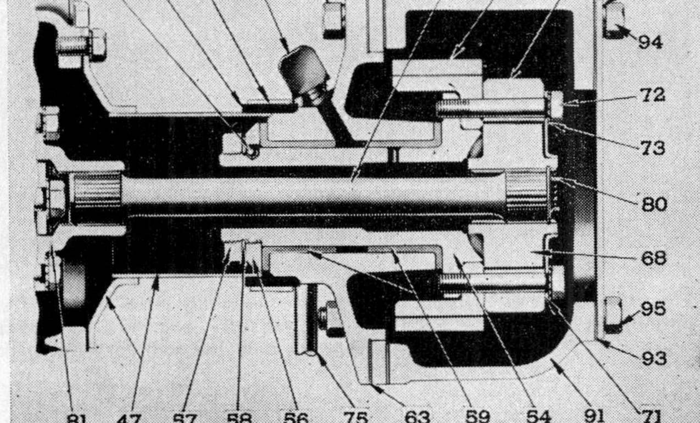

Fig. 0513C—A GM engine blower drive gear and support assembly.

0514A by using a puller attached either to two or four cap screws which can be threaded into gear.

When reinstalling camshaft gear, heat same in oil to facilitate installation. Timing marks on camshaft gear, balance shaft gear, idler gear and crankshaft gear should register as outlined in paragraph 66B and shown in Fig. 0511C or 0512A.

CAMSHAFT

67. Refer to Fig 0514B which shows camshaft installation for a six cylinder model. The two intermediate journals of the series 3-17 engine camshaft rotate in aluminum bearings (14); and front and rear journals rotate in supports, each containing two steel backed bronze bushings. The two-piece intermediate journal aluminum bearings are retained to the journal by lock rings. This construction enables the intermediate bearings to be installed with the camshaft as an assembly after which they are locked in the cylinder block by set screws which are accessible after removing the cylinder head.

67A. To remove camshaft, first remove gear train cover (flywheel housing) as outlined in paragraph 66 and cylinder head as outlined in paragraph 61. Remove both air cleaners, fan blades, fan drive belt, and balance weight cover. Refer to Fig. 0514C. Place a block of wood between the balance weights and remove nut (9—Fig. 0515A) retaining weight to shaft. Remove weight by prying off same as shown, and remove weight positioning Woodruff key and thrust washer. Other thrust washer is located on other side of front journal bearing support. Remove cap screws attaching rear journal bearing support to cylinder block. Remove two set screws which lock intermediate journal aluminum bearings in cylinder block. Set screws are accessible from top surface of block. Withdraw camshaft, intermediate journal bearings and gear assembly rearward and out of cylinder block.

67B. Renew aluminum bearings when maximum running clearance of 0.009 is exceeded or when thickness of bearing shells at their center is less than 0.340.

Front and rear journal support and bushing assemblies are available with bushings which are of 0.010 and 0.020 undersize; whereas bushings only are available in 0.020 undersize. Intermediate bearings with standard outside diameters are available with inside diameters in standard size and undersizes of 0.010-0.020. They are also available with a standard inside diameter and 0.010 oversize on the outside diameter to correct looseness between bearing and bore in cylinder block.

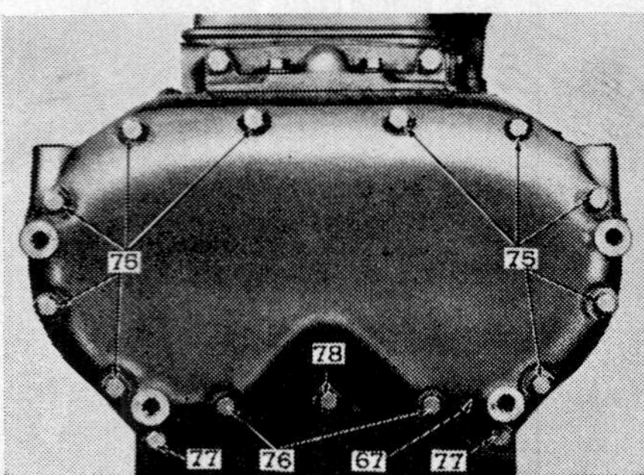

Fig. 0514C — GM engine balance weight cover which is located on front of engine.

75. Cap screw ⅜—24x2⅞ inches
76. Cap screw ⅜—16x3½ inches
77. Cap screw ⅜—16x1⅞ inches
78. Cap screw ⅜—24x1½ inches

Camshaft end play of 0.004-0.011 with a maximum of 0.018 is controlled by two thrust washers located on either side of the number one journal bearing support. Thrust washers which are available in standard thickness of 0.121 and oversizes of 0.005 and 0.010 are installed with the steel side next to the bearing support.

Tighten camshaft gear and balance weight retaining nuts to 300-325 ft. lb. torque.

Journal Diameters:

Front & Rear 1.497
Intermediate 1.498

Fig. 0514B—A 6 cylinder GM engine showing installation of camshaft and balance shaft. A similar installation is used in the type RB engine for the Oliver Super 99 GM tractor. Location of camshaft (1), balance shaft (25), thrust washers (6 and 27) and thrust bearings (3 and 26) will vary according to engine type as shown in Fig. 0499C.

1. Camshaft
3. Front bearing
6. Thrust washers
12. Camshaft thrust shoulder
14. Intermediate bearings (2 used on 3-71 engine)
17. Camshaft rear bearing
18. Camshaft gear (left helix)
25. Balance shaft
26. Front bearing
27. Thrust washers
31. Balance shaft thrust shoulder
32. Balance shaft rear bearing
33. Balance shaft gear (right helix)
40. Balance weights
41. Balance weight hub (not used on 3 and 4 cylinder engines)

Fig. 0515A—A GM engine showing method of removing balance weight assemblies. Weights as illustrated are for a 6 cylinder engine. One-piece weights are used on all 3 and 4 cylinder engines.

1. Camshaft
3. Camshaft front bearing
26. Balance shaft front bearing
40. Balance weight
41. Balance weight hub (not used on 3 and 4 cylinder engines)

Bearings Inside Diameter:
Front & Rear.................1.500
Intermediate 1.501
Running Clearance:
Front & Rear...........0.0025-0.006
Intermediate 0.0025-0.009
Camshaft End Play........0.004-0.018
Exhaust Valve Lift.............0.394

BALANCE SHAFT

69. Refer to Fig. 0514B. The front and rear journals of the balance shaft rotate in supports, each containing two steel backed bronze bushings.

69A. To remove balance shaft, first remove the gear train cover (flywheel housing) as outlined in paragraph 66. Remove the balance weight cover as follows: Remove both air cleaners, fan blades and fan belt and cap screws attaching balance weight cover to front face of engine and lift off the cover.

Place a block of wood between the balance weights and remove nut attaching weight to shaft. Remove weight by prying off same as shown in Fig. 0515A. Remove weight positioning Woodruff key, and cap screws attaching rear journal bearing support to cylinder block. Withdraw balance shaft and gear rearward and out of cylinder block.

Shaft journal recommended running clearance is 0.0025-0.006. Journal support and bushings assemblies are available with bushings which are 0.010 and 0.020 undersize; whereas bushings only are available in 0.020 undersize.

Balance shaft end play of 0.004-0.011 with a maximum of 0.018 is controlled by two thrust washers located on either side of the front journal bearing support. Thrust washers, available in a standard thickness of 0.121 and oversizes of 0.005 and 0.010, are installed with steel side next to bearing support. Tighten balance weight, and balance shaft gear retaining nuts to 300-325 ft. lb. torque.

ROD AND PISTON UNITS

70. Piston and connecting rod units are removed from above after removing cylinder head and oil pan. Number three rod and piston unit can be removed without performing additional work. Removal of either number one or two rod and piston units requires removal of oil pump and pump outlet line. Install cylinder liner hold down clamps to prevent liners from working upward while rotating the crankshaft or removing the piston and rod units. Piston and rod units can be installed as an assembly with the cylinder liner as shown in Fig. 0515C or without the cylinder liner by using conventional methods.

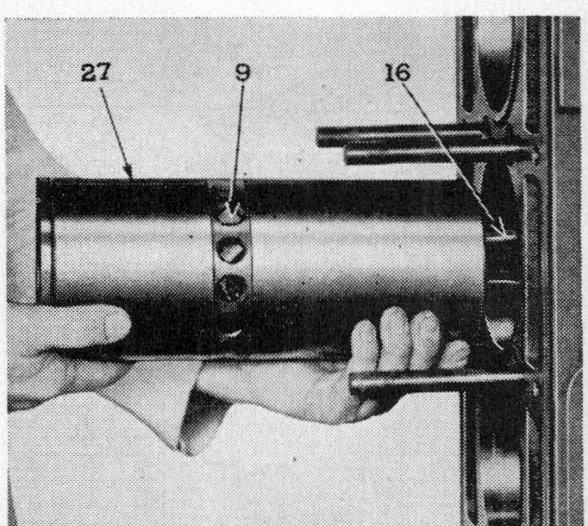

Fig. 0515C—GM engine. Showing method of installing piston and rod assembly with the cylinder liner.

Numbered side of connecting rod and cap should face blower side of engine. Torque the $\frac{7}{16}$ inch connecting rod bolts to 65-75 foot pounds.

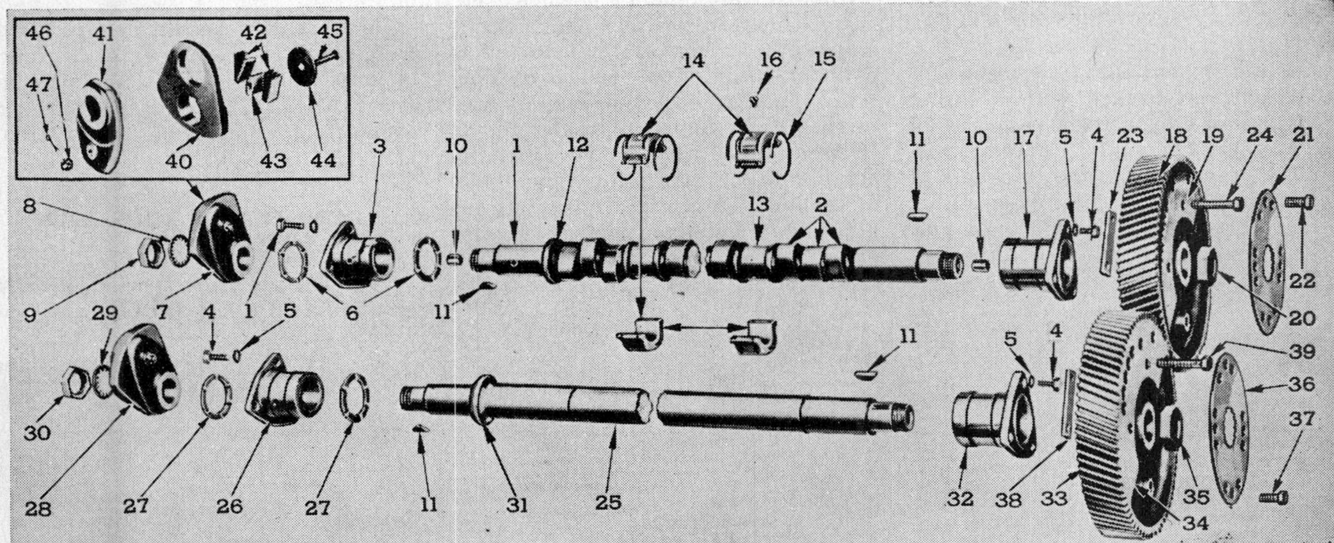

Fig. 0515B—A GM engine showing details of camshaft (1) and balance shaft (25). Camshaft intermediate bearings (14) which are removed when removing the camshaft are locked in cylinder block with set screws (16). Set screws are accessible after removing cylinder head. Spring loaded front balance weights (7 and 28) are used only on 6 cylinder engines whereas one-piece balance weights are used on all other engines.

3. Front bearing
6. Thrust washers
10. Plug
12. Thrust shoulder
15. Lock ring
17. Rear bearing
21. Gear nut retainer
23. Rear balance weights (used on 4 and 6 cylinder engines)
26. Front bearing
27. Thrust washers
31. Thrust shoulder
32. Rear bearing
36. Gear nut retainer

40-47. Spring loaded type of balance weight (used only on 6 cylinder engines)

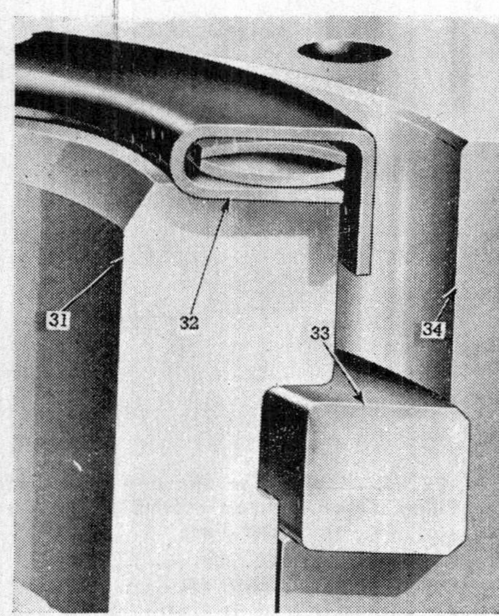

Fig. 0516A—A GM engine cylinder liner installation. Insert (33) controls relationship of top surface of liner flange to machined surface of cylinder block.

32. Compression gasket 34. Cylinder block

PISTONS, LINERS AND RINGS

70A. Cast steel, tin plated pistons equipped with four compression and two oil rings operate in dry slip fit type cylinder liners. Pistons are bushed for the full floating piston pin. Cylinder liners contain twenty air inlet ports of figure 8 design which are radially placed at mid length of sleeve.

70B. **PISTONS AND LINERS.** Pistons are supplied in standard size with a skirt diameter of 4.244 and oversizes of 0.010, 0.020 and 0.030. Refer to paragraph 70G for data on piston pin bushings.

Desired piston skirt clearance of 0.004-0.009 is checked with a spring scale pull of 6 lb., using a ½ inch wide feeler gage. The actual clearance will be 0.001 greater than the thickness of feeler gage. For example: When actual clearance is 0.006, a spring scale pull of approximately 6 lb. will be required to withdraw a 0.005 feeler gage.

70C. Cylinder liners are a slip fit (0.0005 to 0.0025 loose) in the cylinder block and can be removed, in most cases, without the use of a puller. Both the cylinder block bores and inside diameter of the sleeves can be bored and then honed to obtain desired fit of sleeve to block and of piston to sleeve. Inspect cylinder block bores for fit and contact of liner to bore. Check cylinder bore for out-of-round condition with limits of 0.001-0.003, and taper condition with limits of 0.001-0.002. If either of the preceding conditions are beyond maximum, it will be necessary to fit new liners by honing the cylinder block bores with 80 and 120 grit stones. Standard bore inside diameter is 4.6265 to 4.6275.

New liners with standard inside diameters of 4.2495 to 4.2505 are supplied with standard outside diameters of 4.6255 plus oversizes of 0.005, 0.010, 0.020 and 0.030. The oversize liners make it possible to obtain a 0.0005-0.0025 loose fit of the liner when clean-up honing of the cylinder bore is necessary.

Check inside diameter of liner for taper and out-of-round conditions with maximum limits of 0.002. These checks should be with liner installed in cylinder block.

70D. Before reinstalling the cylinder liner, place a liner insert (33—Fig. 0516A) into cylinder block counterbore for liner flange and then install the liner. Cylinder liners are installed dry and without any coating such as oil-mixed aluminum powder, etc. Clamp liner in position and measure distance from top of liner to top of cylinder block. Top surface of liner flange on high block type engines used in the Oliver Super 99 tractor should be from 0.0465-0.050 below surface of cylinder block, and with a maximum of 0.002 difference in height between adjacent liners. Make corrections either by reducing the thickness of the flange insert or by plating the insert.

70E. **RINGS** There are four compression and two oil rings per piston. Refer to Fig. 0516B.

Check ring gap and side clearance against the following limits.

End Gap:
- Compression Rings ..... 0.025 -0.040
- Oil Control Rings....... 0.010 -0.020

Side Clearance:
- 1st Compression .... 0.010 -0.022
- 2nd Compression .... 0.008 -0.015
- 3rd & 4th Comp. .... 0.006 -0.013
- Oil Control 0.0015-0.008

PISTON PINS AND BUSHINGS

70F. Hollow type piston pin which is of the full floating type, rotates in bushed bores of the piston and in similar bushings located in the small end of the connecting rod.

70G. Piston pins are retained in piston by stamped metal retainers (8—Fig. 0516B) and lock rings (2). Since the fit of metal retainers may vary from 0.001 loose to 0.0005 press, it may be necessary to pry out the retainer by punching a hole through its center.

The standard 1.4996-1.500 diameter piston pin is available in 0.010 oversize. Maximum allowable limit of piston pin clearance in either bushing is 0.010.

The split type bushings should be installed in piston with split side at bottom of piston and with inner end of bushing flush with edge of pin boss. Rod bushings should be installed in rod with split side at top of rod, as shown in Fig. 0517A, and with outer edge of bushings flush with edge of rod.

New bushings should be sized after installation to provide a pin clearance of 0.0025-0.0035 in piston bushings, and 0.0015-0.0025 clearance in rod. Connecting rod center to center length is 10.124-10.126 inches.

CONNECTING RODS AND BEARINGS

70H. The precision type, steel backed aluminum rod bearings are renew-

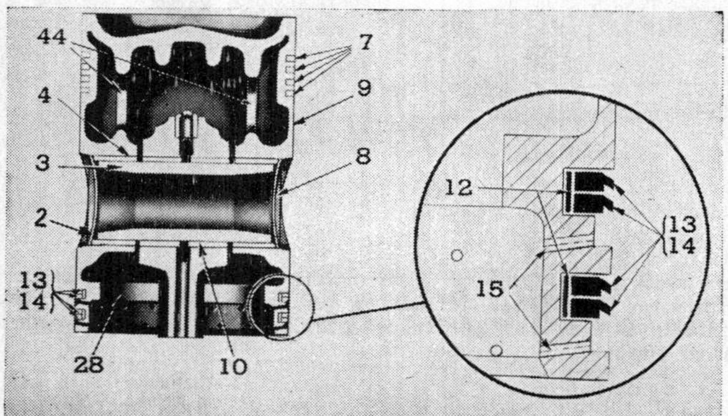

4. Bushing, piston
7. Compression rings
8. Piston pin retainer
9. Piston
10. Bushing, rod
12. Ring expander
13. Oil control rings
14. Oil control rings

Fig. 0516B—A GM engine piston, rings and piston pin assembly.

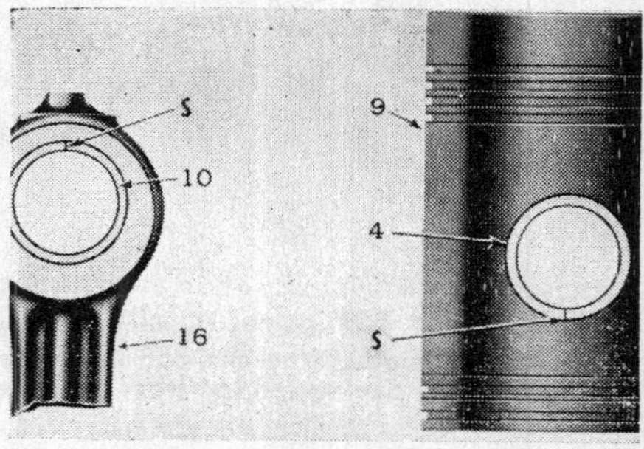

Fig. 0517A—GM engine. Showing location of bushing split (S) in piston pin bushings (4 and 10) for piston and connecting rod.

Fig. 0517B—A GM engine showing the use of a small diameter rod to remove rear main bearing upper shell (3) when crankshaft is installed.

able from below without removing the rods from the engine. Bearing shells are supplied in undersizes of 0.002, 0.010, 0.020 and 0.030.

Check crankpins and bearings for wear, scoring and out-of-round conditions.

Crankpin Diameter 2.7495
Taper, Maximum 0.003
Out-of-Round, Max. 0.003
Bearing Running
  Clearance 0.0015-0.0045
    Renew if Clear-
      ance Exceeds 0.006
End Play .... 0.006 -0.012
Rod, Center to
  Center Length 10.125

Fig. 0518A—A GM engine showing separate thrust washers (6) at rear main bearing to control crankshaft end play.

CRANKSHAFT AND BEARINGS

71. The induction hardened crankshaft rotates in four precision type, steel backed aluminum bearings which are renewable from below without removing the crankshaft. A special bolt or pin inserted in the crankshaft main journal oil hole will facilitate removal of upper shells for main journals 1, 2 and 3. Rear main journal upper bearing shell can be removed with a small diameter curved rod as shown in Fig. 0517B. Identification numbers on main bearing caps face blower side of engine.

Bearing shells are supplied in undersizes of 0.002, 0.010, 0.020 and 0.030. In regrinding crankshafts for use with undersize bearings, all journal fillets must have 0.130 to 0.160 radius between crank check and journal.

71A. Crankshaft can be removed after removing the engine from tractor chassis, flywheel housing, oil pan, rod caps, crankshaft front cover, and main bearing caps.

71B. Crankshaft end play of 0.004 0.011 with a maximum of 0.018 is controlled by varying the thickness of thrust washer halves (6—Fig. 0518A), located on either side of the rear main bearing journal. Grooved face of thrust washers contact thrust surface on crankshaft. Thrust washers are supplied in oversizes of 0.005 and 0.010. If thrust surfaces of crankshaft are worn or ridged excessively, they should be reground. Standard dimensions at rear main bearing thrust washer surfaces, indicating that standard thrust washers of 0.1220 thickness should be used, are shown in Fig. 0518B.

Excessive grooving of the crankshaft where rear oil seal lip contacts the surface can be corrected by installing a seal sleeve on the crankshaft and using an oversize oil seal. For data on crankshaft front and rear oil seals, refer to paragraphs 72 through 72G.

71C. Check shaft and main bearings for wear, scoring and out-of-round conditions.

Main Journal Diameter..........3.500
Running Clear., Mains...0.0015-0.0045
  Renew If Clear-
    ance Exceeds 0.006
Crankpin Diameter 2.7495
Running Clearance,
  Crankpins 0.0015-0.0045
  Renew If Clear-
    ance Exceeds 0.006
Main & Rod Journal
  Out-of-Round, Max. 0.003
Main & Rod Journal
  Taper, Max. 0.003

Fig. 0518B—A GM engine showing standard dimensions of the crankshaft rear main journal thrust surfaces for the use of standard thickness thrust washers.

Crankshaft End Play,
  Recommended 0.004-0.011
Crankshaft End Play
  Control ....No. 4 Brg.
Main Brg. Bolt Tightening
  Torque—Ft.-Lb. 180-190

CRANKSHAFT OIL SEALS

72. **REAR OIL SEAL.** The lip type oil seal (20—Fig. 0519A) for the rear end of the crankshaft is mounted in the flywheel housing. Oil seal can be renewed after removing the flywheel as outlined in paragraphs 72A through 72D.

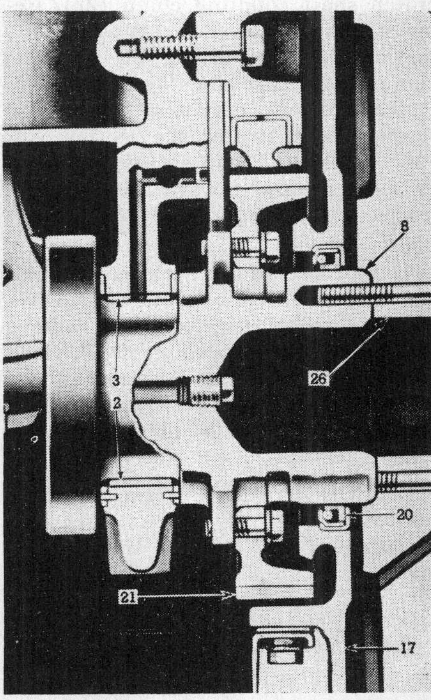

Fig. 0519A—A GM engine showing crankshaft rear oil seal (20) which is located in flywheel housing (17). Remove flywheel to renew seal.

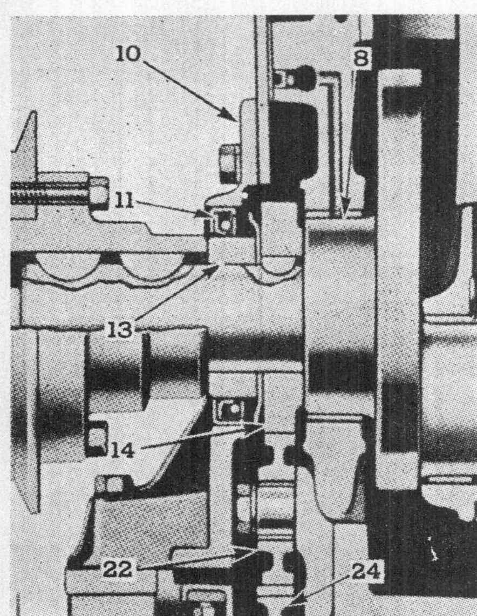

Fig. 0519B—GM engine. Crankshaft front oil seal (11) is located in crankshaft front cover (10). Remove crankshaft pulley and oil seal spacer ring (13) to renew seal.

14. Oil pump drive gear
22. Oil pump drive idler gear
24. Oil pump driven gear

3. Oil pump
4. Pump relief valve
36. Copper gasket
56. Idler gear
73. Regulator valve

Fig. 0520A—A GM engine oil pump installation. Non-adjustable pump relief valve (4) is preset for 100 psi, and non-adjustable oil regulator valve (73) is preset for 45 psi.

72A. If tractor is equipped with pto, remove the drive shaft of same as outlined in paragraph 145H. PTO drive shaft is splined into the flywheel.

72B. Remove storage batteries, clutch housing front cover, and flat dust cover which is attached both to the front frame and to the clutch release bearing carrier. Referring to Fig. 0458, remove master link from clutch shaft coupling chain (22). Remove chain and slide coupling (21) forward after removing the coupling clamp bolt. Disconnect outer end of clutch shifter shaft and loosen set screw which locates the shifter fork (26) on shaft. Bump shifter shaft out of shifter fork toward left side of tractor. Lift clutch shaft (23) rearward and out.

72C. Correlation mark clutch and flywheel, then unbolt clutch from flywheel. Support flywheel and remove flywheel attaching cap screws. Thread two $\frac{7}{16}$ x 4 inch standard thread cap screws into tapped holes provided at flywheel bolt flange to remove flywheel.

72D. Oil seal can be renewed after removing the flywheel. Excessive grooving of the oil seal lip contacting surface on the crankshaft can be corrected by installing a seal sleeve (GM part No. 5193413) and an oversize oil seal (GM part No. 5192776). Seal sleeve is a slip fit on the crankshaft.

72E. **FRONT OIL SEAL.** The lip type oil seal (11—Fig. 0519B) for the front end of the crankshaft is mounted in the crankshaft front cover (10). Oil seal can be renewed after removing the crankshaft pulley as outlined in paragraphs 72F and 72G.

72F. First remove tractor hood and radiator is outlined in paragraph 110. After removing radiator, remove air cleaner to blower pipe. Remove four cap screws attaching oil cooler to mounting bracket in front frame and lay oil cooler aside. Remove fan belt. Remove crankshaft pulley retaining cap screw. Remove pulley by using a puller attached to two cap screws which are threaded into the pulley. Crankshaft front oil seal can be renewed at this time.

72G. The lip of the crankshaft front oil seal does not contact the crankshaft; instead the lip contacts a renewable seal sleeve (13) which is a slip fit on the crankshaft. Check lip con-

Fig. 0520B—A GM engine oil pump.

3. Pump body
4. Pressure relief valve (non-adjustable)
20. Bushing
21. Pump body cover
22. Driven gear
23. Drive gear
24. Driven gear shaft
25. Bushing
36. Copper gasket
37. Relief valve port plug
50. Bushing
56. Idler gear
82. Idler gear support
85. Thrust washer
86. Cap screw

tacting surface of seal sleeve, and renew or reverse the sleeve if necessary.

FLYWHEEL AND RING GEAR

73. Remove flywheel as outlined in paragraphs 72A through 72C.

Ring gear can be renewed after removing the flywheel. To install a new ring gear heat same to 400 degrees F. and install with either side facing the engine. Torque tighten flywheel attaching cap screws to 150-160 ft. lb.

OIL PUMP AND SYSTEM

74. The components which make-up the lubricating system include a gear type pump which is located on the underside of the cylinder block and gear driven by a gear located on the forward end of the crankshaft; an oil pump relief valve (preset to 100 psi) to by-pass excess oil from discharge side of pump to inlet side of pump; an oil pressure regulator (preset to 45 psi) to maintain a constant engine oiling pressure; an oil cooler by-pass valve (preset to 40 psi) to by-pass the oil cooler should it become restricted; an oil cooler; and a partial flow type oil filter.

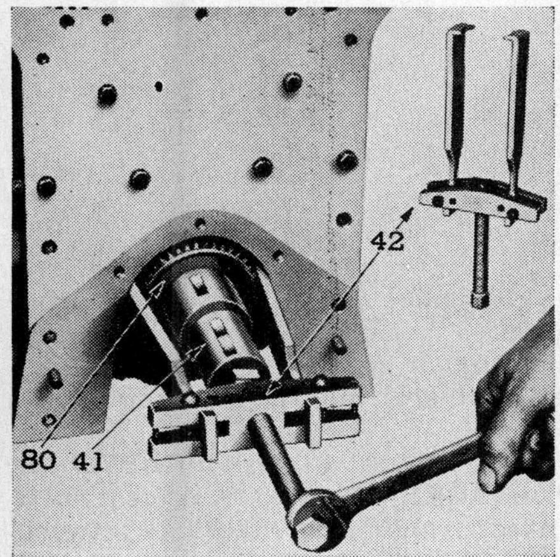

Fig. 0521A—A GM engine showing method of removing oil pump drive gear from crankshaft.

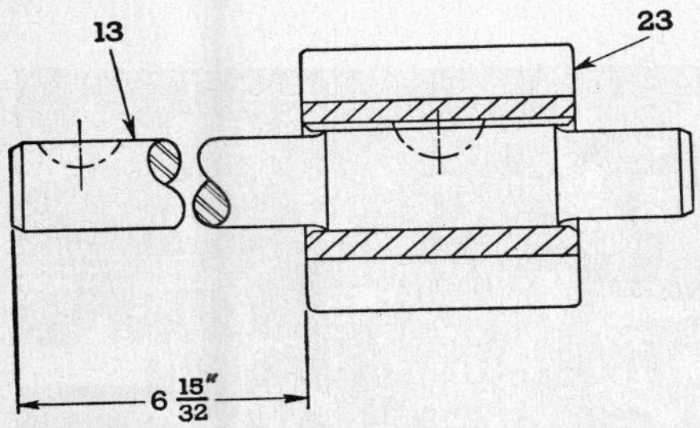

Fig. 0521B—A GM engine showing location of oil pump gear (23) on drive shaft (13).

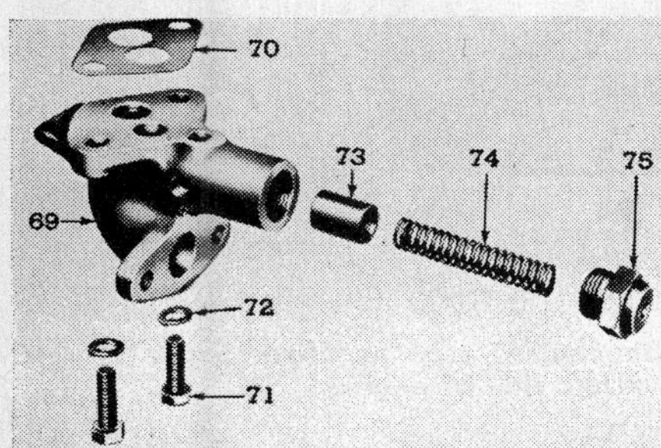

Fig. 0521C—A GM engine non-adjustable type oil pressure regulator which is preset for 45 psi.

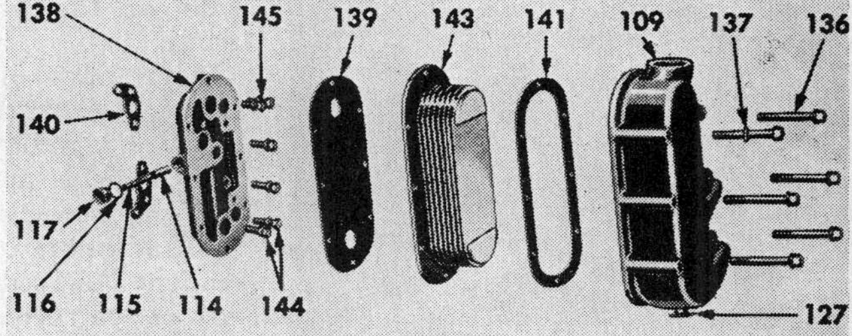

Fig. 0522A—A GM engine oil cooler and oil cooler by-pass valve details. Oil cooler is mounted in the tractor front frame.

74A. **PUMP.** The gear type pump, Fig. 0520A, which is driven by a gear located on the forward end of the crankshaft through an idler gear is mounted on the underside of the cylinder block. To remove pump first remove oil pan. Remove cap screws attaching oil pressure regulator, oil pump and pump intake screen support bracket to engine block.

Shims located between pump mounting base and main bearing cap are used to adjust backlash of pump idler gear to pump drive gear on crankshaft to 0.005-0.020. A 0.005 shim will change the backlash approximately 0.0035.

74B. Pump drive gear located on forward end of crankshaft can be removed after removing crankshaft pulley as outlined in paragraph 72F. Support engine, and remove engine front mounting bolts and crankshaft front cover. Refer to Fig. 0521A and use a similar puller set-up to remove pump drive gear from crankshaft.

Install pump drive gear so that chamfer on gear hub is toward main bearing cap.

74C. Pump disassembly procedure is self-evident. Refer to Fig. 0520B. Use a suitable puller to remove pump driven gear (49) before disassembling the pump.

Bushings (20 and 50) for the drive shaft (two bushings in pump body and one in pump cover), bushings (25) for pump driven gear (two bushings) and a bushing for the idler gear (one bushing) are supplied for service. Final size bushings to provide a free-running fit for the shaft and without perceptable looseness.

Pump body and cover are assembled without a gasket. Pump drive idler gear (56) is installed with flush side of hub and gear teeth facing the idler gear support (82). Install pump gear (23) on drive shaft (13) so that location of gear is $6\frac{15}{32}$ inches from keyway end of shaft as shown in Fig. 0521B. Press pump driven gear (49—Fig. 0520B) on keyway end of shaft to a point where a 0.005 feeler gage can be inserted between the pump body and the end of the gear hub.

74D. **PUMP RELIEF VALVE.** Refer to Fig. 0520A. The non-adjustable pump relief valve, located in the pump body, by-passes oil from the pump outlet to the pump inlet when pump pressure exceeds 100 psi. The relief valve is accessible after removing the oil pan.

74E. **OIL PRESSURE REGULATOR VALVE.** An oil pressure of 45 psi is maintained within the engine, regardless of engine speed and oil temperatures by means of a regulator (73—Fig. 0520A) installed between pump outlet and oil gallery in cylinder block. Removal and disassembly of the non-adjustable oil pressure regulator, Fig. 0521C, is self-evident.

74F. **OIL COOLER BY-PASS VALVE.** A by-pass valve is located between the inlet side of the oil cooler and oil cooler element to by-pass the oil if the element becomes restricted.

The non-adjustable by-pass valve (114—Fig. 0522A), preset for 40 psi, can be removed from oil cooler adaptor cover (138) without disturbing any of the engine components.

74G. **OIL COOLER.** An oil cooler of the heat exchanger type, using engine water as the coolant is mounted in the tractor front frame between the crankshaft pulley and radiator. Water from the outlet side of radiator flows through the oil cooler, which is a cast iron housing surrounding the exchanger core for the oil, to the inlet

Fig. 0522B—GM engine oil cooler which is located in tractor front frame is accessible after removing tractor radiator.

138. Oil cooler cover
OCI. Oil cooler inlet oil line
OCO. Oil cooler outlet oil line
PI. Water pump inlet hose
RO. Radiator outlet hose

side of the engine cooling system pump.

To remove oil cooler from tractor frame, Fig. 0522B, first remove tractor hood and radiator. Oil cooler is attached to a mounting bracket with four cap screws.

Disassembly of the oil cooler is self-evident. Use carbon tetrachloride or any of the recommended steam detergent solutions as a cleaning agent for the oil side of the cooler.

OIL PAN

75. Oil pan can be removed in the conventional manner without disturbing any of the engine or tractor components.

New oil pan gaskets of cork have four punched holes which serve as guides in positioning the gasket on the engine block. All other cap screw holes are punched out when installing the oil pan attaching cap screws. Before installing the oil pan, clean out (remove) punch out cork from the four blind cap screw holes in the flywheel housing and from the four blind cap screw holes in the crankshaft front cover.

BLOWER

76. Refer to Fig. 0523A. The Roots type blower, mounted on the left side of the engine, supplies fresh air needed for combustion and scavenging. Normal air box pressure is 4.2 inches Mercury or 2.06 psi @ 1200 crankshaft rpm, and 7.6 inches Hg. or 3.73 psi at 1600 crankshaft rpm. Blower drive at approximately twice engine speed is supplied by the blower drive gear which is located in the gear train of the engine. Engine accessory units such as the fuel pump (101), water pump (99) and governor weight housing (100) are mounted on the blower and are driven by the blower rotor shafts.

GENERAL INSPECTION

76A. Blower may be inspected for any of the following conditions while it is installed on the engine. To make the inspections remove the blower air inlet housing (78).

76B. **OIL SEAL LEAKAGE.** Damage or leaking oil seals are usually apparent by the presence of oil on blower rotors or inside surfaces of the housing. A further check for leaking oil seals can be made with engine operating at idle speed. A thin film of oil in the rotor compartment, radiating away from the seal (sun burst pattern), is indicative of an oil seal failure.

Remove blower and blower end plates to renew oil seals.

76C. **WORN BLOWER DRIVE.** Wear or damage in the blower drive flexible coupling is indicated by a rattling type of noise, and may be detected as follows: Grasp upper rotor and rotate it in either direction as far as possible; then release the rotor. Rotor can be rotated approximately ⅜ to ⅝ inch when measured at the lobe crown. Spring back when released should be at least ¼ inch. If rotors cannot be moved or if they move too freely, remove blower drive flexible coupling for further inspection.

To remove and inspect blower drive flexible coupling remove flywheel housing, or remove blower and blower drive gear hub support assembly.

76D. **LOOSE ROTOR SHAFTS OR DAMAGED BEARINGS.** A loose shaft causes rubbing or scoring between rotors, and between rotors and end plates. Worn or damaged bearings will cause rubbing at some point between mating rotor lobes or scoring of the

Fig. 0523A—GM engine blower installation. On engines used in Oliver Super 99 GM tractors blower is mounted on left side of engine.

AC. Air cleaner to blower pipe
GT. Governor control housing
OL. Blower drive gear hub oil line
TO. Tank to fuel pump line
9. Tachourmeter drive
10. Fuel pump outlet line
13. Chevron starting aid primer line
48. Blower drive gear hub support
78. Blower inlet housing
92. Flywheel housing cover
99. Water pump
100. Governor weight housing
101. Fuel pump
104. Flywheel housing cover
114. Blower drive seal and clamp

blower housing at the end where the bearings have failed.

Remove blower for complete overhaul.

76E. **EXCESSIVE BACKLASH.** Rubbing or scoring of rotor lobes through their entire length is caused by excessive backlash in blower timing gears.

Remove blower for complete overhaul.

REMOVE AND OVERHAUL

77. **BLOWER DRIVE SHAFT.** Refer to Figs. 0523A and 0523B. To remove blower drive shaft (79), proceed as follows: Remove battery from left side of tractor and disconnect tachourmeter drive cable at drive unit (9). Remove six cap screws attaching flywheel housing cover (92) to flywheel housing and lift off cover. Thread a $\frac{5}{16}$ inch—24 x 2 inch bolt into exposed end of blower drive shaft (79). Remove snap ring (80) from exposed end of blower drive shaft and withdraw the shaft.

77A. If blower drive shaft is broken and the pieces cannot be removed as outlined in preceding paragraph 77, it will be necessary to remove the blower, as outlined in paragraph 80, to permit removal of shaft pieces.

78. **BLOWER DRIVE GEAR AND FLEXIBLE COUPLING.** Refer to Fig. 0523B. Blower drive gear (53) is located in the timing gear train of the engine and is housed by the flywheel housing.

Blower drive gear can be removed either by removing the engine flywheel housing (91) or by removing the blower and blower drive gear hub support (61).

78A. R&R IF FLYWHEEL HOUSING IS REMOVED. Refer to Fig. 0523B. To remove blower drive gear and flexible coupling if flywheel housing is removed proceed as follows: Thread a $\frac{5}{16}$ inch—24 x 2 inch bolt into exposed end of blower drive shaft (79). Remove snap ring (80) from exposed end of blower drive shaft and withdraw the shaft. Remove six cap screws (72) attaching blower drive gear and flexible coupling to blower drive gear hub (54), and lift off drive gear and flexible coupling.

Reinstall flexible coupling so that cam lobes of the drive cam are in line with two oil grooves located in blower drive gear hub.

78B. R&R IF BLOWER IS REMOVED. To remove blower drive gear and flexible coupling if blower is removed, proceed as follows: Remove bearing oil line (OL—Fig. 0523A) from blower drive support. Refer to Fig. 0523B. Remove bolts and cap screws attaching blower drive support (61) to engine rear end plate. Withdraw blower drive support assembly from rear end plate. (There are no timing marks on blower drive gear.) Remove six cap screws attaching drive gear (53) and flexible coupling (67) to blower drive gear hub (54), and lift off gear and flexible coupling.

Reinstall flexible coupling so that cam lobes of the drive cam are in line with two oil grooves located in blower drive gear hub.

78C. OVERHAUL. Refer to Fig. 0524A. With blower drive flexible coupling removed from blower drive gear hub disassemble the flexible coupling by pushing coupling cam (68) out of coupling support (67).

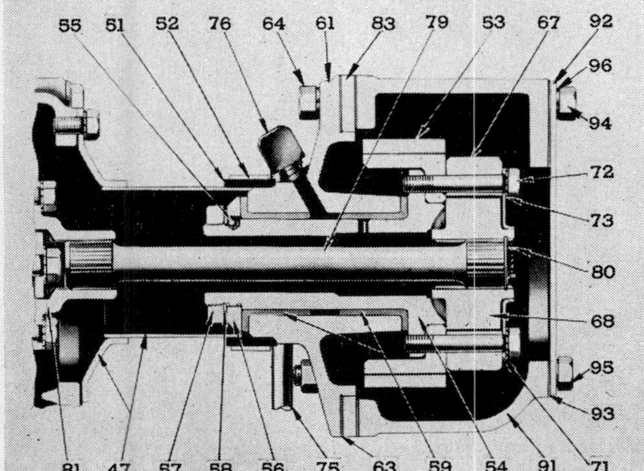

47. Blower drive cover
51. Seal
52. Clamp
53. Blower drive gear
54. Gear hub
55. Lock ball
56. Thrust washer
57. Gear hub nut
58. Lock washer
59. Flanged bushings
61. Support
63. Gasket
67. Flexible coupling drive support
71. Retainer
75. Oil line
76. Oil line fitting
79. Blower drive shaft
80. Snap ring
81. Blower rotor gear hub
83. Engine rear end plate
91. Flywheel housing
92. Flywheel housing cover
93. Gasket

Fig. 0523B—A GM engine blower drive gear and support assembly.

Fig. 0524A — Relationship of GM engine blower flexible coupling drive cam (68) to oil grooves (74) which are located in blower drive gear hub (54).

53. Blower drive gear
67. Flexible coupling drive support
69. Spring pack (21 leaves per pack)

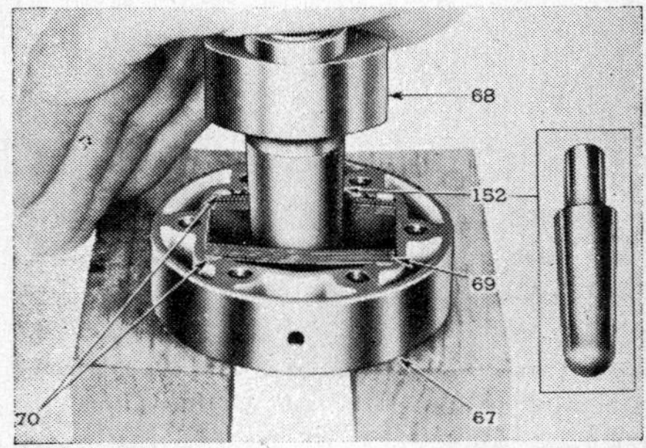

Fig. 0524B—Using a tapered sleeve to install flexible coupling drive cam (68) between spring packs (69).

67. Coupling drive support
70. Spring pack seats
152. Tapered sleeve, GM tool J 1471

Check blower drive shaft serrations in coupling cam for wear and damage. Check spring packs (69) which consist of 21 leaves per pack for breakage. Check spring pack contacting surface of coupling cam for wear.

Reassemble flexible coupling, as shown in Fig. 0524B, by placing the spring packs, each pack containing 21 leaves, in bore of coupling support. A small amount of grease will hold leaves together to form a pack. Install a coupling spring seat (70) in each of the four grooves located in coupling support. Place blower drive cam (68) on end of GM Special tool J1471 or a similar type tapered sleeve; then insert tool between the springs to force them apart while pressing the cam into position.

Backlash between balance shaft gear and blower drive gear is 0.003-0.008.

79. **BLOWER DRIVE GEAR HUB AND SUPPORT.** Refer to Figs. 0523B and 0525B. To renew either blower drive gear hub (54) or the bushings (59), first remove the blower as outlined in paragraph 80; then proceed as outlined in paragraph 78B.

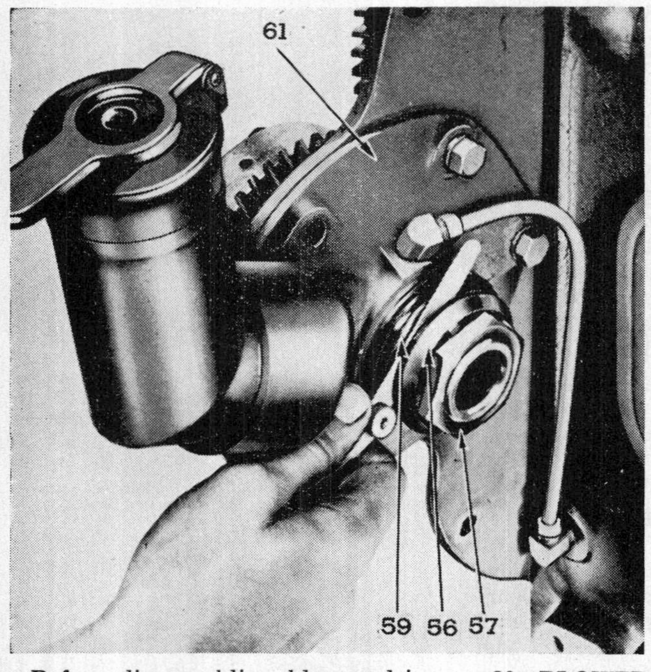

Fig. 0525A—Checking end clearance of GM engine blower drive gear hub assembly by inserting feeler gage between thrust washer (56) and flange of bushing (59). Recommended clearance is 0.005-0.010.

Before disassembling blower drive gear hub support, check clearance between flange of gear hub bearing (59—Fig. 0525A) and thrust washer (56) as shown. If clearance exceeds recommended clearance of 0.005-0.010, it will be necessary to renew the gear hub support and bushings as an assembly. Steel backed babbitt bushings (59) are not available separately for service.

Disassemble drive gear hub support by removing nut (57) from end of drive gear hub. Desired running clearance between hub and bushings is 0.001-0.005. Inside diameter of new bushings is 1.6260-1.6265.

80. **BLOWER R&R.** Refer to Fig. 0523A. To remove blower assembly from engine, proceed as follows: Drain engine coolant. Remove tractor hood as outlined in paragraph 60A.

Remove cover from top of governor control housing (GT), and remove rocker arms cover. Remove link connecting injector rack control tube to governor differential lever. Disconnect throttle booster spring from governor. Remove two cap screws attaching governor control housing to cylinder head, and four cap screws attaching governor weight housing (100) to governor control housing. Pull upper end of control housing away from cylinder head while pushing lower end of control housing toward engine to free the dowels, and lift off governor control housing.

Disconnect fuel lines (TO and 10) at fuel pump (101), and chevron starting aid line (13) from air inlet housing (78). Loosen hose clamps on hose connecting air cleaner pipe (AC) to blower air inlet housing, and slide hose toward the air cleaner.

Detach water pump inlet hose, bypass, and pump outlet connections on the pump.

Remove battery from left side of tractor and disconnect tachourmeter drive cable at drive unit (9). Remove six cap screws attaching flywheel housing cover (92) to flywheel housing and lift off cover. Thread a $\frac{5}{16}$ inch—24 x 2 inch bolt into exposed end of blower drive shaft. Remove snap ring from exposed end of blower drive shaft and withdraw the shaft.

Loosen clamp (114) on blower drive shaft cover seal. Remove four cap screws (CS) attaching blower assembly to cylinder block. Slide blower assembly slightly forward to withdraw blower drive shaft cover from seal, and lift blower assembly off engine.

Fuel pump (101) water pump (99) and governor weight housing (100) can be removed either before or after removing blower assembly from engine.

80A. **BLOWER OVERHAUL.** Refer to Fig. 0526A. Normal wear of blower rotor gears (13 and 14) causes a decrease of rotor-to-rotor clearance between the leading edge of the upper rotor lobes and the trailing edge of the lower rotor lobes. Clearance between the opposite sides of the rotor

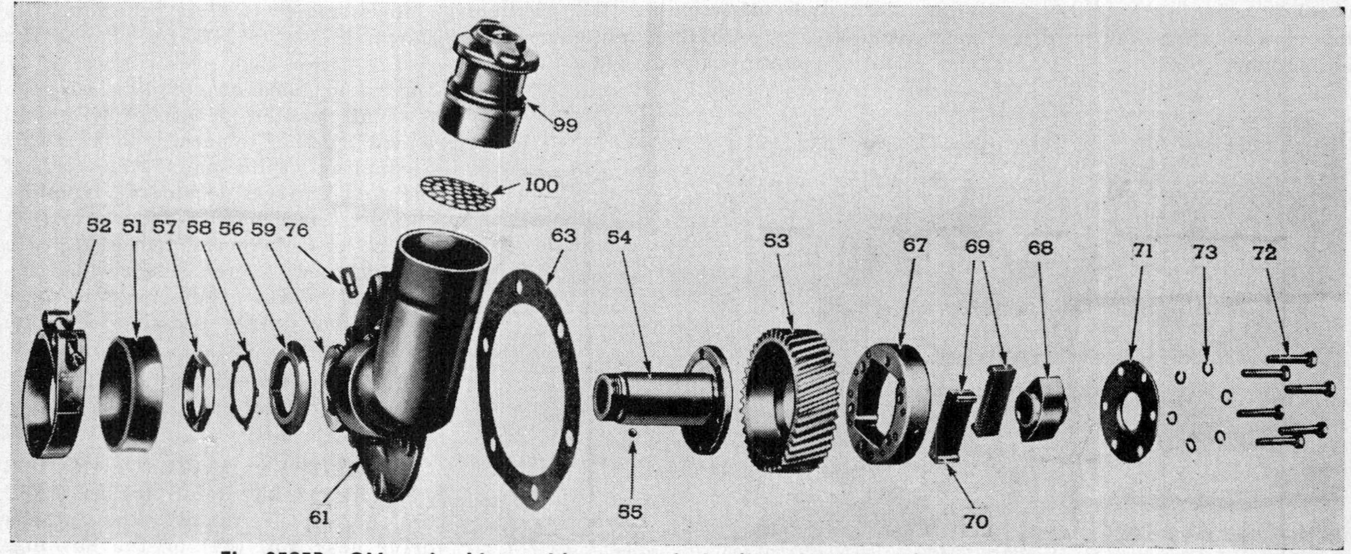

Fig. 0525B—GM engine blower drive gear unit details and relative location of components.

51. Drive cover seal
52. Clamp
53. Blower drive gear
54. Blower drive gear hub
55. Lock ball
56. Thrust washer
57. Nut
58. Lock washer
59. Bushings
61. Support
63. Gasket
67. Coupling drive support
68. Flexible coupling drive cam
69. Spring pack
70. Spring seat
71. Retainer
76. Oil line fitting
100. Screen.

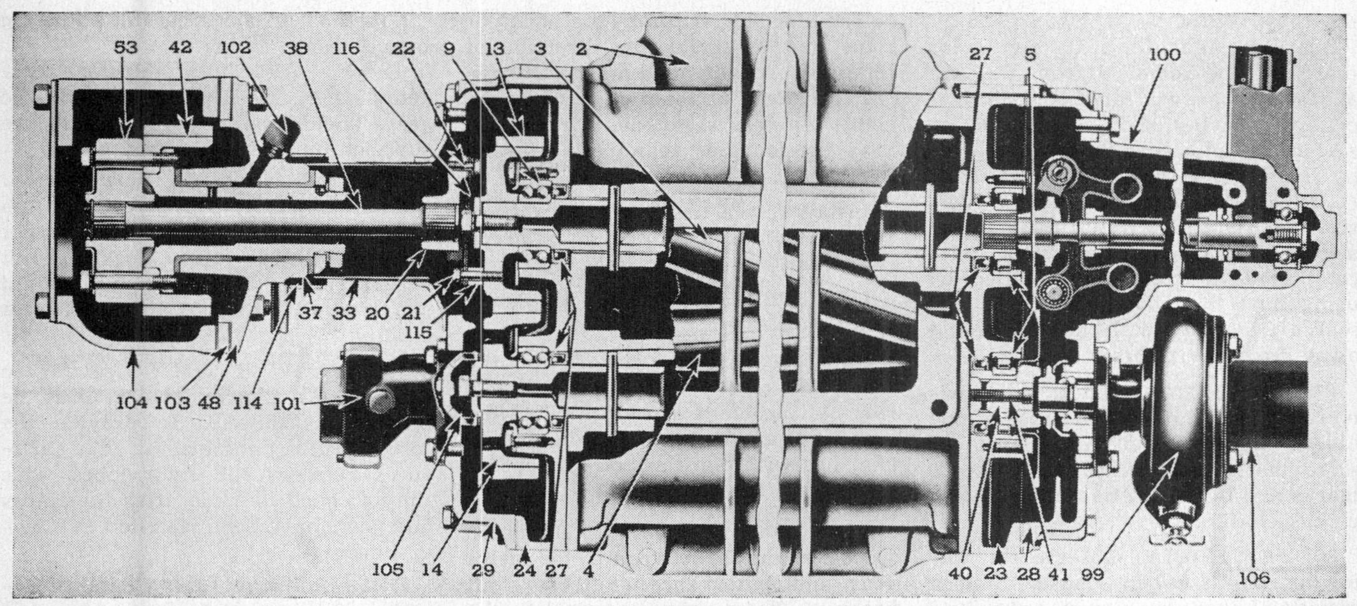

Fig. 0526A—GM engine blower and drive assembly with fuel pump (101), governor weight housing (100) and water pump (99) attached to blower.

2. Blower housing
3. Upper rotor (right helix)
4. Lower rotor (left helix)
5. Roller bearing (front)
9. Ball thrust bearing (rear)
13. Upper rotor gear (right helix)
14. Lower rotor gear (left helix)
20. Rotor drive gear hub
23. End plate (front)
24. End plate (rear)
27. Oil seal
28. End plate cover (front)
29. End plate cover (rear)
33. Drive shaft cover
37. Cover seal
38. Blower drive shaft
40. Water pump drive coupling and slinger
41. Allen screw
42. Blower drive gear
48. Support
53. Flexible coupling drive assembly
102. Oil line fitting
103. Engine rear end plate
104. Flywheel housing
105. Fuel pump drive coupling
114. Cover seal clamp
115. Plate to gear spacer
116. Rotor drive hub plate

lobes is increased correspondingly. While rotor lobe clearances can be corrected, rotor drive gear backlash cannot be corrected. If backlash of blower rotor drive gears exceeds 0.005, it will be necessary to renew both gears as a matched pair.

80B. DISASSEMBLY. With blower removed from engine, and governor weight housing, fuel pump and water pump removed from blower end plate covers, proceed as follows: Remove 10 cap screws attaching each blower end plate cover to blower end plates.

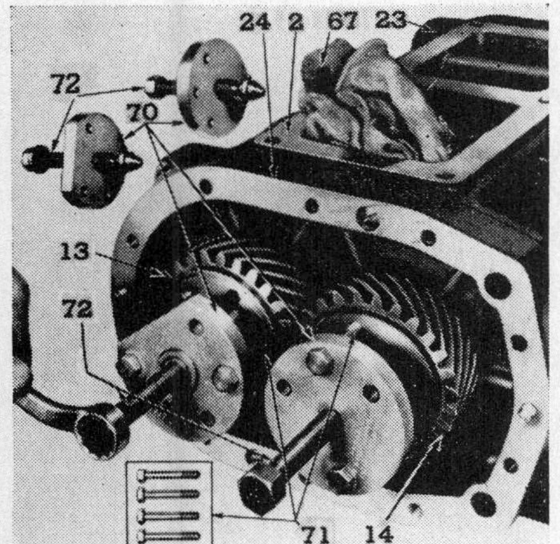

Fig. 0526B—Showing puller set-up to remove blower rotor timing gears. A clean rag (67) placed between the rotors prevents their turning when removing the gears.

13. Upper rotor gear (right helix)
14. Lower rotor gear (left helix)

Remove Allen screw (41) retaining water pump drive coupling (40) to rotor shaft. Pull coupling from shaft by threading a ½ inch—20 cap screw into tapped hole of coupling.

Remove rotor gear hub drive plates (116), gear hub (20) and spacers (115) from drive end of upper rotor shaft. Remove cap screws retaining rotor drive gears on rotor shafts. Place a clean cloth between rotors to prevent their turning when pulling the rotor gears. Remove rotor gears (13 and 14), using GM special tool J1682-CB or equivalent, as shown in Fig. 0526B. The two gears must be pulled simultaneously from the rotor shafts. Shims (17—Fig. 0527B) located under the gears are used to obtain correct rotor timing (trailing and leading rotor edge clearances of the rotor lobes). Keep shims with their respective gears.

Remove rotor shaft bearing retainers (6 and 10). Remove two fillister head screws (26) attaching the front end plate to blower housing, and loosen two fillister head screws attaching the rear plate approximately three turns. Attach GM special tool set J1682-CB or equivalent as shown in Fig. 0527A to push simultaneously both rotor shafts out of their rear bearings.

Use a similar tool set-up to remove the front end plate and bearings from

3. Upper rotor (right helix)
4. Lower rotor (left helix)
23. End plate (front)
24. End plate (rear)
70. GM tool set J 1682-CB

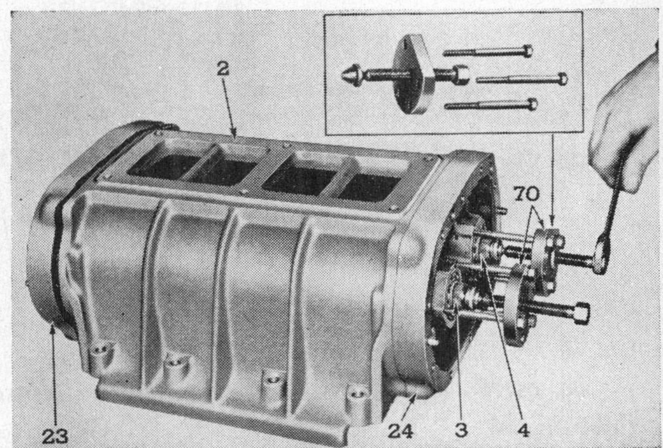

Fig. 0527A—Showing method and tools used in removing rotor shafts from rear bearings, and front end plate from housing on a GM blower.

the rotor shafts.

Bearings and oil seals for rotor shafts can be removed from the end plates and renewed at this time. Excessive grooving of oil seal contacting surface on rotor shaft can be corrected by installing a seal sleeve wear ring (GM part No. 5192439) and an oversize seal (GM part No. 5192438). Seal sleeve wear rings are a shrink fit to rotor shaft and can be installed by using heat to expand the rings. Oil seals are installed with their lip facing away from the rotors.

80C. INSPECTION. Check the blower rotors, and internal surfaces of the blower housing and end plates for score marks and burrs. Minor score marks and burrs can be removed with crocus cloth.

80D. ASSEMBLE. Observe the following points when reassembling a blower. Lobes on upper rotor and teeth on its drive gear form a right-hand helix. The lower rotor and gear have left-hand helices. Helix can be determined by holding the gear or rotor and observing the direction of the twist. A right-hand helix will have the lobes or gear teeth pointing to the right; whereas a left-hand helix will have the lobes or gear teeth pointing to the left.

Install rotors in blower housing so that the blind or omitted serration on rotor shafts and gears face up.

Install oil seals with their lip facing away from the rotors.

Roller bearings are installed at the front end of the rotor shaft. Double row ball bearings are installed at the rear end of the rotor shaft. Install bearings with markings on race facing away from the rotors.

Insert same thickness of shims, as were removed, between rotor drive gear and inner race of rotor shaft bearings. These shims are used to obtain correct rotor timing (trailing and leading edge clearances of the rotor lobes). Refer to next paragraph 80E for method of checking and adjusting blower rotor timing.

Check backlash of blower rotor gears. If backlash exceeds 0.005, the gears should be renewed. Gears are supplied only as a matched pair.

After adjusting the blower rotor timing, install rotor drive gear hub (20) and hub plates (116) on upper rotor drive gear (13). Check run-out of hub splines. Splines in hub should run true within 0.005 total indicator reading.

80E. BLOWER TIMING. Refer to Fig. 0528A. Blower rotors must be timed or positioned to provide the correct trailing and leading edge clearances between the rotor lobes. This timing check is made after the rotors and gears have been installed in blower housing.

Rotor timing or rotor clearance adjustment is varied by moving one of the rotor drive helical gears in or out on the rotor shaft relative to the other gear. Moving the gears in or out on the

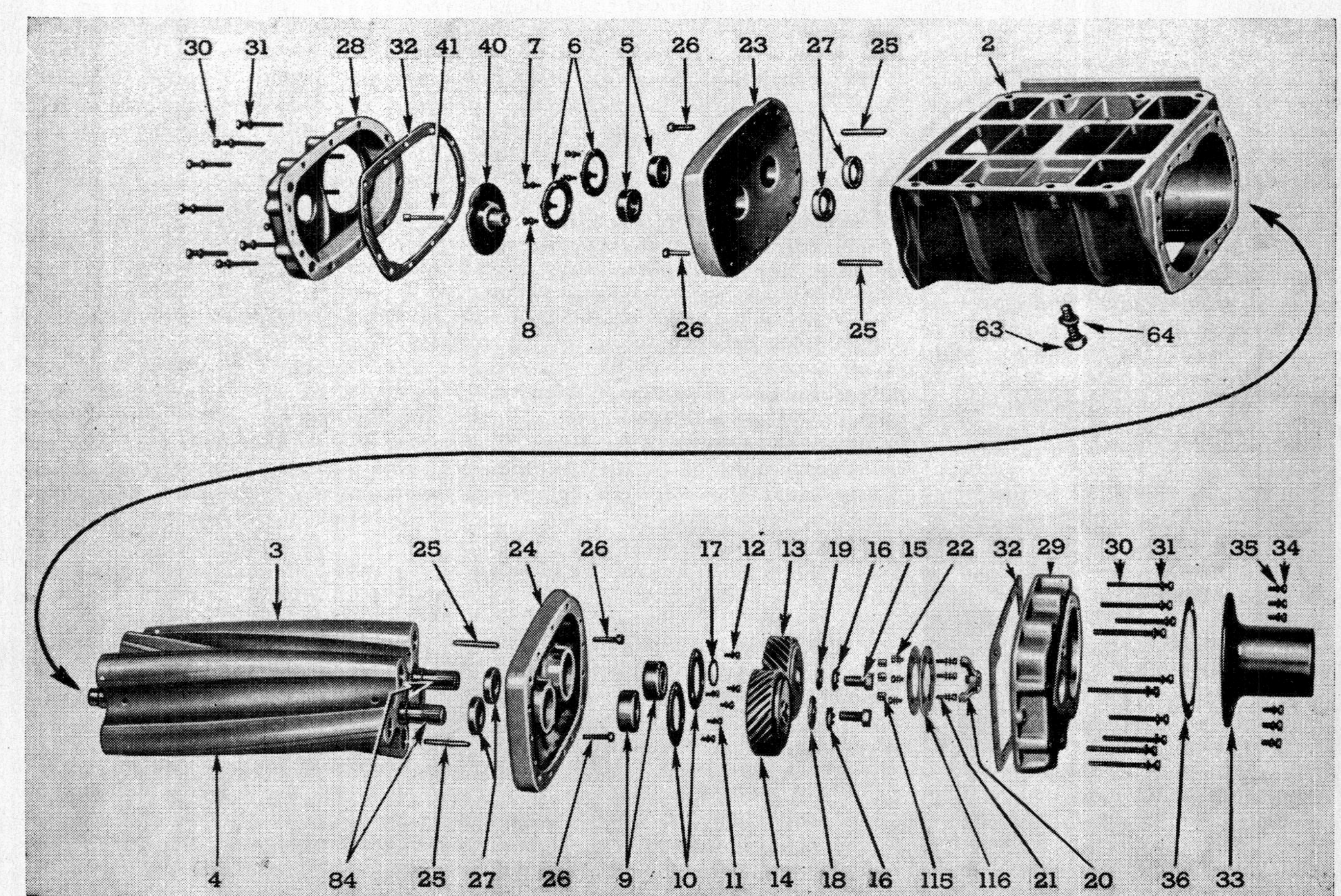

Fig. 0527B—GM engine blower details and relative location of components. Shims (17) control rotor clearances (timing).

2. Blower housing
3. Upper rotor (right helix)
4. Lower rotor (left helix)
5. Roller bearing (front)
6. Bearing retainer
9. Ball thrust bearing (rear)
10. Bearing retainer
13. Upper rotor gear (right helix)
14. Lower rotor gear (left helix)
17. Shim (for timing rotors)
18. Fuel pump drive coupling disc
20. Rotor drive gear hub
23. End plate (front)
24. End plate (rear)
25. Dowel pin
27. Oil seal
28. End plate cover (front)
29. End plate cover (rear)
32. Gasket
33. Drive shaft cover
36. Gasket
40. Intermediate drive shaft and coupling
41. Allen screw
84. Blower rotor shaft
115. Spacer
116. Rotor drive hub plate

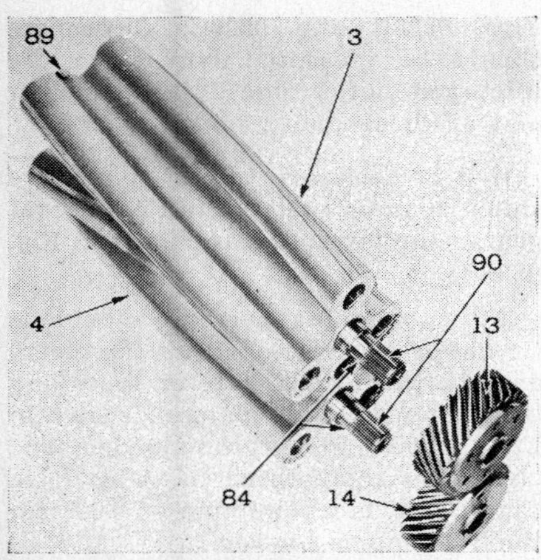

Fig. 0527C—Lobes on upper rotor (3) and teeth on its drive gear (13) form a right hand helix. Lower rotor (4) and gear (14) have left hand helices.

20. Rotor gear drive hub
23. End plate (front)
24. End plate (rear)
91. Blower outlet side
92. Leading edge for upper rotor
96. Trailing edge for upper rotor

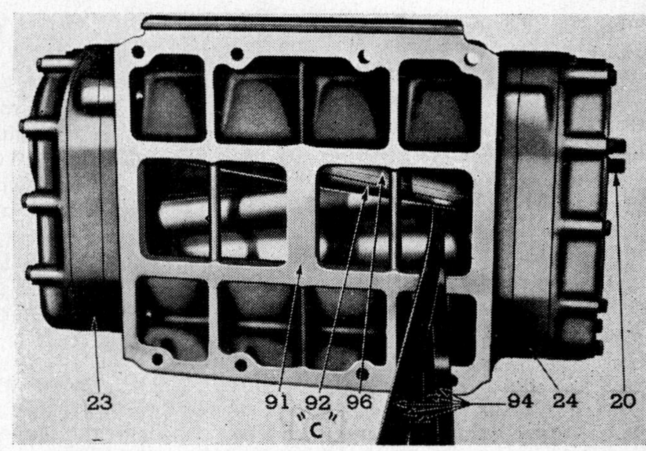

Fig. 0528B—Method of checking rotor clearance with feeler gage ribbons. Measurements should be made both from inlet and outlet sides of blower.

B
A
GEAR END
AIR INLET SIDE

BLOWER FOR RB AND LB ENGINE MODELS

BLOWER FOR RA, RB, LA AND LB ENGINE MODELS

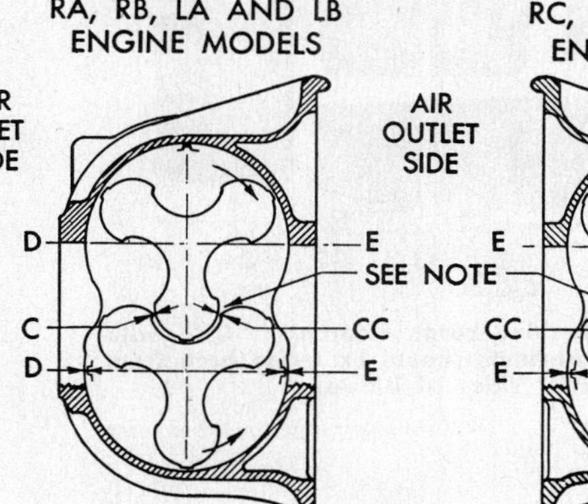

BLOWER FOR RD AND LD ENGINE MODELS

BLOWER FOR RC, RD, LC AND LD ENGINE MODELS

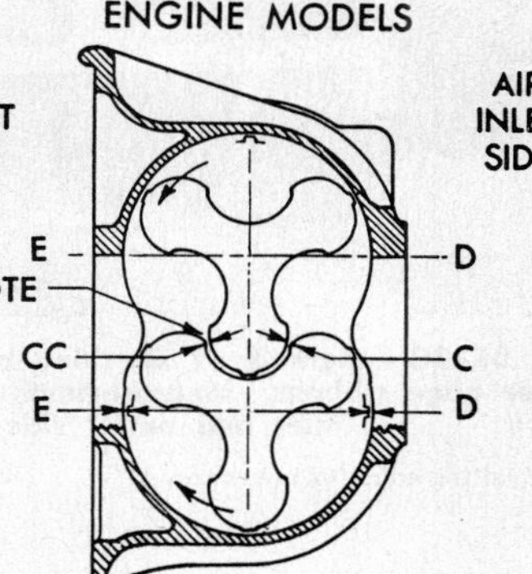

		A	B	C	CC	D	E
3-71	MIN.	.007	.007	.012	.002	.016	.004
	MAX.				.006		
4-71	MIN.	.007	.009	.014	.002	.016	.004
	MAX.				.006		
6-71	MIN.	.007	.014	.014	.002	.016	.004
	MAX.				.006		

TIME ROTORS TO DIMENSIONS ABOVE

NOTE: TIME ROTORS TO DIMENSION ON CHART FOR CLEARANCE BETWEEN TRAILING SIDE OF UPPER ROTOR AND LEADING SIDE OF LOWER ROTOR (CC) FROM BOTH OUTLET AND INLET SIDE OF BLOWER

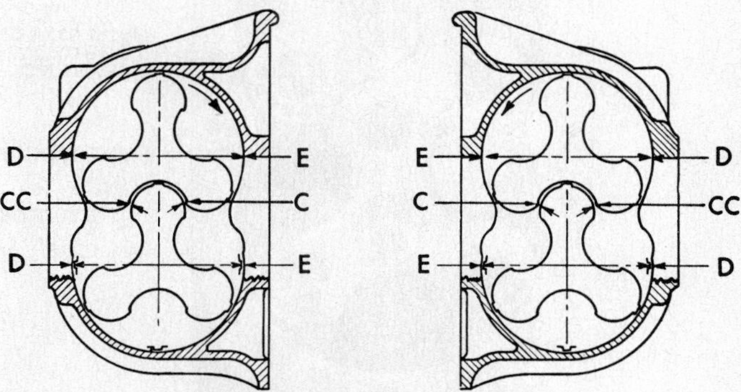

ALL VIEWS FROM REAR OF ENGINE

Fig. 0528A—Timing dimensions (rotor to rotor clearances and rotor to end plate clearances) for GM engine blowers. For GM model RB engines, as used in the Oliver Super 99 GM tractors, make clearance measurements as shown for engine models RA, RB, LA and LB.

rotor shafts is accomplished by removing or adding shims between the gear and rotor shaft bearing. A 0.003 shim will change the rotor clearance 0.001. Shims are available in thickness of 0.002, 0.003, 0.005 and 0.010.

If the upper gear which has a right-hand helix is moved away from the rotor (adding shims), the upper rotor which has a right hand helix will turn counter-clockwise when viewed from the gear end. Adding shims under the upper gear will increase the clearance (C) which is measured at leading edge of upper rotor lobes and trailing edge of lower rotor lobes. Recommended minimum clearance measured at point (C) is 0.012 for the 3-71 engine blower.

If the lower gear which has a left-hand helix is moved away from the rotor (adding shims), the lower rotor which has a left helix will turn clockwise when viewed from the gear end. Adding shims under the lower gear will increase the clearance (CC) which is measured at trailing edge of upper rotor lobes and leading edge of lower rotor lobes. Recommended clearance at point (CC) is 0.002-0.006.

80F. The clearances between rotor lobes should be checked with ½ inch wide feeler gage ribbons as shown in Figs. 0528B and 0528C. Clearances should be measured both from the inlet and outlet sides of the blower and at three points on each lobe.

If it is necessary to add or remove shims, remove and reinstall both rotor gears simultaneously as shown in Fig. 0526B.

After obtaining correct rotor lobe timing or clearances, check the clearance between ends of rotor lobes and blower housing end plates as shown in Fig. 0529A. The recommended end clearance, represented as A & B in Fig. 0528A, is a minimum of 0.007 for the 3-71 engine blower.

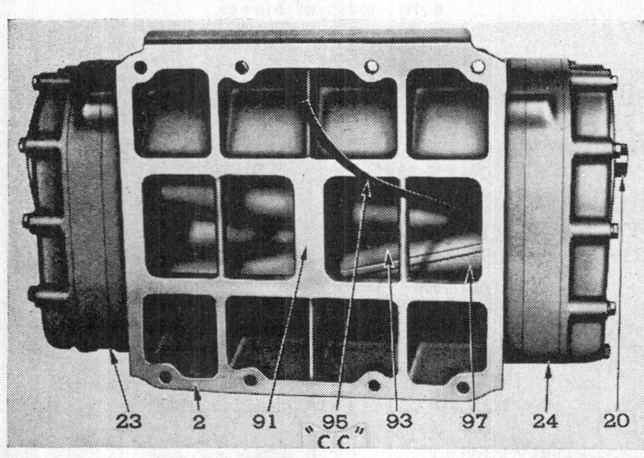

Fig. 0528C—Method of checking rotor clearance "CC" with feeler gage ribbons. Measurements should be made both from inlet and outlet sides of blower.

93. Trailing edge for lower rotor
97. Leading edge for lower rotor

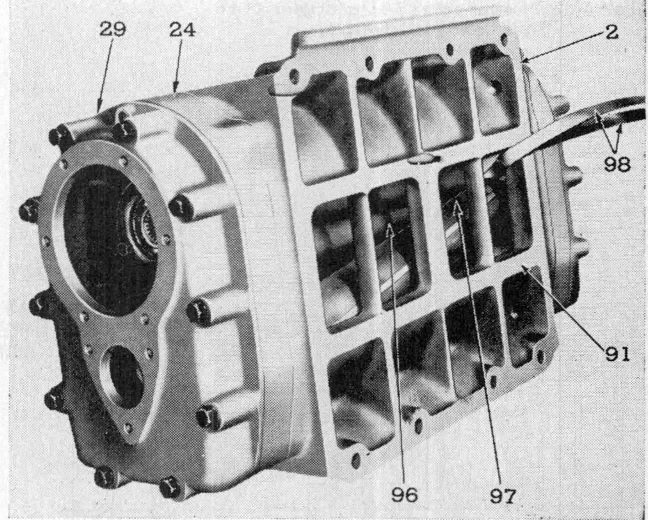

Fig. 0529A—Method of checking end clearance between rotors and blower end plates.

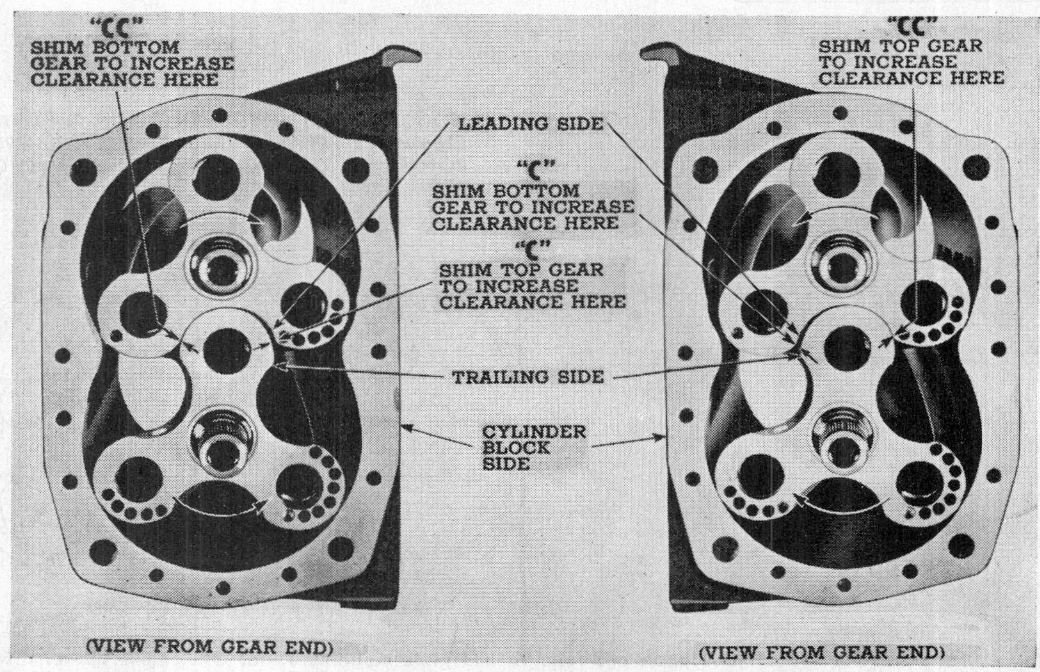

Fig. 0528D—Showing proper location of shims to obtain correct rotor lobe clearances "C" and "CC".

DIESEL FUEL SYSTEM (GM Diesel 71 Series)

85. Main components of the Diesel fuel system are: Fuel supply tank; a gear type fuel pump, mounted on the blower and driven by the lower rotor shaft of the blower; inlet and outlet fuel manifolds located on the cylinder head to connect fuel supply pump to injectors and injectors to fuel supply tank; unit type fuel injectors; and an injector rack control tube which transfers governor action to the injector racks for metering control of the fuel.

TROUBLE SHOOTING

85A. The following data, supplied by General Motors Detroit Diesel Engine Division, should be helpful in locating trouble on the GM-71 series engine.

85B. **HARD STARTING.** Cranking speed, engine compression and engine air supply are okay.

Check for no fuel which could be caused by air leaks, faulty pump, or obstruction in fuel line. Also could be caused by injector racks not in full fuel position.

85C. **NO FUEL OR INSUFFICIENT FUEL.** Check for air leaks in fuel system, fuel flow obstruction, faulty fuel pump or missing restriction fitting located in return fuel manifold. For the latter, check system pressure as outlined in paragraph 99B.

85D. **UNEVEN RUNNING OR FREQUENT STALLING.** Check for below normal coolant temperature, no fuel or insufficient fuel, improper injector timing, incorrect injector rack control setting, leaking injector spray tips, faulty governor adjustments or binding in injector rack control tube.

85E. **DETONATION.** The trouble can be caused either by oil in the inlet supply of air for the engine, by low engine coolant temperature or by faulty injection. Oil in the engine air supply can be traced either to restricted air box drains, to a defective gasket between blower and engine, to defective blower oil seals, or to excessive oil supply in air cleaner oil cups. Check injectors for incorrect timing, for leaking check valve, for enlarged spray tip holes or for a damaged spray tip.

85F. **LACK OF POWER.** Check governor, injector rack control, injector timing and exhaust valve clearance for correct adjustment. Also check for insufficient fuel or insufficient air supply.

85G. **EXHAUST SMOKE ANALYSIS.** Black or gray color of the exhaust smoke indicates incomplete burning of fuel. Check for insufficient air supply, excess fuel or irregular fuel distribution, lugging engine and incorrect grade of fuel.

Blue color of the exhaust smoke indicates fuel or lube oil not being burn-

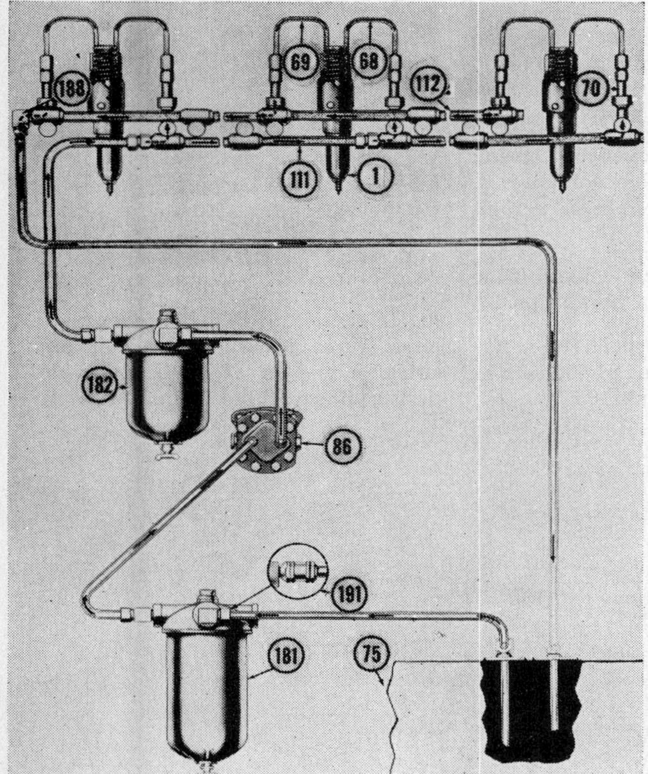

Fig. 0529B—Schematic diagram of a typical fuel system.

1. Injector
68. Inlet pipe
69. Outlet pipe
70. Fuel connector
86. Fuel pump
111. Inlet manifold
112. Outlet manifold
181. Fuel strainer
182. Fuel filter
188. Restriction elbow
191. Check valve (industrial engines only)

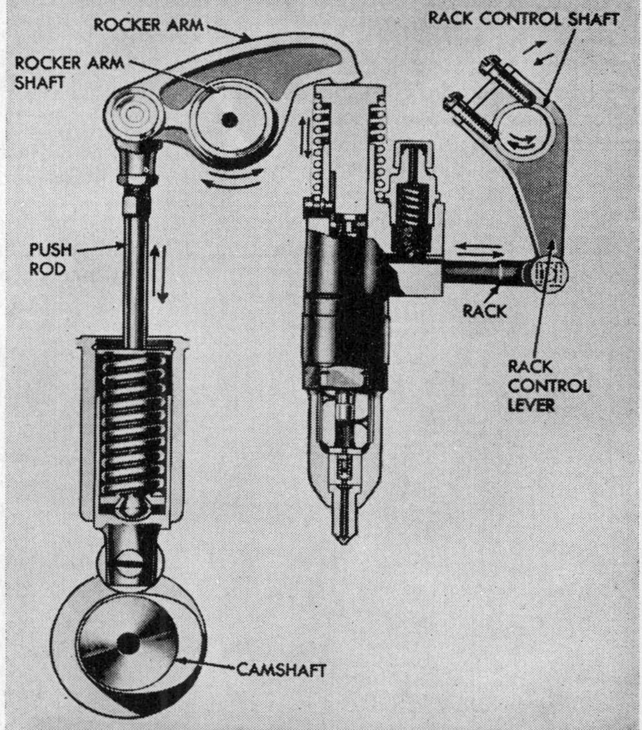

Fig. 0529C—GM 3-71 engines used in the Oliver Super 99 tractors are equipped with 70MM high valve unit type injectors.

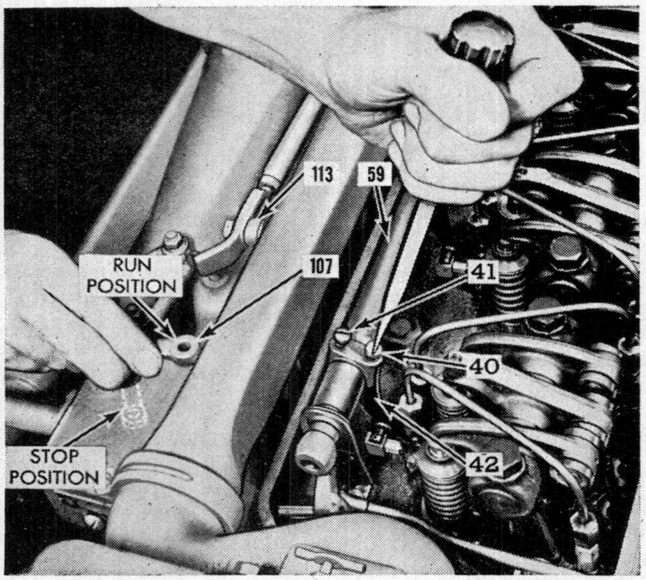

42. Control tube lever
59. Injector rack lever control tube
107. Governor control lever
113. Variable speed control lever

Fig. 0530A—Position of injector racks are set with speed control lever in full-speed position. First synchronize position of No. 1 injector rack with the governor; then synchronize the position of all remaining racks to the No. 1 injector rack.

ed in the cylinder; that is, fuel or lube oil is being blown through the cylinder during the scavenging period. Check for internal fuel or lube oil leaks and for excessive oil supply in air cleaner cups.

White color of the exhaust smoke usually indicates a mis-firing cylinder. Check for faulty injectors, low compression, and low cetane fuel.

TUNE-UP

86. The following adjustments are necessary to completely tune-up a GM series 71 engine. Governor adjustments as listed apply to the variable speed type governor which is used on the GM 3-71 engine for the Oliver Super 99 tractor. Adjustments should be made after the engine has reached operating temperature.

A. Adjust exhaust valve tappet clearance to 0.009 hot.

B. Adjust injector timing on engine used in Oliver tractors to 1.460 inches for 70MM injectors with special tool J1853. If injectors are of 60MM type the timing dimension is 1.484 inches. Timing dimension for 80MM injectors is 1.460 inches. Refer to paragraph 90.

C. Adjust variable governor spring plunger setting (governor gap) to 0.005-0.007 as outlined in paragraph 103B.

D. Adjust injector rack control tube levers as outlined in paragraph 87.

E. Adjust engine idle speed to a no-load crankshaft speed of 500 rpm as outlined in paragraph 103C.

INJECTOR RACK CONTROL TUBE

87. **ADJUSTMENT.** Refer to Fig. 0530A. The position of the injector racks must be correctly set in relation to the governor. This relationship is established by adjusting the injector rack control tube lever for No. 1 cylinder first, and then synchronizing the position of all remaining injector rack control tube levers to the No. 1 injector rack.

To adjust the injector racks, proceed as follows: Remove tractor hood and engine rocker arms cover. Back out buffer screw on governor until it extends about ⅝ inch from housing. Loosen, approximately three turns, adjusting and lock screws (40 & 41) on all injector rack control levers (42). Be sure that all levers swing freely on rack control tube. Place governor variable speed control lever (113) in full-speed position (rearward) and governor control lever (107) in the stop position.

Starting with No. 1 injector rack control tube lever, rotate adjusting screw (40) until screw just bottoms. Slowly move governor control lever (107) toward the run position, but do not force it past any point where resistance to movement is encountered. If resistance to movement is encountered hold governor control lever at this point, and back out the lever adjusting screw until governor control lever (107) can be moved without resistance. Lock the adjustment with lock screw (41).

Check the adjustment of No. 1 injector rack control tube lever as follows: Move governor control lever to extreme end of governor cam while noting the resistance required to move

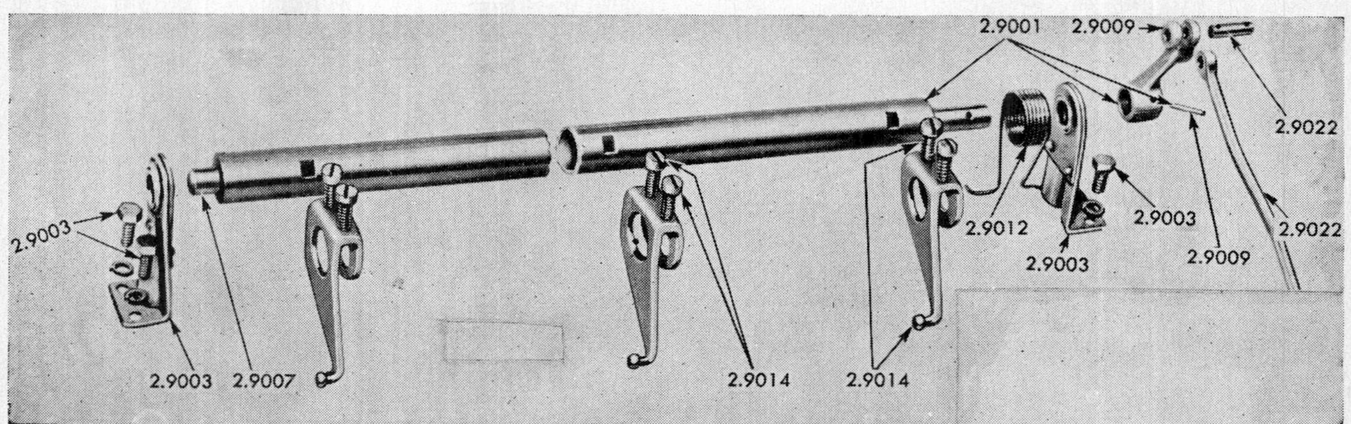

Fig. 0530B—Injector rack control tube and control tube levers assembly.

Fig. 0531A—To locate a misfiring cylinder, make the injector inoperative by holding down the injector follower guide (7) with a screw driver.

the lever. If a step-up in resistance is encountered, the injector rack is too tight. Correct this condition by slightly loosening the adjusting screw. Recheck adjustment.

If governor control lever can be moved to extreme end of governor cam without encountering any step-up in resistance, continue the check as follows: Hold governor control lever in full run position (extreme end of governor cam) and check for free movement at injector rack coupling by applying light finger pressure to coupling end of rack. If rack cannot be moved, the adjustment is okay. If injector rack can be moved with light finger pressure, the rack lever adjustment is too loose. Correct this looseness by slightly tightening the adjusting screw (40). Recheck the adjustment.

After the No. 1 injector rack control tube lever is correctly positioned, do not change this adjustment while adjusting the remaining injector rack levers.

87A. Adjust remaining rack levers as follows: Place and hold governor control lever in full run position (extreme end of governor cam). Rotate adjusting screw (40) on No. 2 control tube lever while checking for rotary movement at injector rack coupling. When rack coupling loses its rotary movement, lock the adjustment.

Check adjustment of No. 2 rack lever with that of No. 1 lever by comparing looseness or tightness of rack couplings. If No. 1 rack coupling feels loose, No. 2 rack is too tight. If No. 2 rack feels loose, it should be tightened. Do not change adjustment of No. 1 injector rack lever.

Adjust all remaining rack levers in a similar manner and always check the adjustment by comparing it with No. 1 injector rack.

88. **REMOVE AND REINSTALL.** To remove injector rack control tube and levers assembly, proceed as follows: Remove tractor hood and engine rocker arms cover. Remove governor control housing cover, and remove link which connects injector rack control tube to governor differential lever. Remove injector rack control tube bracket cap screws using a $\frac{7}{16}$ inch thin-wall socket. Lift control tube assembly up and away from cylinder head.

88A. When reinstalling the control tube assembly, position control tube spring so that the control tube will return injector racks to no fuel position. After installing the control tube assembly and before connecting the control tube lever to governor differential lever linkage, check for binding conditions between injector rack and rack operating lever on control tube. Changing position of control tube brackets on cylinder head or shifting control tube rack levers on control tube will sometimes correct a binding condition.

INJECTORS

89. The 70MM high valve injectors as used in the Oliver Super 99 tractor GM 3-71 Diesel engine are of the unit type, and are made by General Motors. The unit type fuel injectors, one for each cylinder, combines in a single unit all of the parts necessary to meter and inject fuel, create high fuel pressure, atomize fuel, and by-pass fuel through the injector body. By-passed fuel serves as a coolant and eliminator of air pockets.

89A. **LOCATING A MISFIRING CYLINDER.** If one engine cylinder is misfiring, it is reasonable to suspect a faulty injector. Generally, a faulty injector can be located by making each injector inoperative. As in checking spark plugs in a spark ignition engine, the faulty injector is one which, when it is made inoperative, least affects the running of the engine.

With engine running at idle speed, make the injectors inoperative, one at a time, by holding down the injector follower guide (7—Fig. 0531A) with a screw driver as shown.

89B. A misfiring cylinder because of dirt under the spray tip valve may be corrected by flushing the injector as follows: With engine operating, set throttle in idle position. Loosen, approximately six turns, both adjusting screws (40 & 41) on the rack control

G. Timing gage
LN. Lock nut
PR. Push rod
RA. Rocker arm
2. Injector body
7. Follower guide

Fig. 0531B—Time injectors (synchronizing the no fuel position of each injector) by positioning the injector follower guide to a certain distance above the injector body. Timing dimension for 60 MM injectors is 1.484 inches; for 70 and 80 MM injectors, 1.460 inches.

lever of the injector to be flushed. Move the injector rack control lever into the full fuel position. This will cause a maximum amount of fuel to be forced through the injector. During this time the engine will detonate, but do not allow detonation to continue for more than a few seconds. If detonation does not occur, continued flushing of the injector will not correct a misfiring cylinder.

Remove and test the condemned injector as outlined in paragraphs 91 through 92E.

90. **TIMING INJECTORS.** This adjustment synchronizes the no fuel position of each injector by positioning the top surface of each injector follower guide (7) a certain distance, depending on the size of the injector, above the machined face of the injector body (2—Fig. 0531B). For Oliver tractors equipped with GM 3-71 engines using 70MM injectors this measurement is 1.460 inches and is made with a GM special timing gage J1853. For 60MM injector applications this measurement is 1.484 and is made with timing gage J1242. For 80MM injectors use data and gage applying to the 70MM injectors.

To time the injectors proceed as follows: Place governor control lever in no fuel (off) position. Rotate engine crankshaft until both exhaust valves are opened fully. Position injector timing gage so that one end of gage enters the drilled hole located in top surface of injector body. Loosen push rod lock nut (LN). Adjust injector rocker arm (RA) by lengthening or shortening rocker arm push rod (PR) until bottom face of timing gage head will just touch the top of injector follower guide (7). Tighten push rod lock nut and recheck the timing dimension.

Check and adjust the timing of each injector in a similar manner.

91. **R&R INJECTORS.** Refer to Fig. 0532A. To remove any or all injectors, proceed as follows: First remove tractor hood and engine rocker arms cover. Remove fuel lines (68 & 69), connecting inlet and outlet fuel manifolds to injector body. Cover each fuel fitting with a shipping cap to prevent dirt from entering the injector and fuel lines.

Rotate engine crankshaft so that the clevis pins of the three rocker arms for one cylinder are on the same plane. Remove rocker arms shaft support bolts, then swing rocker arms assembly over and away from the valves and injector.

Remove injector hold-down clamp (36) and special washer (38). Using GM tool J1227-A or equivalent placed under the injector body, free the injector from its seat. Lift injector from its seat while disengaging the injector rack from the control lever.

91A. After installing the injector, torque tighten the injector clamp retaining nut to 20-25 ft. lb.

Time injectors as outlined in preceding paragraph 90, and adjust injector rack control tube levers as outlined in paragraphs 87 and 87A.

92. **INJECTOR BENCH TEST.** Refer to Fig. 0534. The unit type injector must pass five different bench tests. The five tests are: Rack freeness, binding plunger, spray tip orifice, valve opening pressure, and holding pressure. The latter three tests require an injector test fixture.

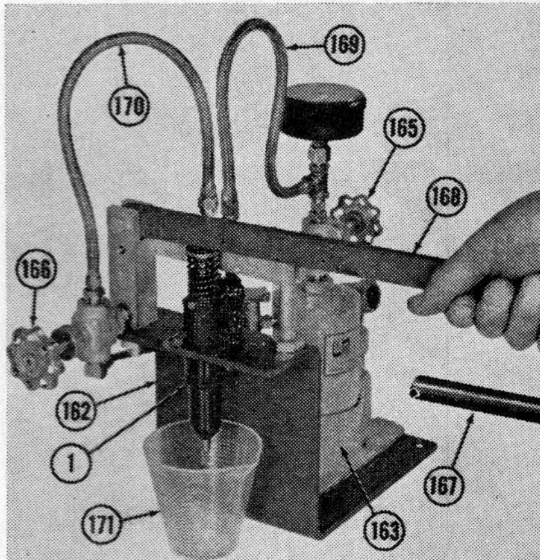

Fig. 0533A—Test fixture set-up for popping the injector to check spray tip orifices.

169. Fuel supply line 170. Fuel return line

The extreme pressure of the injector spray is dangerous and can cause the fuel to penetrate the human flesh. Avoid this source of danger when checking injectors by directing the spray away from your person.

92A. RACK FREENESS TEST. If the various parts of the injector have been assembled correctly or are free of defects and dirt, the injector rack (24) will move through its full length by its own weight.

Hold injector in a horizontal plane with coupling end of rack facing up. Now, turn injector about its long-axis until coupling end of rack points downward. The rack should move through its full length by its own weight.

92B. BINDING PLUNGER TEST. To check for a binding plunger (17) in

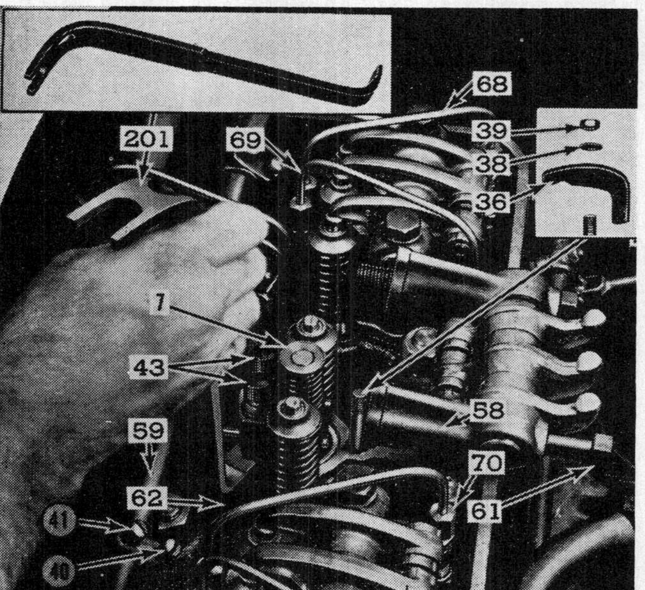

Fig. 0532A—Method of removing an injector with GM tool J 1227-A.

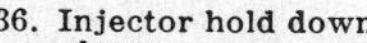

36. Injector hold down clamp
59. Injector rack lever control tube
68. Fuel inlet pipe
69. Fuel outlet pipe

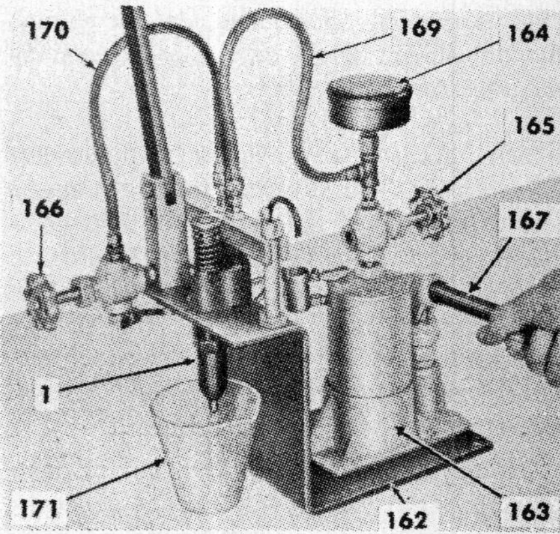

Fig. 0533B—Test fixture set-up for testing injector valve opening pressure and valve holding pressure.

169. Fuel supply line 170. Fuel return line

plunger bushing (18) install injector spring side up in a holding fixture or test fixture. Depress plunger (17), then release it while noting its action on the return stroke. Perform this test with injector rack in full fuel position, no fuel position and midway between these two positions.

A sticking plunger will show up as a slow return when the depressed plunger is suddenly released. A free plunger will return with a definite snap.

92C. SPRAY TIP ORIFICE TEST. This test is called the popping test. Refer to Fig. 0533A. Mount injector in a test fixture. Keep the injector supplied with fuel at 10-15 psi gage pressure. Operate the injector test fixture popping handle (168) at 60 strokes per minute while observing the spray pattern. An equal amount of fuel should be discharged in a fogged condition from each of the seven orifices.

92D. VALVE OPENING PRESSURE TEST. This test is made to determine the condition of the valve sealing surfaces of items (29, 30, 32, 33 & 34—Fig. 0534) and the condition of spray tip valve spring (31). The test should not be confused with the actual injection pressure because after the valve opens the fuel pressure continues to increase until injection is completed.

To make the valve opening pressure test, mount injector in a test fixture. Operate fuel supply pump on test fixture with even strokes while noting gage pressure reading when injector sprays fuel. Injector should start spraying fuel between 350 to 850 psi. A gage reading of less than 350 psi indicates either dirty or defective valve sealing surfaces.

92E. HOLDING PRESSURE TEST. This test is made for three reasons. The first reason is to determine the condition (clearance) of plunger (17—Fig. 0534) and bushing (18). The other

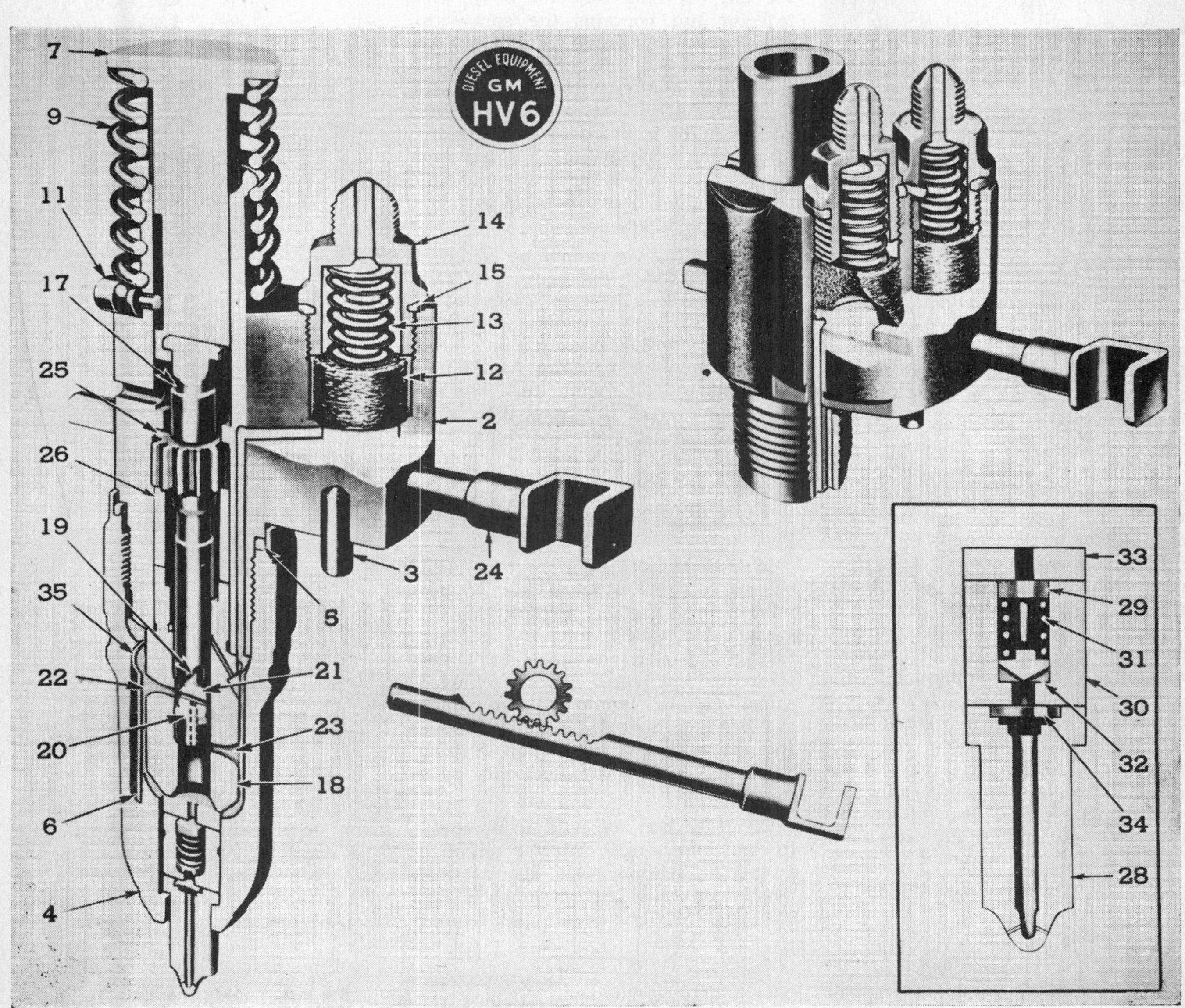

Fig. 0534—A GM engine high valve type injector.

2. Injector body
3. Dowel
4. Injector nut
5. Rubber seal ring
6. Spill deflector
7. Injector follower
9. Plunger spring
11. Stop pin
12. Filter
13. Spring
14. Cap
15. Gasket, copper
17. Plunger
18. Plunger bushing
19. Upper helix on plunger
20. Lower helix on plunger
21. Metering recess
22. Upper port
23. Lower port
24. Rack
25. Gear
26. Gear retainer
28. Spray tip
29. Spray tip valve
30. Valve cage
31. Valve spring
32. Valve stop
33. Valve seat
34. Check valve
35. Fuel supply chamber

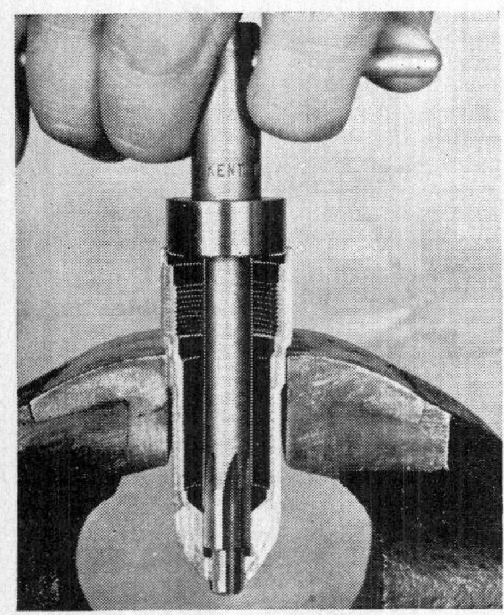

Fig. 0535A—Cleaning carbon from spray tip seat in injector nut with GM tool J4986-1.

reasons are: To determine if there is internal leakage between plunger bushing (18) and injector body, and to determine if there are any internal leaks at the seal ring, spray tip or injector body fuel line fittings.

Refer to Fig. 0533B. To make the hold pressure test, operate fuel supply pump on test fixture until gage pressure is just below the popping or minimum valve opening pressure of 350 psi. Lock this pressure in the injector by turning pump valve (165) to the closed position; then observe pressure drop on gage.

The time required for gage pressure to drop from 350 psi to 150 psi should not exceed 50 seconds for a new injector or 35 seconds for a used injector. If pressure drop exceeds these limits, locate the leak as follows: Thoroughly dry outside of injector by blowing off excess fuel with compressed air. Operate fuel pump occasionally to maintain testing pressure while looking for moist spots of fuel. A leak at the rack opening in injector body indicates either excessive clearance between plunger and bushing or a defective fit between bushing and injector body. A dribble or drop of fuel at the spray tip orifices indicates faulty sealing surfaces in the valve parts.

93. **INJECTOR OVERHAUL.** The GM unit type injector can be partially disassembled either for inspection and cleaning of the spray tip and valve parts, or for inspection of the plunger, plunger bushing, rack and gear.

Unless the shop is equipped with the necessary tools and a test fixture, servicing of injectors should not be attemped. For shops equipped to service GM unit type injectors, the following data can be used and applied.

93A. SPRAY TIP. Refer to Fig. 0536B. To disassemble the injector spray tip, proceed as follows: Carefully clamp injector body in a vise or special fixture, and remove injector nut (4) from injector body. The injector nut contains the spray tip (28), valves and valve seats (items 29, 30, 32, 33 & 34), valve spring (31) and spill deflector (6). Removal of the injector nut will release the plunger bushing (18) from injector body. Handle injector spray tip, bushing and the related valves with care as the sealing action between each part depends on a lapped surface.

If the spray tip cannot be removed from the injector nut because of carbon, proceed as follows: Place injector nut with open side on a wood surface. Using hollow brass tubing placed on the shoulder of spray tip, bump spray tip out of injector nut. The inside diameter of the brass tube must be large enough so that tube contacts only the outside edge or shoulder of the spray tip.

Wash all injector parts in clean fuel oil.

93B. Since sealing action of the various valve parts depends on a lapped mirror finish surface, carefully inspect these parts with a magnifying glass. Slight imperfections such as burrs, scratches and stains can be removed effectively by lapping. Parts which are damaged, scored, pitted or chipped should be renewed. Injector bushing and plunger are supplied only as a matched set.

Clean carbon deposits from spray tip seat and bore in injector nut with a special reamer, GM special tool J4986-1 or equivalent, as shown in Fig. 0535A. Carefully operate the reamer

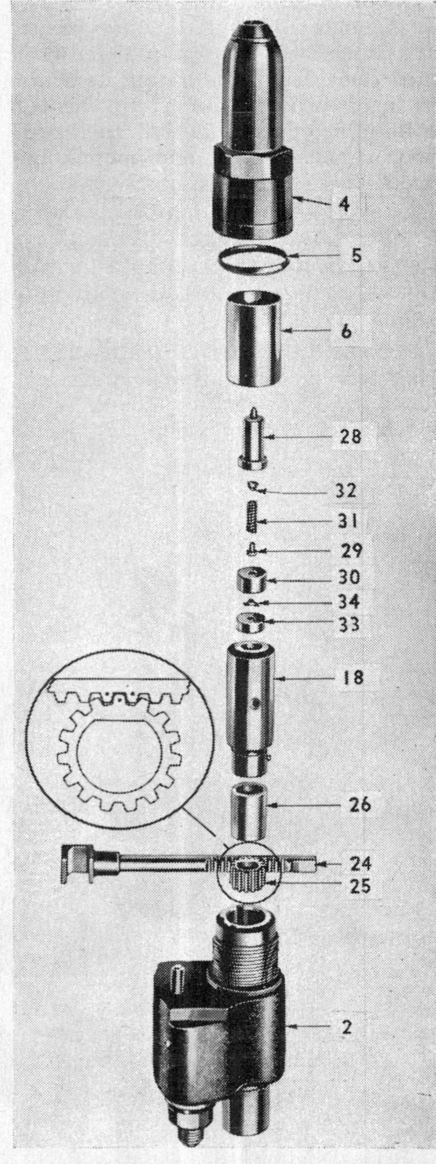

Fig. 0536B—Injector rack, gear, and spray tip details and relative location of parts. Refer to Fig. 0534 for legend.

during this cleaning process so as to prevent removal of any metal.

The inside of the injector spray tip can be cleaned with a spray tip reamer, GM Special tool J1243 or equivalent, as shown in Fig. 0535B. Clean the seven orifices in the spray tip with 0.007 diameter wire. Before using the wire, remove any sharp burrs on the wire with a fine grade abrasive stone. Carefully inspect the spray tip orifices.

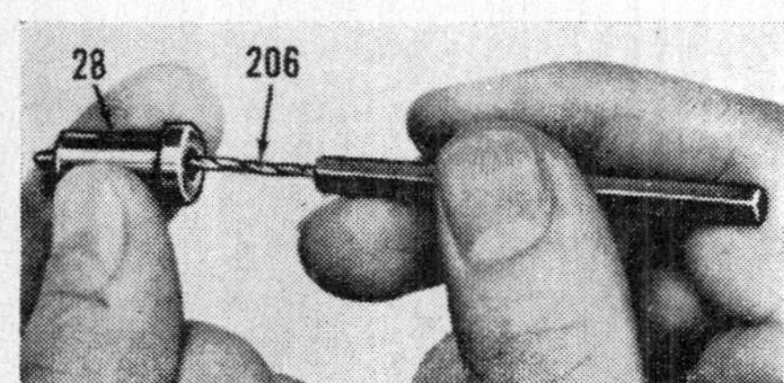

Fig. 0535B—Reaming injector spray tip with GM tool J 1243.

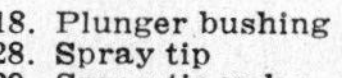

18. Plunger bushing
28. Spray tip
29. Spray tip valve
30. Valve cage
33. Valve seat
34. Check valve

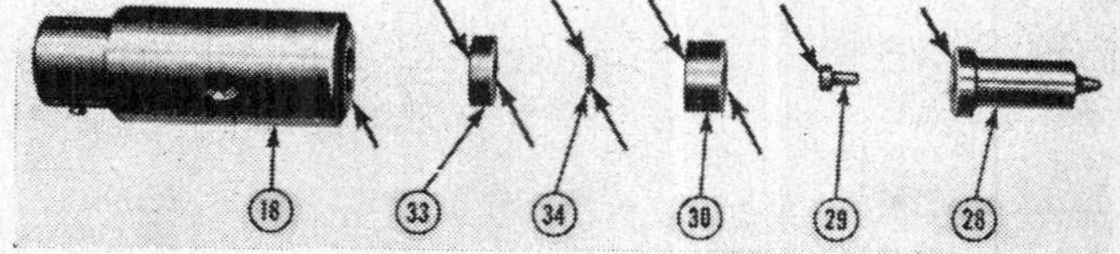

Fig. 0536A—Arrows indicate injector sealing surfaces which may require lapping.

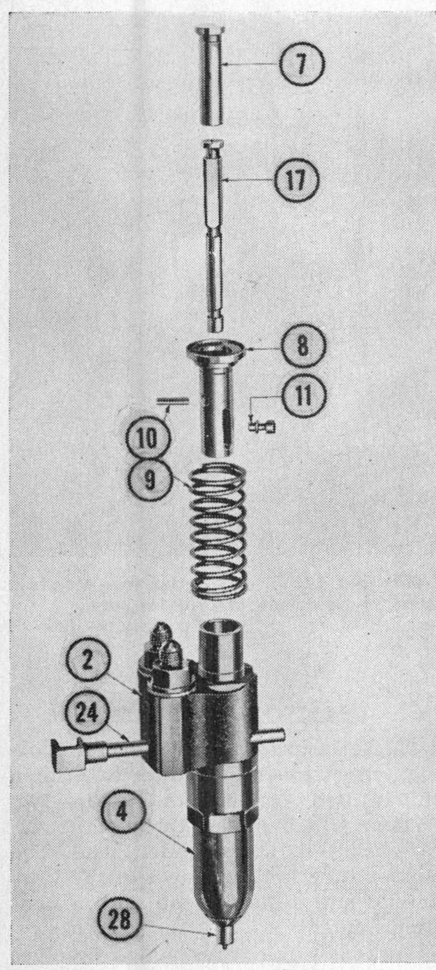

Fig. 0537A— Non high valve type injector plunger and follower details. Follower (7) and guide (8) are integral on high valve type injectors.

2. Injector body
4. Injector nut
9. Plunger spring
10. Follower pin (used on non high valve types only)
11. Stop pin
17. Plunger
24. Rack
28. Spray tip

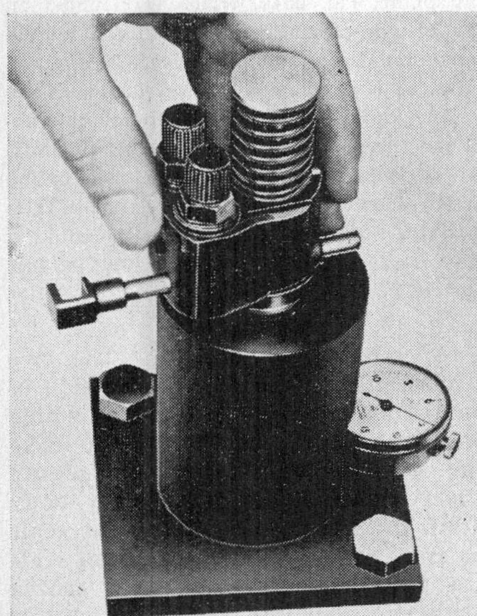

Fig. 0537B—Checking concentricity of injector spray tip to injector nut with GM tool J 5119.

Sealing surfaces of the injector which may require lapping are indicating in Fig. 0536A. These surfaces should be lapped on a suitable lapping block with lapping compound or Carborundum H40, 600 grain size Norton Alundum or equivalent. Light pressure lap the sealing surfaces, using a figure 8 motion, until surface is flat. Then, wash the part in clean fuel oil and dry lap the surface of the part to a mirror finish. Dry lapping or mirror finishing should be done on a different lapping block than that used for abrasive lapping. Thoroughly wash and inspect the parts frequently during the lapping process.

93C. Reassemble spray tip and related valve parts to the injector body in the order as shown in Fig. 0536B. Install a new injector nut to body gasket on injector body. Place recessed side of valve seat (30) over check valve (34). Lubricate the injector nut threads and carefully place the nut, containing spill deflector, over the assembled valve parts. Hand tighten injector nut while rotating end of spray tip to make certain that none of the valve parts have shifted. Torque tighten the injector nut to 55-65 ft. lb.

After torque tightening the injector nut, check the outside surface of the spray tip for concentricity to the outer surface of the injector nut as shown in Fig 0537B. Spray tip and injector nut must be concentric within 0.008 total indicator reading. If indicator reading exceeds 0.008, loosen injector nut and recenter the spray tip until the reading is within the recommended runout.

93D. COMPLETE DISASSEMBLY. Refer to Fig. 0537A. The injector can be completely disassembled as follows: Support injector in a holding fixture or vise. Compress plunger spring (9) by hand, then raise the spring from stop pin (11) with a screw driver and withdraw the pin as shown in Fig. 0538B. Remove follower guide (8—Fig. 0537A) which is used only on non high valve injectors, follower (7), plunger spring (9) and plunger (17) from injector body (2). On non high valve injectors remove the follower pin (10) to separate the previously mentioned parts.

Remove injector nut, spray tip, valve parts and plunger bushing from injector body as outlined in paragraph 93A.

Bump lower side of injector body on a piece of wood to jar gear retainer (26—Fig. 0536B) and gear from injector body. Pull rack (24) out of injector body.

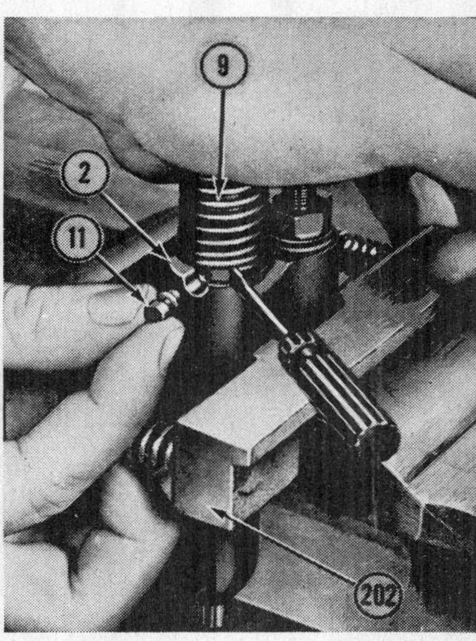

Fig. 0538B—Showing method of removing stop pin (11) to remove injector plunger.

Remove the two fuel filter caps (14—Fig. 0539), filters (12) and springs (13) from injector body.

Carefully handle and thoroughly wash all injector parts with a soft bristle brush in clean fuel oil. Inspect and if necessary lap the injector spray tip and valves as outlined in paragraph 93B.

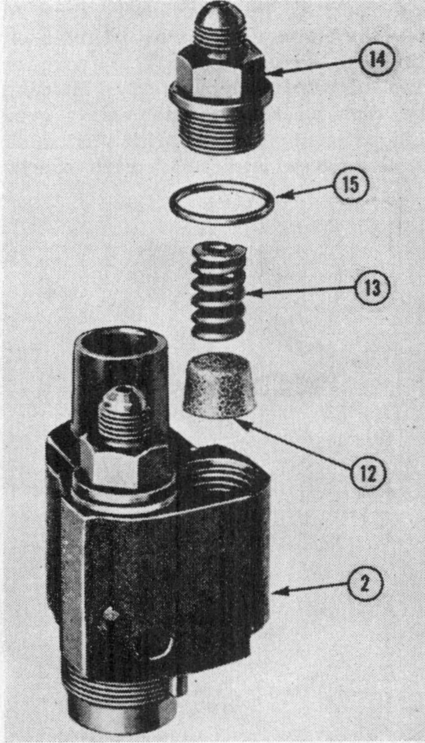

Fig. 0539—Details of injector filters, spring, and caps.

2. Injector body
12. Filter
13. Spring
14. Filter cap
15. Gasket, copper

Fig. 0540—Primary fuel supply pump which is mounted on the blower is driven by the lower rotor shaft of the blower.

CP. Chevron starting aid primer line
PI. Pump inlet line
PO. Pump outlet line
TR. Governor control rod
95. Fuel pump
109. Fuel pump drain

Fig. 0543—Inlet fuel manifold (111) and outlet fuel manifold (112) are located on right side of cylinder head.

114. Restricted fitting
128. Secondary fuel filter

Inspect and clean both the plunger bushing and plunger with clean fuel oil and tissue. Do not attempt to polish these highly finished surfaces. If for any reason either of these parts require renewal, both the plunger and bushing should be renewed as a matched pair.

93E. Reassemble the injector gear and rack as follows: Insert rack in injector body so that marked teeth on rack are visible from gear retainer side of the injector body. Holding the rack in this position, slide gear (25) into engagement with gear rack so that gear and rack teeth marks are registered as shown in Fig. 0536B. Install gear retainer (26) to hold gear and rack in proper mesh.

Reassemble plunger bushing, spray tip and injector nut to injector body as outlined in paragraph 93C.

Reassemble the plunger, spring and related parts to injector body by reversing the disassembly procedure and with reference to Figs. 0534 and 0537A. Install new filters (12) with concave side of filter facing bottom of hole in injector body.

After reassembling the injector, make a bench test, involving five tests, as outlined in paragraphs 92 through 92E.

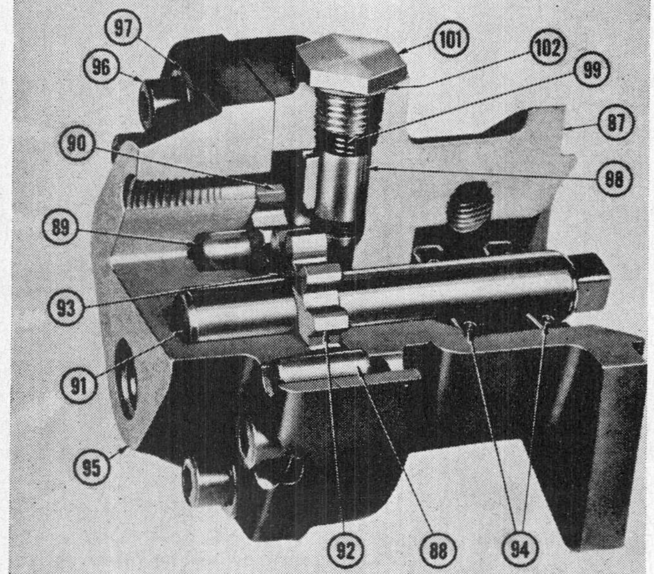

93. Ball (gear retaining)
94. Oil seals
98. Relief valve
99. Spring
102. Gasket

Fig. 0541—A GM engine primary fuel supply pump. Install oil seals (94) with lips facing drive end of shaft (91).

INJECTOR HOLE TUBES

95. To insure efficient injector cooling, each injector is inserted into a thin-walled copper tube (40—Fig. 0505A) which passes through the water jacket in the cylinder head. The copper tube is flared-over at the lower end and sealed at top with a neoprene ring.

To renew the tubes which involves removal of the cylinder head, refer to paragraphs 61C and 61D.

PRIMARY FUEL SUPPLY PUMP

99. The GM 71 series Diesel engine is equipped with a gear type primary fuel supply pump, Fig. 0540, which is mounted on the blower assembly and is driven by the lower rotor shaft of the blower. This pump provides a continuous flow of fuel from the supply tank through the fuel chambers within the injectors and then, back to the tank. A restriction (114—Fig. 0543) placed in the outlet elbow of the outlet fuel manifold (112) provides sufficient resistance to maintain a fuel pressure of 60-65 psi throughout the fuel system.

99A. **SEAL LEAKAGE.** The two tapped holes located in lower side of pump body carry off fuel oil which passes the pump shaft seals. If leakage exceeds one drop per minute when checked at fuel pump body drain tube (109—Fig. 0540), the seals must be renewed. Pump shaft oil seals are installed with their lip facing drive end of pump.

99B. **FUEL FLOW CHECK.** To check fuel flow, disconnect fuel return line at supply tank. Operate engine at 1200 rpm and measure the fuel flow for a period of one minute. At least one half gallon of fuel should flow from the return line in one minute.

If fuel flow is insufficient check for air leaks in fuel system by immersing end of fuel return line in clean fuel oil. Air bubbles indicate an air leak on the suction side of the pump. Also check for a restricted fuel strainer on suction side of pump, for a restricted fuel filter on outlet side of pump or for an inoperative pressure relief valve which is located in the pump body. Pressure relief valve can be removed without removing the pump assembly.

99C. **REMOVE & REINSTALL.** To remove fuel pump proceed as follows: Disconnect suction and pressure lines from pump. Remove three cap screws which attach pump to blower housing and lift pump from blower. Copper type flat washers are used with the three attaching cap screws.

Non-adjustable pump pressure relief valve can be removed without removing the pump assembly from blower housing. Pressure relief valve opens at 60-65 psi.

99E. **DISASSEMBLE & OVERHAUL.** Refer to Figs. 0541 and 0544. Disassembly of the pump is self-evident after an examination of the unit.

Pump shaft (91) rotates in unbushed bores of the pump body. Install pump driven gear (90) with slot in face of gear facing pump cover. Install the two lip type oil seals (94) with lips facing drive end of pump. The pump body and cover are assembled without a gasket.

70. Fuel connector
111. Inlet manifold
112. Outlet manifold
113. Fitting
114. Restricted elbow
117. Lock nut
118. Washer, copper
119. "T" connector

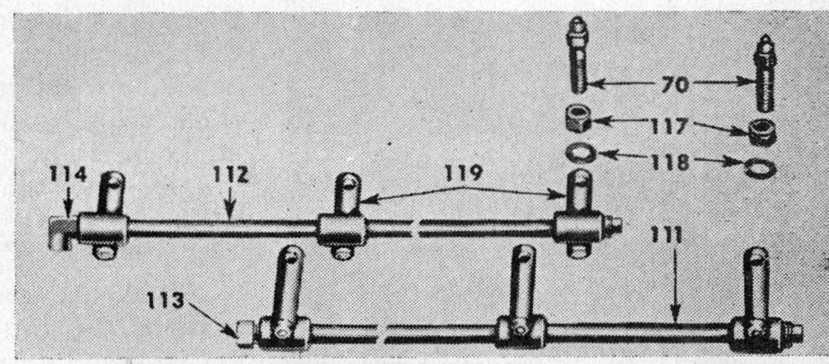

Fig. 0544A—GM engine fuel manifold details.

FUEL MANIFOLDS

100. Refer to Figs. 0543 and 0544A. Two fuel manifolds, one an inlet (111), the other an outlet (112), are mounted on the right side of the cylinder head. These fuel manifolds plus short detachable pipes complete the flow of fuel from the supply pump to the injectors and from the injectors to supply tank. A restriction placed in the outlet fuel manifold elbow (114) provides sufficient resistance to maintain a fuel pressure of 60-65 psi in the fuel circuit. The restricted elbow fitting can be identified by a letter "R"

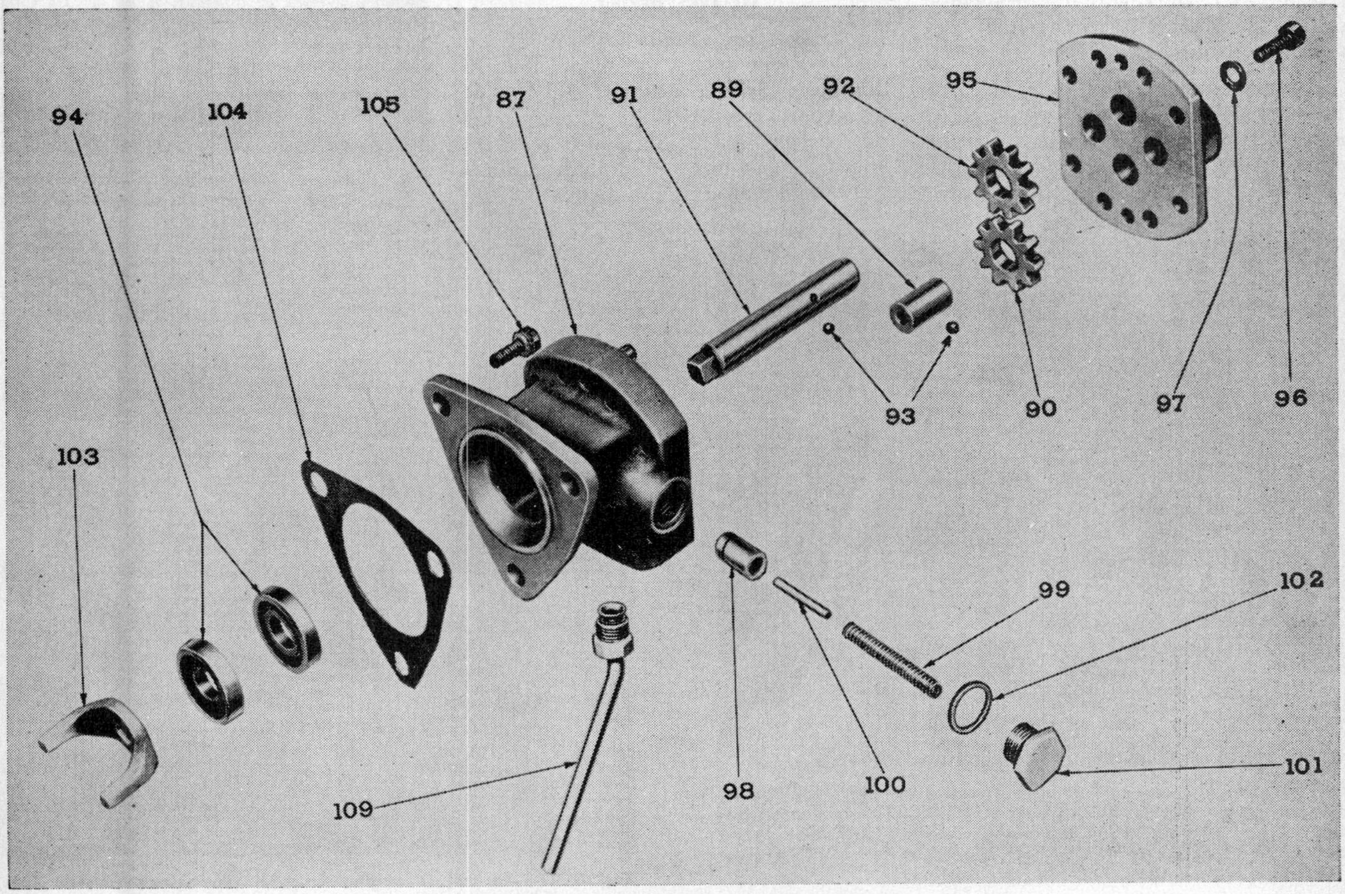

Fig. 0544—GM engine fuel pump details.

87. Pump body
89. Shaft, driven
90. Gear, driven
91. Shaft, drive
92. Gear, drive
93. Ball, gear retaining
94. Oil seals
95. Cover
98. Relief valve
99. Spring
100. Pin
102. Gasket
103. Coupling
104. Gasket
109. Fuel pump drain line

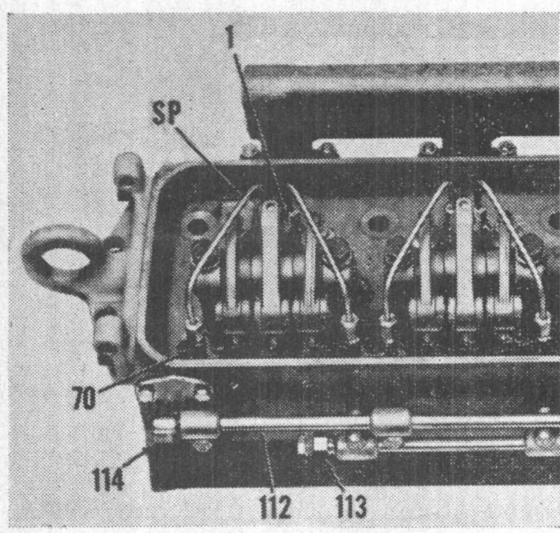

Fig. 0546A—Fuel manifold assembly installation. Refer to Fig. 0544A for legend.

stamped on the body of the fitting.

100A. To remove either fuel manifold, proceed as follows: Remove tractor hood, engine rocker arms cover and secondary fuel filter (128). Remove inlet and outlet pipes connecting injectors to fuel manifold. Remove cap screws attaching manifold "T" connectors (119) to cylinder head. Remove fuel connectors (70) that seat into tapered seats which are located in the manifold "T" connectors. Remove manifolds.

When reinstalling fuel manifolds, install and tighten fuel connectors (70) before installing cap screws attaching manifold "T" connectors to cylinder head. After installing the fuel manifold check for fuel leakage where the "T" connecters enter the cylinder. head. Fuel leakage at this point indicates a faulty seal between the fuel connectors and "T" connectors.

DIESEL STARTING AID

101. A chevron type ether priming system, utilizing ether filled cartridges, is used as a starting aid in cold weather. This element is used in lieu of the manifold pre-heater which is used on some of the other Oliver Diesel powered tractors. Control of the ether injector is located on a sub-panel, bolted to the instrument panel. Servicing procedures are conventional.

GOVERNOR
(GM Diesel 71 Series)

A flyweight, variable speed centrifugal governor, Fig. O548A, which is located on the left side of the engine is used on the GM engine for the Oliver tractor. Governor receives its drive from the upper rotor shaft of the blower.

ADJUSTMENT

103. The three main adjustments on the variable speed governor are: A. Governor Spring Plunger Setting or Governor Gap, B. Engine Idle Speed Adjustment and C. Maximum No-load Speed Adjustment. A fourth adjustment, considered minor, is the throttle booster spring adjustment. All governor adjustments should be made with engine at operating temperature.

103A. **HIGH SPEED ADJUSTMENT.** High speed no-load crankshaft rpm of 1840 is obtained by varying the number of shims (102) which are located between variable speed spring (101) and spring retainer (104). Add shims, available in thicknesses of 0.015 and 0.078, to increase the speed.

Shims are accessible after removing tractor hood, governor control housing cover (3), and two cap screws attaching variable speed spring hous-

Fig. 0548—A GM engine variable speed type governor installation.

BS. Booster spring
WBP. Water by-pass
2. Control housing
3. Cover
57. Buffer screw
59. Weight housing
90. Variable speed spring housing
107. Control lever
108. Lever retracting spring
113. Speed control lever

Fig. 0546B—GM engine variable speed type governor spring plunger setting (governor gap) of 0.005-0.007 is measured between spring plunger (103) and plunger guide (105) when speed control lever (113) is at one-half speed or better. Capscrew (28) controls this setting.

2. Control housing
23. Differential lever
27. Operating shaft lever
29. Lock nut
87. Injector rack control tube to governor link
90. Variable speed spring housing
92. Idle speed adjusting screw
103. Spring plunger
105. Spring plunger guide

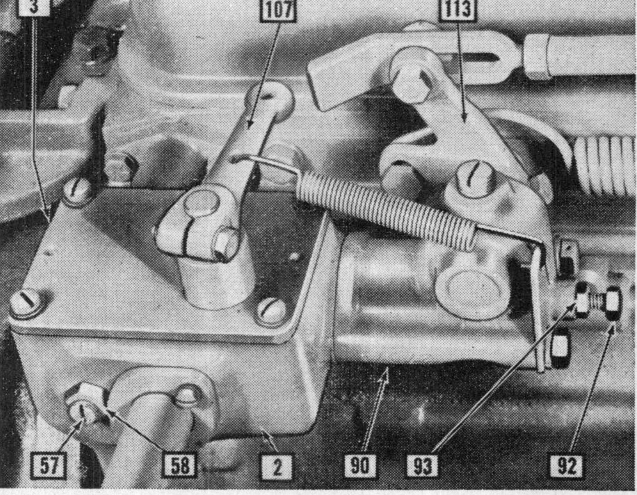

Fig. 0547—To adjust engine no-load idle speed to 450 rpm, place speed control lever (113) in idle position (forward) and rotate screw (92) in or out.

2. Control housing
3. Cover
57. Buffer screw
90. Variable speed spring housing
107. Control lever

ing (90) to governor control housing (2).

103B. **GOVERNOR SPRING PLUNGER SETTING (TORQUE CONTROL).** To obtain proper performance and full engine power, the governor spring plunger (103—Fig. 0546B) must be adjusted correctly in relation to the spring plunger guide (105). This is referred to as the governor gap adjustment.

To adjust governor gap, first remove tractor hood and governor control housing cover. Place speed control lever (113) at one-half speed position or better so as to place spring tension against the spring plunger. Governor gap adjustment of 0.005-0.007 between spring plunger guide (105) and spring plunger (103) can be obtained by rotating the gap adjusting screw (28) which is located in the operating shaft lever.

After adjusting the governor gap, check adjustment of injector racks as outlined in paragraphs 87 and 87A.

103C. **IDLE SPEED ADJUSTMENT.** Refer to Fig. 0547. Start and warm-up engine. Place variable speed control lever (113) in the idle position, making sure it is all the way forward. If engine surges excessively, rotate buffer screw (57) just enough to eliminate the surge. Rotate idle speed adjusting screw (92) to obtain recommended no-load crankshaft speed. Set engine idling speed 10 to 20 rpm below recommended speed of 500 rpm. If engine does not surge or roll, increase idling speed to recommended 500 rpm with buffer screw.

Recheck adjustment of buffer screw (57) as follows: Accelerate engine by moving variable speed control lever (113); then quickly return the engine to idle speed while observing the action of the injector racks and control tube. Injector racks and control tube should cease movement after one or two surges.

103D. **THROTTLE BOOSTER SPRING.** Refer to Fig. 0548. Adjust tension of throttle booster spring (BS) until variable speed control lever (113) can be moved rearward to the full throttle position with ease.

R&R AND OVERHAUL

104. Refer to Fig. 0548. Governor assembly consists of two main sub-assemblies which

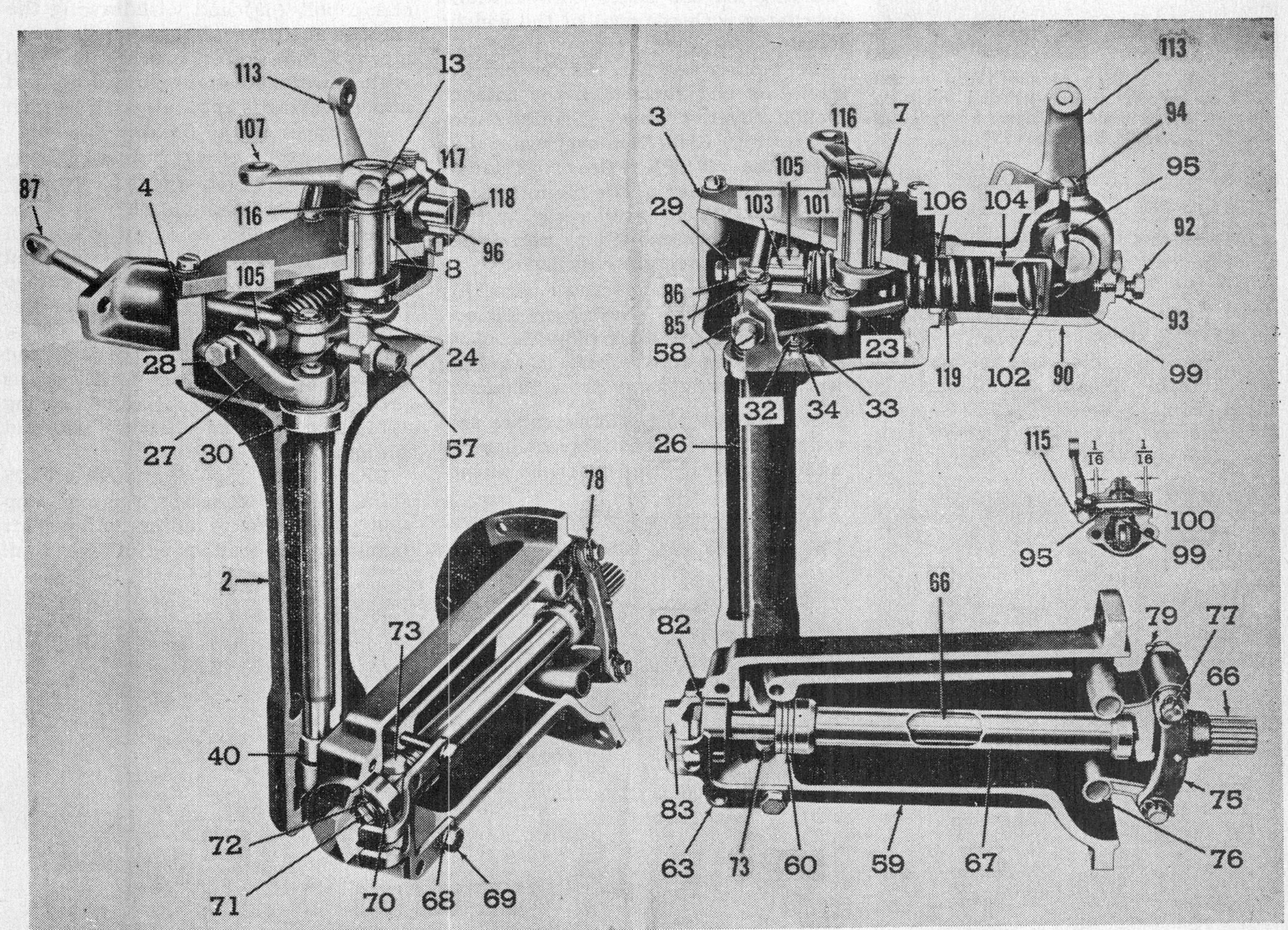

Fig. 0548A—A variable speed type governor is installed on the GM engines used in the Oliver Super 99 GM tractor.

2. Control housing
3. Cover
4. Gasket
7. Throttle shaft
8. Needle bearing
23. Differential lever
24. Washer
26. Operating shaft
27. Operating shaft lever
28. Gap adjusting screw
30. Bearing
32. Bearing retaining screw
40. Oilite bushing
57. Buffer screw
59. Weight housing
60. Riser thrust bearing
63. Cover
66. Weight shaft
67. Riser
70. Bearing
73. Operating shaft fork
75. Weight carrier
76. Weight
77. Weight pin
79. Snap ring
82. Gasket
83. Cap
86. Spring retainer
87. Injector rack control tube to governor link
90. Variable speed spring housing
92. Idle speed adjusting screw
96. Needle bearing
99. Spring lever
101. Variable speed spring
102. Shims
103. Spring plunger
104. Spring retainer
105. Spring plunger guide
106. Spring retainer stop
107. Control lever
113. Speed control lever
116. Retaining ring
117. Washer
119. Spring retainer stop spacer (not used in Oliver application)

Fig. 0549—Location of the sharp corner (FC) on balance weight housing which interferes with complete removal of governor weight housing (59).

are: A. Weight Carrier Housing (59), and B. Control Housing (2).

104A. **REMOVE.** To remove the governor control housing (2) which is also called the governor tower, first remove tractor hood and engine rocker arms cover. Disconnect throttle booster spring (BS), retracting spring (108) and control linkage from variable speed control lever (113). Remove governor control housing cover (3); and remove linkage (87—Fig. 0548A) by disconnecting it from injector rack control tube lever, and differential lever (23). Remove two cap screws attaching governor control housing to cylinder head, and four cap screws attaching governor control housing to weight housing (59). Pull upper end of control housing away from cylinder head, and at the same time push lower end of control housing toward the engine to free the dowels. Lift off control housing. Governor weight housing is still attached to the blower.

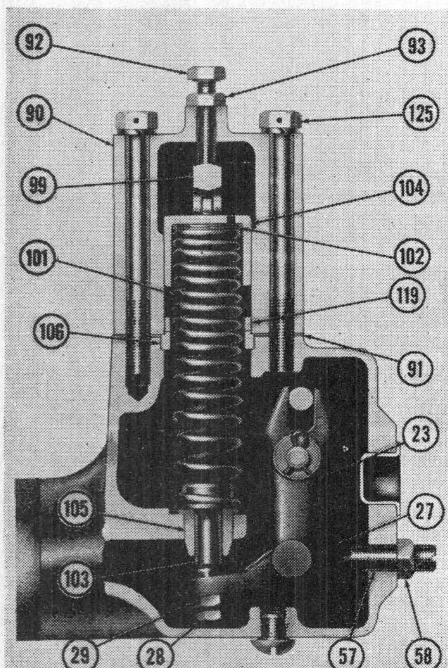

Fig. 0550 — Details of governor variable speed spring and lever.

104B. Governor carrier weight housing (59) can be removed after removing the governor control housing either by removing the blower, or modifying the air cleaner bracket mounting pad surface which is located on the balance weight cover. Modification of this surface involves filing or grinding off the sharp corner which interferes with removal of the weight housing.

To modify the air cleaner bracket mounting pad surface on the balance weight cover, proceed as follows: Refer to Fig. 0549. Remove water bypass tube (WBP). File or grind off the sharp corner of the mounting pad surface (FC) to produce a ½ inch radius, or just enough to permit removal of the weight housing.

Remove six cap screws attaching governor weight housing to blower. Separate weight shaft (66—Fig. 0554) from blower rotor shaft by pulling weight housing away from blower.

105. **OVERHAUL.** Herewith are procedures for overhaul of the components which make up the governor assembly.

105A. CONTROL HOUSING COVER. Refer to Fig. 0550. Needle bearings (8) for throttle shaft (7) can be renewed after removing control lever clamp bolt (89) and withdrawing the throttle shaft. Install new needle bearings so that lower bearing is flush with lower end of bearing boss, and upper bearing is approximately ⅛ inch below upper end of bearing boss.

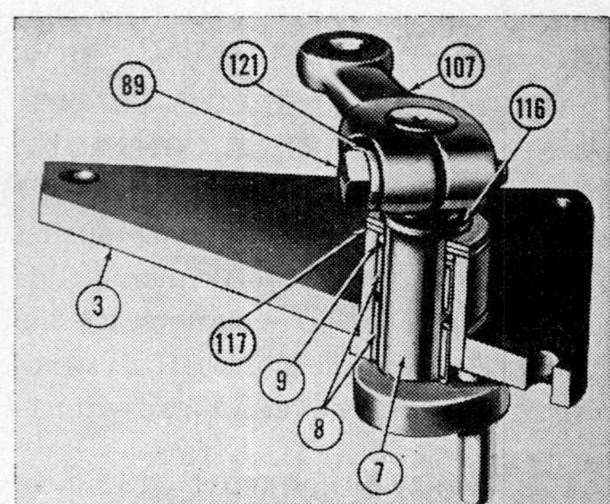

Fig. 0552—A GM engine variable speed type governor control housing cover.

7. Throttle shaft
8. Needle bearing
9. Seal ring
107. Governor control lever
116. Retaining ring
117. Seal ring washer

105B. SPRING HOUSING AND SPEED CONTROL LEVER SHAFT. Refer to Figs. 0550 and 0551. It is not necessary to remove governor control housing to service the units contained in the spring housing. First, remove governor control housing cover. Remove two cap screws attaching spring housing (90) to control housing; and remove spring retainer (104), shims (102), spring retainer stop (106), spring (101), speed spring plunger (103) and plunger guide (105).

Shims (102) (approximately a 0.325 shim pack) and spring retainer stop (106) are used to adjust crankshaft maximum no-load speed of 1840 rpm.

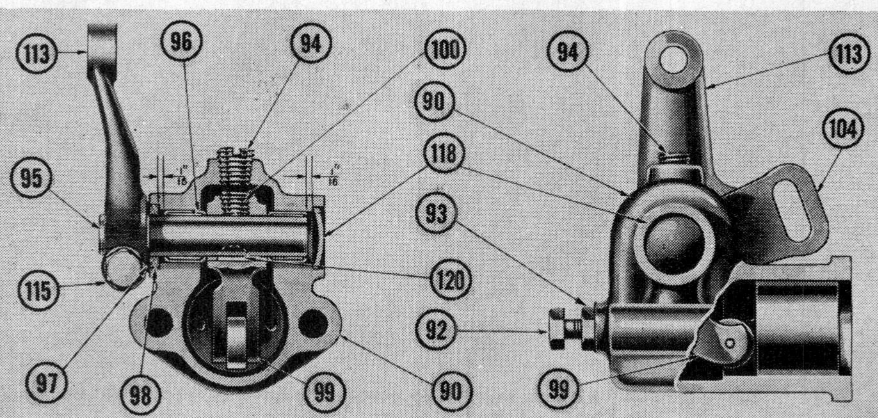

Fig. 0551—Details of variable speed spring housing.

23. Differential lever
27. Operating shaft lever
28. Governor gap adjusting screw
57. Buffer screw
90. Spring housing
92. Idle speed adjusting screw
95. Spring lever shaft
96. Needle bearings
97. Washer
98. Packing
99. Spring lever
100. Set screw
101. Variable speed spring
102. Shims
103. Spring plunger
104. Spring retainer
105. Spring plunger guide
106. Spring retainer stop
113. Speed control lever
119. Spring retainer stop spacer (not used in Oliver application)
120. Woodruff key

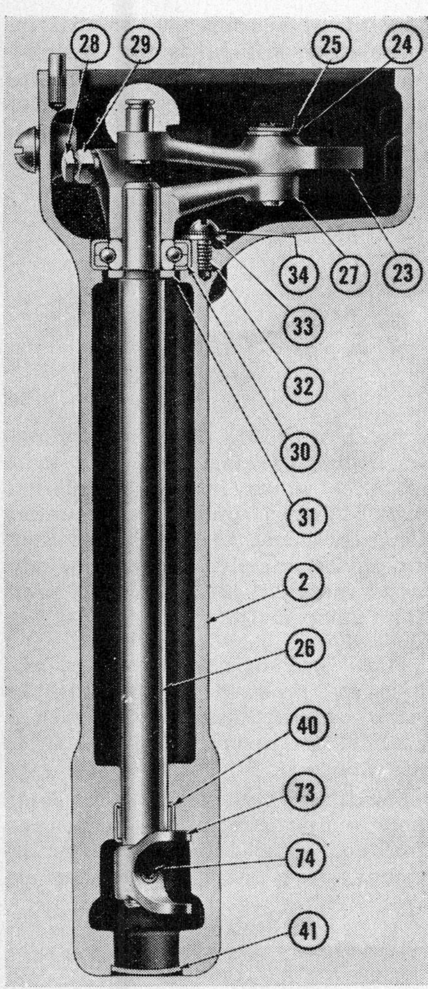

Fig. 0553—Details of variable speed type governor control housing.

Needle bearings (96) for spring lever shaft (95), and spring lever (99) can be renewed as follows: First, remove threaded plug (94) from top of spring housing. Working through plug opening, remove set screw (100) from spring lever (99). Remove speed control lever (113) from shaft (95) and Woodruff key. Support housing and press on speed control lever end of spring lever shaft (95). Press shaft, expansion plug (118) and needle bearing from housing, thus releasing spring lever (99) from shaft.

Install needle bearings (96) approximately $\frac{1}{16}$ inch below counterbore in spring housing as shown in Fig. 0551. The same thickness of shims (102) should be installed as were removed. Variable speed spring (101) should be installed with closely wound coils in speed spring retainer (104).

If crankshaft maximum no-load speed of 1840 rpm cannot be obtained and injector timing and injector rack setting are in correct adjustment, it will be necessary to vary the thickness of shims (102). Add shims to increase the speed.

2. Control housing
23. Differential lever
24. Washer
25. Retainer
26. Operating shaft
27. Operating shaft lever
28. Governor gap adjusting screw
30. Bearing
32. Bearing retaining screw
40. Oilite bushing (current production models)
41. Expansion plug
73. Operating shaft fork
74. Set screw

105C. CONTROL HOUSING. Refer to Fig. 0553. First, remove control housing from weight housing as outlined in paragraph 104A, and spring housing as outlined in paragraph 105B. To remove operating shaft (26), first remove differential lever (23) from operating lever (27). Working through opening in lower side of control housing (2) bump out expansion plug (41) from housing. Use a ⅛ inch Allen wrench to loosen Allen screw (74) which retains operating fork (73) to lower end of operating shaft. Remove operating shaft upper bearing retaining screw (34); then press operating shaft assembly out of operating fork (73). Operating shaft lever (27) is a press fit on operating shaft.

Oilite bushing (40) and ball bearing (30) can be renewed at this time.

105D. WEIGHT CARRIER HOUSING. Refer to Fig. 0554. First, remove weight carrier housing as outlined in paragraph 104B. Start disassembly of weight carrier housing by removing governor weight housing cap (83) and bearing retainer screw (71). Insert a long $\frac{5}{16}$ inch—24 thread cap screw in threaded end of weight shaft. Press on head of this cap screw to remove weight shaft assembly from weight housing. Weight carrier is a press fit

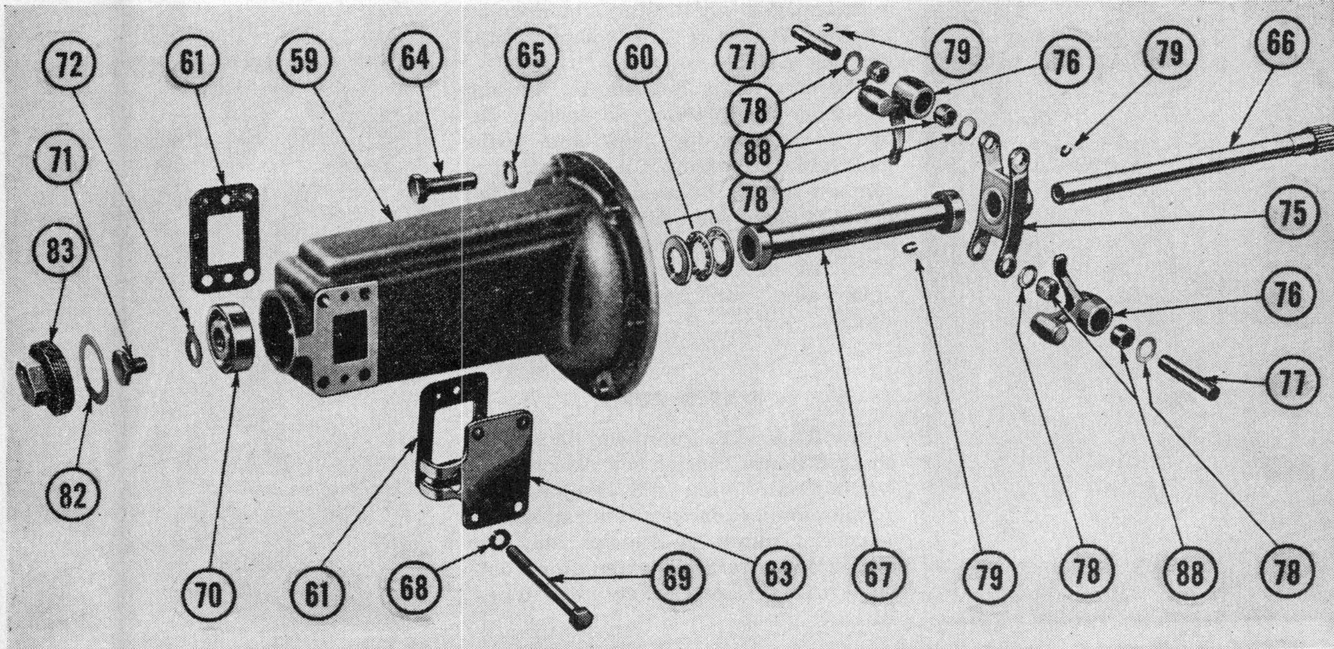

Fig. 0554—Details of variable speed type governor weight housing.

59. Weight housing
60. Riser thrust bearing
61. Gasket
63. Cover
66. Weight shaft
67. Riser
70. Bearing
75. Weight carrier
76. Weight
77. Weight pin
79. Retainer
88. Needle bearings

on weight shaft (66). If weights are to be removed from weight carrier, place correlation marks on weights and carrier for correct reassembly.

Weight shaft bearing (70) is a thrust bearing designed with a certain amount of looseness between the cone and cup. Weights operate on needle bearings (88). Renew weight pins (77) if worn more than 0.002 out-of-round.

When reassembling the weight carrier housing assembly install governor riser thrust bearing (60) so that the bearing race which has the larger inside diameter contacts the operating shaft fork. Install weight shaft bearing (70) with bearing numbers facing weight housing cap (83).

COOLING SYSTEM (GM Diesel 71 Series)

RADIATOR

110. Procedure for removal of radiator which involves removal of the hood is as follows: Loosen front grille center strap bolts, and remove two cap screws attaching front grilles to support. Lift off grilles. Detach both headlights and disconnect the wires. Remove hood rear strap, muffler, precleaners, two wiring harness clips from right side of hood, and lift hood from tractor. Remove radiator top hose and disconnect lower hose from oil cooler. Remove radiator to front frame cap screws, two on each side, and shield to radiator shell screws, two on each side. Lift off radiator.

Radiator filler neck cap is of the pressure type which is rated at 4 psi.

Fig. 0555—Fan belt tension is adjusted by pivoting fan pulley shaft bracket (FSB).

BWC. Balance weight cover
FMS. Pulley shaft bracket mounting support

FAN AND FAN BELT

111. To renew fan belt, remove air cleaner located on right side of engine.

Adjust tension of fan belt by means of fan bracket (FSB—Fig. 0555).

111A. To remove fan blades, first remove air cleaner located on right side of engine. Remove four bolts attaching fan to fan pulley hub extension, and lift off fan blades.

111B. To renew fan pulley or the prelubricated type ball bearings for the fan shaft, first remove fan blades, fan belt and three cap screws attaching fan support bracket (FSB) to fan mounting support (FMS).

Disassembly is self-evident after examining the unit.

THERMOSTAT

112. Cooling system is equipped with a thermostat which starts to open at 170 deg. F. and is fully opened at 185 deg. F.

Thermostat is located in housing (T—Fig. 0556) which is located between radiator top hose and water manifold. To renew thermostat, first remove tractor hood and radiator top hose. Remove four cap screws attaching thermostat housing to water manifold. Lift off thermostat housing water outlet elbow and remove the thermostat.

WATER PUMP

113. **REMOVE.** Pump can be removed as follows: Disconnect oil cooler to water pump hose, and by-pass hose at the pump. Remove three cap screws attaching pump to blower, and two cap screws attaching water pump outlet packing flange to cylinder block. Lift off pump.

113A. **OVERHAUL.** Disassembly procedure is as follows: Remove pump cover (15—Fig. 0557) and pump body drain plug (18). Working through drain plug opening, bump out taper pin (37) which retains impeller to shaft. Support pump on mounting flange end, and press impeller shaft out of the impeller and pump body. Drive coupling (24) is also a press fit on pump shaft.

113B. Prelubricated bearings and shaft are renewed as an assembly. Pump seal seat insert (27) which is available for service is a press fit in pump body.

Install water slinger (21) on pump shaft so that slinger is $\frac{5}{16}$ inch away from end of shaft bearing. After installing pump shaft and bearing as-

Fig. 0556 — GM engine water pump is mounted on the blower.

T. Thermostat housing
BP. By-pass
WP. Water pump Assembly

sembly in pump body and before installing the drive coupling, stake pump shaft bearing in bearing bore at three or four places.

Position impeller on shaft so that pin hole in shaft is midway between two blades on the impeller; then press impeller on shaft until end of impeller hub is 0.052 to 0.072 below the machined surface for pump body cover. Rotate pump shaft and check for recommended clearance of 0.005-0.045 between impeller blades and pump body. Insert a 0.184 inch drill in pin hole of shaft and drill through hub of impeller; then install a taper pin through the shaft and impeller.

2. Pump body
3. Impeller
4. Seal washer
5. Seal
6. Spring
8. Guide
10. Seal clamp ring
12. Retaining cup
13. Gasket
15. Pump cover
18. Drain valve
21. Slinger
22. Bearing and shaft assembly
23. Bearing and shaft assembly
24. Drive coupling
25. Slinger
26. Stake points for bearing
27. Seat insert
37. Pin

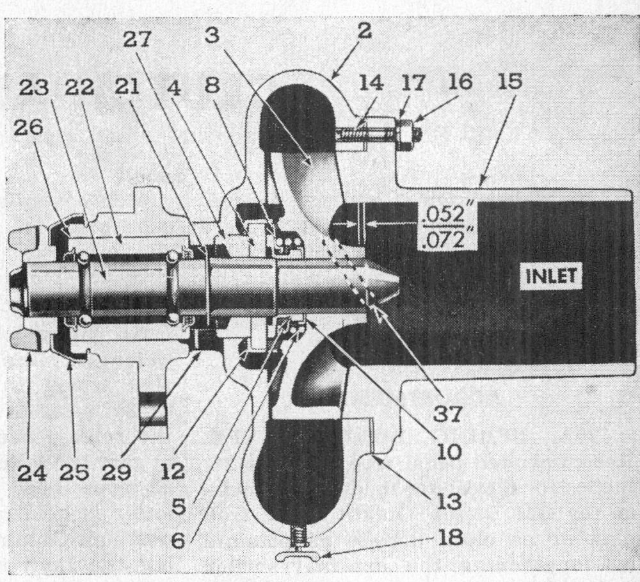

Fig. 0557—Details of GM engine water pump.

ELECTRICAL SYSTEM (GM Diesel 71 Series)

GENERATOR AND REGULATOR

115. GM 3-71 engines as installed in the Oliver Super 99 tractors are equipped with a Delco-Remy model 1100316, 12 volt, third brush type generator. Cold output is 20.0 amperes @ 14.0 volts @ 2300 generator rpm. Brush spring tension is 28 ounces. Field current draw is 1.58-1.67 amperes @ 12 volts @ 80 deg. F.

A model 1118791 Delco-Remy regulator is used.

STARTING MOTOR

116. GM 3-71 engines as installed in Oliver Super 99 tractors are equipped with a Delco-Remy model 1108801, 12 volt motor, fitted with a Dyer drive and a D-R model 1118095, 12 volt starter solenoid.

Tested at no-load, the starter current draw should be 115 amperes @ 11.6 volts @ 7000 rpm. Minimum locked torque should be 20 pounds feet @ 570 amperes @ 2.3 volts. Brush spring tension is 36-40 ounces.

Solenoid specifications are: Current consumption for both windings, 49.0-55.0 amperes @ 10.0 volts; and for hold-in winding, 11.0-13.0 amperes @ 10.0 volts. Relay on solenoid point opening voltage is 8.5 volts; and point closing voltage, 3.5-4.2 volts.

Fig. 0558—Installation view of generator and starting motor on a GM 3-71 engine.

CLUTCH AND CONTROLS

125. Main clutch is a spring loaded, single plate Borg & Beck 12E on 6 cylinder HC and Diesel; Rockford 14-TT on GM Diesel. An over-center type Rockford is available optionally on all models.

ADJUSTMENT

125A. **SPRING LOADED TYPE.** Recommended pedal free travel is $2\frac{1}{4}$ inches for 6 cylinder models; $2\frac{9}{16}$ inches for GM Diesel. Obtain stated free play or as close thereto as possible by lengthening the external shifter rod at the clevis end of same.

125B. **OVER-CENTER TYPE.** Correct adjustment is when a force of 30-35 pounds is required to engage clutch into over-center position. Measure the force with a spring scale hooked to upper end of shifter handle. Access to clutch adjusting ring Figs. 0575 and 0576 is obtained by removing batteries and clutch housing front cover and dust cover. Loosen clutch adjusting ring lock and rotate ring clockwise to "tighten" the adjustment.

OVERHAUL

125C. Herewith is procedure for removing both types of clutches from the tractor. If pto is continuous type remove the drive shaft of same as per paragraph 145J.

Remove battery or batteries and clutch housing front cover (31—Figs. 0457 and 0457A). On six speed tractors, remove the flat dust cover (40) by removing the screws which attach it to the front frame and to the clutch release bearing carrier. On four speed tractors, remove the screws retaining the clutch release bearing carrier to the carrier support.

Refer to Figs. 0458 and 0577. Remove chain (22) after extracting master link. Slide clutch shaft coupling (21) forward after loosening clamp bolt. Disconnect outer end of clutch shifter shaft, loosen set screw which retains fork (26) to shifter shaft then bump shaft out of fork toward left side of tractor. On spring loaded clutches the clutch shaft can now be removed. Unbolt clutch cover from flywheel and lift out the clutch or clutch and clutch shaft.

On GM Diesels with spring loaded clutch, removal will be facilitated if clutch cover is first secured to pressure plate (to compress springs) with 3 cap screws ($\frac{3}{8}$ x $2\frac{1}{4}$) screwed into pressure plate through the holes between release lever adjusting nuts.

On spring loaded installations only, before installing dust cover make sure that travel limit screw (28B — Fig. 0458) is set to prevent over-travel of the release bearing. There should be a gap of $\frac{1}{2}$ inch between underside of screw head and support when the latter is held upright in the installed position. Obtain this setting by rotating the limit screw.

125D. **OVERHAUL SPRING LOADED TYPE.** Procedure for disassembly is contained in the Standard Units Manual. Checking standards are as follows: All parts numbers are Borg-Warner numbers.

Borg-Warner Model	**B & B 12E**	**Rock. 14TT**
Cover Assembly	361216	165360
Pressure Plate	305293	M5246-1
Number of Pressure Springs	16	15

BORG & BECK 12E. Of the 16 pressure springs, 8 are No. 3814 painted purple for service. These should show a pressure of 130-140 pounds when compressed to a length of $1\frac{11}{16}$ inches. The 8 other heavier, uncolored springs are No. 3951. These should show a pressure of 150-160 pounds at length of $1\frac{11}{16}$ inches.

Lever setting is $2\frac{5}{32}$ inches using $\frac{11}{32}$ (0.340) inch keystock instead of the lined plate with clutch assembled.

ROCKFORD 14TT. Pressure springs No. 505-2 should show pressure of 170-180 pounds when compressed to their working height of $1\frac{13}{16}$ inches. Lever setting is $1\frac{15}{16}$ inches measured from the bearing surface of each lever to the friction surface of the **pressure plate.** Using Borg-Warner fixture, lever height is $1\frac{5}{8}$ inches with 3K4 sleeves (0.701) in place.

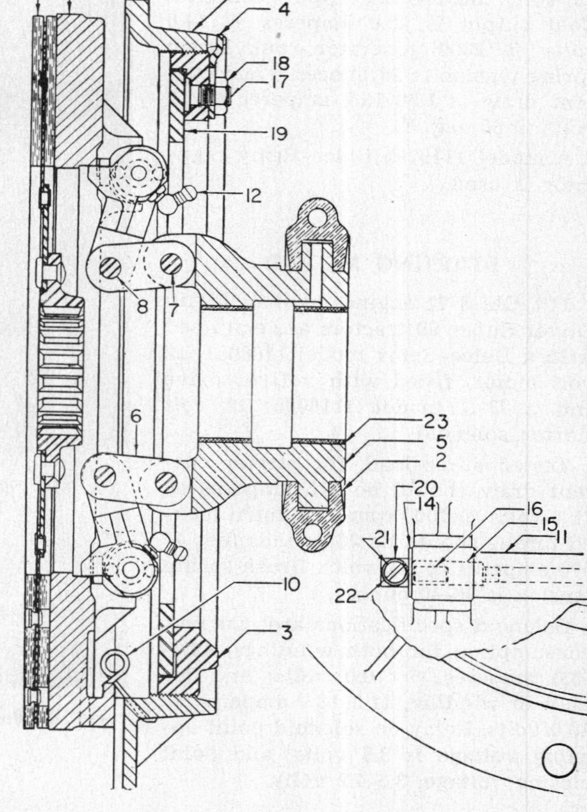

Fig. 0575—Over-center type clutch as used optionally on 6 cylinder Super 99 tractors.

1. Driven (lined) member
2. Release bearing
3. Adjusting ring
4. Adjusting lock
5. Sleeve
6. Connecting link
9. Pressure plate
10. Return spring
11. Camshaft
13. Back plate
14. Retainer plate
15. Roller
16. Roller
19. Adjusting ring plate
20. Cam plate
23. Sleeve bushing

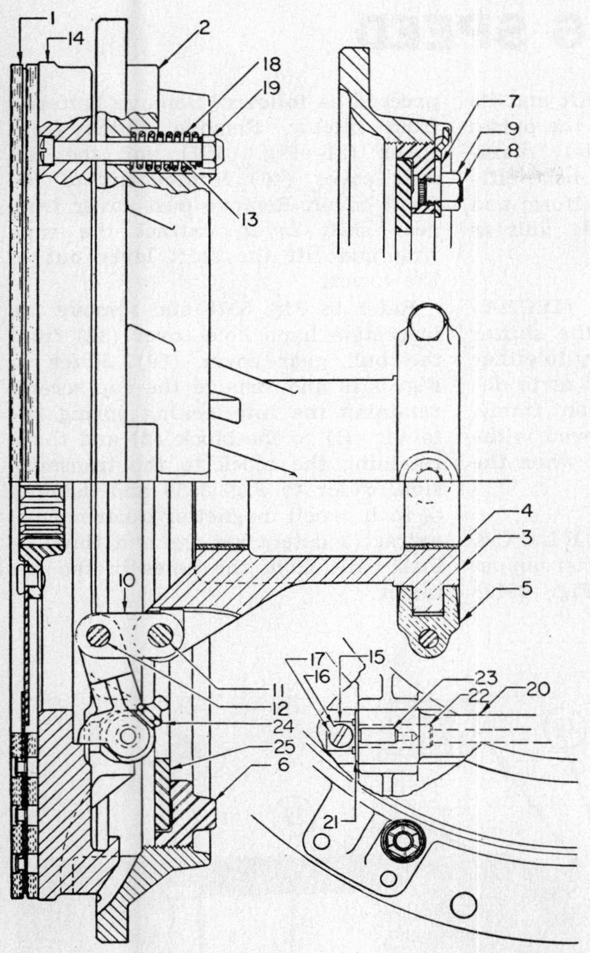

Fig. 0576—Over-center type clutch as used optionally on Super 99 GM Diesel tractors.

1. Driven (lined) member
2. Back plate
3. Sleeve
4. Sleeve bushing
5. Release bearing
6. Adjusting ring
7. Adjusting lock
10. Connecting link
14. Pressure plate
15. Cam block
18. Return spring
20. Camshaft
21. Retainer plate
22. Roller
23. Roller
25. Adjusting ring plate

Fig. 0577A—Rear view six cylinder model showing spring loaded Borg-Beck 12E clutch. Pressed steel cover shown at (31) in Fig. 0457 must be removed for access to starter mounting bolts (SB). Wheel axle shaft is (WS).

S. Clutch retaining screws
1. Pilot bushing
21. Coupling half
22. Coupling chain
23. Clutch shaft (hollow)
24. Shifter shaft
26. Shifter fork
27. Bushing for shaft 24
28. Release bearing carrier
29. Support for carrier 28
31. Housing front cover
36. Dust shield
40. Dust cover
40A. PTO drive hub
44. Shifter rod
53. Pedal return spring
58. Release bearing

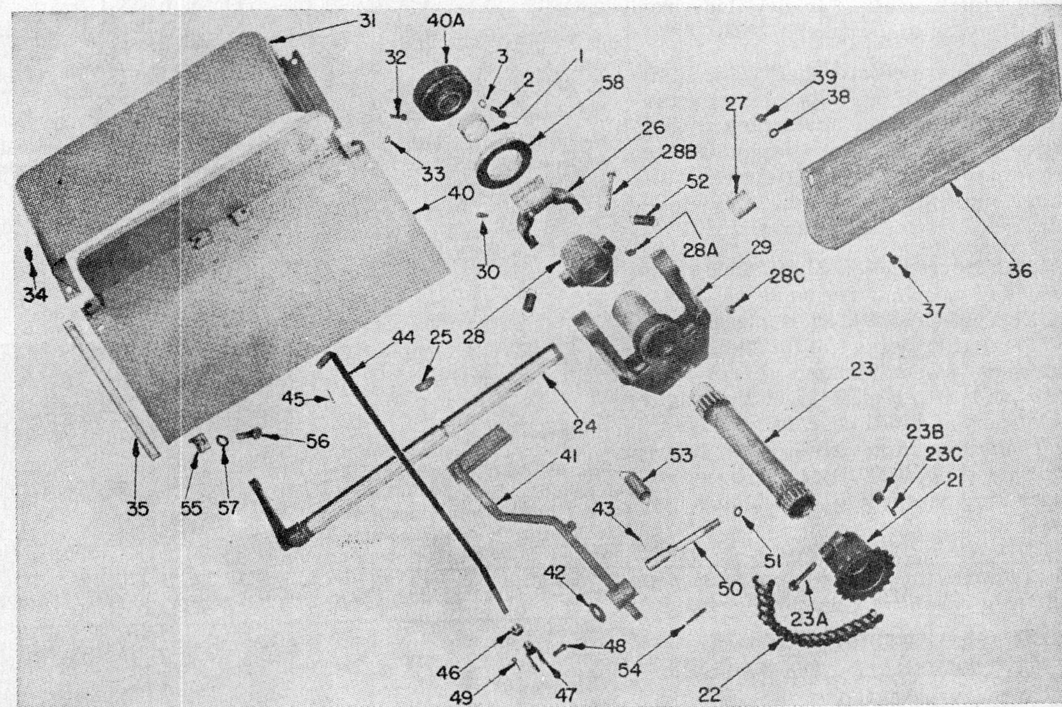

Fig. 0577—Components of main clutch control system. Dust shield (36) is mounted vertically to steel adaptor plate at rear of engine block.

TRANSMISSION 6 SPEED

126. For 4 speed transmission refer to paragraphs 129 through 129D.

The transmission housing which is called the rear frame, is divided into two compartments by a wall in the casting. In the front compartment are the transmission gears and shafts. On the rear side of the wall are the bevel gear end of the transmission bevel pinion shaft, the bevel ring gear, differential, and bull gears and bull pinions.

All shafts except the reverse idler are carried on adjustable type roller bearings.

OVERHAUL

127. **GUIDE TO LOCALIZED REPAIRS.** Although most transmission repair jobs do and should include complete disassembly there are exceptions. These exceptions are the infrequent instances where the location of a worn or failed part is such that the resultant metal cuttings are not likely to get into the bearings or between gear teeth. Information contained in paragraphs 127A through 127D is intended as a general guide only, not a procedure, for completing localized repairs.

127A. Input Shaft. This shaft and its bearings shown at (25) in Fig. 0581 can be removed without disturbing shifter forks or rails after removing the transmission top cover, clutch dust cover, bull gear cover, pto drive shaft if so equipped, and the main clutch shaft. For detailed procedure refer to paragraph 128J.

127B. Intermediate Shaft. This shaft (24—Fig. 0586) and its bearings can be removed without disturbing shifter forks and rails after removing the transmission top cover, clutch dust cover and bull gear cover. For detailed procedure refer to paragraph 128F.

127C. Bevel Pinion Shaft. This shaft (10) and bearings for same as shown in Figs. 0586 and 0589 is cleared for removal after doing the following preliminary work: Remove bevel ring gear and differential assembly, right axle shaft and bull gear, left bull gear, transmission input shaft and intermediate shaft. For detailed procedure refer to paragraphs 128K through 128N.

127D. Gear Shifter Forks & Rails. To remove the various parts of the shifter mechanism refer to paragraphs 128A through 128D.

128. GENERAL. The procedure outlined in paragraphs 128A through 128M may not be the fastest but it is arranged in a sequence that permits localized repair of each shaft and its bearings. If it is known at the outset that unit is to be completely disassembled it will be slightly more efficient to remove fenders, platform and bull gear cover as a single unit as shown in Fig. 0499B.

128A. **GEAR SHIFTING MECHANISM.** To remove all of the shifter rails and forks it is necessary to either remove the engine flywheel or to detach the rear frame from front frame. The inner rail can be removed without disturbing the flywheel when the differential is out.

128B. POPPET INTERLOCK BLOCK. To remove the shifter poppet and interlock block (2—Fig. 0579) proceed as follows: Remove batteries from tractor. Remove clutch front cover (31—Fig. 0457) and the flat dust cover (40) located aft of the front cover. Remove dust cover from gear shift lever, extract the snap ring and lift the shift lever out of the tower.

Refer to Fig. 0578 and remove the inspection hand hole cover (48) from the bull gear cover (39). Refer to Fig. 0579 and remove the cap screws retaining the interlocking spring retainer (R) to the block (2) and those retaining the block to the transmission. Refer to Fig. 0580 and using a ¼ inch pencil magnet or other means, extract 3 detent springs and three $\frac{7}{16}$ inch balls from the holes in the retainer.

Fig. 0578—Top view Super 99 chassis with engine removed. Not shown are the clutch front cover located immediately forward of the transmission top cover (44), and the clutch dust cover which is located directly above support (29) to which it is bolted.

L. Clip
3. Wheel axle carrier
7. Gear shift tower
19. PTO housing
23. Clutch shaft
29. Carrier support
39. Bull gear cover
44. Transmission cover
48. Hand hole cover

Fig. 0579—View of gearshift interlock block (2) containing 7 shifter detent and interlock balls. It can be removed without removing transmission top cover. Item (R) is retainer, (14) is cover and (P) is Welch plug.

Working through the hand hole cover opening slide the upper shifter rail (U) back (rearward) until it is disengaged from block, then lift block off case wall. Two ½ inch diameter interlocking balls are mounted in a cross passage in the block and two additional $\frac{7}{16}$ balls are in the center vertical hole. Be careful not to lose these when block is withdrawn from transmission.

When reinstalling the block place top shifter rail (rod) in rearward position and two lower rails in forward position. Insert the two ½ inch balls into block via the shift rail holes. While holding both balls in center of block passage, slide the block over the two lower rails. Install block to case gasket and install the block to transmission retaining cap screws finger tight. With top rail towards front of tractor insert two $\frac{7}{16}$ balls into center vertical hole in block. Slide upper rail into block and with all rails in neutral position drop one $\frac{7}{16}$ inch ball in each of the 3 holes in top of block. Insert a detent spring in each of the 3 holes then install the retainer.

Check rails for binding. If rails are free tighten all cap screws securely.

128C. RAILS AND FORKS. To remove all rails and forks first remove the interlock block as outlined in paragraph 128B then proceed as follows: Remove fuel tank rear support and lay instrument panel on fuel tank.

Remove cap screws retaining the transmission top cover (44—Fig. 0578) and withdraw cover by tilting front edge up and out toward left side of tractor. Remove set screws (13—Fig. 0581) from the three shifter forks. Working through opening at hand hole inspection cover (48—Fig. 0578) in bull gear cover (39) rotate upper shift rail to position where a wrench can be used on lug retaining set screw (14—Fig. 0582) and remove set screw. Remove set screw from lugs on the two other rails and remove lugs from rails.

Forks for the two bottom rails can be removed at this time by sliding the outer and inner lower rails rearward and withdrawing forks out through top opening in transmission. To remove center rail fork it will be necessary to remove flywheel or detach front frame from rear frame or remove the transmission intermediate shaft.

128D. To remove all of the shifter rails from the transmission, either remove the clutch from the flywheel as per paragraph 125C and flywheel from crankshaft or, detach front frame from rear frame. The lower inner rail can be removed without disturbing flywheel by removing the differential.

Install all rails with their notches at front. Install upper (center) fork and lower right fork with their hubs at front; left fork hub at rear. After rails and forks are installed, reinstall interlock block as per last half of paragraph 128B.

128F. **INTERMEDIATE SHAFT.** To remove this shaft (24—Fig. 0586) proceed as follows:

Remove clutch front cover (31—Fig. 0457) and the flat dust cover (40) located aft of the front cover. Remove fuel tank rear support and lay instrument panel on tank. Remove the cap screws retaining the transmission

Fig. 0580—Showing gearshift interlock block with retainer removed. Top shifter rail (rod) is (U). Use a magnet to remove detent balls from block.

Fig. 0581—Shifter forks but not the shifter rails, can be removed without removing flywheel. Screws (13) retain forks to shifter rails. Item (25) is the input shaft.

Fig. 0582 — View of shifter rail lugs (L) with bull gear cover removed. Lugs can be removed without disturbing bull gear cover by working through inspection hole cover opening in bull gear cover.

top cover (44—Fig. 0578) and withdraw cover by tilting front edge up and out toward left side of tractor.

Remove gear shift tower from bull gear cover and drop arm from steering gear. Remove platform, fenders and bull gear cover as a single assembly as shown in Fig. 0499B after removing the necessary fastenings; or cover alone after removing seat cover bolts, cover to platform bolts and cover to frame bolts.

128G. Refer to Figs. 0583, 0584 and 0585. Remove oil cup (70) from wall of transmission, bearing cover (57) from front wall and castellated nut from front end of intermediate shaft. Bump front end of shaft several times to loosen bearings, then extract the snap ring (60) from rear end of shaft. Bump shaft forward out of gears and out of rear bearing cone or, pull shaft forward using a puller with reaction legs. Rear bearing cup remains in transmission, front cone remains on shaft.

When reinstalling shaft place rear bearing cone in rear cup, enter shaft from front while threading gears and spacer on to shaft in the order shown in Fig. 0585. Start rear bearing cone on to shaft. Use a short piece of pipe or tubing and a pry bar to buck up the cone while bumping shaft rearward into the cone. Install snap ring to rear end of shaft. Tighten and secure shaft nut with cotter pin. Adjust shaft bearings to 0.000-0.002 end play by varying the shims. Because shims are used without sealing gaskets coat them with castor oil to obtain sealing effect. Coat sealing surface of oil cup (70—Fig. 0583) with rubber cement before installing.

128H. **REVERSE IDLER.** Reverse idler (53—Fig. 0587) can be removed after the intermediate shaft is out. Remove lock wire (LW) from Roll-Pin and with a straight drift bump pin down and out of boss. Bump idler shaft forward and remove shaft and gear.

The steel-backed bronze bushing in gear should be renewed when running clearance exceeds 0.006. After Roll-Pin is installed secure it with lock wire.

128J. **INPUT SHAFT.** To remove the input shaft which is joined to the rear end of the clutch shaft it is necessary to first remove transmission top cover, bull gear cover etc., as outlined in paragraph 128F then proceed as follows (refer to Figs. 0584 and 0583):

Remove rear half of shaft coupling from front of input shaft (25), also bearing retainer (27) sleeve and "O" ring. Remove oil cup (71) from rear wall of transmission. Bump shaft forward until front bearing cup is out of front wall. Using a puller move front bearing cone about one inch towards front end of shaft. Move shaft forward then swing rear end of shaft toward left side of cover opening and lift out the assembly. Rear bearing cup remains in transmission, bearing cones remain on shaft. Procedure for further disassembly is apparent.

Beginning with tractor serial 521300, an improved input shaft entered production. New shaft can be installed in place of old by changing the bearing cones and oil seal.

Fig. 0583—To remove bevel pinion shaft from 6 speed transmission it is necessary to remove both bull gears and the differential (D).

L. Shifter lugs
14. Retainer ring
70. Oil cup
71. Oil cup

Fig. 0584—Input shaft (25) of 6 speed transmission with shifter forks removed. An improved version is used in tractors after serial 521299.

11. Bevel pinion shaft bearing carrier
25. Input shaft
27. Input shaft bearing carrier
57. Bearing cover for intermediate shaft

Fig. 0585—Correct order of assembly of intermediate shaft for 6 speed transmission.

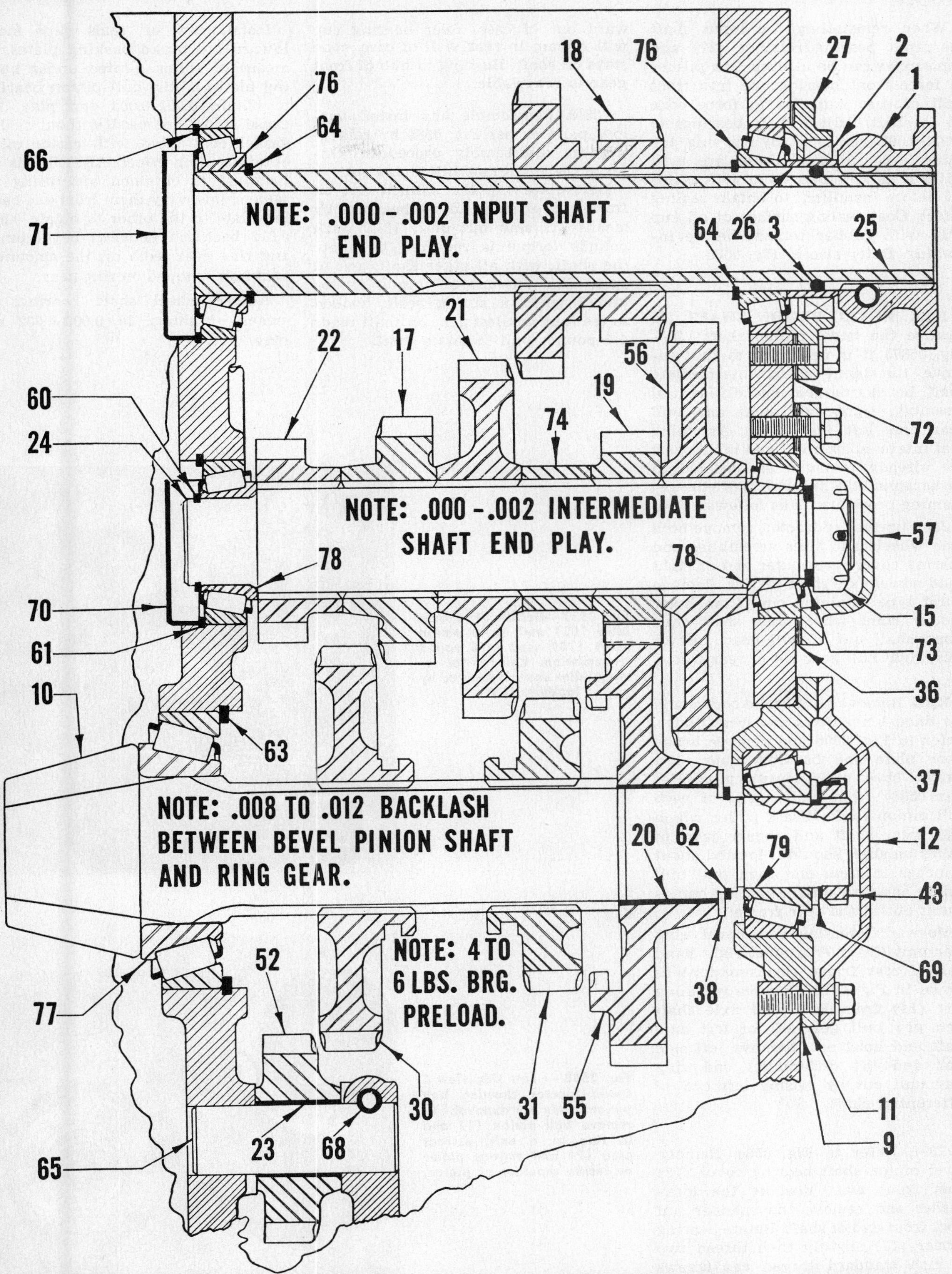

Fig. 0586—Section through 6 speed transmission. All shafts can be removed without removing the shifter forks. Mesh position of bevel pinion shaft is fixed. Removal of this shaft requires removal of right hand wheel axle shaft and bull gear, left bull gear and the differential. Input shaft (71) is slightly different beginning with tractor serial 521300. Refer to Fig. 0589 for parts legend.

When reinstalling the input shaft the front bearing retainer (27) and cap screws can be utilized as a pusher to force front bearing cup into front wall of case and also to force cone on to shaft. Adjust shaft bearings to 0.000-0.002 end play by varying the shims (72). Because shims are used without gaskets coat them with castor oil before installing, to obtain sealing effect. Coat sealing surface of oil cup (71) with rubber cement before installing. Refer also to Fig. 0586.

128K. **BEVEL PINION SHAFT.** To remove the bevel pinion shaft (10—Fig. 0587) it is necessary to first remove the input shaft, intermediate shaft, bevel ring gear and differential assembly, right axle shaft and bull gear and left bull gear. Assuming that intermediate shaft and input shaft are already removed as outlined in paragraphs 128F through 128J, the remaining procedure is as follows:

Jack up rear of tractor, remove both rear wheels and tires assemblies and bearing cover from outer end of right hand wheel axle shaft carrier. Remove spiral type retaining ring from inner end of right wheel axle shaft then bump shaft out of bull gear. Lift or hoist right bull gear out of rear frame.

128L. Remove both brake covers and the lined inner disks and wear plates. Refer to Fig. 0606 and remove brake wear plate and backing plate (18). Extract sheet metal closure plug (P—Fig. 0588) from hollow end of each bull pinion shaft. Use a puller which will enter shaft and engage jaws of puller on shaft shoulder located about 7 inches in from end then pull bull pinion and bull pinion bearing cup as a unit out of the rear frame.

Move differential and bevel ring gear unit (D) over against right hand wall of rear frame (transmission) as shown in Fig. 0612. Remove retaining ring (14) from left hand axle shaft then pry bull gear off of left axle shaft and hoist out. Remove left side gear and lift differential and ring gear unit out by raising left end of differential shaft.

128M. Refer to Fig. 0584. Remove bevel pinion shaft bearing cover (12) from front wall, unstake the lockwasher and remove the spanner nut from front end of shaft. Rotate bearing carrier (11) slightly then thread two ½ inch standard thread cap screws into tapped holes of same to force carrier and front bearing out of transmission. Bump front end of shaft rearward out of case. Rear bearing cup will remain in rear wall of case, cone stays on shaft. Bushing in hub of front gear is renewable.

128N. Reassemble and install various parts as per Fig. 0589 by reversing the disassembly procedure. Observe these points while doing so:

Before locking the splined nut on front end of shaft adjust bearings by means of same nut until 10-15 inch pounds torque is required to rotate the shaft, with all other shafts out of case. This amount of torque is obtained when a spring scale hooked to teeth of smallest gear on shaft reads 4-6 pounds pull to rotate shaft.

Install new oil seals (lips facing inward) to brake backing plates. By means of shims located under backing plates adjust bull pinion bearings to obtain 0.001-0.003 end play. Because shims are used without sealing gasket coat them with castor oil to obtain sealing effect. AFTER this adjustment is obtained and using the same shims, vary them from one backing plate to the other to obtain 0.008-0.012 backlash between bevel pinion and ring gear teeth or, the amount of backlash stamped on ring gear.

Adjust wheel shaft bearings by means of shims, to 0.000-0.002 end play.

Fig. 0587—Details of reverse idler (53) and bevel pinion shaft (10) used in 6 speed transmission. Roll-Pin for reverse idler shaft is secured by lockwire (LW).

Fig. 0588 — Left side view 6 speed tractor showing bull pinion ready for removal. To remove bull pinion (1) and bearing as a unit, extract plug (P) and engage puller on inside shoulder of pinion.

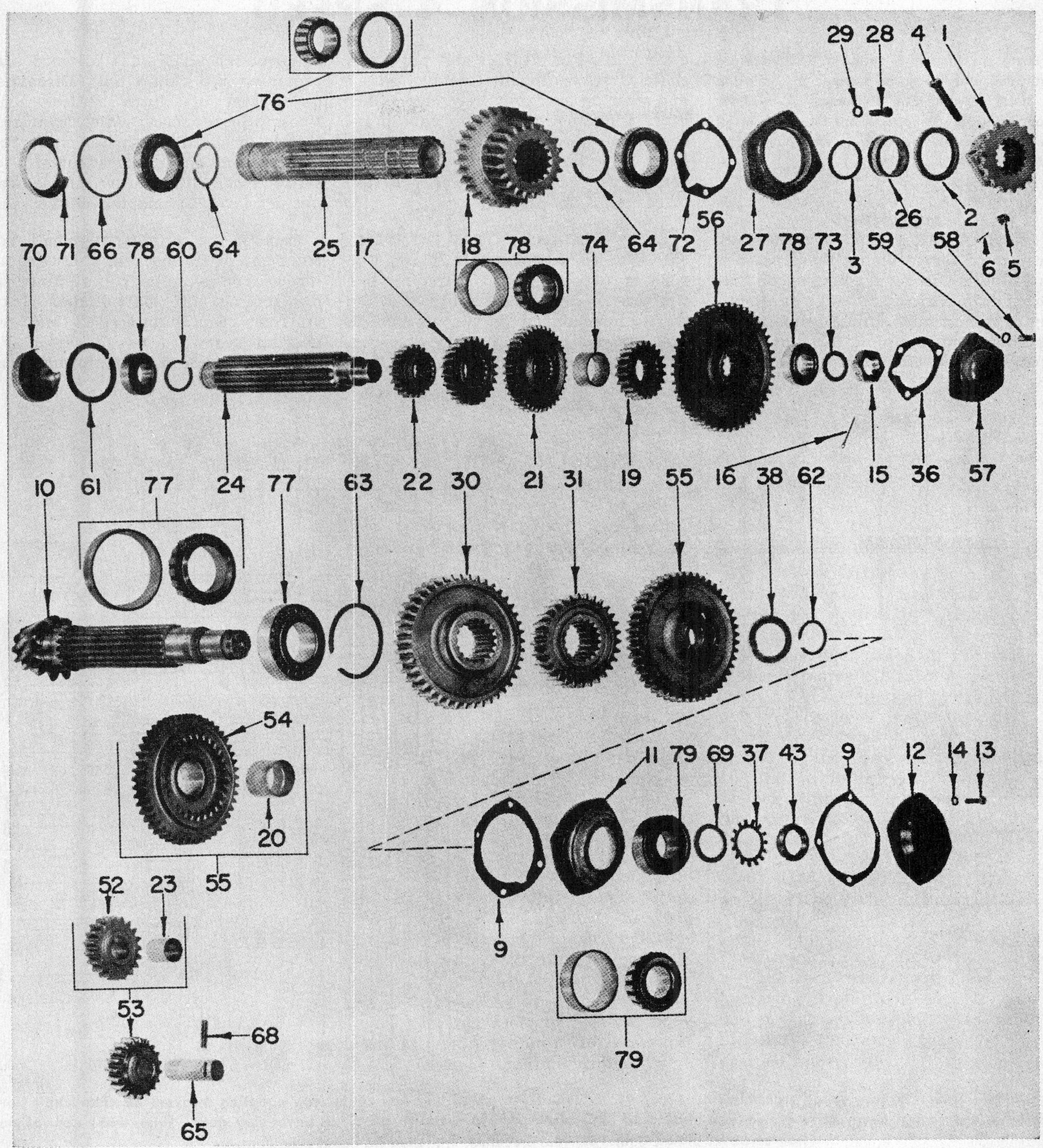

Fig. 0589—Components of 6 speed transmission with adjustable bearings for all shafts except reverse idler.

1. Coupling, rear half
2. Oil seal
3. "O" ring
10. Bevel pinion shaft
11. Bearing carrier
12. Bearing cover
17. Intermediate second gear
18. Input sliding gear
19. Intermediate first gear
20. Idler gear bushing
21. Intermediate high gear
22. Intermediate reverse gear
23. Reverse idler bushing
24. Intermediate shaft
25. Input shaft
26. Seal sleeve
27. Bearing retainer
30. Reverse, second, fourth sliding gear
31. First, third, fifth, sixth sliding gear
36. Shim
37. Lockwasher
38. Locating washer
43. Lock nut
52. Reverse idler gear
55. Pinion idler gear
56. Intermediate low gear
57. Bearing cover
60. Snap ring
61. Snap ring
62. Snap ring
63. Snap ring
64. Snap ring
65. Idler shaft
66. Snap ring
68. Roll pin
69. Retainer washer
70. Bearing oil cup
71. Bearing oil cup
72. Shim
73. Retainer washer
74. Spacer
76. Bearing
77. Bearing
78. Bearing
79. Bearing

TRANSMISSION 4 SPEED

Early production 6 cylinder tractors were equipped with the same 4 speed transmission as used in the late production 4 cylinder models 90 and 99. Procedure for overhaul of 6 speed transmissions begins with paragraph 128.

OVERHAUL

129. **SHIFTER FORKS AND RAILS.** Procedure for servicing these parts is as follows: Disconnect battery or batteries from wiring. Remove fuel supply tank and tank support and lay instrument panel on platform or other resting place. Remove clutch front cover (31—Fig. 0457A) and pto drive shaft if so equipped. Remove pulley and steering gear unit from belt pulley carrier. Remove belt pulley carrier assembly as shown in Fig. 0594. Refer to Fig. 0595 then remove cap screws retaining the shifter assembly to top of case and lift off.

129A. **SLIDING GEAR (INPUT) SHAFT.** To remove this shaft, Fig. 0596 or (31—Fig. 0590), first remove the pulley carrier assembly and the shifter mechanism as per paragraph 129. Remove coupling chain (22—Fig. 0458) after extracting the master link. Slide clutch shaft (23) forward after removing the clamp bolt. Disconnect outer end of clutch shifter shaft. Loosen the set screw which locates the shifter fork (26) to shifter shaft, then bump shaft out of fork toward left side of tractor. Withdraw release bearing carrier (29) and the clutch shaft.

Remove the coupling rear half from front end of sliding gear shaft and cap screws from front bearing retainer and pry retainer off shaft. Using a drift on rear gear, bump shaft and rear bearing rearward as shown in Fig. 0596 and withdraw shaft through gears

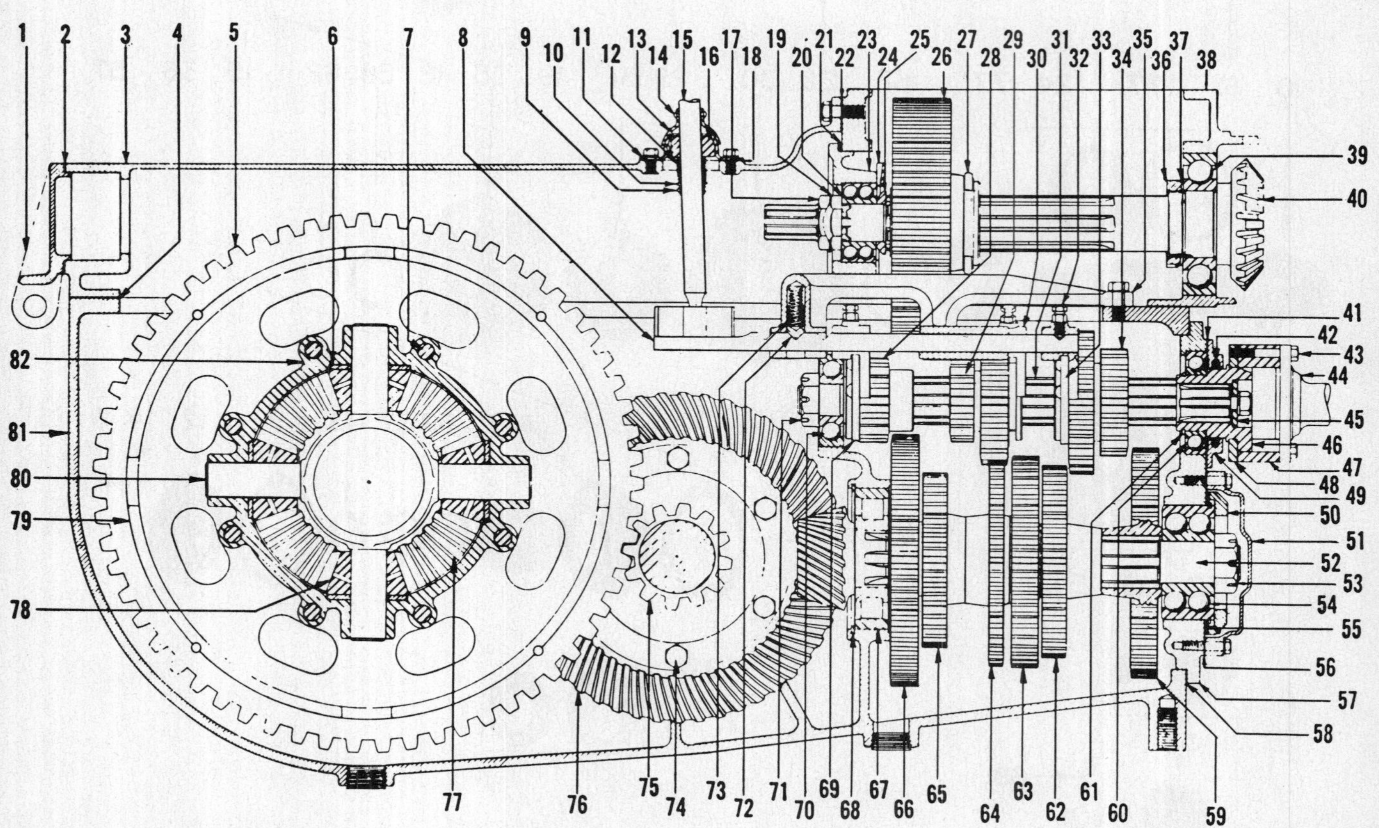

Fig. 0590—Series 99 four speed transmission. Super 99 six cylinder 4 speed unit has chain type coupling retained by clamp bolt but is otherwise the same. Bevel drive pinion and shaft sold separately or as a set with matching bevel ring gear. Shims (55) control position of bevel pinion gear. Correct backlash is etched on bevel ring gear.

3. Bull gear cover
5. Bull gear
6. Differential pinion
8. Shifter rails
10. Spring stop
11. Shift lever bearing
12. Shift lever spring
13. Shift lever cap
21. Shim
23. Bearing cage
24. Retainer
25. Thrust washer
26. Pulley drive gear
27. Shifter fork
28. Sliding gear
29. Sliding gear
30. Center fork
33. Front fork
34. Sliding gear
35. Rail support
38. Pulley carrier
40. Pulley gear and shaft
42. Felt washer
44. Clutch shaft
46. Clutch coupling
49. Retainer
50. Retainer
52. Bevel pinion shaft
55. Bearing shim
57. Front end plate
59. Stationary gear
61. Snap ring
62. Stationary gear
63. Stationary gear
64. Oiler gear
65. Reverse gear
66. Stationary gear
68. Snap ring
69. Rear fork
72. Rail poppet
73. Poppet spring
75. Bull pinion and shaft
76. Bevel ring gear
77. Differential gear
78. Differential pinion
80. Pinion carrier
81. Main frame
82. Differential housing (left)

Fig. 0594—Removing belt pulley carrier from 4 speed tractors. Unit will be easier to handle if pulley is removed before removing carrier unit.

Fig. 0596—Bumping the spline (input) shaft of Super 99 four speed transmission backward prior to removal of shaft.

Fig. 0595—Removing gear shifters and forks assembly from Super 99 four speed transmission.

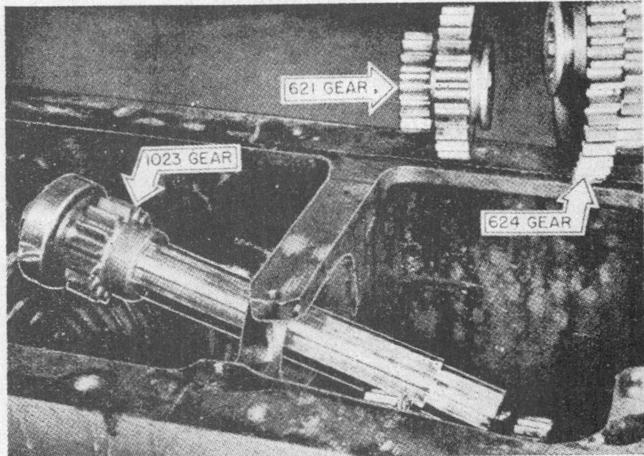

Fig. 0598 — Withdrawing of spline (input) shaft from Super 99 four speed transmission.

Fig. 0595A—Super 99 GM tractor showing supercharger on the three cylinder engine.

and rear top opening as shown in Fig. 0598.

129B. **REVERSE IDLER GEAR.** After the sliding gear shaft is out the procedure for removing reverse idler gear is self-evident after referring to Fig. 0599.

129C. **BEVEL PINION SHAFT.** To remove this shaft shown at (52) in Fig. 0590, first remove the pulley carrier, shifter unit, sliding gear shaft and reverse idler gear as outlined in paragraphs 129 through 129B then proceed as follows:

Remove bearing cover and bearing retainer from transmission front end cover as shown in Figs. 0601 and 0602. Remove snap ring from bearing and shims from behind snap ring and tie the shims to the snap ring for identity. Remove dowels from transmission front cover using a ⅝—11 nut for a puller then remove front cover as shown in Fig. 0600. Bump pinion shaft forward until rear bearing and gear have cleared the bearing bore in case then lift the assembled unit out of the case.

Procedure for disassembling the bevel pinion shaft is self-evident after referring to Figs. 0590 and 0603. The oiler gear bushing (26—Fig. 0603) should have 0.004-0.007 running clearance on hub of adjacent gear. If running clearance exceeds 0.010 renew the bushing which may require sizing after installation. Make sure that oil holes in bushing register with oil holes in gear. If outside diameter of hub of adjacent gear is less than 2.866 renew the gear.

129D. When reassembling the original or a new bevel pinion shaft use the shims (49—Fig. 0603) under the front bearing retainer to bring the heels of the bevel pinion teeth flush with the toes of the teeth on the bevel ring gear.

If the bevel pinion to ring gear backlash is now within the limits of 0.006-0.012, the job is completed. If measured backlash is not within the stated limits it should be adjusted as outlined in paragraph 129E.

129E. To adjust backlash, proceed as follows: By means of the shims under each bull pinion bearing carrier, adjust bearings so that bevel ring gear rotates freely but has minimum end play. After bearings are so adjusted, vary the same shims until the amount of backlash shown stamped on end of pinion (or 0.006-0.012 backlash) is obtained.

Fig. 0599—Removal of reverse idler gear from Super 99 four speed transmission.

Fig. 0600 — Super 99 four speed transmission with front cover removed showing bevel pinion shaft and gears assembly.

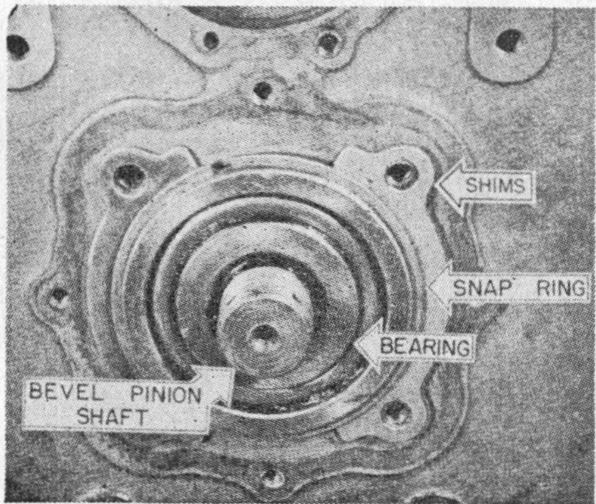

Fig. 0601—Half shims are used behind the bevel pinion shaft front bearing snap ring in the Super 99 four speed transmission. Shims control mesh position of pinion.

Fig. 0602 — Removing bevel pinion shaft front bearing retainer from transmission front cover on Super 99 four speed tractor.

DIFFERENTIAL AND MAIN DRIVE BEVEL GEARS

DIFFERENTIAL (6 Speed)

(Type used with 4 speed transmissions begins with paragraph 131.)

The differential is combined with the main drive bevel ring in the rear compartment of the transmission case which is called the rear frame. Also mounted on the differential as shown in Fig. 0606 are the two spur gear type bull pinions which drive the final drive bull gears. Beginning with serial 521300, a four pinion differential is used.

130. **REMOVE AND INSTALL.** Procedure for removal of differential, bevel ring gear and bull pinions as a single assembly is as follows: Remove gearshift tower and drop arm from steering gear. Disconnect foot accelerator linkage on tractors so equipped. Remove platform, fenders and bull gear cover as a single unit as shown in Fig. 0499B or, the cover alone by removing seat cover bolts also the cover to platform and cover to rear frame bolts.

Remove both rear wheels and tires assemblies and the bearing cover from outer end of right hand wheel axle shaft carrier. Remove spiral type retaining ring from inner end of right wheel axle shaft then bump shaft out of bull gear. Hoist gear out of rear frame. Remove both brake covers and wear plates (18—Fig. 0606).

Extract sheet metal closure plug (P—Fig. 0588) from each bull pinion shaft. Use a puller which will enter the shaft and engage jaws of puller on shaft shoulder located about 7 inches in, then pull bull pinion and bull pinion bearing cup as a unit out of the rear frame.

Move differential and bevel ring gear unit (D) over against right hand wall of rear frame (transmission) as shown in Fig. 0612. Remove retaining ring (14) from left hand axle shaft then pry bull gear off left hand axle shaft and hoist out. Remove left differential side gear and lift differential

Fig. 0603A — Brakes on 4 speed and 6 speed tractors are adjusted by turning the nut (A). Item (13) is brake cover.

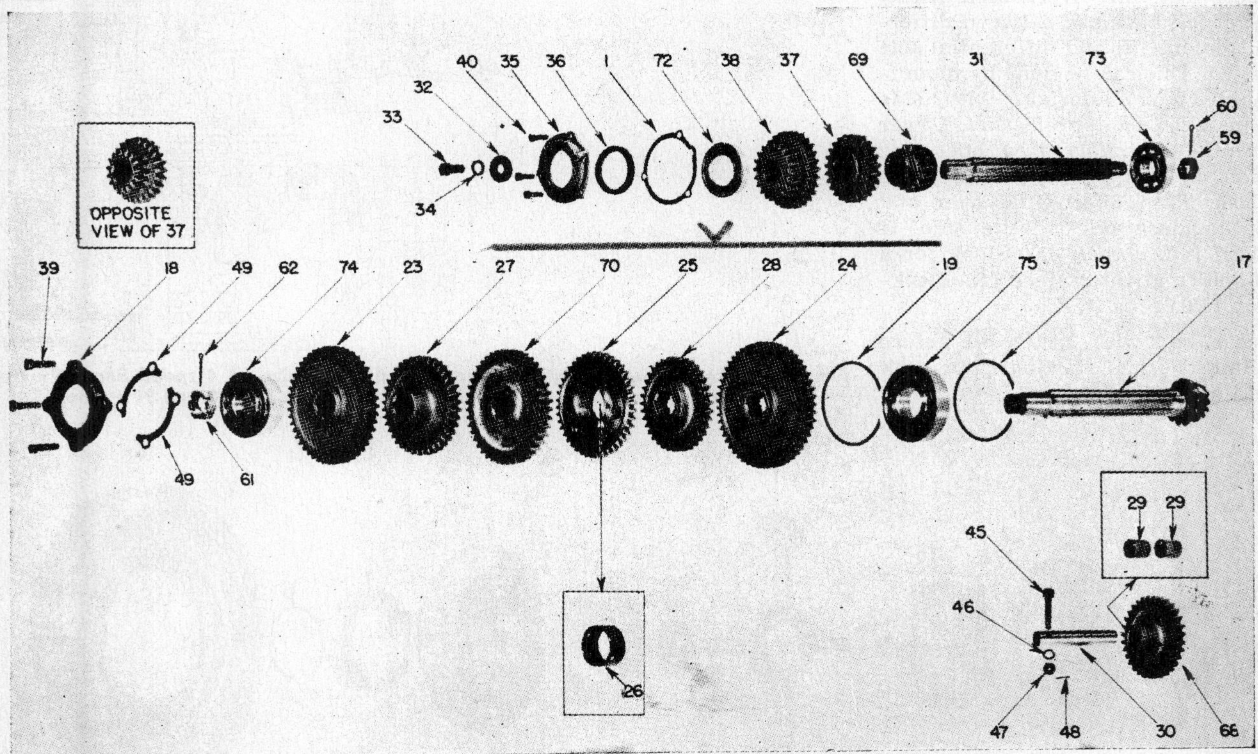

Fig. 0603—Components of 4 speed transmission used in early production Super 99 six cylinder tractors.

17. Bevel pinion shaft
18. Pinion bearing retainer
19. Snap ring
25. Oiler gear
28. Reverse gear
29. Reverse idler bushing
30. Reverse idler shaft
31. Spline shaft
49. Bevel pinion adjusting shim
66. Reverse idler gear

and ring gear unit out by raising left end of differential shaft.

130A. When reinstalling the differential observe the following points: Assemble new oil seals to backing plates with lips of seals facing inward. No gaskets are provided between backing plates and transmission; obtain sealing effect by coating shims with castor oil. Adjust bearings by means of shims (4—Fig. 0606) to obtain 0.002-0.003 end play, making sure that bevel gear teeth are not bottomed on bevel pinion teeth. AFTER this adjustment is obtained, and using the same shims, vary them from one backing plate to the other to obtain 0.008-0.012 backlash between teeth of bevel gears, or, the amount of backlash stamped on ring gear.

Adjust axle shaft bearings to 0.000-0.002 end play. Refer to caption Fig. 0586 for data on later four pinion differential.

130B. **OVERHAUL.** Observe the following points when overhauling the removed differential: To remove pinions (5) extract snap rings (12) from pins (11) then use puller in threaded holes to extract pinion pins from spider.

When riveting ring gear heat rivets to orange color.

Before installing overhauled differential check backlash between differential pinions (5) and differential side gears (6). This can be done by mounting bull pinion upright with side gear installed on top of same. Lower differential assembly into operating position on bull pinion and measure backlash. Turn differential over and check backlash on opposite side. If backlash is less than 0.006 look for a worn spider at thrust face (S). Backlash up to 0.018 is O. K.

Coat shims with castor oil before installing. Normal starting shim pack is 0.094 on each side.

DIFFERENTIAL (4 Speed)

(Six speed type begins with paragraph 130.)

On these tractors the differential is combined with the single final drive bull gear in the rear portion of the transmission case. The bull gear itself is riveted to either the right half of the differential housing or to a separate hub called the spider. A bolted-together assembly located ahead of the bull gear combines the single bull pinion, and main drive bevel ring gear. These tractors are not ordinarily equipped with pinion shaft brakes.

131. **R & R AND OVERHAUL.** Refer to Fig. 0605. To remove the differential and bull gear assembly first block up the rear of tractor and remove the rear wheels. Remove wheel guards and platforms assembled. Remove pto if so equipped and bull gear cover from top of transmission. Disconnect brake linkage if tractor is equipped with rear wheel brakes. Support bull gear in hoist. Remove both rear wheel carriers with their axle shafts, as assemblies, from transmission. Hoist bull gear and differential unit out of transmission.

131A. Procedure for bench overhaul of the assembly is self-evident after an examination of Fig. 0604. The bull gear is available only as an assembly with its spider or right hand differential case to which it is riveted. Carrier bearings (12) complete with their outer races will remain on the differential case and can be renewed after removing the snap rings (11).

MAIN DRIVE BEVEL GEARS (6 Speed)

(Four speed type begins with paragraph 134.)

132. **BEVEL PINION.** The bevel pinion is also the transmission sliding gear shaft and is available separately from the bevel ring gear. The mesh (fore and aft) position of pinion is not adjustable. To remove or renew the bevel pinion follow the procedure outlined in paragraphs 128K through 128N.

133. **BEVEL RING GEAR.** Bevel ring gear is available separately from the bevel pinion. To remove or renew the bevel ring gear independently of the bevel pinion follow the procedure outlined in paragraphs 130 through 130B.

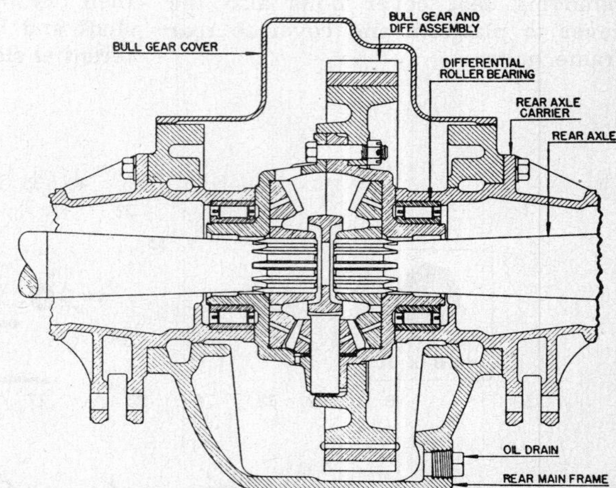

Fig. 0605—Rear view through differential and final drive of 4 speed Super 99 tractor. This design is same as used on late 4 cylinder models 90 and 99 tractors.

Fig. 0604—Components of differential and bull gear unit used on Super 99 tractors with 4 speed transmissions.

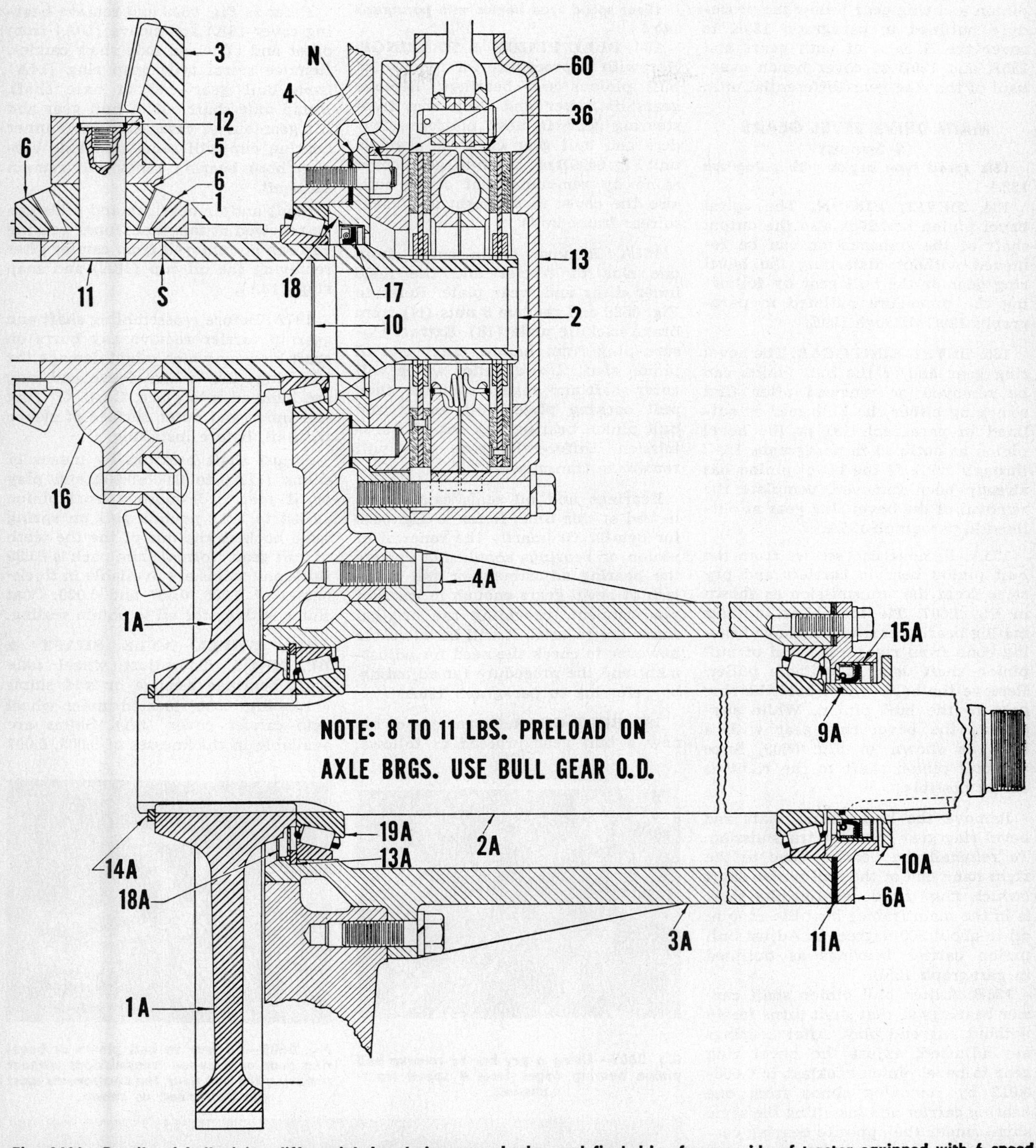

Fig. 0606—Details of bull pinion differential, bevel ring gear, brakes and final drive for one side of tractor equipped with 6 speed transmission. Beginning with serial 521300, the differential is of the four pinion type. The four pinion type can be used in place of the two pinion unit by also changing the bevel ring gear and side gears.

1. Bull pinion
1A. Bull gear
2. Expansion plug
2A. Wheel axle shaft
3. Bevel ring gear
3A. Axle carrier
4. Shims
4A. Shaft carrier screw
5. Differential pinion
6. Side gear
6A. Bearing cover
9A. "O" ring
10. Differential shaft
10A. Oil seal sleeve
11. Differential pin
11A. Shim
12. Retaining ring
13. Brake cover
13A. Snap ring
14A. Snap ring
15A. Oil seal
16. Differential spider
17. Oil seal
18. Brake backing plate
18A. Oil cup
19A. Bearing
36. Actuating disc
60. Brake wear plate

133A. **BEVEL PINION & RING GEAR.** To remove or renew the bevel pinion and ring gear follow the procedure outlined in paragraph 128K to cover the R & R of both gears and 130A and 130B to cover bench overhaul of the ring gear-differential unit.

MAIN DRIVE BEVEL GEARS (4 Speed)

(Six speed type begins with paragraph 132.)

134. **BEVEL PINION.** The spiral bevel pinion which is also the output shaft of the transmission can be removed without disturbing the bevel ring gear or the bull gear by following the procedure outlined in paragraphs 129C through 129E.

135. **BEVEL RING GEAR.** The bevel ring gear and/or the bull pinion can be removed or renewed after first removing either the bull gear as outlined in paragraph 131 or the bevel pinion as outlined in paragraphs 129C through 129E. If the bevel pinion has already been removed, complete the removal of the bevel ring gear as outlined in paragraph 135A.

135A. Remove cap screws from the bull pinion bearing carriers and pry same from the transmission as shown in Fig. 0607. Tie the shims to their mating bearing carriers. Remove bearing cone from right hand end of bull pinion shaft using a suitable puller. Remove the bolts which hold the ring gear to the bull pinion. While supporting the bevel ring gear with a hoist as shown in Fig. 0609, force the bull pinion shaft to the right as far as possible.

Remove the bull pinion shaft and bevel ring gear from the transmission. To reinstall the bearing cone to the right hand end of the bull pinion shaft (which must be done after the shaft is in the main frame) heat the cone in oil to about 300 degrees F. Adjust bull pinion carrier bearings as outlined in paragraph 135B.

135B. Adjust bull pinion shaft carrier bearings so that shaft turns freely without any end play. After bearings are adjusted, adjust the bevel ring gear to bevel pinion backlash to 0.006-0.012 by removing shims from one bearing carrier and installing the same shims under the opposite bearing carrier.

135C. **BEVEL PINION & RING GEAR.** To renew the bevel pinion and ring gear, follow the procedure as outlined in paragraphs 129C through 129E for the pinion, and paragraph 135A through 135B for the bevel ring gear.

BULL GEARS, PINIONS, WHEEL AXLES (6 Speed)

(Four speed type begins with paragraph 140.)

136. **BULL PINION & BEARINGS.** Herewith is procedure for renewal of bull pinions and bearings: Remove gearshift tower and drop arm from steering gear. Remove platform, fenders and bull gear cover as a single unit Fig. 0499B or, the bull gear cover alone by removing seat cover bolts also the cover to platform and cover to rear frame bolts.

136A. Remove brake cover from one side of tractor and the lined inner disks and wear plate. Refer to Fig. 0588 and remove 6 nuts (N) from brake backing plate (18). Extract closure plug from hollow outer end of pinion shaft. Use a puller which will enter shaft and engage shoulder, then pull backing plate, bull pinion and bull pinion bearing cup out of transmission. Differential side gear will remain in transmission.

Bearings and oil seals can be renewed at this time. Refer to Fig. 0606 for details. Ordinarily the renewal of pinion or bearings should not change the bearing adjustment or the backlash of bevel gears enough to warrant readjustment, providing the original shims are re-used. It will be advisable however to check the need for adjustment and the procedure for adjusting, by referring to paragraph 130A.

137. **BULL GEAR.** To remove or renew a bull gear proceed as follows: Remove bull gear cover as per first section of paragraph 130.

Refer to Fig. 0606 and remove bearing cover (6A) and sleeve (10A) from outer end of wheel axle shaft carrier. Remove spiral type snap ring (14A) from bull gear end of axle shaft. Bump axle shaft out of bull gear and lift gear out of case. Axle shaft inner bearing cup will remain in axle carrier, both bearing cones will remain on shaft.

Shaft and/or bearings and seals can be renewed at this time. Inner bearing cup can be removed from carrier after removing the oil cup (18A) and snap ring (13A).

137A. Before reassembling shaft and gear to carrier remove any burrs on shaft keyway which might damage the "O" ring (9A) when sleeve and ring are bumped into place. Coat "O" ring and inner and outer surface of sleeve with oil, before installing.

Adjust shaft bearings by means of shims (11A) to 0.000-0.002 end play or, if gear is de-meshed from pinion adjust to 8-11 pounds pull on spring scale hooked into one of the the teeth of bull gear. Normal shim pack is 0.039 thick and shims are available in thicknesses of 0.005, 0.007 and 0.020. Coat shims with castor oil to obtain sealing.

138. **WHEEL AXLE SHAFT & BEARINGS.** To adjust wheel axle shaft bearings remove or add shims (11A—Fig. 0606) located under wheel axle carrier cover (6A). Shims are available in thicknesses of 0.005, 0.007

Fig. 0607—Using a pry bar to remove bull pinion bearing cages from 4 speed transmission.

Fig. 0609—To remove bull pinion or bevel ring gear on 4 speed transmissions without removing the bull gear, the components must be positioned as shown.

Fig. 0608—Correct arrangement of components of bevel pinion shaft for 6 speed transmission.

and 0.020. Coat shims with castor oil to obtain sealing effect.

138A. To remove or renew wheel axle shaft or bearings follow the procedure for renewal of bull gear as per paragraph 137.

139. **AXLE CARRIER (HOUSING).** To remove or renew the axle carrier (3A—Fig. 0606) proceed as for removing bull gear as per paragraph 137 and R & R carrier from transmission (rear frame) by R & R of cap screws (4A) retaining carrier to rear frame.

BULL GEARS, PINIONS AND WHEEL AXLES (4 Speed)

(Six speed type begins with paragraph 136.)

These tractors are equipped with only one bull pinion and one bull gear. The bull pinion is bolted to the bevel ring gear. The diameter of the bolt flange on the bull pinion is such that the pinion cannot be removed unless either the bull gear or the bevel pinion shaft of the transmission is first removed.

140. **BULL PINION AND BEARINGS.** The bearings for the single bull pinion can be removed or renewed without removing the bull gear by removing the top rear cover from the transmission and the pinion bearing carriers from the transmission. A suitable puller can then be used to extract the bearing cones from the pinion shaft.

To remove or renew the bull pinion first remove the single bull gear as outlined in paragraph 141, then proceed as follows: Remove cap screws from the bull pinion bearing carriers and pry same from the transmission. Bump the bevel ring gear and bull pinion to the left as far as possible then hoist the unit from the transmission housing. Bearings can be easily renewed at this time.

141. **BULL GEAR.** Refer to Fig. 0605. To remove or renew the single bull gear which is integral with the differential, proceed as follows: Block up the tractor and remove rear wheels. Remove wheel guards and platform assembled. Remove pto if so equipped and bull gear cover from top of transmission. Disconnect brake linkage from brakes if tractor is equipped with rear wheel brakes. Support bull gear in hoist while removing both rear wheel carrier and axle shaft assemblies from transmission housing. Hoist the bull gear and differential assembly from the tractor. The differential carrier bearings complete with their outer races will remain on the differential case portion of the bull gear and can be renewed after removing the snap rings (11—Fig. 0604).

142. **WHEEL AXLE SHAFT & BEARINGS.** Rear wheel must be off tractor when adjusting the bearings for one wheel axle shaft as follows: Vary the shims under the bearing cap (36—Fig. 0611) until axle rotates freely but without end play. To do an accurate job of checking the adjustment, the oil seal should be removed from the cap. Apply a light coat of sealing compound or castor oil to shims to prevent oil leakage.

142A. To remove or renew one wheel axle shaft or shaft bearings, refer to Fig. 0611 and remove rear wheel. Apply brakes and hold in applied position to keep brake splines in alignment. Remove bearing cap from outer end of wheel carrier housing and using a suitable puller extract the shaft, outer bearing cup and both bearing cones as a single unit. The inner cup remaining in the carrier can be extracted with a suitable puller entered from outer end.

Lip of the spring loaded oil seal should face inward. It should be noted from the illustration that a spacer (43) is interposed between the axle shaft and the seals. Be careful when reinstalling shaft to avoid damaging inner oil seal (IS). Renew rubber seal ring interposed between the inner face of the wheel hub and the outer face of the spacer on some tractors. Adjust the wheel bearings as outlined in paragraph 142.

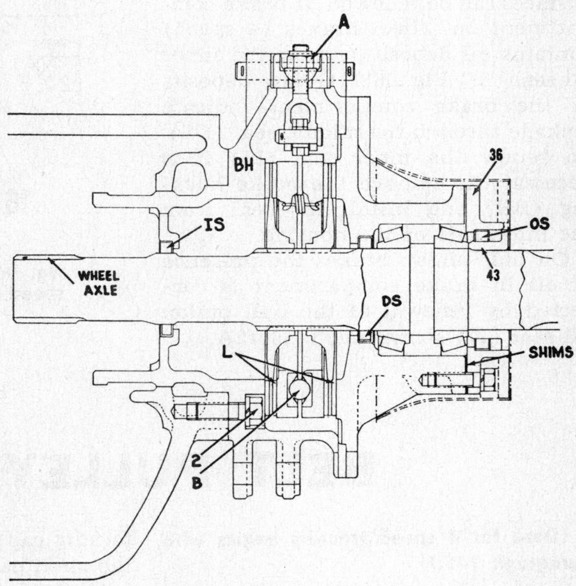

Fig. 0611—Sectional view of wheel axle shaft and disc brakes used on 4 speed Super 99 transmission. Brakes are adjusted by rotating the nut (A).

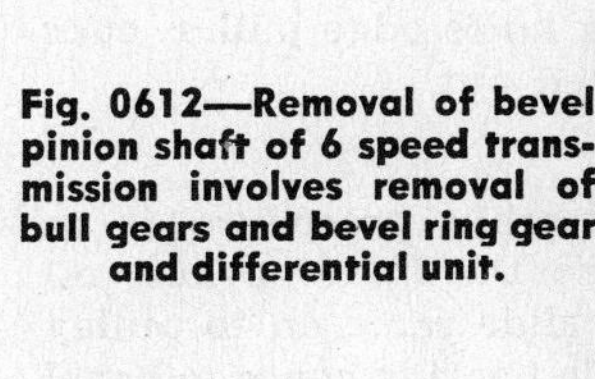

Fig. 0612—Removal of bevel pinion shaft of 6 speed transmission involves removal of bull gears and bevel ring gear and differential unit.

BRAKES

Double disc type brakes located on the rear wheel axle shafts are used on tractors equipped with 4 speed transmissions; similar brakes mounted on the bull pinion shafts are used on 6 speed transmissions.

143. **ADJUSTMENT.** To "take up" on the brakes, rotate nut (A) shown Figs. 0612A and 0606. All of the drum the same to the other brake. Equalize the brakes by backing off on the tight brake. Synchronize the pedals by varying the length of the pedal rods.

143A. **DISCS, DRUMS AND SEALS.** The procedure for removing the lined discs is self-evident after referring to Figs. 0612A and 0606. All of the drum surfaces can be renewed. If brake compartment on wheel brakes (4 speed) contains oil deposits, renew the inner oil seal (IS) Fig. 0611. Grease deposits in the brake compartment indicate leakage through the middle seal (DS). To renew the inner seal (IS) it is necessary to remove the brake housing (BH) and install the seal from the inner side of the housing.

On bull pinion brakes the presence of oil in brake compartment is corrected by renewal of the bull pinion oil seal (17) shown in Fig. 0612A.

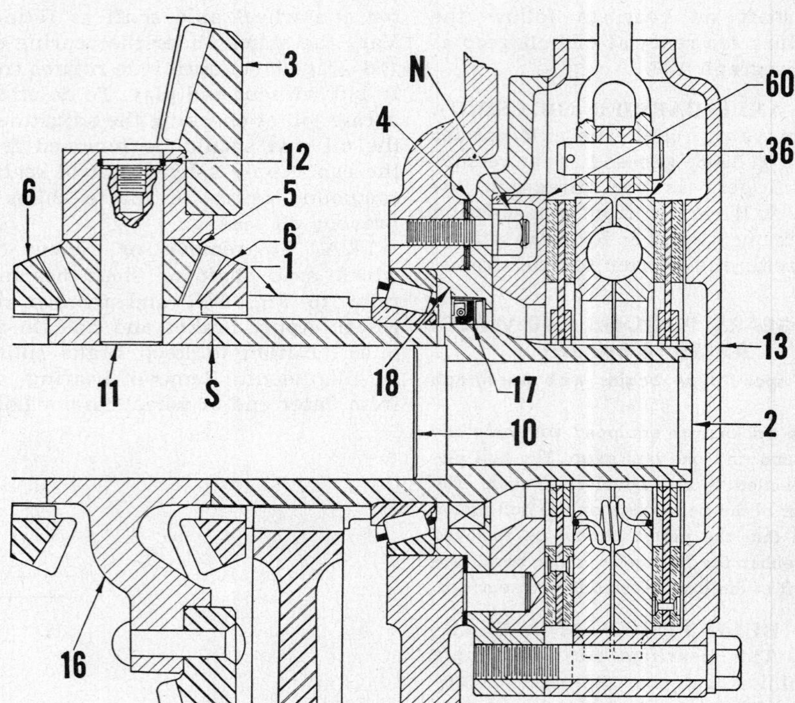

Fig. 0612A—Section through bull pinion and disk brakes on one side of 6 speed tractor. Brakes are adjusted by turning nut (A) shown in Figs. 0603A, 0611.

BELT PULLEY AND PTO (6 Speed)

(Data for 4 speed tractors begins with paragraph 146.)

145. The continuous type pto unit mounted on the rear face of the main rear frame (transmission) as shown in Fig. 0613 is driven from the engine flywheel by a long drive shaft which extends through the hollow clutch shaft and hollow transmission input shaft. As will be seen in Fig. 0614 the pto housing contains the long drive shaft (43) also an over-center clutch, reduction gears and external output pto shaft (39). The belt pulley element (18) when used, is mounted in place of the cover (16) shown in Fig. 0613.

Paragraphs 145A through 145K cover the servicing of the complete unit and are arranged in an order of coverage which begins with the belt pulley portion of the system, proceeds to the clutch and long drive shaft and ends with the external pto shaft.

BELT PULLEY CARRIER

145A. **R & R AND OVERHAUL.** Procedure for removal of the belt pulley is conventional. Pulley shaft and bearing carrier are removed as an assembly after removing the 4 carrier retaining cap screws. Be careful to wire the shim pack together as these shims control the mesh position of the pulley shaft bevel pinion.

145B. Refer to Figs. 0614 and 0615. To disassemble the already removed carrier, mount shaft flange in vise and remove nut (33). Press or pull bevel gear (8) off of shaft. Pull shaft out of carrier (18). Bearing cups will remain in carrier, cones remain on gear and shaft. Oil seal (21) will be damaged in removal. Cones may be renewed using a knife edge puller, cups may be bumped out.

145C. Reassembly procedure is as follows: Grease the lip of a new oil seal (21) and slide same on to pulley shaft. Assemble bearing cones to bevel gear (8) and to pulley shaft. Insert Woodruff key into pulley shaft and assemble both bearing cups to the carrier (18). Place carrier on press bed as shown in Fig. 0615. Insert a suitably sized split type collar "K" as

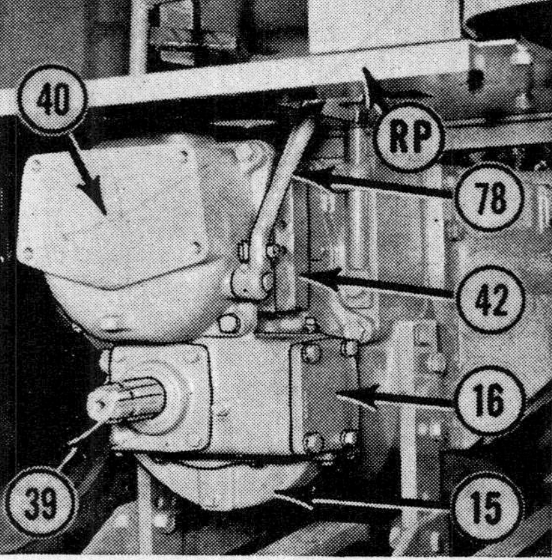

Fig. 0613—Continuous type pto is mounted on rear face of rear main frame as shown.

RP. Rear platform
15. Pan
16. Cover for pulley mounting
39. PTO shaft
40. Clutch housing cover
42. Hand hole cover
78. Operating lever

shown, or two pieces of ¼ inch key stock between inner face of shaft flange and outer face of seal as shown. Now by pressing downward on the pulley end of shaft, force the oil seal (21) squarely into place flush with end of carrier.

Turn the assembly upside down and press the bevel gear on to shaft until shaft has only about 0.010 end play in bearings. Now mount pulley end of shaft vertically in a vise and with a spring scale measure amount of pull (torque) required to rotate carrier around shaft. This amount of pull is the drag of the oil seal and the value should be noted. Install and tighten the gear nut (33) until the spring scale pull required to rotate carrier is 7-10 pounds **higher** than the seal drag reading. Lock the nut by staking.

If shims have become lost or if a new bevel pinion has been installed it will be advisable to check the mesh position of the pinion and to reset if necessary, as outlined in paragraph 145D.

145D. MESH POSITION OF PINION. To check mesh position without removing pto housing from tractor proceed as follows: Remove the bearing cover (28—Fig. 0614) and with the pulley carrier (18) and gasket shims (22) removed, measure the distance (with a depth micrometer) "B" from outer race of bearing (98) to end face of pto housing as shown in Figs. 0614 and 0617. With pinion assembled to carrier (lay or clamp straightedge to end face of bevel pinion) measure the distance "A" from end face of pinion to gasket contacting shoulder surface of the carrier (18) as shown. Record these two measurements.

Note the plus or minus stamped on the face of the pinion and record it. Now add .010 to dimension "B" and subtract the total from dimension "A". To the remainder after the subtraction, add the plus value stamped on the pinion, or if the stamped value has a minus sign, subtract it. This total represents the thickness of the shim pack (22—Fig. 0614) which when installed will automatically locate the proper mesh position of bevel pinion.

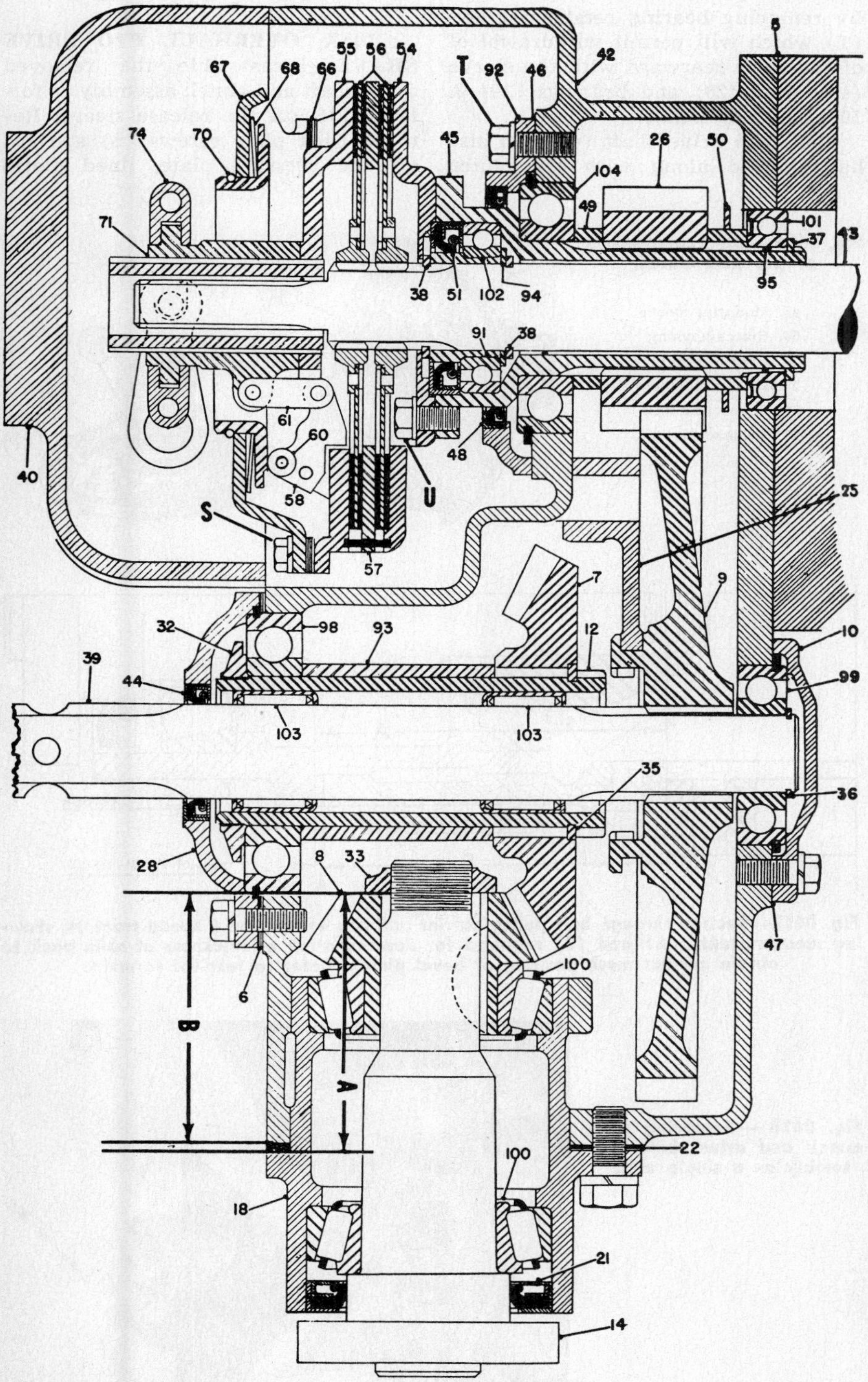

Fig. 0614—Sectional view of combined continuous type pto and belt pulley drive. Shaft (43) passes through hollow transmission and clutch shafts to engage the engine flywheel. Refer to Fig. 0619 for parts legends.

Fig. 0615—Oil seal (21) assembled to shaft (14) is installed into (18) as a single unit using split collar (K) or keystock as shown.

EXAMPLE: Dimension "A" is 4.018
Dimension "B" is 3.926
"B" plus .010 is 3.936

Pinion face marked +0.008

So: 4.018
3.936
———
0.082 plus 0.008 is 0.090 which is shim pack thickness

(if stamped value on pinion was −0.008 the shim pack would be 0.082 minus 0.008 or 0.074)

Therefore under the above stated conditions with a pinion stamped "+0.008" we would install a shim pack 0.090 thick to set the pinion at the correct mesh position.

The same results can be obtained without calculation if the checking is done when the pto shaft (39) is out of the housing. By this method the pto shaft rear bearing (98) is installed to pto housing and bumped inward (snap ring removed) until only about ⅛ inch of it remains in the pto housing. Now push the pinion and carrier assembly (minus the shims) into the pto housing until the end face of pinion contacts the bearing. Using feeler blades measure the gap at shim pack location (22) and install a shim pack of the same thickness. This will locate the pinion correctly.

After pinion mesh position is established adjust backlash of bevel gears (7 & 8) to 0.005-0.010 by varying the shims (6) located under bearing cap (28).

PTO MECHANISM

145E. **ADJUST PTO CLUTCH.** If clutch slips under load it should be "tightened" as follows: Remove shift lever (78—Fig. 0613) and clutch cover (40). Remove one pair of clutch back plate screws (S—Fig. 0614) and one shim pack (66) therefrom. Remove 2 or 3 shims from pack and reinstall shim pack and screws. Do the same to the remaining two pairs of screws. Reinstall the removed parts and test. If clutch is too tight as manifested by difficulty in bringing it into engagement again remove the stated parts and install a shim to each of the 3 packs.

145F. **CLUTCH OVERHAUL.** The pto clutch can be disassembled and overhauled without removing the drive shaft (43—Fig. 0614) or the pto housing from the rear frame. Procedure is as follows: Remove housing cover (40) and release lever. Remove hand hole covers from each side of housing.

145G. Refer to Figs. 0614 and 0616. Remove clutch back plate screws (S) and lift back plate and release linkage off the shaft. The lined plates (driven members 56) can now be withdrawn. To remove clutch cover (54) remove screws (U). Oil seal (51) can now be renewed. Any further work in the clutch compartment will require removal of the pto drive shaft (43).

145H. **R & R PTO DRIVE SHAFT.** If the pto clutch has been previously removed as in paragraphs 145F and 145G the drive shaft can be withdrawn by removing bearing retainer screws (T) which will permit withdrawal of of the shaft rearward with the sleeve (45), gear (26) and bearings (101 & 104) as a single unit.

145J. The drive shaft (43) can also be removed along with the clutch assembly as a single unit. This procedure which would be followed when it is known that the clutch requires no attention is as follows: Remove clutch housing cover (40) and release lever. Remove hand hole cover from each side of housing. Working through hand hole cover openings remove the 4 bearing retainer cap screws (T—Fig. 0614) and pull shaft and clutch assembly rearward as shown in Fig. 0618.

145K. **OVERHAUL PTO DRIVE SHAFT.** Disassemble the removed drive shaft and clutch assembly as follows: Lift off the release sleeve. Remove back plate screws (S) and lift off the pressure plate, lined plates

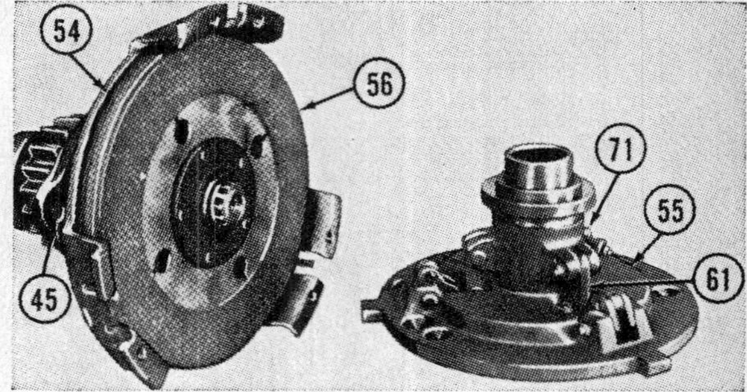

Fig. 0616—Main components of pto disk clutch.

45. Mounting sleeve
54. Clutch housing
55. Pressure plate
56. Driven members
61. Connecting link
71. Release sleeve

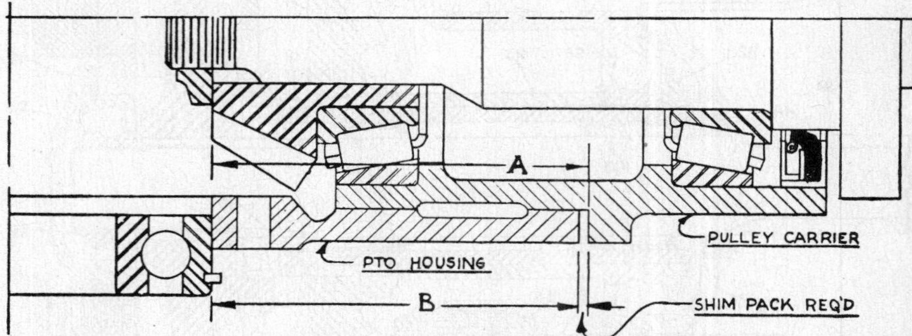

Fig. 0617—Section through belt pulley carrier and pto housing on 6 speed tractors, showing measurements (A) and (B) required for determination of thickness of shim pack to obtain correct mesh position of bevel pinion. Refer to text for formula.

Fig. 0618 — Removing pto clutch and drive shaft assembly as a single unit.

and clutch housing (54). Bump front end of drive shaft (43—Fig. 0614) which will dislodge bearing (102) and seal (51) from the seal mounting sleeve (45). Remove both snap rings (38) and washer (94) then bump or press bearing (102) and seal journal sleeve (91) rearward from the drive shaft. Bearings (101 & 104) and the spur gear (26) can be pressed off mounting sleeve (45) after removing snap ring (37). See Figs. 0621 & 0624.

145L. **R&R COMPLETE PTO UNIT.** To remove complete pto unit housing, drain lubricant and remove rear platform shown at (RP) in Fig. 0613. Remove clutch release lever (78) and the safety shield. Remove 6 cap screws attaching unit to rear frame. Support unit with hoist then withdraw from tractor as shown in Fig. 0620. Removal will be facilitated by using two guiding studs screwed into cap screw holes in rear frame.

145M. **R&R AND OVERHAUL PTO (EXTERNAL OUTPUT) SHAFT.** To remove the pto shaft (39—Fig. 0619) first remove the complete unit housing as described in paragraph 145L. Procedure for overhauling the shaft when pto unit is off the tractor is as follows: Remove oil pan from bottom of pto housing and bearing covers (10 & 28). Remove snap ring (36) from front end of shaft and the shifter poppet (detent) screw, spring and ball from the housing. Bump pto shaft (39) rearward out of the front bear-

Fig. 0620—Removing pto clutch and belt pulley housing and drive shaft assembly from tractor.

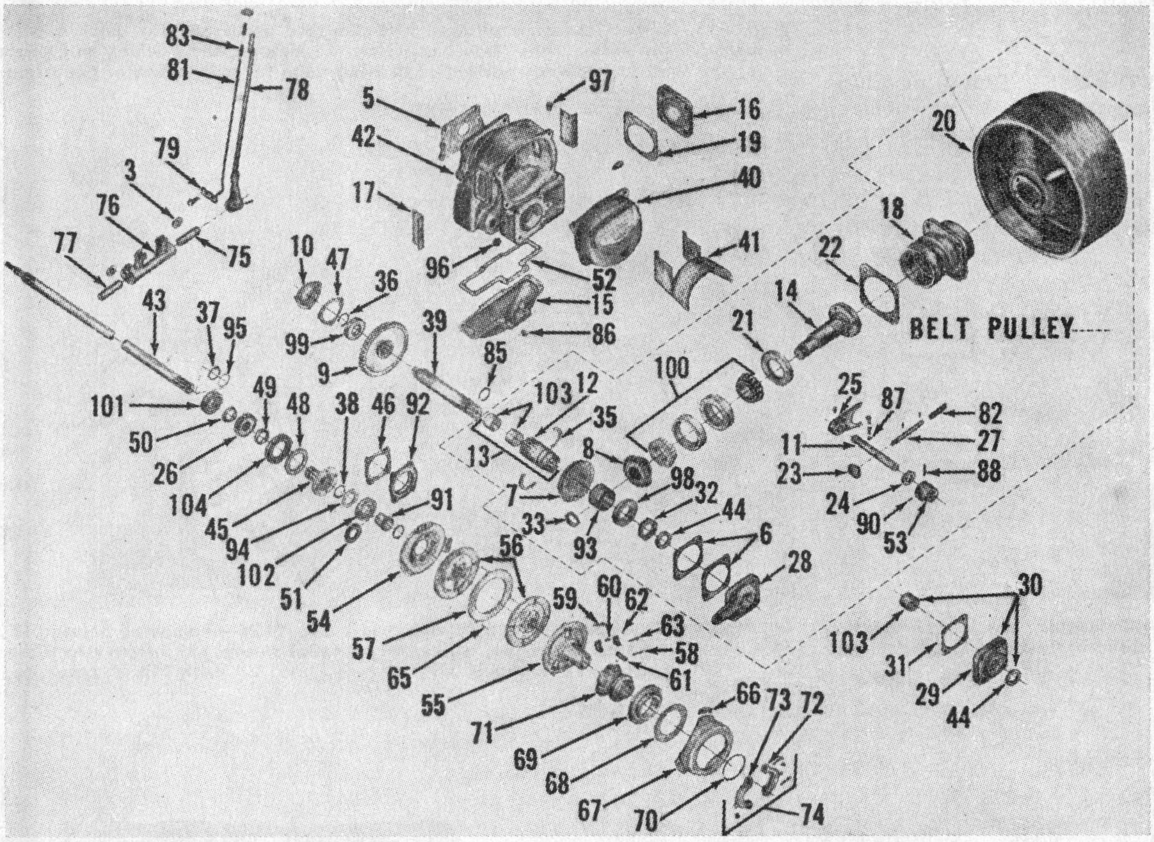

Fig. 0619—Components of the pto and combined belt pulley unit shown in Fig. 0614.

4. Poppet spring
5. PTO housing gasket
6. Shim
7. Pulley drive gear
8. Pulley pinion
9. PTO driven gear
10. Bearing cover
13. Pulley drive shaft
14. Pulley shaft
17. Hand hole cover
18. BP bearing carrier
21. Oil seal
22. Shim
24. Oil seal
25. Shift fork
28. Bearing cover
35. Split ring
36. Snap ring
37. Snap ring
38. Snap ring
39. PTO shaft
40. Clutch housing cover
41. Shield
42. PTO housing
43. PTO drive shaft
44. Oil seal
45. Mounting sleeve
48. Oil seal
49. Spacer
50. Spacer
53. Shift rod collar
54. Clutch housing
55. Pressure plate
56. Driven member
57. Clutch center plate
60. Clutch lever roller
61. Connecting link
66. Shim pack
67. Back plate
68. Pressure washer
69. Washer plate
74. Release bearing
75. Clutch release shaft
76. Clutch release fork
77. Release pivot
78. Release lever
91. Oil seal sleeve
92. Bearing retainer
93. Spacer
95. "O" ring
97. Breather
98. Bearing
100. Bearing
101. Bearing
102. Bearing
103. Needle bearing

ing (99), large spur gear (9) and needle bearings (103). Lift spur gear and shaft from the housing.

Remove lock screw from shifter fork (25) and withdraw shifter fork and rail, and front bearing (99) from pto housing. Remove nut (32) from rear end of pulley drive shaft (12), bump rear end of shaft forward and extract the split rings (35). Remove bevel gear (7) and spacer (93) from shaft and pulley drive shaft and rear bearing (98) from the pto housing. If the needle bearings (103) are to be renewed use a driver similar to the one shown in Fig. 0623. Outer end of needle bearing at threaded end of shaft should be located ¼ inch in from end; bearing at opposite (front) end should be located ⅝ inch in from front end of shaft. Shaft nut (32) should be locked by staking using a pin punch.

145N. **BACKLASH.** Before installing the various covers to the pto housing make sure that the bevel gear (7) on pulley drive shaft has .005-.010 backlash as shown in Fig. 0622. Obtain specified lash by varying the shims (6) shown in Fig. 0614. Mesh position of bevel pinion (8) is controlled by shims (22).

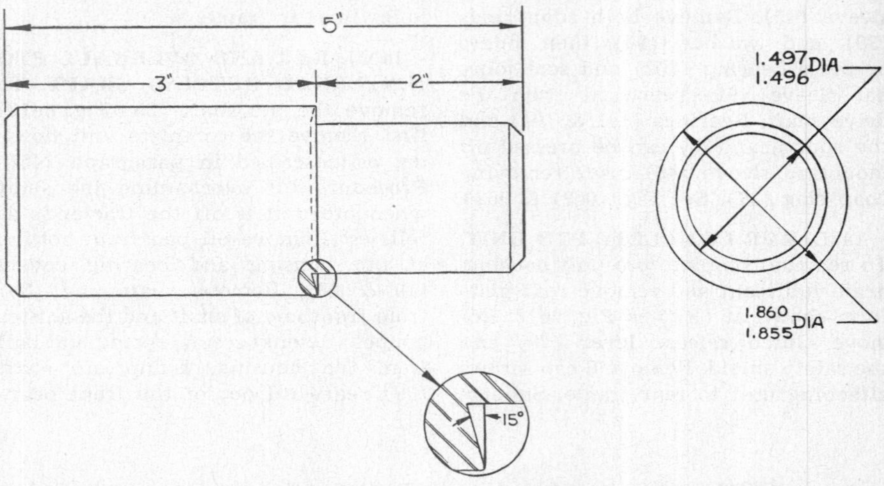

Fig. 0623—Dimensions of a piloted drift designed especially for quick and safe renewal of caged type needle roller bearings. Note 15 degree angle which will concentrate the load on shoulder portion of bearing cage for prevention of distortion.

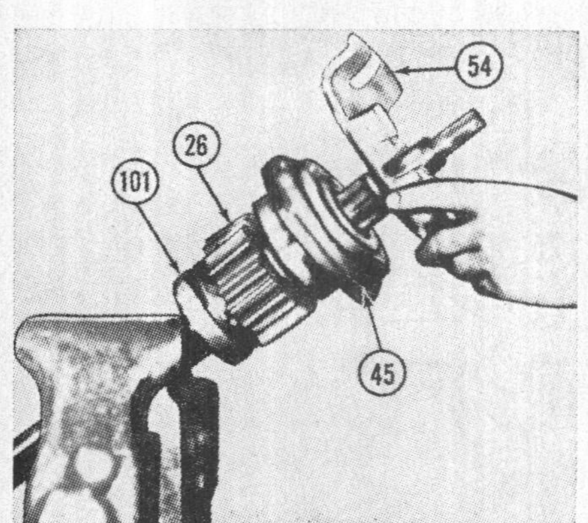

Fig. 0621—Removing pto clutch housing (54) from mounting sleeve (45). Drive gear is (26).

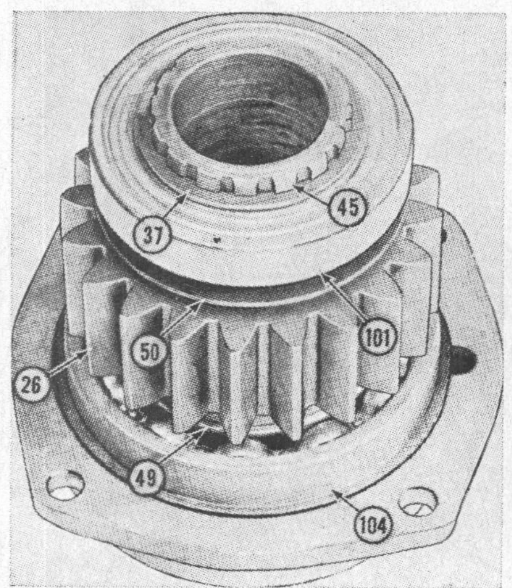

Fig. 0624 — Subassembly containing pto clutch mounting sleeve (45), drive gear (26) spacers (49) and (50).

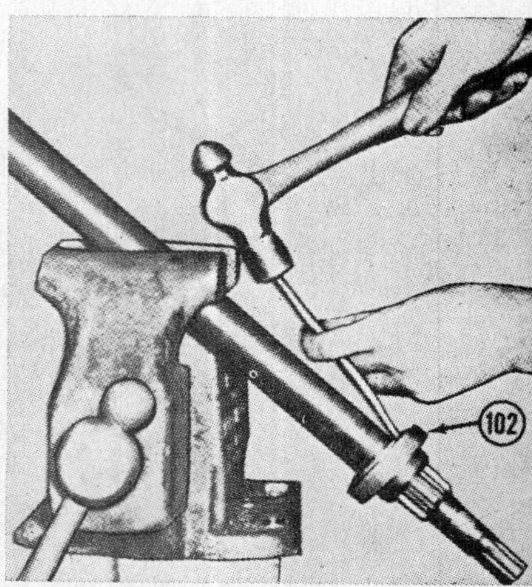

Fig. 0626—Removing bearing (102) and oil seal sleeve (91) from pto drive shaft used with "live" type pto.

Fig. 0622—Checking backlash of pto belt pulley bevel gears (6 speed tractors) which is controlled by shims.

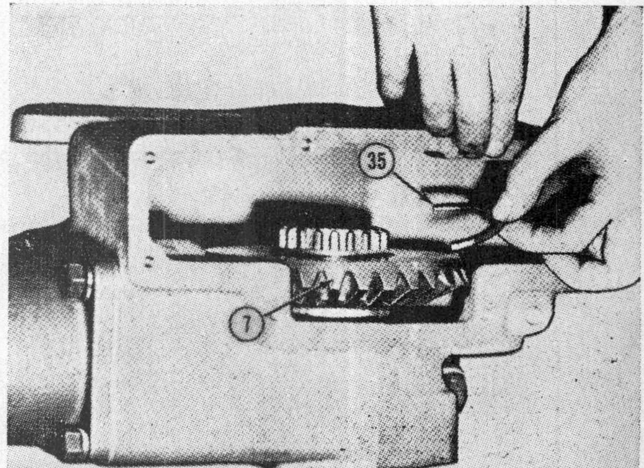

Fig. 0625—Inserting split lock rings (35) used (6 speed tractors) to locate the bevel gear (7) on hollow pulley shaft shown at (12) in Fig. 0614.

BELT PULLEY AND PTO (4 Speed)

(Data for 6 speed tractors begins with paragraph 145.)

BELT PULLEY

146. **PULLEY AND SHAFT.** Refer to Fig. 0627. To remove only the belt pulley and shaft, proceed as follows: Remove nut (10) from end of pulley shaft and pull pulley off the shaft. Remove oil seal retainer from same end of shaft and bearing cage (1) from opposite end. Tie the shims (34) together. Bump the pulley shaft to the left out of the carrier. Procedure for disassembly and overhaul are self-evident. The straight bevel gear (5) can be purchased separately. Adjust backlash of bevel gear teeth by varying the shims (34) under the bearing cage as outlined in paragraph 146C.

146A. **R&R PULLEY CARRIER.** To remove the complete pulley carrier assembly from the tractor proceed as follows: Disconnect battery or batteries from wiring. Remove fuel supply tank and tank support and lay instrument panel on platform or other resting place. Remove clutch front cover (31—Fig. 0457A) and pto drive shaft if so equipped. Remove pulley and steering gear unit from belt pulley carrier housing. Remove locating dowels from front frame also the retaining cap screws and hoist carrier from tractor as shown in Fig. 0594.

146B. **OVERHAUL.** Procedure for overhauling the pulley carrier after it has been removed from the tractor is self-evident after referring to Figs. 0627 and 0628. The straight bevel gears can be purchased separately. After unit is assembled but before installing it to the tractor, check and adjust the mesh position and backlash of the bevel gears as outlined in paragraph 146C.

146C. **MESH AND BACKLASH.** Correct mesh position is with heel (large) end of drive bevel gear flush with toe end of other gear as shown in Fig. 0627. Correct backlash is 0.005-0.010. Obtain flush mesh position by varying the shims (20) and obtain the desired 0.005-0.010 backlash by varying the shims (34) located under the pulley shaft bearing cage. Bearings are non-adjustable ball type.

PTO UNIT

146D. The power take-off is of the same general design as used on the series 70 tractors except that the pto external shaft is spline coupled to the rear end of the pulley drive shaft (DS) shown in Fig. 0627. The service and overhaul procedures are self-evident.

1. Bearing cage
2. Retainer
4. Spacer
5. Pulley drive pinion
6. Pulley shaft
9. Oil seal
13. Bevel gear and shaft
17. Pulley drive gear
18. Thrust washer
20. **Shims**
21. Bearing cage
25. Retainer
26. Shifter fork
27. Shifter arm
28. Shifter shaft
29. Shifter stop
31. Shifter lever
32. Pulley carrier
33. Carrier cover
34. **Shims**

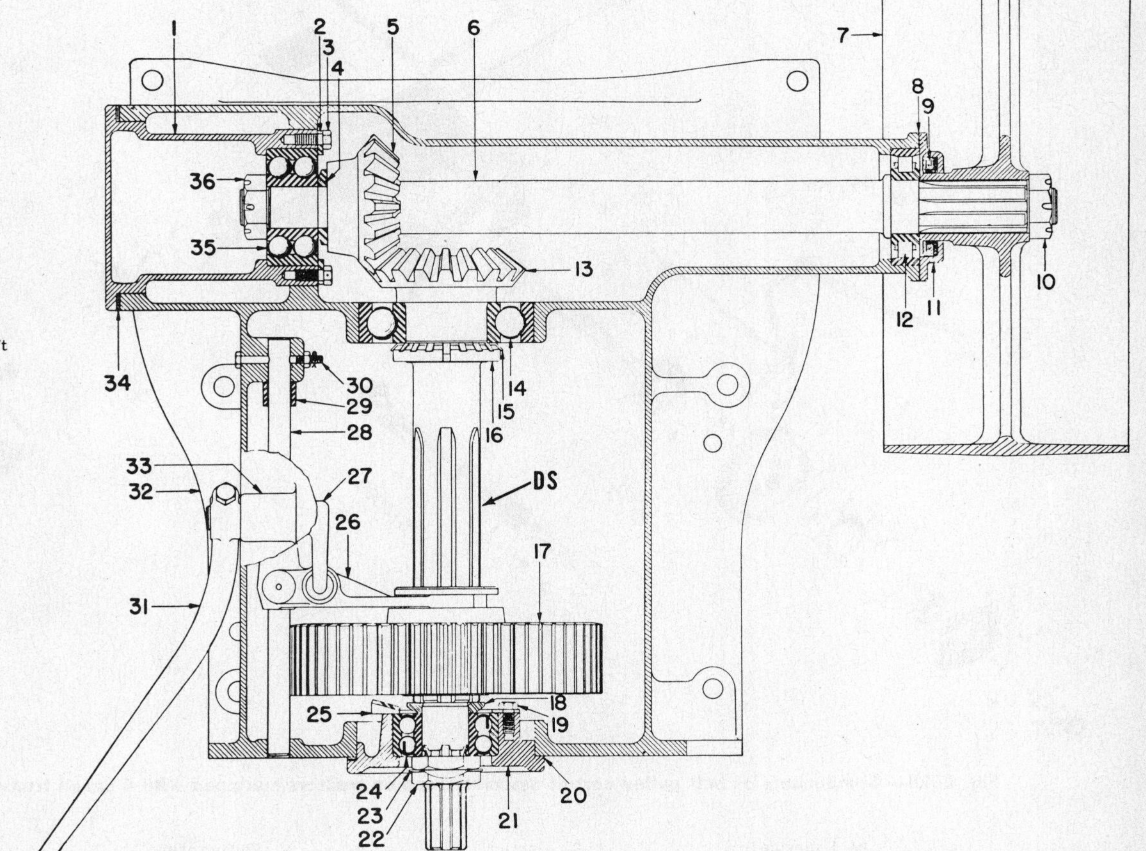

Fig. 0627—Sectional top view of belt pulley carrier and pto drive unit used on models having 4 speed transmission. This unit forms the front top cover of the transmission.

HYDRAULIC SYSTEMS

BRIEF DESCRIPTION

147. Two types of hydraulic systems are available for the Super 99 tractors. The non-depth control system comprises an engine mounted, engine driven, Vickers pump, a fluid reservoir mounted on the side of the chassis, 4 way control valve mounted on the reservoir and a remote double acting two hose work cylinder conforming to ASAE standards.

In the depth control system, one assembly comprising the Vickers pump, control valve and reservoir is mounted on the rear face of the transmission or rear face of continuous pto clutch housing. On tractors without pto, the pump is driven by a long drive shaft splined into the engine flywheel. On pto equipped tractors the pump is joined to the pto clutch shaft by a conventional drive coupling. The work cylinder used in depth control systems is called the "Hydro-Stop" type and is fitted with three hoses to provide hydraulic (non-manual) depth control.

Both systems use SAE 10W engine oil as the operating fluid. Reservoir capacity is 1¾ gallons for the depth control system; 4.3 gallons for the non-depth control system which has the engine mounted pump. Operating pressure is 1250 psi maximum for both systems.

148. **TROUBLE SHOOTING.** Internal leakage arising from leaking valves, seals or gaskets is manifested by failure of the unit to control the ground engaging implement. Under such conditions the implement will either settle to the ground from the raised position or the soil penetration depth will slowly increase from a fixed position due to the natural ground suction of the tool.

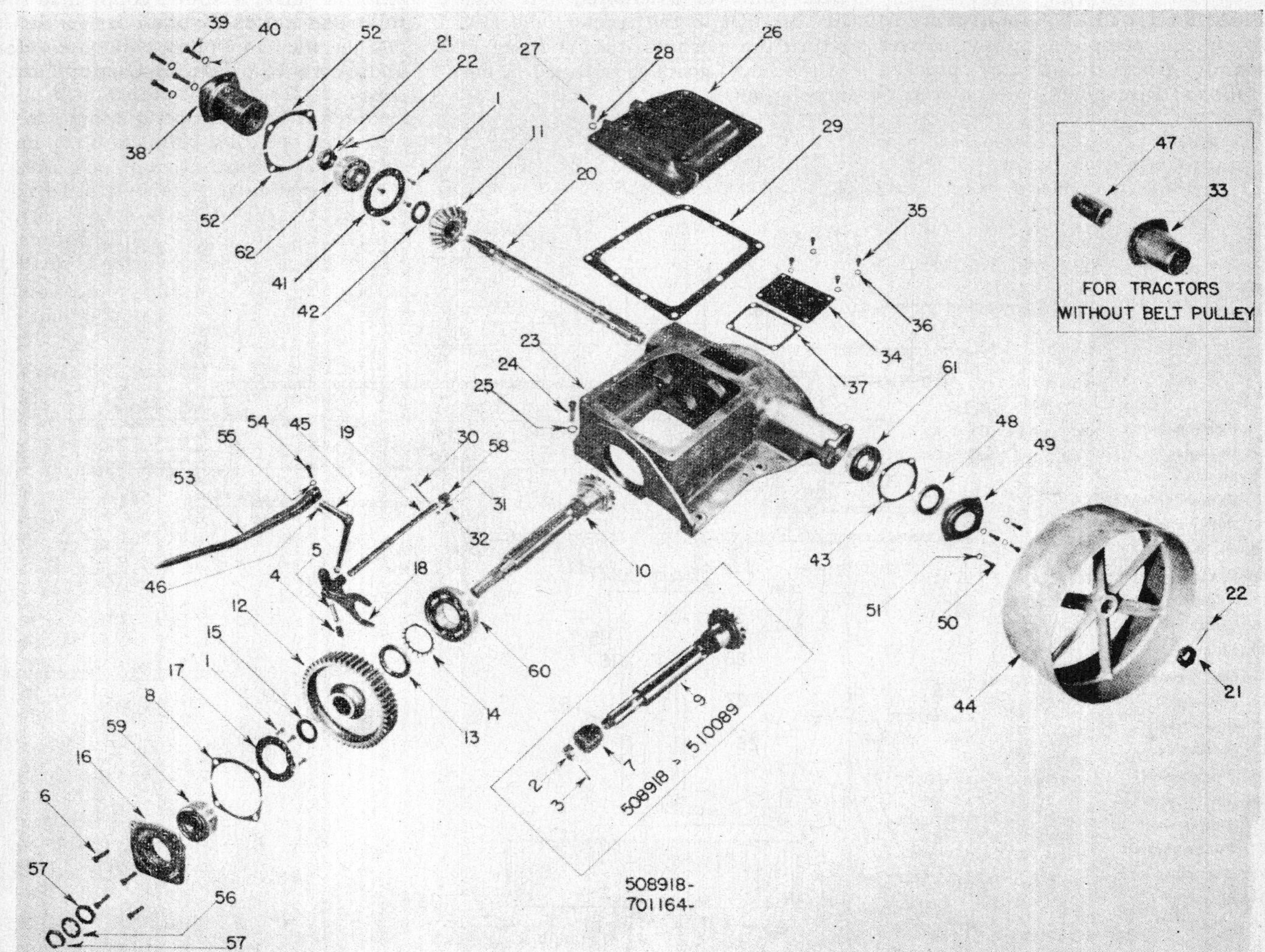

Fig. 0628—Components of belt pulley carrier assembly used on tractors equipped with 4 speed transmission.

4. Poppet
5. Poppet spring
7. Coupling
8. Shim
9. Pulley gear
10. Pulley gear
11. Pinion
15. Thrust washer
12. Drive gear
20. Pulley shaft
23. Carrier
26. Carrier cover
38. Bearing cage
41. Bearing retainer
42. Spacer
45. Shifter arm
47. Spacer
48. Oil seal
49. Bearing cover
52. Shim
58. Shifter stop
59. Ball bearing
60. Ball bearing, front
61. Ball bearing, right
62. Ball bearing, left

A check of the pressure relief valve unseating pressure can be made by connecting a high reading pressure gauge anywhere into the outlet (pressure) side of the system. If valve (33—Fig. 0634 does not unseat in the range of 1235-1270 psi remove it for inspection. Renew any ridged, grooved, worn or otherwise damaged parts. Valve is not adjustable.

DEPTH CONTROL SYSTEM

(Non-depth system begins with paragraph 150.)

149. **HYDRO-STOP WORK CYLINDER.** Two sizes of cylinder are available. Bore diameter for both sizes is 4 inches, stroke is 8 inches or 16 inches.

To overhaul a removed cylinder refer to Fig. 0629 and proceed as follows: Remove rod clevis (4) and gland retainer (12) by unscrewing. Remove Tru-Arc snap ring (6). Gland (2) and piston and rod assembly can now be withdrawn from the cylinder. Access to the piston rod chevron packing (14) is obtained by unscrewing the holder (9). Quill (10) is removed from cylinder base by unscrewing the Allen head screw (18). Install all new "O" rings and other seals.

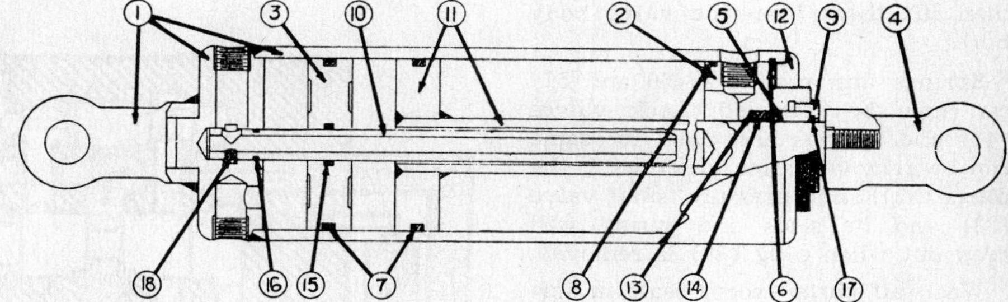

Fig. 0629—Hydro-Stop hydraulic work cylinder used on some Super 99 tractors has 4 inch bore and is available with stroke of 8 inches or 16 inches. Depth stop is hydraulically controlled.

1. Base and clevis
2. Gland
3. Piston stop
4. Clevis rod
5. Bearing
6. Snap ring
7. "O" ring
8. "O" ring
9. Packing holder
10. Quill
11. Piston and rod
12. Gland retainer
13. Backer packing
14. Packing
15. "O" ring
16. "O" ring
17. Wiper
18. Allen screw

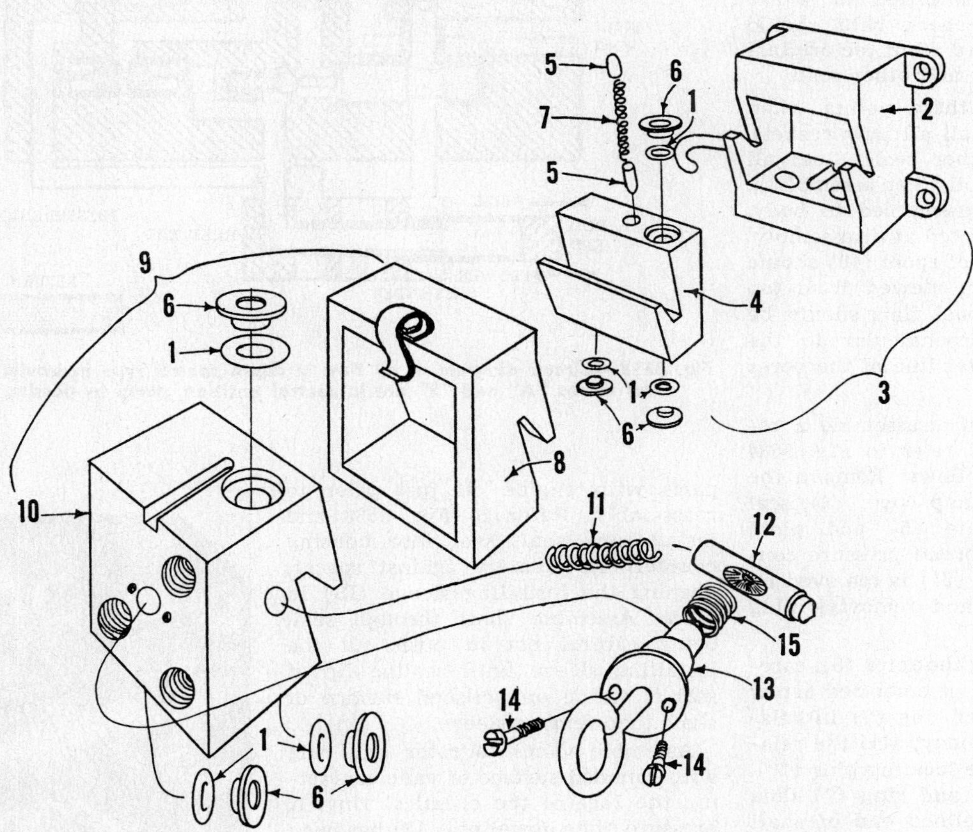

Fig. 0630—Multiple disconnect coupling for hydraulic system hose.

1. "O" ring
2. Housing
3. Coupling half
4. Slide
5. Slide stop pin
6. Seal washer
7. Stop pin spring
8. Housing assembly
9. Half assembly
10. Slide
11. Lock pin spring
12. Lock pin
13. Cam assembly
14. Screw 6-32
15. Cam spring

149A. **CONTROL VALVE.** Valve body assembly (28—Fig. 0634) can be removed from the pump by removing the wire locked attaching screws.

Overhaul a removed valve as follows: Remove the screws which attach the two plates (47) to valve body. Withdraw from bores the control spool caps (45) and stop spool cap (53). Seals and wipers (35, 44, 46 and 52) can be renewed at this time.

Spool valves (43 & 49) are marked with the letter "T" on a flat adjacent to one of the spool collars. Carefully note the installed position of spools then lift them from the valve body bores.

Springs and retainers (50 and 51) can now be removed. Lock valves (41) and springs and balls for same can be removed after unscrewing the plugs (42). High pressure relief valve (33) and its seals and spring will drop out when plug (38) is removed.

Wash all parts except seals in solvent. If body bores show scoring or other damage discard the body. Discard any spool valves or lock valves which are scored, pitted or worn. Pitted or ridged check balls should be renewed. Discard all of the original gaskets, "O" rings and other seals.

149B. Observe these points when reassembling: Install all new gaskets, "O" rings and other seals. Coat all parts with engine oil when assembling. After spools are assembled to body, position them as noted at disassembly. Letter "T" on flat of spool (49) should face upward when viewed from top of valve body. Spool flats should be approximately perpendicular to the perpendicular center line of the bores in body.

149C. **PUMP.** To disassemble a removed pump unit refer to Fig. 0634 and proceed as follows: Remove the screws (2) and pump cover (3), seal (6), pressure plate (5) and plate spring (4). The normal pressure control (relief) valve (21) is removed by pulling pin (20) and removing plug (23).

Before removing the rotor (8), carefully note position of embossed arrow on outer surface of ring (7) in relation to the pump body, also the relative position of the locating pins (10). Remove pins (10) and ring (7) then pull rotor from splined end of shaft (14).

To inspect pump shaft and bearings, remove snap ring (15) and pull shaft (14) and bearing (63) as a unit from pump body. Bearing (62) will remain in body. Wash bearings and check for "lumpiness". To remove inner bearing (62) from body, first extract the seal (13) and spacer (12) then working from rotor end of body, engage outer race of bearing with a drift and bump it out.

After washing all parts inspect rubbing surfaces of same for scoring or pitting. All vanes should slide freely in their rotor slots. Discard all seals. Discard any ridged, scored or pitted rubbing parts.

149D. Observe the following points when reassembling the pump: Coat all parts with engine oil just prior to reassembly. Refer to Fig. 0633 and install new shaft seal into housing counterbore squarely against spacer, keeping the installing sleeve (IS) in place. Assemble shaft through seal, being careful not to push out the installing sleeve until sealing lip of seal is riding on polished surface of shaft then remove sleeve.

Assemble vanes to rotor and ring with rounded surface of vane contacting the race of the elliptical ring. If pressure plate dowel pin (17) has been removed it should be inserted before assembling the valve (21) to pressure plate. Pump cover screws (2) should be torqued to 65-75 foot pounds.

149E. **PUMP TEST SPECIFICATIONS.** Required for a capacity test of the pump are means for maintaining oil temperature at 90—110F, a shut off valve for pressure side of circuit, means for driving the pump and a gauge capable of reading pressures to 1500 psi. With such a set up the pump should deliver 12.5 gallons per minute at 1725 pump rpm when shut off valve is turned to position where 1000 psi pressure is registered. At zero gauge pressure (head) and same rpm the delivery should be at least 16 gallons per minute.

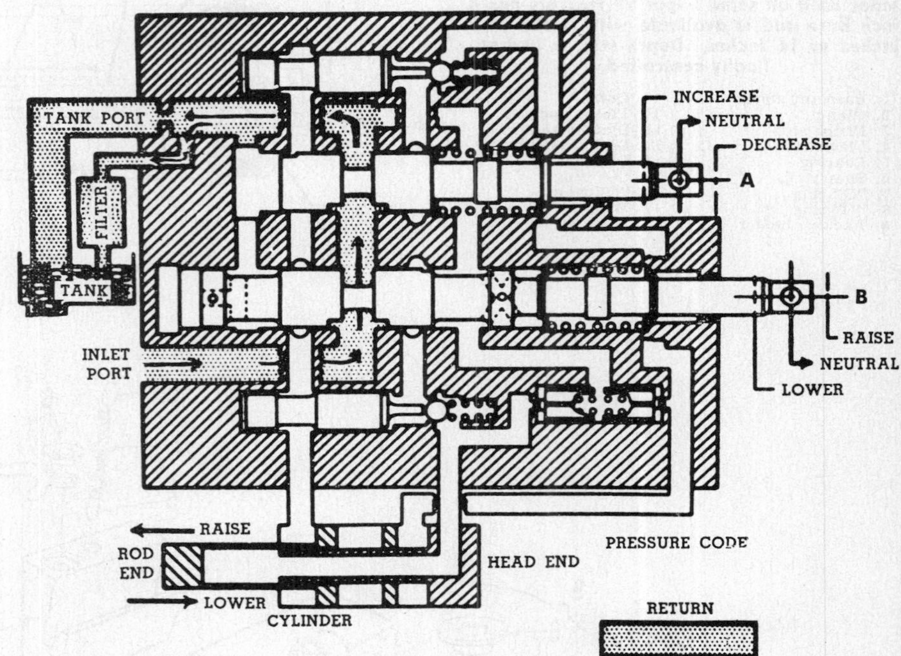

Fig. 0632—Circuit diagram of oil flow in depth control type hydraulic system when control valves "A" and "B" are in neutral position, pump by-passing to reservoir.

Fig. 0633—Details of shaft seal used on Vickers hydraulic pump. Seal installing sleeve (15) is furnished with each new seal and must be removed after seal is installed.

NON-DEPTH CONTROL SYSTEM

(Depth control type begins with paragraph 149.)

150. **WORK CYLINDER.** A double acting cylinder of 4 inch bore, 8 inch stroke is furnished as standard equipment with the system, but a cylinder of similar bore having 16 inch stroke is optional at extra cost. Both cylinders conform to ASAE Standards.

150A. **CONTROL VALVE.** Procedure for removal of this valve which is mounted on top of the fluid reservoir is conventional and self-evident.

To disassemble and overhaul the removed valve refer to Fig. 0635 and proceed as follows: Remove valve bonnet (2) and cap screw (3) washer, center spring (5) and collar (6) from end of valve spool. Remove control handle and valve spool from housing. Be careful when removing "U" cup seals (16) to prevent damaging collar surfaces of spool.

Light pitting and scoring can be removed from housing bore by honing if smooth surface can be restored without enlarging bore diameter more than 0.005. If valve housing requires oversizing beyond 0.005 reject it and install a new one.

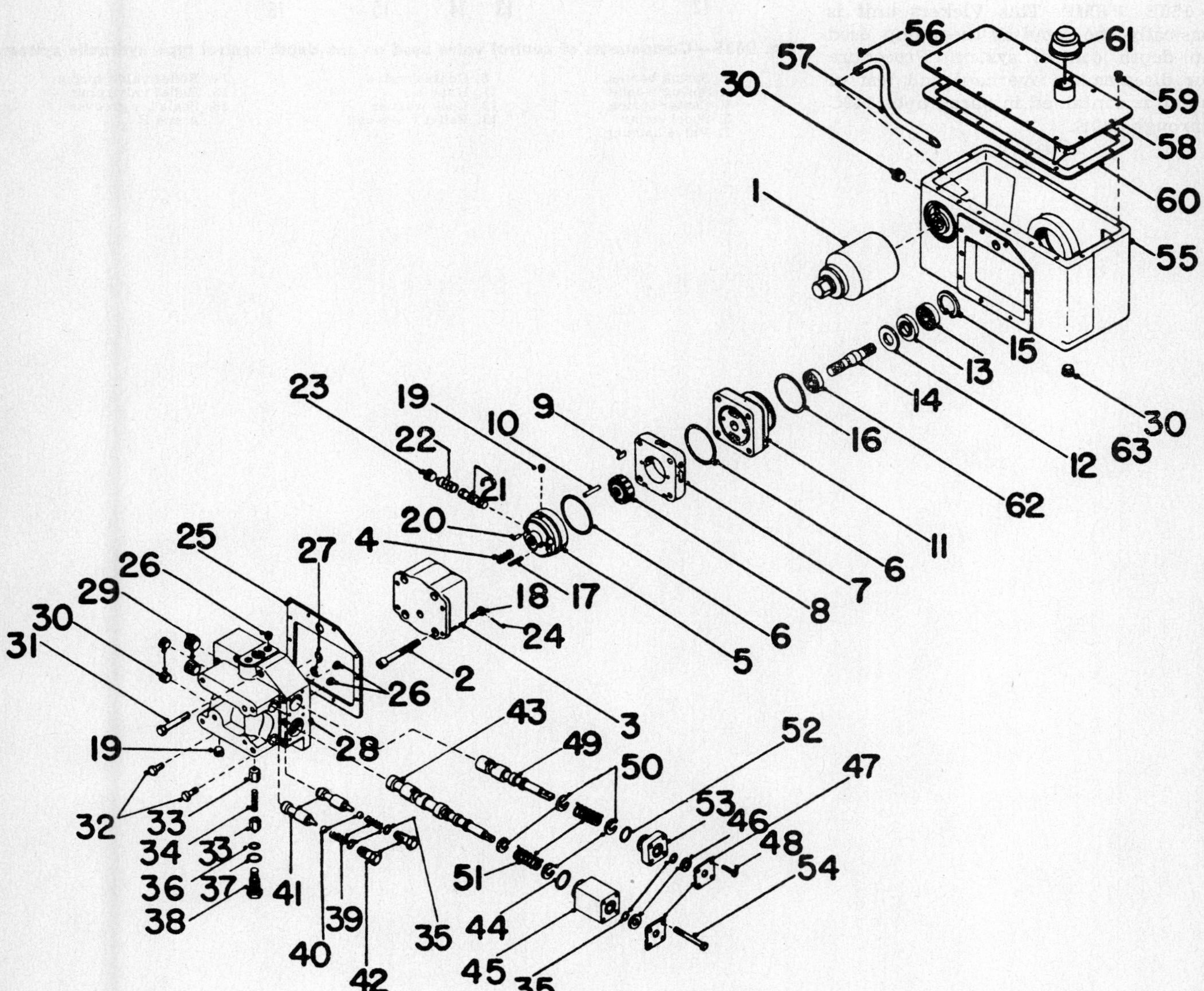

Fig. 0634—Exploded view of the Oliver 7AS-2043-A hydraulic system depth control type. Pump shown here is also used on non-depth control system, the control valve for which is shown in Fig. 0635.

1. Oil filter
2. Pump cover screw
3. Pump cover
4. Spring for plate 5
5. Pressure plate
6. "O" ring
7. Pump ring
8. Pump rotor
9. Vane for rotor 8
10. Dowel pin
11. Pump body
12. Shaft spacer
13. Seal for shaft 14
14. Pump shaft
15. Snap ring
16. "O" ring
17. Plate dowel pin
18. Lock screw
19. Plug
20. Dowel pin
21. Pump control valve
22. Valve spring
25. Gasket
27. "O" ring
28. Valve body
33. Relief valve poppet
34. Relief valve spring
35. "O" ring
36. "O" ring
37. "O" ring
39. Plunger spring
40. Plunger ball
41. Valve plunger
43. Implement control spool
44. "O" ring
45. Control spool cap
46. Spool wiper
47. Spool plate
49. Stop adjustment spool
50. Spring retainer
51. Spool spring
52. "O" ring
53. Stop spool cap
55. Reservoir
56. Groov-pin
57. Drain tube
60. Gasket
62. Pump bearing, inner
63. Pump bearing, outer

Install all new seals. Recommended method for installation of seals (16) is to enter spool until only the groove (A) at bonnet end extends out of body bore. Install seal ring into groove, coat outer surface of seal with engine oil then push spool and seal into bore until groove (B) is exposed at opposite end of valve housing. Install seal ring to groove (B) in a similar manner. Be careful to avoid putting a twist into seal rings when they enter the valve bore.

150B. **PUMP.** This Vickers unit is basically the same as the pump used on depth control system. Procedure for disassembly overhaul and test of pump is contained in paragraphs 149C through 149E.

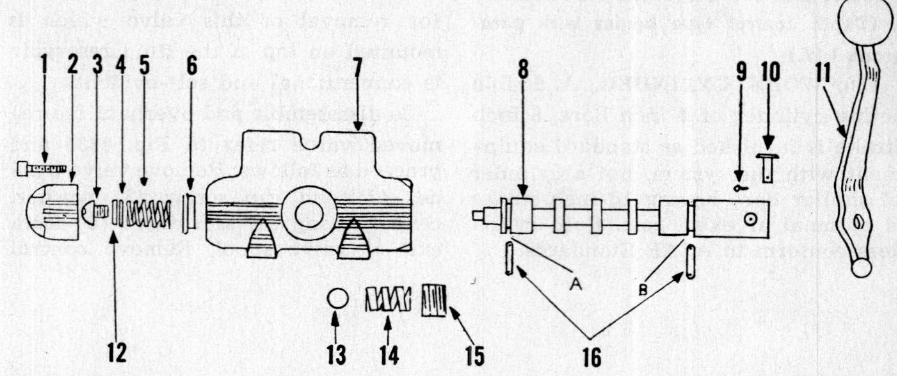

Fig. 0635—Components of control valve used on non-depth control type hydraulic system.

2. Spring bonnet
4. Spring washer
5. Center spring
6. Spool collar
7. Valve housing
8. Control valve
11. Handle
12. Lock washer
13. Relief valve ball
14. Relief valve spring
15. Relief valve plug
16. Seals for grooves A and B

OLIVER

SHOP MANUAL SUPPLEMENT

Series ■ Super 99GMTC ■ 950 ■ 990 ■ 995

This supplement provides methods for servicing the Super 99GMTC, 950, 990 and 995 tractors which, except for the differences covered in this supplement, are serviced in the same manner as the Super 99 and Super 99 GM models covered in the previous section of this book or in I & T SHOP SERVICE MANUAL No. 0-7.

IDENTIFICATION

Model 950 tractors have the tractor serial and specification numbers stamped on a metal plate which is affixed to left side of clutch dust cover. Engine serial number is stamped on right rear mounting flange of engine.

Models Super 99GMTC, 990 and 995 have the tractor serial and specification numbers stamped on a metal plate which is affixed to the left side of the clutch dust cover. Engine serial number appears in two places: On a metal plate affixed to engine valve cover and on left side of cylinder block.

INDEX (By Starting Paragraph)

In the index which follows, the first column lists the tractor components, the 2nd & 4th columns list the beginning paragraph number located either in this supplement or in the previous section covering the Super 99 tractor in which the desired service information will be found. The 3rd & 5th columns list which section of this manual contains the desired paragraph number.

	Model 950 HC & D		Models Super 99GMTC-990-995	
	Paragraph Number	Manual Section	Paragraph Number	Manual Section
BELT PULLEY	145	Super 99	145	Super 99
BRAKES	143	Super 99	143	Super 99
CARBURETOR	29	Super 99	...	...
CLUTCH				
Engine clutch adjustment	125A	Super 99	125A	Super 99
Engine clutch R&R (except 99GMTC & 995)	125C	Super 99	125C	Super 99
Engine clutch R&R (99GMTC & 995)	...	...	28	This Supplement
Engine clutch overhaul	125D	Super 99	125D	Super 99
Power take-off clutch	145E	Super 99	145E	Super 99
COOLING SYSTEM				
Pump	46	Super 99	113	Super 99
Radiator	45	Super 99	110	Super 99
Water nozzles	...	...	61E	Super 99
DIESEL SYSTEM				
Injector rack	...	...	87	Super 99
Injector hole tubes	...	...	61C	Super 99
Fuel manifold	...	...	100	Super 99
Energy cells	41	Super 99	...	...
Fuel supply pump	43	Super 99	99	Super 99
Fuel system bleeding	37	Super 99	...	...
Governor adjust	38	Super 99	103	Super 99
Nozzles testing	40A	Super 99	92	Super 99
Nozzles R&R	40B	Super 99	91	Super 99
Nozzle servicing	40C	Super 99	93	Super 99
Injection pump R&R	36F	Super 99	91	Super 99
Injection pump timing	36	Super 99	90	Super 99
Trouble shooting	32	Super 99	85A	Super 99
Starting aid	42	Super 99	101	Super 99
DIFFERENTIAL				
Overhaul	130B	Super 99	130B	Super 99
Remove & reinstall	130	Super 99	130	Super 99
ELECTRICAL				
Distributor	26	This Supplement	26	This Supplement
Generator & regulator	26	This Supplement	26	This Supplement
Starting motor	26	This Supplement	26	This Supplement
ENGINE				
Assembly R&R	10	Super 99	60	Super 99
Balance Shaft	...	...	69	Super 99
Blower	...	...	76	Super 99
Cam followers	15	Super 99	64	Super 99
Camshaft	19	Super 99	67	Super 99
Connecting rods & brgs.	23	Super 99	70H	Super 99
Crankshaft & bearings	24	Super 99	71	Super 99
Cylinder head	11	Super 99	61	Super 99

Index (Cont.)

CONDENSED SERVICE DATA

	950 HC & D	Super 99GMTC, 990, 995
GENERAL		
Engine Make	Own	GM
Engine Model	950	3-71
Cylinders, Number of	6	3
Cylinder Bore—Inches	4	4¼
Stroke—Inches	4	5
Displacement—Cubic Inches	302	213
Compression Ratio—Non-Diesel	6.2	.....
Compression Ratio—Diesel	16.0	17.0
Pistons Removed From	Above	Above
Main Bearings, Number of	4	4
Cylinder Sleeves	Dry	Dry
Forward Speeds	6	6
TUNE-UP		
Firing Order	1-5-3-6-2-4	1-3-2
Valve Tappet Gap—Inlet	0.009 cold	None
Valve Tappet Gap—Exhaust	0.016 cold	0.009 hot
Valve Face and Seat Angle	45	30
Generator, Distributor and Starter, Make	Delco-Remy	Delco-Remy
Ignition Distributor Model	1112561	.....
Distributor Breaker Gap	0.022	.....
Distributor Timing High Idle	28°	.....
Flywheel Mark Distributor High Idle Timing	IGN	.....
Flywheel Mark Distributor Retard Timing	4½° BTDC	.....
Spark Plug Size	18mm	.....
Injection Pump, Make and Type	Bosch PSB	GM70
Nozzle Opening Pressure	1750	350-850
Plug Electrode Gap	0.025	.....
Carburetor Make	MS	.....
Carburetor Model	TSX581	.....
Engine Low Idle RPM—Battery Ignition	400-500	.....
Engine Low Idle RPM—Diesels	675-725	500
Engine High Idle RPM—Non-Diesels	2000	.....
Engine High Idle RPM—Diesels	2000	(5)
Engine Governed RPM—Loaded	1800	(6)
Belt Pulley RPM @ 1800 Engine RPM	1021	1021
SIZES—CAPACITIES—CLEARANCES (Clearances in Thousandths)		
Crankshaft Journal Diameter	2.625	3.500
Crankpin Diameter	2.625	2.750
Connecting Rod Center to Center Length—Inches	6¾	10⅛
Camshaft Journal Diameter—Nominal	1.750	1.500
Piston Pin Diameter	1.250	1.500
Valve Stem Diameter—Nominal	0.375	0.344
Compression Ring Width	0.125	0.125
Oil Ring Width	0.187	0.187
Main Bearings, Clearance	.5-3	1.5-3
Rod Bearings, Clearance	.5-1.5	1.5-3
Piston Skirt Clearance	(1)	4-9
Camshaft End Play	(2)	4-18
Crankshaft End Play	5-12	4-18
Camshaft Bearings Running Clearance	(3)	(7)
Cooling Systems—Gallons	5½	5½
Crankcase Oil, Refill—Quarts	6	11
Crankcase Oil, if Filter is Changed	7	13
Transmission and Differential—Quarts	32	32
Hydraulic System, Maximum—Quarts	(4)	(4)
Add for Belt Pulley and PTO—Quarts	4	4
Torque Converter	.....	(8)
TIGHTENING TORQUES (In Foot-Pounds)		
Oil Line Cylinder Head Bolts	96-100	.....
Cylinder Head, Non-Diesel—Foot-Pounds	91-100	.....
Cylinder Head, Diesel—Foot-Pounds	112-117	165-175
Main Bearings	108-112	180-190
Connecting Rods	87-92	65-75
Manifold Nuts and Rocker Shaft Brackets	25-27	.....
Flywheel	67-69	150-160

(1) 5-10 lbs. pull on spring scale using ½ x 0.003 feeler with blade positioned 90 degrees from pin. (2) Spring loaded. (3) No. 1 journal 0.0015-0.003; all others 0.0025-0.0045. (4) From 9 to 11¼ quarts, depending on cylinder sizes. (5) Super 99GMTC, 1840; 990, 1920; 995, 2120. (6) Super 99GMTC, 1675; 990, 1800; 995, 2000. (7) Front and rear, 0.0025-0.006; intermediate, 0.0025-0.009. (8) Fill with 6 quarts, start engine and add to bring level to "Full" mark.

POWER STEERING

Models So Equipped

10. Power steering for all models so equipped is a linkage booster type. The three basic components of these systems are the pump, control valve and hydraulic cylinder. On the Super 99GMTC model tractors, the control valve is an integral part of the hydraulic cylinder while on models 950, 990 and 995 the control valve is part of the drag link.

LUBRICATION AND BLEEDING

All Models

11. Fluid capacity for the power steering system is approximately 1½ quarts. It is recommended that Automatic Transmission Fluid, Type A, or SAE10W oil mixed in the ratio of 16 parts oil to one part of Oliver additive 102 082-A be used in the power steering system. To fill and bleed the system, remove reservoir cover, or filler cap, then add fluid to bring fluid level to full mark. Run engine at approximately 1000 rpm and cycle system from one side to the other until all air is freed from system. Refill reservoir.

TROUBLE-SHOOTING

All Models

12. The accompanying table lists troubles which may be encountered in the operation of the power steering system. The procedure for correcting most of the troubles is evident; for those not readily remedied, refer to the appropriate subsequent paragraphs.

SYSTEM OPERATING PRESSURE

13. A pressure test of the hydraulic circuit will disclose whether the pump or some other unit in the system is malfunctioning. To make such a test proceed as follows:

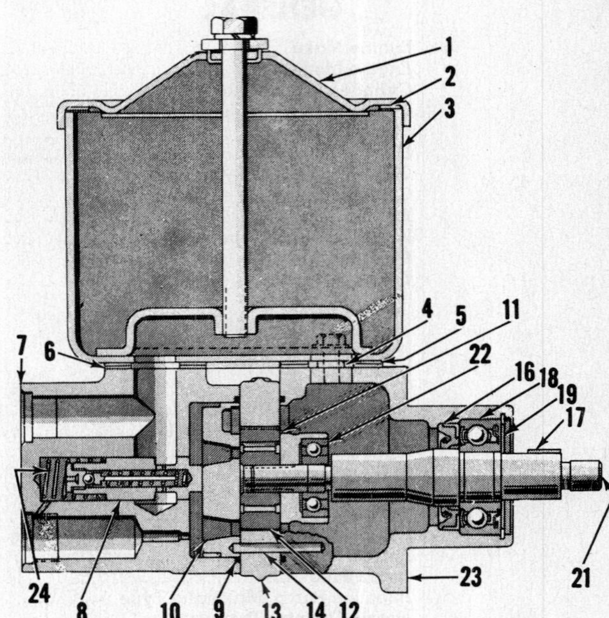

Fig. OS2—Sectional view of vane type power steering pump used on Super 99GMTC and 950. Malfunctions of the relief valve are corrected by renewing the flow control and relief valve assembly.

1. Reservoir cover
2. Gasket
3. Reservoir
4. Spacer gasket
5. Body gasket
6. Cover gasket
7. Pump cover
8. Flow control valve
9. Seal gasket
10. Pressure plate
11. Vane
12. Rotor
13. Ring
14. Locating pins
16. Shaft seal
17. Key
18. Outer bearing
19. Snap ring
21. Shaft
22. Inner bearing
23. Body
24. Flow control valve spring

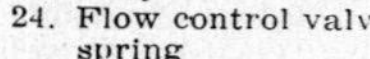

Connect pressure test gage and a shut-off valve in series with the pump pressure line as shown in Fig. OS1. Note that the shut-off valve is connected in the circuit between the gage and the power unit. Open the shut-off valve and run the engine at idle speed until oil is warmed. With engine running at high idle speed, turn the front wheels of the tractor to extreme right or extreme left position and note gage reading which should be 750-900 psi on series Super 99GMTC and 950; 950-1050 psi on series 990 and 995. Note: To avoid heating the steering fluid excessively, do not hold wheels in this test position for more than 30 seconds.

If the pressure is higher than specified, the relief valve is probably stuck in the closed position. If the pressure is less than specified, turn the wheels to straight ahead position; at which time, the gage reading should drop considerably. Now slowly close the shut-off valve and retain in closed position only long enough to observe the gage reading. Pump may be seriously damaged if valve is left in closed position for more than 10 seconds. If gage reading increases to the specific ranges, with valve closed, the pump and relief valve are satisfactory and the trouble is located in the control valve, cylinder or the connections.

While a low pressure reading with shut-off valve closed indicates the need of relief valve or pump repair, it does not necessarily mean the remainder of the system is in good condition. After overhauling the pump or renewing the relief valve, recheck the pressure reading to make sure the power unit and connections are in satisfactory operating condition.

Refer to the following paragraphs for data concerning the flow control and relief valve.

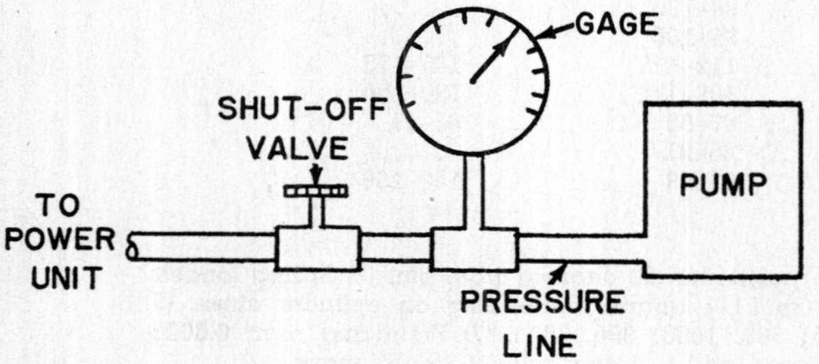

Fig. OS1 — To test the power steering system operating pressure, a shut-off valve and pressure gage can be installed in series with the pump pressure line.

FLOW CONTROL AND RELIEF VALVE

Series Super 99GMTC-950

14. A sectional view of the pump, flow control valve and relief valve are shown in Fig. OS2. To remove the flow control and relief valve, first remove pump, then remove reservoir cover (1) and reservoir (3) Remove the cap screws retaining cover (7) to pump body and carefully remove cover. Withdraw flow control valve assembly (8) from cov-

POWER STEERING SYSTEM TROUBLE-SHOOTING CHART

	Loss of Power Assistance	Power Assistance in One Direction Only	Unequal Turning Radius	Erratic Steering Control	Foaming Out of Reservoir	Wheels Shimmy
Binding, worn or bent mechanical linkage	★		★	★		★
Insufficient fluid in reservoir	★					★
Faulty pump drive belt	★			★		
Filter restricted or improperly installed	★			★		
Low pump pressure	★					
Faulty control valve	★			★		
Damaged or restricted hose or tubing	★	★			★	
Pump to valve hose lines reversed	★					
Wrong fluid in system	★			★	★	★
Improperly adjusted tie rods			★	★		
Air in system	★			★	★	★
Faulty flow control valve in pump				★		
Faulty pump	★			★		
External fluid leaks	★					
Faulty wheel bearings						★

er and remove spring (24). The pressure relief valve is contained within the flow control valve, opening pressure is non-adjustable and component parts are not available separately. Therefore, if the system pressure test indicates a malfunctioning relief valve, renew the complete flow control and relief valve assembly. Make certain that flow control valve slides freely in the cover bore with absolutely no binding tendency. O. D. of valve can be polished lightly with crocus cloth, but any deep scoring of valve or cover will necessitate renewal of both parts.

When reassembling, tighten the cover retaining screws to a torque of 25-30 Ft.-Lbs. and make certain the pump shaft does not bind. When installing the reservoir, be sure the reservoir gaskets and spacer gaskets are in position and tighten the retainer screws to a torque of 3½-4-Ft.-Lbs. Fill and bleed system as outlined in paragraph 11.

Series 990-995

14A. An exploded view of the pump is shown in Fig. OS4. To remove the flow control and relief valve, first drain pump and remove from generator. Use care not to damage drive shaft seal as it passes over the splined end of generator shaft. Remove reservoir and screen, then while holding the valve cover plate, remove lock wire. Retainer, spring and the flow control and relief valve can now be removed.

The internal pressure relief ball contained inside the flow control (relief valve) spool need not be removed if the system pressure test showed normal operation pressure. If, however, the test pressure was not within the normal 950-1050 psi range, remove the hexagon head plug and ball from spool and clean. If damaged parts are found, install completely new assembly as separate parts are not catalogued.

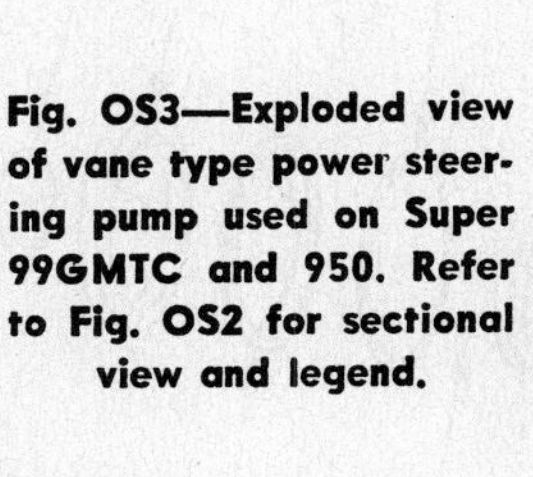

Fig. OS3—Exploded view of vane type power steering pump used on Super 99GMTC and 950. Refer to Fig. OS2 for sectional view and legend.

PUMP

Series Super 99GMTC-950

15. **R&R AND OVERHAUL.** The procedure for removing the pump is evident. To disassemble the pump, refer to Figs. OS2 and OS3 and proceed as follows:

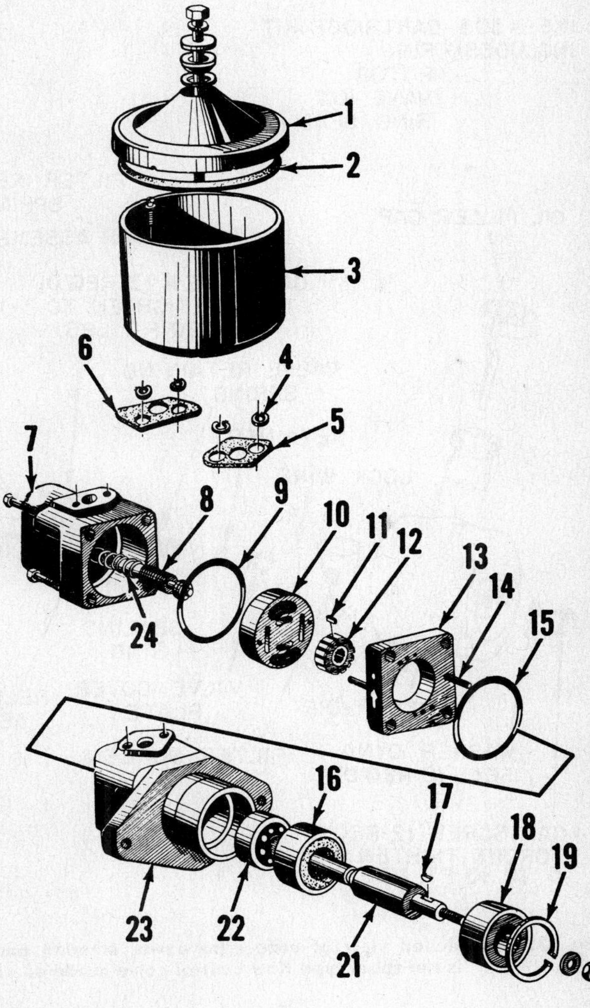

Remove reservoir cover (1) and reservoir (3). Remove the cap screws retaining cover (7) to pump body and carefully remove cover. Withdraw flow control valve assembly (8) from cover and remove spring (24). The system pressure relief valve is contained within the flow control valve and component parts are not available separately. Therefore if there are any manifestations of a faulty relief valve, renew the complete flow control and relief valve assembly. Remove pressure plate (10) and carefully note the position of cam contour ring (13). Remove locating pins (14) and ring (13). Slide the vanes (11) from slots in rotor, then remove rotor (12) from drive shaft.

Remove drive pulley, extract key (17) from shaft and remove snap ring (19). Gently tap shaft (on rotor end) out of pump body. The need and procedure for further disassembly is evident.

Thoroughly clean all parts except bearings in a suitable solvent and renew any which are damaged or worn. Check the finished surfaces of rotor (12), vanes (11) and ring (13) for roughness or irregular wear. Vanes must slide freely in the rotor slots but must not be excessively loose. Vane contacting surface of ring (13) must not be excessively scored. If any irregularities in these parts cannot be removed with crocus cloth, install a new cartridge repair kit which includes seals (9 & 15), vanes (11), rotor (12) and ring (13).

Machined face of pressure plate (10) must not be deeply scored. Slight irregularities, however, can be removed by lapping providing not more than 0.005 thickness of metal is removed. Refer to paragraph 14 for information concerning the flow control and relief valve. Shaft bearings (18 & 22) must be renewed if they feel "lumpy." Renew shaft (21) if seal contacting surface is grooved.

16. When reassembling, dip all parts in power steering oil and proceed as follows: Seat inner bearing (22) in housing by using a driver which contacts the outer race. Seat outer bearing (18) on shaft by using a driver which contacts the inner race. Install seal (16) with lip of same facing rotor end of pump housing. Install shaft and bearing assembly into housing (tap gently if necessary), then install snap ring (19).

Install seal ring (15), then locate ring (13) so the cam contour is in its original position and install locating pins (14). Install rotor and vanes so that rounded end of vanes contact the cam ring contour. Install spring (24) and flow control valve (8) in cover, then position seal ring (9) and pressure plate (10). Install cover and tighten the retaining screws to a torque of 25-30 Ft.-Lbs. Turn pump shaft several revolutions by hand to make certain that it does not bind.

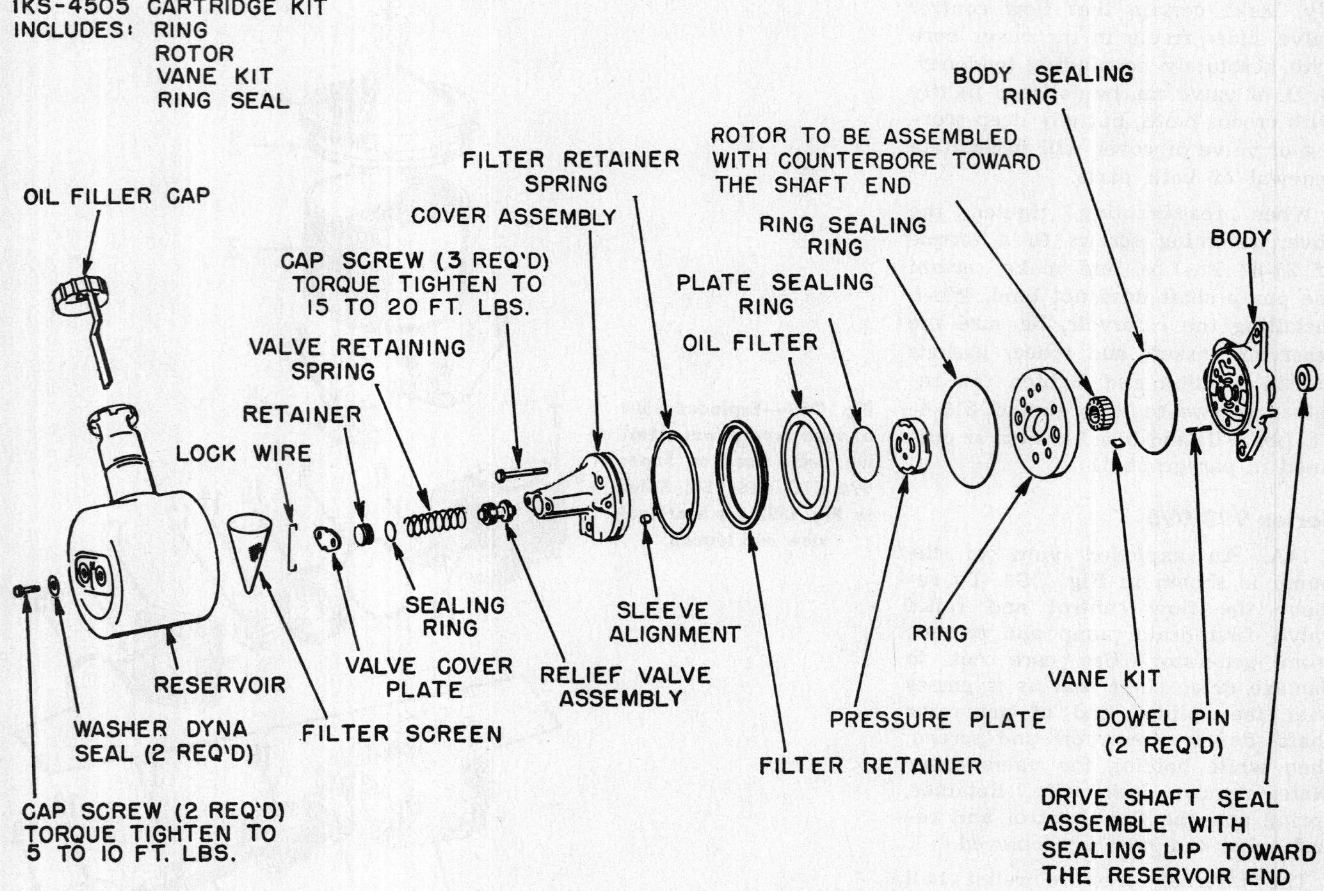

Fig. OS4—Exploded view of vane type power steering pump used on models 990 and 995. Item marked "relief valve assembly" is the spool type flow control valve inside of which is mounted the spring loaded ball type relief valve.

Install spacers (4) and gaskets (5 & 6). Install reservoir and tighten the retaining screws to a torque of 3½-4 Ft.-Lbs.

Reinstall pump on tractor, then fill and bleed the power steering system as outlined in paragraph 11.

Series 990-995

17. **R&R AND OVERHAUL.** Procedure for removal of the pump from the generator is self-evident. When sliding pump off and on generator be careful not to damage the drive shaft seal as it passes over the splined end of the generator shaft.

An exploded view of the pump is shown in Fig. OS4. Disassembly procedure is as follows: Remove first the reservoir and filter screen. When removing the relief valve, hold cover plate from flying, due to pressure of valve retaining lock wire, then remove plug, spring and valve. Remove cover mounting screws and lift cover off the body assembly. To remove filter assembly, extract the filter retaining spring from groove in cover assembly.

To remove ring, rotor and vanes, carefully lift the pump ring from dowel pins and lift rotor and vanes assembly from body. Drive shaft seal in body can be removed with a brass rod.

If pump ring, rotor or vanes are worn or scored, renew all of these parts as an assembly. If scoring or scratching of pressure plate face does not exceed 0.005, it can be corrected on a fine surface grinder or by lapping. Similar correction can be applied to face of pump body if irregularities are not deeper than 0.0015. Make sure the small pressure sensing hole in body face at inner end of pressure port is not restricted.

The internal pressure relief ball contained inside the relief valve assembly spool need not be removed if system pressure test showed normal operating pressure. If, however, the test pressure was not within the normal 950-1050 psi range, remove the hexagon head plug and ball valve from the spool and clean. If damaged parts are found, install complete new assembly as no separate parts are catalogued.

18. **REASSEMBLY.** Refer to Fig. OS4 and observe the following points when reassembling the pump: Coat inside surfaces with clean, Type "A" automatic transmission oil, or SAE 10W engine oil. Renew all seals and allow them to soak at least 20 minutes in the fluid before installing them. Lip of drive shaft seal in pump body should face reservoir end of pump. The small counterbore on rotor should face the mounting flange of the pump body. Rounded ends of rotor vanes should face outward. Apply a coating of light grease to rotor contacting surface of pressure plate to aid centering of rotor during pump installation. Install the alignment sleeve.

Install a new "O" ring in groove inside the cover. Add the filter retaining spring, filter retainer with flat side against spring and the oil filter to outside of cover. If relief valve spool does not move freely in bore of cover, polish it carefully with fine crocus cloth, being careful not to round the spool land edges. Hexagon end of flow control valve faces outward (rearward). If a substitute drive shaft is available use it to rotate pump drive shaft. Pump rotor should rotate without binding. Correct the cause of binding before installing pump to tractor. Reinstall pump on tractor, then fill and bleed power steering system as outlined in paragraph 11. Pump should deliver a minimum of 1.3 gallons per minute at 850 rpm and 700 psi.

14. Back-up ring
15. Retainer
16. Wiper ring
17. Piston rod seal
20. Housing seal
21. Centering spring
22. Washer
23. Pilot lock nut
24. Nut
25. Housing and spool assy.
28. Clamp
32. Ball socket housing
33. End plug
34. Safety plug
35. Seat
36. Spring
37. Washer
39. Spool mounting shaft
40. Shaft seal
41. Retainer
44. Safety plug
45. Relay link spacer cover
46. Seat
47. Housing and pin assy.
50. Self-locking bolt
54. Pilot nut seal
55. Housing seal

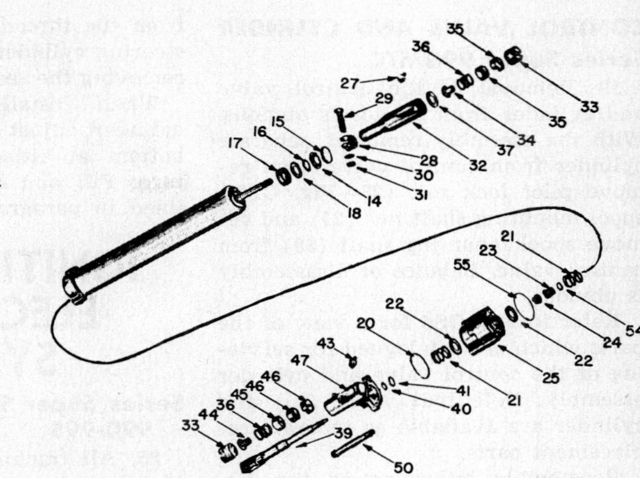

Fig. OS5—Exploded view of power steering cylinder and control valve used on Super 99GMTC tractors.

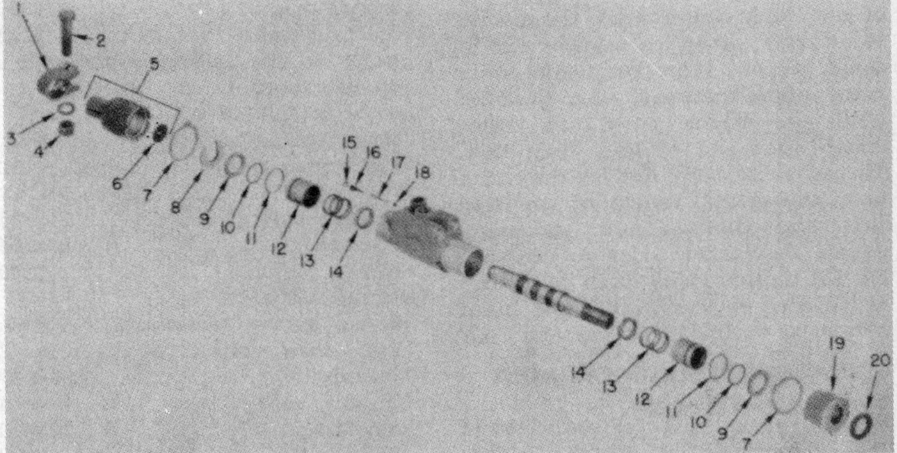

Fig. OS6—Exploded view of the drag link mounted type power steering control valve used on models 950, 990 and 995.

1. Clamp
5. Reducer
6. Expansion plug
7. Lock ring
8. Washer
9. Washer
10. "O" ring
11. "O" ring
12. Gland
13. Spring
14. Reaction ring
15. "O" ring
16. Plug
17. Relief valve spring
18. Steel ball
19. End cover
20. Dust seal

CONTROL VALVE AND CYLINDER

Series Super 99GMTC

19. Removal of the control valve and cylinder from tractor is obvious. With the assembly removed, separate cylinder from control valve, then remove pilot lock nut (23—Fig. OS5), spool mounting shaft nut (24) and remove spool mounting shaft (39) from control valve. Balance of disassembly is obvious.

Refer to Fig. OS5 for a view of the parts which are catalogued for servicing of the control valve and cylinder assembly. Note that valve (25) and cylinder are available as separate replacement parts.

Reassemble by reversing the disassembly procedure and adjust cylinder so that it will not bottom at either full left or right turn. Fill and bleed system as outlined in paragraph 11.

POWER CONTROL VALVE AND DRAG LINK

Series 950-990-995

20. Procedure for removal of control valve unit is self-evident, however, prior to removal, measure the overall length of the drag link and control valve assembly. To remove the end covers, straighten both legs of each lock ring (Fig. OS6) and unscrew the covers, using a strap type wrench or water pump pliers. If the valve body or the mating valve spool are scored, rusted or corroded, install a new valve assembly. Install all new "O" rings.

When reassembling, insert the valve spool into the valve body with threaded end of spool nearest the front of the tractor. Install new end cover lock rings if available. One leg of each lock ring engages the slot in valve body; other leg engages slot in cover. Screw short drag link onto valve until bottomed, then back-off until one thread is visible. Place clamp over end of long drag link. Thread valve body to long drag link until the overall length of the drag link and control valve assembly equals that taken prior to removal. Do not tighten clamp bolts until unit is installed on tractor. Fill and bleed system as outlined in paragraph 11.

POWER STEERING CYLINDER

Series 950-990-995

21. Removal and reinstallation of the power steering cylinder is obvious.

The only service that can be performed on the power steering cylinder is the renewal of the piston rod "O" ring and oil seal and their retaining rings. This is accomplished by removing the ball socket end assembly from the threaded end of the power steering cylinder piston rod and then removing the seal retaining snap ring.

When installing power steering cylinder adjust same so it will not bottom at either full left or right turn. Fill and bleed system as outlined in paragraph 11.

IGNITION AND ELECTRICAL SYSTEM

Series Super 99GMTC-950-990-995

25. All tractors are equipped with 12-volt systems. The non-diesel tractors use one 12-volt battery while all diesel tractors use two 6-volt batteries connected in series. In both cases, the positive terminal of battery is grounded.

26. Electrical units are Delco-Remy and their applications are as follows:

Series Super 99GMTC

Generator 1100316
Regulator 1118791
Starting Motor 1108801

Series 950HC

Generator, manual steering. 1100349
Generator, power steering. 1100322
Regulator 1118791
Starting motor 1113072
Distributor 1112561

Series 950 Diesel

Generator, manual steering. 1100349
Generator, power steering. 1100322
Regulator 1118791
Starting Motor 1113602

Series 990-995

Generator, manual steering. 1102023
Generator, power steering. 1100322
Regulator 1118896
Starting motor 1114022

Refer to the following tables for specifications.

Generator 1100349
Brush spring tension .... 28 oz.
Field draw, volts 12
Amps 1.58-1.67
Output, volts .... 14 cold
Amps .... 20 cold
RPM 2300

Generator 1100322
Brush spring tension .... 28 oz.
Field draw, volts 12
Amps 1.58-1.67
Output, volts .... 14 cold
Amps .... 20 cold
RPM 2300

Generator 1102023
Brush spring tension .... 28 oz.
Field draw, volts 12
Amps 1.48-1.62
Output, volts .... 14 cold
Amps .... 25 cold
RPM 2000

Generator 1100316
Brush spring tension .... 28 oz.
Field draw, volts 12
Amps 1.58-1.67
Output, volts .... 14 cold
Amps .... 20 cold
RPM 2300

Regulator 1118791
Ground polarity .... Positive
Cut-out relay
Air gap 0.020
Point gap 0.020
Closing voltage, range ... 11.8-14.0
Adjust to 12.8
Voltage regulator
Air gap 0.075
Setting volts, range 14.0-15.0
Adjust to 14.4

Regulator 1118896
Ground polarity .... Positive
Cut-out relay
Air gap 0.020
Point gap 0.020
Closing voltage, range... 11.8-13.5
Adjust to 12.6
Voltage regulator
Air gap 0.075
Setting volts, range..... 13.8-14.8
Adjust to 14.3
Current regulator
Air gap 0.075
Setting range, amps....... 23-27
Adjust to, amps 25

Starting Motor 1113072
Brush spring tension .... 48 oz.
No load test
Volts 11.5
Amps 50
RPM 6000
Lock test
Volts 3.3
Amps 500
Torque-ft.-lbs. 22

Starting Motor 1113602
Brush spring tension .... 35 oz.
No load test
Volts 11.6
Amps 90
RPM 8000
Lock test
Volts 2.2
Amps 600
Torque-ft.-lbs. 20

Starting Motor 1114022
Brush spring tension .... 35 oz.
No load test
Volts 11.5
Amps 105
RPM 6000
Lock test
Volts 2.3
Amps 600
Torque-ft.-lbs. 19

Starting Motor 1108801
Brush spring tension......36-40 oz.
No load test
Volts 11.6
Amps 115
RPM 7000
Lock test
Volts 2.3
Amps 570
Torque-ft.-lbs. 20

Distributor 1112561
Breaker contact gap..........0.022
Breaker arm spring
pressure ....17-21 oz.
Start advance ....0-2° @ 275
Intermediate advance ...5-7° @ 400
Intermediate advance ..9-11° @ 800
Maximum advance ..14-16° @ 1300

Advance data given is in distributor degrees and distributor rpm. Double listed values for flywheel degrees and rpm.

IGNITION TIMING

Model 950 HC

27. To time ignition of model 950 HC, set the distributor point gap to 0.022, remove timing hole cover from engine rear plate, then rotate engine until the IGN mark on flywheel is centered in the hole. Mark the IGN symbol with chalk or paint to make it more visible. Install a timing light, start engine and run at high-idle, no-load rpm of 2000. Direct timing light close to the timing hole cover, loosen distributor clamp and rotate distributor until IGN mark aligns with pointer of timing hole. Tighten distributor clamp. Reinstall timing hole cover.

CLUTCH

28. For clutch information on all models except models Super 99GMTC and 995, refer to Super 99 models covered in previous section of this manual. On 995 and Super 99GMTC models, use the information in paragraph 39 of this supplement to remove clutch although bear in mind that it will not be necessary to remove fuel tank. Overhaul of the clutch is conventional.

TORQUE CONVERTER

30. The torque converter used in the 995 Lugmatic can also be installed in the Super 99GM series tractors.

This four-element unit is mounted in the tractor main frame ahead of the transmission and depending upon load conditions, functions as a torque converter or a fluid coupling. At tractor stall, or when starting a load, a torque ratio of 2.5 to 1 is available from the converter in addition to the ratios provided by the six-speed transmission.

Two types of torque converter drives have been used. On tractors prior to serial number 69 643-900 the torque converter received its drive from the engine through a drive ring fitted with drive (spring) pins. On tractors serial number 69 643-900 and up, the drive is taken by a ring gear. When working on the early type, they should be converted to the later type by installing a conversion package supplied by the Oliver Corporation under part number 102 669-AS.

A brief explanation of the converter operation is as follows:

FIRST CONVERTER PHASE. The first converter phase is when both stators are being held stationary and the greatest torque multiplication is taking place. During this phase, the stators are being held stationary because the oil leaving the turbine strikes the stators in a manner that attempts to turn the stators in the direction opposite to the rotation of the pump and turbine. However, because the stators are mounted on cams (over-running clutches) and can rotate only in one direction, vortex flow of oil is redirected so that it enters the pump in the same direction as its rotation and thus aids the pump and results in torque multiplication.

SECOND CONVERTER PHASE. During this phase, the first stator is free-wheeling. The first stator begins free-wheeling as soon as the output load demands lessen. That is, when the rpm of the turbine increases in relation to the rpm of the pump. As the turbine rpm increases, the oil leaving the turbine strikes the blades of the first stator in such a manner that it no longer tends to lock-up but starts to free-wheel in the direction of the turbine rotation. At this time, three things are happening: Torque multiplication is decreasing, turbine rpm is increasing and the vortex oil flow is lessening.

FLUID COUPLING PHASE. The torque converter is in the fluid coupling phase at the time when the second stator also begins to free-wheel. This means that as turbine rpm approaches the pump rpm, the vortex oil flow further decreases and permits the second stator to free-wheel in conjunction with the first stator. As the turbine and pump approach equal rpms, the input torque and output torque are nearing the one-to-one ratio and the torque converter functions as a fluid coupling.

However, as the output torque demands increase, the converter automatically adjusts the torque input deman to the torque output demand and first one stator, then the other becomes stationary (locks-up) to produce torque multiplication.

TROUBLE SHOOTING

33. The following trouble shooting chart will give some of the troubles that may be encountered in the torque converter and corrections for same are evident.

1. LOW CONVERTER "OUT" PRESSURE. Could be caused by:
   - a. Low oil level. Refer to paragrph 38.
   - b. Oil line leakage.
   - c. Defective charging pump. Refer to paragraph 36.
2. HIGH OIL TEMPERATURE. Could be caused by:
   - a. Low oil level. Refer to paragraph 38.
   - b. High oil level. Refer to paragraph 38.
   - c. Low coolant level in cooling system.
   - d. Clogged or dirty oil cooler.
   - e. Operating tractor at inefficient converter range. Shift to lower gear.
   - f. One or both stators locked. Refer to paragraph 37.
   - g. Stators installed without rollers. Refer to paragraphs 40 and 41.

Fig. OS7 — Oil pressure and temperature checks can be made at either of the oil passages shown.

Fig. OS8 — Drain converter by removing plug at lower right front of clutch housing.

3. HIGH ENGINE SPEED AT CONVERTER STALL. Could be caused by:
   a. Low oil level. Refer to paragraph 38.
   b. Low converter "out" pressure. Refer to paragraph 36.
   c. High oil temperature. Refer to paragraph 37.

4. LOW ENGINE SPEED AT CONVERTER STALL. Could be caused by:
   a. Low engine output. Tune or overhaul engine.
   b. Converter element interference. Refer to paragraphs 40 and 41.

5. LOSS OF POWER. Could be caused by:
   a. Stators installed without rollers. Refer to paragraphs 40 and 41.
   b. Low converter "out" pressure. Refer to paragraph 36.
   c. Low engine rpm at converter stall. Tune or overhaul engine or overhaul torque converter.

TESTING

34. Assuming the engine and other components of the tractor are in satisfactory condition, the converter operation can be checked as follows:

35. STALL CHECK. Place tractor transmission in sixth gear and lock brakes. Operate engine at full throttle and engage clutch. Desired engine rpm at converter stall is 1500 rpm. CAUTION: Do not exceed a converter "OUT" oil temperature of 250 degrees F. during this operation as damage to converter seals and break-down of oil could occur.

36. OIL PRESSURE CHECK. To make an oil pressure check, install a gage capable of registering at least 50 psi in either of the two oil passages shown in Fig. OS7. Place transmission in sixth gear and lock the brakes. Operate engine at full throttle and engage clutch. When converter stalls, the "OUT" oil pressure should show a minimum of 24 psi. Do not stall converter longer than necessary and in no case should the oil temperature be allowed to exceed 250 degrees F.

37. OIL TEMPERATURE AND STATOR CHECK. The operating temperature of the converter "OUT" oil can be determined by installing a thermometer at either of the two oil passages shown in Fig. OS7. The maximum converter "OUT" oil temperature allowable is 250 degrees F. A rapid drop from a high to a low temperature would indicate satisfactory operation of the oil cooler.

A check for locked stators can be made as follows: Stall the converter as outlined in the stall check test until the "OUT" oil reaches 230 degrees F., then disengage tractor main clutch and allow engine to continue running at full throttle while observing the thermometer. The temperature should start to drop in 15 seconds. Assuming the oil cooler to be operating satisfactorily, a slow temperature drop indicates the probability of locked stator or stators. A rapid temperature drop indicates normal stator operation.

37A. If the tests outlined in paragraphs 34 through 37 indicate a malfunctioning torque converter, remove and overhaul same as outlined in paragraphs 39 through 41.

PREVENTIVE MAINTENANCE AND LUBRICATION

38. As the converter has only four major moving parts which run in oil, maintenance is relatively simple. The oil level should be checked daily and the unit observed for oil leaks.

When checking oil level, have the converter at or near operating temperature and the engine running at low idle speed. Be sure dip stick bottoms and add oil, if necessary, to bring level to "full" mark. Do not overfill.

Recommended torque converter lubricant is "Type C-1" hydraulic fluid, however, if this is not available "Type C" hydraulic fluid or SAE 10 Heavy Duty oil meeting military specification (MS) 2104-A can be used.

Torque converter fluid should be drained every 500 hours of operation or when fluid shows signs of contamination or any effects of high operating temperatures. Drain fluid from converter by removing plug at front of clutch compartment on right underside of tractor as shown in Fig. OS8. Make initial refill with six quarts of fluid, start engine and run until converter fluid is warmed. Check fluid level and add oil, if necessary, to bring level to "full" mark on dip stick.

REMOVE AND REINSTALL

Models Super 99GMTC-995

39. The following remove and reinstall procedure will apply to the Super 99GMTC tractor as well as the 995 Lugmatic. The primary difference occurring will be in the removal, and installation of the hood and these detail differences will be apparent as the work proceeds.

To remove the torque converter from tractor, proceed as follows: Remove muffler, air cleaner caps and rear hood strap. Remove grille, disconnect wires and remove headlights, if so equipped; then unbolt hood from frame, radiator and baffle plate. Lift off hood. Shut off fuel and disconnect fuel return line from top front of fuel tank. Remove fuel supply line.

Disconnect lines and wiring loom from bottom of fuel tank. Remove the two nuts from fuel tank front mounting studs and the tank rear retaining strap. On 995 models, loosen the two lower instrument panel cap screws and on all models, remove throttle bellcrank from left rear of fuel tank. Lift off fuel tank.

Disconnect control rod from lower end of throttle lever and move same forward out of the way. Drain the torque converter unit, then remove the "OUT" and "RETURN" lines. Disconnect the tachometer drive cable and the torque converter temperature sending unit (bulb) and move them forward out of the way. On early tractors with one-piece converter cover, remove the tee and nipple from converter, then remove the converter cover and the clutch dust cover. Plug the two holes in converter. Note: Late model tractors have a two piece torque converter cover which can be removed without disturbing the tee and nipple.

Depending on how the tractor is equipped, remove the following components as outlined below:

1. Power Take-Off Only. Remove the pto clutch cover and the two side covers from the pto housing. Remove the cap screws from the pto bearing retainer and pull the pto clutch and shaft assembly from tractor.

2. Power Take-Off and Hydraulic Unit. Disconnect rods from the control levers and remove the hydraulic manifold. Do not lose the three "O" rings and four lever spacers. Attach a hoist to support the pto and hydraulic assembly, then remove the cap screws retaining assembly to rear main frame and pull the entire unit from tractor.

3. Hydraulic Unit Only. Disconnect rods from control levers and remove cap screws from hydraulic manifold. Do not lose the three "O" rings and four lever spacers. Remove the rear platform with manifold and hoses attached. Attach a hoist to support the hydraulic unit, remove top cover from unit and remove the cap screws which retain unit to tractor. Pull hydraulic unit and spacer from pilot, then pull pilot and shaft assembly from tractor.

4. Non-Depth Control Hydraulics. Drain reservoir and disconnect hoses from reservoir and control valve. Disconnect control rod from control valve, then unbolt and remove reservoir.

5. Hydra-Electric or Hydraulic Control Unit. Drain unit and disconnect hydraulic hoses. On Hydra-Electric units, disconnect wiring as necessary. Unbolt and remove unit from side of tractor.

Note: Additional clearance will be obtained if the hydraulic pump and attached hoses are removed at this time.

Fig. OS9 — View showing torque converter unit being removed from tractor. Unit shown has early pin type drive ring.

Remove the hand hole cover next to the gear shift tower and the hand hole covers from each side of the main frame at clutch cover compartment.

Work through the hand hole next to the shift tower and remove the internal snap ring which holds clutch shaft in transmission input shaft, however, before completely releasing snap ring position a piece of pipe, or some suitable object, in clutch shaft to prevent snap ring from falling into transmission. Use a small pry bar and separate clutch shaft from input shaft. Move clutch shaft rearward and remove the two piece plastic bushing. Jack-up one rear wheel of the tractor, move clutch shaft rearward and engage one of the dogs of the clutch shaft between teeth on bevel ring gear. Turn the tractor wheel forward and pull the clutch shaft rearward past the bevel ring gear as far as it will go. Note: The bull gears will prevent complete removal of shaft. Temporarily replace hand hole cover.

NOTE: On tractors equipped with a two lever gear shift, it will be necessary to loosen the shifter poppet block and slide the high-low range rod to the rear to allow the clutch shaft to clear the shift arm.

Disconnect the clutch shifter rod from the shifter shaft. Loosen the lock screw in the clutch shifter fork and remove the two clutch release bearing return springs from the carrier and retainer.

Block the release bearing carrier support and drive the shifter shaft from the fork, remove key and slide the shaft from the frame as far as possible. (The rear tire will prevent complete removal of the shaft.)

Remove the fork, clutch release bearing carrier and support assembly from the tractor. Compress (unload) the clutch using three ⅜-16 X 2½ inch cap screws and flat washers to fasten the clutch cover and pressure plate together. Remove the clutch assembly and driven plate from the converter flywheel.

Install a ½-13 inch eyebolt in the tapped hole atop the converter housing and attach a hoist. Remove the drain tube assembly from the converter housing.

Remove the cap screws retaining the converter to the engine flywheel housing and remove the converter from tractor. See Fig. OS9.

OVERHAUL

Models Super 99GMTC-995

40. To overhaul the removed torque converter, remove flywheel, refer to Fig. OS10 and proceed as follows: Remove sump plate (40) and gasket (39). Remove relief valve plug (32), spring (38) and ball (37). Remove the lock bolts from output shaft retainer (47) and pull the output shaft assembly from converter. Use caution not to damage the two hook type seals on

Fig. OS10—Exploded view of an early type torque converter showing component parts and their relative positions. Later type converters have a ring gear drive instead of the pin type drive ring (25) shown.

1. Charging pump
4. Snap ring
7. Converter pump hub
8. Ball bearing
9. Converter pump
11. "O" ring
12. Hook seal ring
13. Stator spacer
15. Freewheel race
16. Stator
17. Stator
18. Roller
19. Roller spring
20. Spacer washer
21. Turbine
22. Ball bearing
23. "O" ring
24. Cover
25. Drive ring
26. Pin
29. Housing
31. Plug
32. Plug
35. Oil return tube
37. Ball
38. Spring
39. Gasket
40. Sump plate
45. Plug
46. Gasket
47. Bearing retainer
48. Oil seal
49. Output shaft
50. Plug
51. Ball bearing
52. Snap ring
53. Hook type ring
54. Hook type ring

the output shaft. Remove retainer gasket and the eight seal rings from housing. Remove the hook seal rings and snap ring from output shaft, then press shaft from bearing. Pry seal (48) from retainer and inspect bearing. If bearing renewal is required, press bearing from retainer.

Remove the two counter-sunk retaining cap screws. Insert two of removed lock bolts through the converter housing and thread into charging pump body. Support assembly and tap lock bolts until pump and turbine assembly is loose, then remove bolts and pull assembly from housing. Support machined face of charging pump on wood blocks to prevent damage, then unbolt and remove pump cover (24) from converter pump (9). Remove turbine (21), pull first stator (17) about half-way off and install a cylinder of shim stock to hold rollers and springs in position, then complete removal of first stator. Lift-off spacer (20) then remove stator (16) in the same manner that stator (17) was removed. Remove the free-wheel roller race (15). Spread snap ring (4) and lift same from shaft. DO NOT work the snap ring off as damage to shaft could result. Remove spacer (13) and converter pump assembly (9). Straighten lock plates and remove pump hub (7) from converter pump. Remove the hook seal (12) from pump hub and if necessary, the bearing (8). Remove "O" ring seal (11) from pump mounting flange. Remove the three counter-sunk screws from the ground sleeve of the charging pump and separate the assembly. Mark the pump gears with chalk, or some other marking material, prior to removal to insure that they can be reinstalled in the same position. Note: Do not use a punch to mark charging pump gears.

With torque converter disassembled, carefully clean and inspect all parts. Inspect housings, castings and machined surfaces for wear, scratches and/or scoring. Small defects can be removed with crocus cloth or a soft stone. Renew those parts showing wear, scratches and/or scoring. Pay

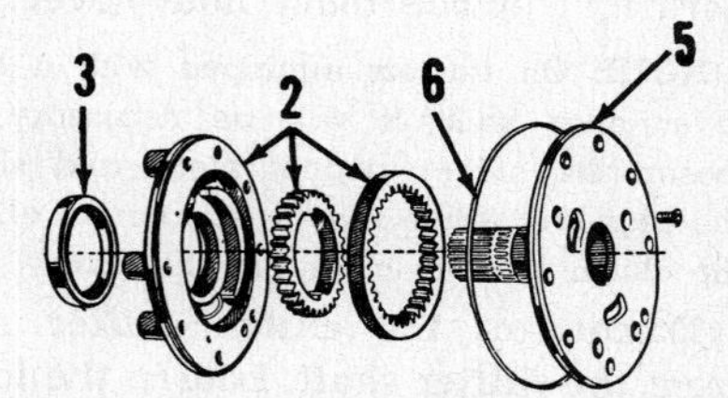

Fig. OS11 — Exploded view of charging pump assembly. Refer to paragraph 41 of text when installing oil seal (3).

2. Body & gear assy.
3. Oil seal
5. Ground sleeve
6. Ring seal

particular attention to machined surfaces for damage that could cause oil leakage. Be sure all oil passages are open and clean. Check all bearings and races for excessive wear, chipping, cracks or rough bearing rotation. If a defective bearing is found, pay close attention to the bearing bore and shaft. Inspect charging pump gears and body and if excessive wear or damage is found, renew the pump gears and body.

Reassembly of the torque converter is the reverse of disassembly, however, keep the following points in mind.

41. Use all new gaskets and seals and when renewing oil seal (3—Fig. OS11), first measure the depth of seal counterbore in charging pump body. If depth of bore is approximately $\frac{5}{16}$-inch the seal may be pressed in until it bottoms. However, if depth of bore is approximately $\frac{7}{16}$-inch, the seal should be pressed in until top of seal is 1 41/64 inches (1.64) from machined inner surface of pump body. Seal is installed with lip toward pump gears.

If springs and rollers were removed from stators, install springs with open end next to roller and towards inside diameter of stator. Springs and rollers can be held in position with heavy grease. Stators can be installed by placing them on chamfer of freewheel race and rotating them in the direction of freewheel travel.

When installing the converter assembly into the housing, install two headless guide bolts in the charging pump assembly, then lower housing assembly onto converter assembly making sure the holes for the counter-sunk screws are aligned. Tap around the housing to force the seal ring on charging pump to enter the bore, then slide into position and install the three counter-sunk screws.

Reinstall the torque converter in tractor by reversing the removal procedure. However, if a new engine, or a new torque converter is being installed, check the torque converter flywheel alignment as follows: Slide the clutch shaft, with the two-piece plastic bushing installed, through the transmission input shaft and into the pilot bearing in the flywheel. Mount a dial indicator on clutch shaft, then turn the clutch shaft and sweep the flywheel face. Total indicated run-out must not exceed 0.010. Now position indicator to sweep the outside diameter of the input shaft at the oil seal land. Turn shafts to check run-out. The total indicated run-out must not exceed 0.005. Shim engine mounts to obtain alignment.

Fill torque converter with oil, start engine and check unit for leaks.

OIL COOLER

42. An oil cooler for the torque converter fluid is incorporated into the engine cooling system and is positioned between the engine outlet elbow and the radiator top tank. Removal for cleaning and/or renewal is obvious.

Fig. OS12 — The Hydra-Electric, or Hydraulic Control unit is mounted on right side of tractor as shown.

HYDRAULIC LIFT SYSTEM

Series 950-990-995

Beginning with serial number 80 826-900 the 900 series tractors were equipped with a Hydra-Lectric or Hydraulic Control unit in place of the rear mounted depth stop hydraulic system. This unit is mounted on the right side of the tractor just forward of the right fender as shown in Fig. OS12.

Service procedures for the Hydra-Lectric unit are very similar to those for the Hydra-Lectric unit of the 770 and 880 series tractors. The basic changes being a different base mounting plate in which an oil filter of the type used on the 550 series tractors is installed and a relocated relief valve. Relief valve pressure on this unit is set at 1300 psi.

These units are also available for installation on the Super 99 series tractors.

The Hydraulic Control unit is almost identical with the Hydra-Lectric unit except that it does not include the electrical units. For servicing, use the information given for the Hydra-Lectric unit but ignore the electrical system.

TROUBLE-SHOOTING

50. **TESTING.** Trouble with the "Hydra-Lectric" unit may be divided into two categories: Electrical and mechanical. If there is doubt as to whether the trouble is electrical or mechanical after checking the fuse in the electrical circuit make the check outlined in paragraph 51, which involves operating the lift system manually and without the use of the electrical circuit. Refer also to paragraph 52 through 58.

51. The selector control valves located inside the reservoir can be operated manually by operating the control levers. If the lift system will not function when it is manually operated, the trouble is mechanical and there is no need to check the electrical system.

Some causes of faulty operation are outlined in the following paragraphs.

52. RELIEF VALVE BLOWS DURING RETRACTING STROKE. Malfunction in restrictor valves or valves are improperly adjusted. Self-sealing couplings are improperly connected.

53. RELIEF VALVE BLOWS DURING EXTENDING STROKE. Excessive load or binding linkage can be the cause of the trouble. Stuck relief valves or restricted relief valve passage are other possibilities.

54. RELIEF VALVE BLOWS WHEN CYLINDER IS RETRACTED OR EXTENDED. Selector valve sticking, binding solenoid linkage, weak or damaged selector valve centering spring. Friction stop collar on work cylinder too close to piston rod yoke.

55. UNIT WILL NOT LIFT. Insufficient oil in reservoir. Worn pump. Relief valve not seating. Leaking "O" ring or packing on work cylinder piston. Selector control valve solenoid linkage improperly adjusted. Selector control valve leaking. Interlock valve seals leaking. Selector valve body gasket leaking. Pump to reservoir seals leaking. Thermal relief valve leaking.

56. OIL FOAMS OUT OF RESERVOIR BREATHER. Wrong type of hydraulic operating fluid. Relief valve not properly adjusted. Oil level too high. Air leak at pump inlet. External leaks.

57. SLUGGISH LIFTING ACTION. Worn pump or obstructed pump inlet passage. Air in system.

58. UNIT WILL NOT HOLD LOAD. Leaking "O" ring or packing on work cylinder piston. Damaged selector control valve.

SYSTEM MAINTENANCE

59. **HYDRAULIC FLUID.** Recommended operating fluid for the hydraulic lift system is SAE 10W engine lubricating oil plus Oliver 102 082-A additive mixed in the ratio of 16 parts motor oil to one part additive. Fill reservoir and hydraulic system to full level mark indicated on the bayonet type gage which is attached to the reservoir air breather.

After draining and refilling the system, it will be necessary to bleed the system as outlined in paragraph 61.

60. **AIR BREATHER.** The air breather should be cleaned daily.

61. **BLEEDING LIFT SYSTEM.** The presence of air in the system will be evidenced by a retarded and spongy operating condition. The hydraulic power lift system will also require bleeding whenever the system is drained and refilled.

To bleed the system, loosen one hose connection at each cylinder. Start the engine and operate the hand control lever through one cycle of cylinder operation. Then tighten the hose connection. Repeat this procedure on the other hose for the same cylinder. Continue this procedure until there is a continuous flow of oil without air bubbles at the loosened connection. Replenish the oil supply in the reservoir.

62. **FLUSHING THE SYSTEM.** Once each year or every 1000 hours of operation, the system should be drained

and flushed with engine flushing oil. Operate the system for about 5 minutes on flushing oil then completely drain the reservoir, hoses and cylinders. Refill reservoir and bleed as outlined in paragraph 61.

PUMP

63. **R&R AND OVERHAUL.** After draining the hydraulic unit and disconnecting hoses, the removal of the engine driven pump is obvious.

To disassemble a removed pump unit refer to Fig. OS13 and proceed as follows: Remove cap screws attaching pump cover (2) and lift off cover. Lift off compression spring (3), and pressure plate (4). Before removing rotor (15) carefully note position of embossed arrow on outer surface of ring (6) in relation to pump body. Remove ring (6); then, pull rotor (15) and vanes (16) from splined end of shaft (11).

To inspect pump shaft bearings (9 and 12) or renew seal (10), remove snap ring (13), pull shaft and bearing (12) as a unit from pump body. Bearing (9) will remain in pump body. To remove inner bearing (9) from pump body first extract seal (10). Using a drift, bump bearing (9) out of pump body.

Thoroughly wash all parts, except bearings, and inspect their rubbing surfaces for wear, scoring and pitting. All vanes should slide freely in their rotor slots.

Observe the following points when reassembling the pump. Carefully oil all parts with engine oil prior to reassembly. Install new shaft seal into housing counterbore with lip facing toward inside of pump. Assemble shaft through seal until lip of seal is riding on polished surface of shaft.

Assemble vanes to rotor and ring with rounded ends of vanes contacting the race of the ring .

CONTROL VALVES

64. Procedures for the removal, disassembly and overhaul of the selector control valve and related parts, interlock valve, relief valve, thermal relief valve, restrictor valve, momentary restrictor valve, and free flow valve are outlined in the following paragraphs 65 through 69.

65. **SELECTOR CONTROL VALVE & INTERLOCK.** Selector valve controls the direction of oil flow to and from the work cylinder. It also contains an interlock system for locking the oil in the cylinder. The solenoid operated spool type selector control valve, body and solenoid assembly can be removed from the hydraulic lift unit as follows: Drain hydraulic fluid from the unit by removing the pipe plug located at the right front side of the reservoir mounting base. Remove reservoir cover and disconnect the wires from the solenoid unit. Remove reservoir housing from mounting base. The selector control valve, body and solenoid can be removed after removing the attaching cap screws.

66. INTERLOCK VALVE. Interlock valve (31—Fig. OS14) which is located in selector control valve body can be removed after removing selector control valve, body and solenoid assembly as outlined in paragraph 65.

Remove the interlock valve retaining steel end plate (22) from each end of the body (33). Using pliers, remove the two interlock valve guides (25) and springs (28). Remove the ¾-inch steel balls. Remove the interlock valve seats (29) by bumping on the end of the interlock plunger (31) with a ⅜ inch brass rod and hammer. Remove the interlock plunger (31).

Renew "O" rings and check the condition of the valve seats and ball valves. Renew valve seats and/or ball valves if they show signs of wear, grooving and/or pitting. Renew the interlock plunger if the ends are chipped or upset.

Reassemble the interlocking valve assembly by reversing the disassembly procedure. An interlock valve can be reseated by placing the ¾-inch ball on the seat and tapping it lightly with a hammer and soft drift.

67. RELIEF VALVE. The poppet type relief valve is located in hydraulic unit mounting base. Relief valves are equipped with a renewable steel seat (60).

The relief valve can be removed for inspection after removing the relief valve guide (55), spring (58) and shims. Check the line contact of both the valve and seat, and if same are grooved, it will be necessary to renew the valve, or valve and seat. Note: Relief valve pressure is set at the factory to 1300 psi. and should not require further adjustment. However, shims (57) are available should adjustment become necessary.

68. THERMAL RELIEF VALVE. Each selector control valve body is equipped with a poppet type safety or thermal relief valve assembly (32) to protect the hoses against expansion of the oil arising from temperature increases when the oil is locked in the cylinder hoses. The non-adjustable thermal relief valve is designed to open when the pressure within the cylinders exceeds 3500 psi.

The thermal relief valve can be removed after draining the reservoir and removing the reservoir cover and housing.

69. RESTRICTOR VALVE. A restrictor valve is inserted between the reservoir and hose connection to restrict the flow of fluid from the work cylinder on the retracting stroke. The adjustable type restrictor valve consists of a valve body, ball check valve, and a needle valve or only a spring loaded plunger.

To retard the lowering speed, loosen the adjusting screw Palnut then rotate the guide in a clockwise direction.

Adjust restrictor valve so that implement will lower at the maximum rate without bouncing.

Method of removal and disassembly of the restrictor valve is self-evident.

Note: Hydra-Lectric or Hydraulic Control units that are used to operate single acting cylinders must have a momentary restrictor assembly installed in either the right front, or

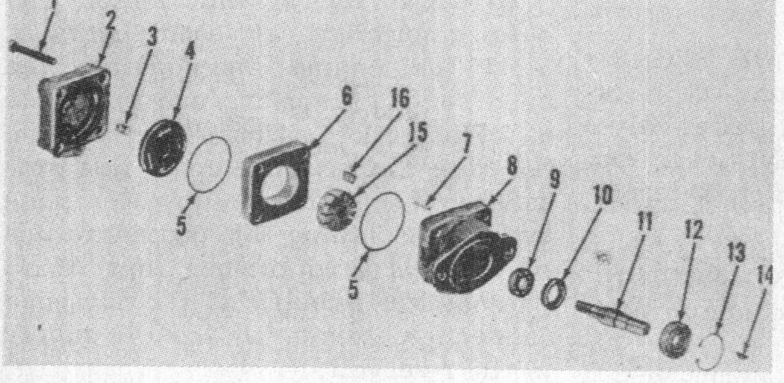

Fig. OS13 — Exploded view of hydraulic lift pump showing component parts and their relative positions.

2. Pump cover
3. Spring
4. Pressure plate
5. Seal ring
6. Pump ring
7. Dowel
8. Pump body
9. Ball bearing
10. Oil seal
11. Pump shaft
12. Ball bearing
13. Snap ring
14. Woodruff key
15. Rotor
16. Vane

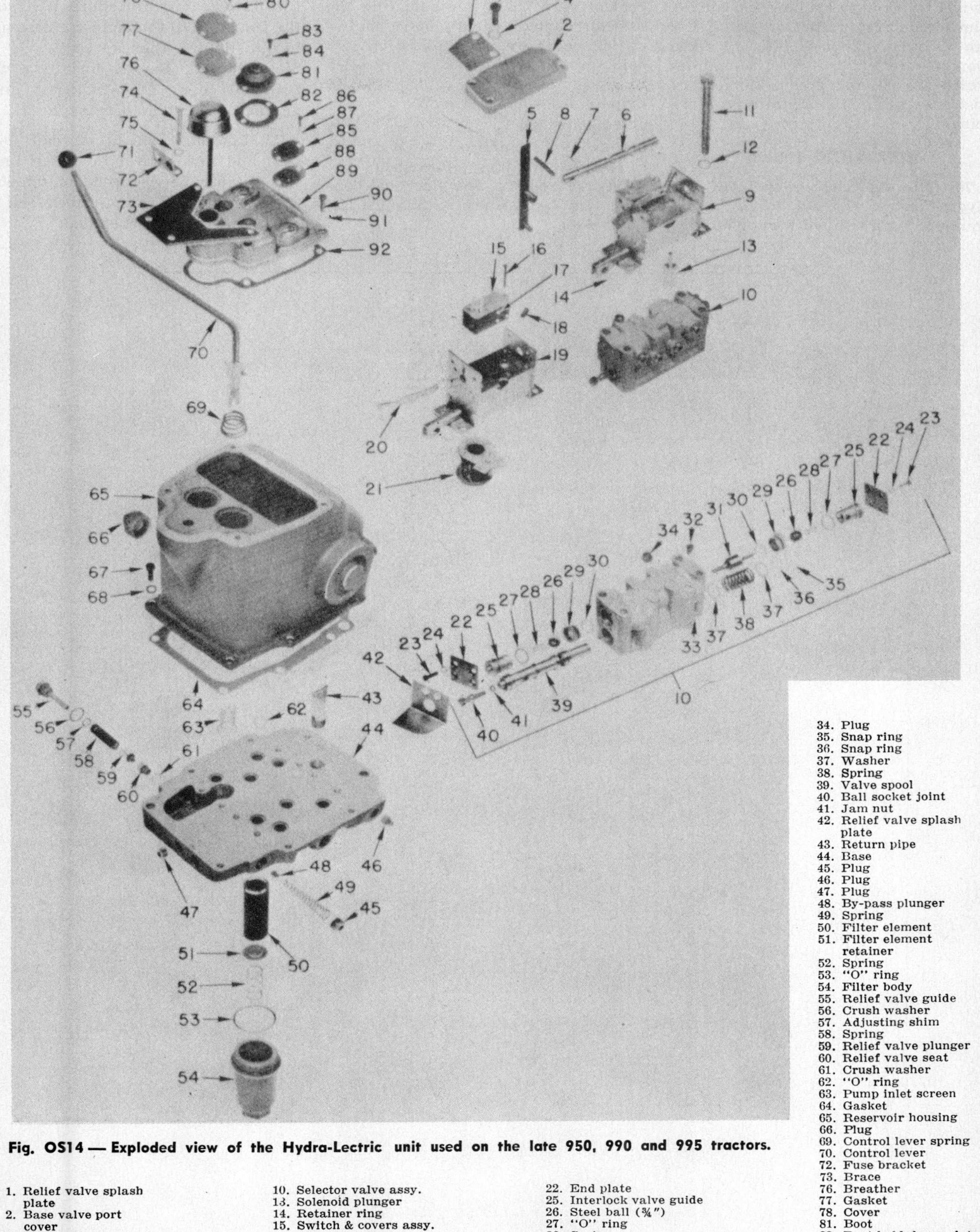

Fig. OS14 — Exploded view of the Hydra-Lectric unit used on the late 950, 990 and 995 tractors.

1. Relief valve splash plate
2. Base valve port cover
5. Spool lever
6. Solenoid rail
7. Pin
8. Pin
9. Solenoid & switch assy.
10. Selector valve assy.
13. Solenoid plunger
14. Retainer ring
15. Switch & covers assy.
17. Switch button
18. Spring
19. Frame & top plate assy.
20. Lever
21. Coil assembly
22. End plate
25. Interlock valve guide
26. Steel ball (¾")
27. "O" ring
28. Spring
29. Seat
30. "O" ring
31. Interlock plunger
32. Thermal relief valve
33. Valve body
34. Plug
35. Snap ring
36. Snap ring
37. Washer
38. Spring
39. Valve spool
40. Ball socket joint
41. Jam nut
42. Relief valve splash plate
43. Return pipe
44. Base
45. Plug
46. Plug
47. Plug
48. By-pass plunger
49. Spring
50. Filter element
51. Filter element retainer
52. Spring
53. "O" ring
54. Filter body
55. Relief valve guide
56. Crush washer
57. Adjusting shim
58. Spring
59. Relief valve plunger
60. Relief valve seat
61. Crush washer
62. "O" ring
63. Pump inlet screen
64. Gasket
65. Reservoir housing
66. Plug
69. Control lever spring
70. Control lever
72. Fuse bracket
73. Brace
76. Breather
77. Gasket
78. Cover
81. Boot
82. Boot hold-down plate
85. Wiring connector opening cover
88. Gasket
89. Reservoir cover
92. Gasket

left rear port opening of the mounting base in place of the regular free flow swivel or 90 degree elbow and nipple. The momentary restrictor assemblies are painted yellow to identify them from the restrictor swivel assemblies which are painted green. Momentary restrictor assemblies require no servicing.

ELECTRICAL UNITS

70. The electrical system of the Hydra-Lectric Control for a model equipped with two work cylinders includes the following: Two double-acting solenoids located on the selector control valve body for actuating the selector control valves; hand control switch; work cylinder follow-up stop collar which is electrically controlled by a magnetic type of control; two 20 ampere fuses and wiring harness.

Solenoids are a part of the solenoid and switch assemblies and some parts of these assemblies are furnished separately for repairs. The electrical contacts can be cleaned with a contact point file.

71. Reassemble the selector control valve, body and solenoid by reversing the disassembly procedure. Reinstall the assembly to the mountng base and torque tighten the cap screws to 18 foot pounds. Before installing the reservoir cover, check solenoid linkage as outlined in paragraph 72.

72. Loosen lock nut on ball joint (40) and turn swivel in or out until the distance between end of solenoid housing and roll pin in each end of rail (6) is equal at each end.

NOTES

OLIVER

Series ■ Super 55 ■ 550

Previously contained in I & T Shop Service Manual No. 0-11A

SHOP MANUAL

OLIVER

MODELS SUPER 55HC - SUPER 55 DIESEL - 550 HC - 550 DIESEL

IDENTIFICATION

Tractor serial number and tractor specification number plate is located on the left hand side of the tractor center frame. Engine serial number is stamped on right side of engine rear flange.

INDEX (By Starting Paragraph)

CONDENSED SERVICE DATA

GENERAL

Engine Make	Own
Cylinders	4
Bore—Inches (Prior Ser. No. 72 832)	3.5
Bore—Inches (After Ser. No. 72 831)	3.625
Stroke—Inches	3.75
Displacement—Cubic Inches (Prior to Ser. No. 72 832)	144
Displacement—Cubic Inches (After Ser. No. 72 831)	155
Compression Ratio, Diesel	15.5
Compression Ratio, HC (Prior Ser. No. 72 832)	*7.3
Compression Ratio, HC (After Ser. No. 72 831)	7.75
Pistons Removed From	Above
Main Bearings, Number of	3
Main Bearings Adjustable?	No
Rod Bearings Adjustable?	No
Cylinder Sleeves Type	Wet
Forward Speeds	6
Reverse Speeds	2
Generator Make	Delco-Remy
Generator Model	Refer to paragraph 75
Starter Make	Delco-Remy
Starter Model	Refer to paragraph 76

TUNE-UP

Firing Order	1-2-4-3
Valve Tappet Gap, Inlet, Cold	0.009-0.011
Valve Tappet Gap, Exhaust, Cold (Non-Diesel)	0.015-0.017
Valve Tappet Gap, Exhaust, Cold (Diesel)	0.019-0.021
Inlet and Exhaust Valve Face Angle	44½
Inlet and Exhaust Valve Seat Angle	45
Ignition Distributor Make	Delco-Remy
Ignition Distributor Model	1112555
Distributor Breaker Gap	0.022
Distributor Timing Retard Degrees	TDC
Distributor Timing Full Advance Degrees	28
Flywheel Mark Indicating:	
Retarded Timing Distributor	TDC
Full Advanced Timing—Distributor	IGN-HC
Distributor Governor Advance. Refer to paragraph 77 this manual	
Spark Plug Make, Non-Diesel	AC, Auto-Lite, Champion
Model for Gasoline	85S Com., BT-8, 8 Com.
Electrode Gap	0.025
Carburetor Make	Zenith, Marvel-Schebler or Carter
Carburetor Model	Refer to paragraph 34
Carburetor Float Setting—Inches	Refer to paragraph 34
Injection Pump Make	Bosch

*On tractors prior to 1957, compression ratio is 7.0.

TUNE-UP—*(Continued)*

Injection Pump Model	See paragraph 42A
Injector (Nozzle) Make	Bosch or CAV
Injector (Nozzle) Type	Pintle
Injection Timing—Degrees	26° BTDC
Injection Timing Mark on Flywheel	FP
Nozzle Opening Pressure—PSI	1750
Engine Low Idle RPM, Non-Diesel	350-400
Engine Low Idle RPM, Diesel	650-700
Engine High Idle RPM	2200
Engine Loaded RPM	2000
Belt Pulley RPM @ 2200 Engine RPM	1451
PTO RPM @ 2200 Engine RPM	749

SIZES—CAPACITIES—CLEARANCES

(Clearances in Thousandths)

Crankshaft Journal Diameter—Inches	2.2495
Crankpin Diameter—Inches	2.250
Connecting Rod C to C Length—Inches	6.75
Camshaft Journal Diameter, Front	1.7495
Camshaft Journal Diameter, second and third	1.7485
Piston Pin Diameter—Inches	1.2495
Valve Stem Diameter—Inlet	0.3725
Valve Stem Diameter—Exhaust	0.3715
Compression Ring Width	1/8
Oil Ring Width (Early 1/4) Latest	3/16
Main Bearings Running Clearance	0.5-3.5
Rod Bearings Running Clearance	0.5-1.5
Piston Skirt Clearance	See Paragraph 21
Crankshaft End Play	0.004-0.009
Camshaft End Play	Spring and Plunger
Camshaft Bearing Running Clearance	Refer to paragraph 19A.
Cooling System—Gallons	3.5
Crankcase Oil—Quarts	4.0
Transmission and Differential—Quarts	20.0
Hydraulic System—Quarts	9.0
With External Valve and One Cylinder—Quarts	11.0
Belt Pulley—Pints	1.0

TIGHTENING TORQUE FOOT POUNDS

Cylinder Head, Non-Diesel	92-100
Cylinder Head, Diesel	112-117
Cylinder Head Oil Screw	96-100
Main Bearing Caps	88-92
Connecting Rod Bolts	46-50
Flywheel Cap Screws	67-69

Oliver "Super 55"

FRONT AXLE SYSTEM

1. Refer to Figs. O650 and O651 for views of the three-piece adjustable front axle and the fixed tread front axle.

STEERING KNUCKLES

2. The knuckle bushing and/or steering knuckle should be renewed when the diametral clearance between the two parts is greater than 0.020. New bronze type bushings are presized and will not require final sizing if they are properly installed.

Recommended toe-in of front wheels is $\frac{3}{16}$-inch.

AXLE OR AXLE CENTER MEMBER

3. To remove the front axle or axle center member, proceed as follows: Remove tractor hood and grille. Unbolt and remove radiator shell. On models with the fixed tread front axle, also remove the radiator. Raise front of tractor, block-up under front main frame, then remove bolts attaching stay (radius) rods to axle or axle center member.

On models with adjustable front axles, unbolt axle extensions from axle main member and swing extensions, steering knuckles and wheels, as units, away from axle center member.

On models with fixed tread front axle, unbolt steering arms from steering knuckles and remove steering knuckles and wheels.

Remove axle pivot pin locking cap screw; then, use a sleeve and stud type puller to pull pivot pin from the front main frame. Axle or axle center member can now be removed.

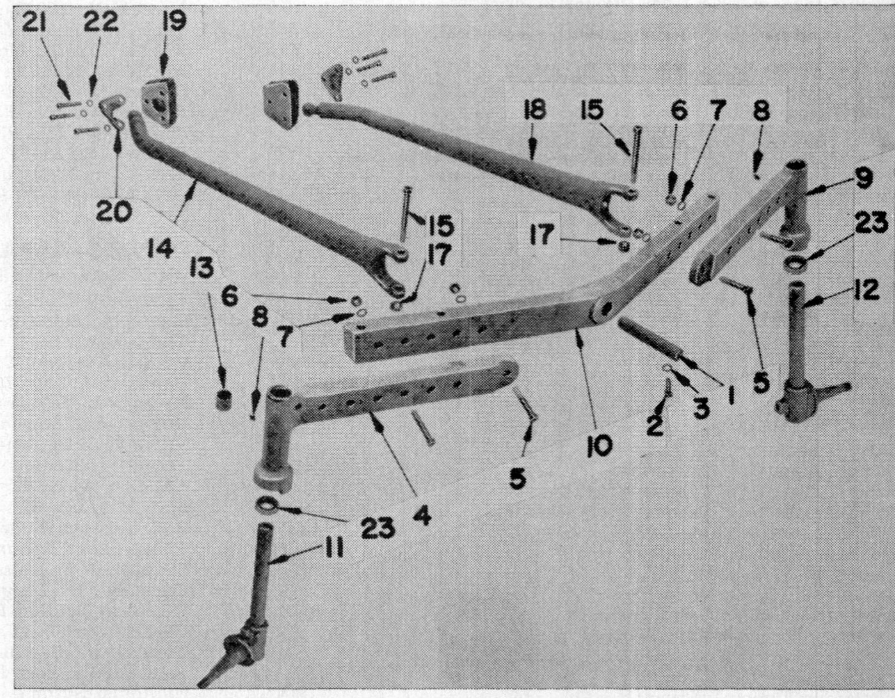

Fig. O651—Deails of adjustable front axle assembly. Front axle center member (10) is not bushed for pivot pin (1).

1. Pivot shaft
2. Pivot pin locking cap screw
4. Axle extension, right
9. Axle extension, left
10. Axle center member
11. Steering knuckle and shaft
12. Steering knuckle and shaft
13. Bushing
14. Stay rod
18. Stay rod
19. Stay rod ball socket
20. Ball socket cap
23. Thrust bearing

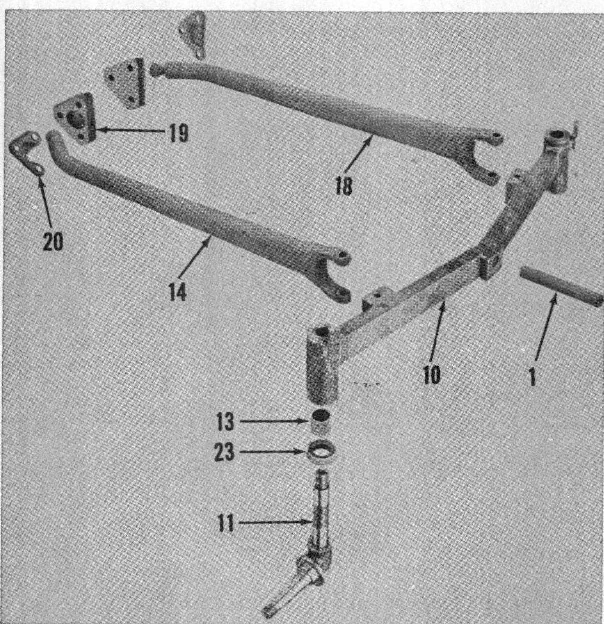

Fig. O650—Details of the fixed tread front axle. Refer to Fig. O651 for legend.

AXLE PIVOT PIN

4. The axle center member is not bushed for the axle pivot pin. To remove axle center member pivot pin proceed as follows: Remove tractor hood, and grille. Remove five machine screws and nuts attaching radiator shell to hood front support. Remove two cap screws attaching radiator shell to sides of front main frame, and four cap screws attaching radiator shell to top of front main frame. Lift off radiator shell. Support tractor under front main frame, and remove axle pivot pin locking cap screw (2—Fig. O651) from lower side of main frame.

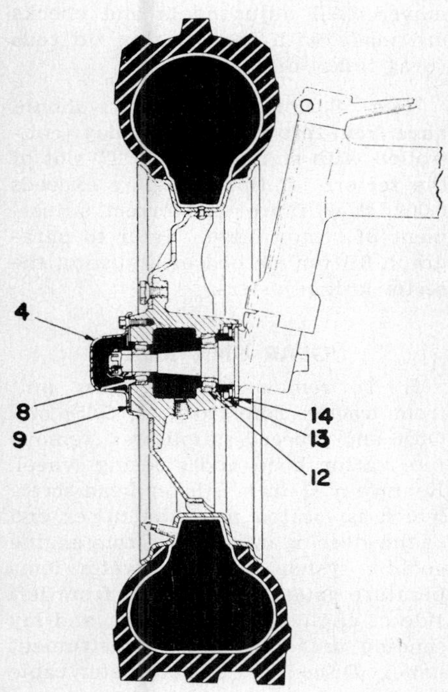

Fig. O651A—Details of front wheel hub and bearings assembly. Item (13) is oil seal; (14) is the seal wearing cup.

Fig. O653—Front axle main member (10) may be removed without disturbing the radiator. Front main frame (FF) attaches to radiator with two cap screws.

Use a sleeve and stud (½-inch NC thread) type puller to pull pivot pin out of front main frame.

AXLE SUPPORT (Front Main Frame)

5. The front axle support (FF—Fig. O653) is the tractor front main frame. To remove frame, proceed as follows: Remove front axle center member pivot pin. Raise front end of tractor and block-up under tractor center frame. Remove two cap screws attaching radiator to top of front frame. Slightly raise the radiator and support it in this raised position or remove it from the tractor. Remove four special bolts attaching engine to front main frame. Support front main frame and remove four cap screws attaching it to center housing. Attach sling and remove the frame from the tractor.

MANUAL STEERING SYSTEM

(For Data on Power Steering Refer to Oliver Appendix I)

For data on power steering refer to page 57.

The steering gear, used on the Oliver Super 55 and 550 tractors, is a Saginaw screw and recirculating ball nut type, mounted on top of the tractor center frame cover. The gear unit is provided with three adjustments which control end play of the wormshaft, end play of the lash adjuster screws, and mesh of the sectors.

GEAR UNIT ADJUST

6. All adjustments with the exception of the end play adjustment of the lash adjuster screws can be made with the gear unit installed on the tractor. End play adjustment of the lash adjuster screws can be made only after removing the sectors from the gear unit.

Before making any adjustments when gear unit is installed on the tractor, disconnect both tie rods (drag links) at the gear sector shaft pitman arms.

6A. **WORMSHAFT (SCREW) END PLAY.** Refer to Figs. O654, O655 and O656. To check and adjust wormshaft end play, proceed as follows: Loosen, both right and left side sector lash adjuster screws (37 or 37B), two full turns. Pull up and push down on steering wheel. If this check shows that end play is present, adjust to zero end play by varying the thickness of shims (29) which are located between gear housing (21), and cover and tube assembly (25).

Adjustment is correct when the pull, required to keep the steering wheel

Fig. O654—Ready for removal of steering gear unit, right side view. Lash adjuster (37B) controls the mesh of left sector (pitman) gear with right sector gear. Shims (29) control end play of worm (nut) shaft.

B. Battery support
P. Pitman arm
TR. Tie rod (drag link)
25. Wormshaft tube
29. Shims

in motion after it passes the center or mid-point of its rotation, is 3½ to 5 pounds measured at a 9-inch radius from the center.

It is not necessary to remove the instrument panel to gain access to these shims. All work, necessary either to add or remove shims, can be performed by working through the large holes (H—Fig. O656A) which are located in the battery support. On Diesel engine models, it will be necessary to remove one of the two batteries to uncover the holes in the battery support.

Shims are available in thicknesses of 0.002, 0.005, 0.010, and 0.030.

6B. **BACKLASH ADJUSTMENT.** Before adjusting sector mesh, make sure that wormshaft bearings are correctly adjusted as outlined in preceding paragraph 6A. Also, make sure that both steering tie rods (drag links) are disconnected at the sector shaft pitman arms, and that the right and left side lash adjusters (37—Fig. O655 are loosened two full turns.

Rotate steering wheel to mid or wheels straight ahead position. Using a screwdriver, rotate the sector adjuster located in side cover (37A—Fig. O656) until all backlash between worm nut and sector is removed.

After adjusting the sector which meshes directly with the ball nut, rotate the steering wheel to the mid or wheels straight ahead position. Rotate the opposite sector adjuster screw on right side of gear housing until all backlash between both sectors is removed. All adjustments and checks are made with the steering tie rods (drag links) disconnected.

Note: Sector lash adjusters should have from zero to 0.002 end play (controlled with shims) in the "T" slot of the sectors. If this end play exceeds 0.002, it will prevent correct adjustment of sector mesh. Refer to paragraph 8B for method of adjusting the sector lash adjusters.

GEAR UNIT R&R

7. To remove steering gear unit from tractor, refer to Figs. O654 and O656 and proceed as follows: Remove the tractor hood and steering wheel. Remove a ¼-inch fillister head screw which is located near the upper end of the steering tube (25). Drain engine cooling system. Remove water temperature gauge sending unit from left side of engine cylinder block, and lay sending unit and tube on instrument panel. Disconnect tachourmeter cable

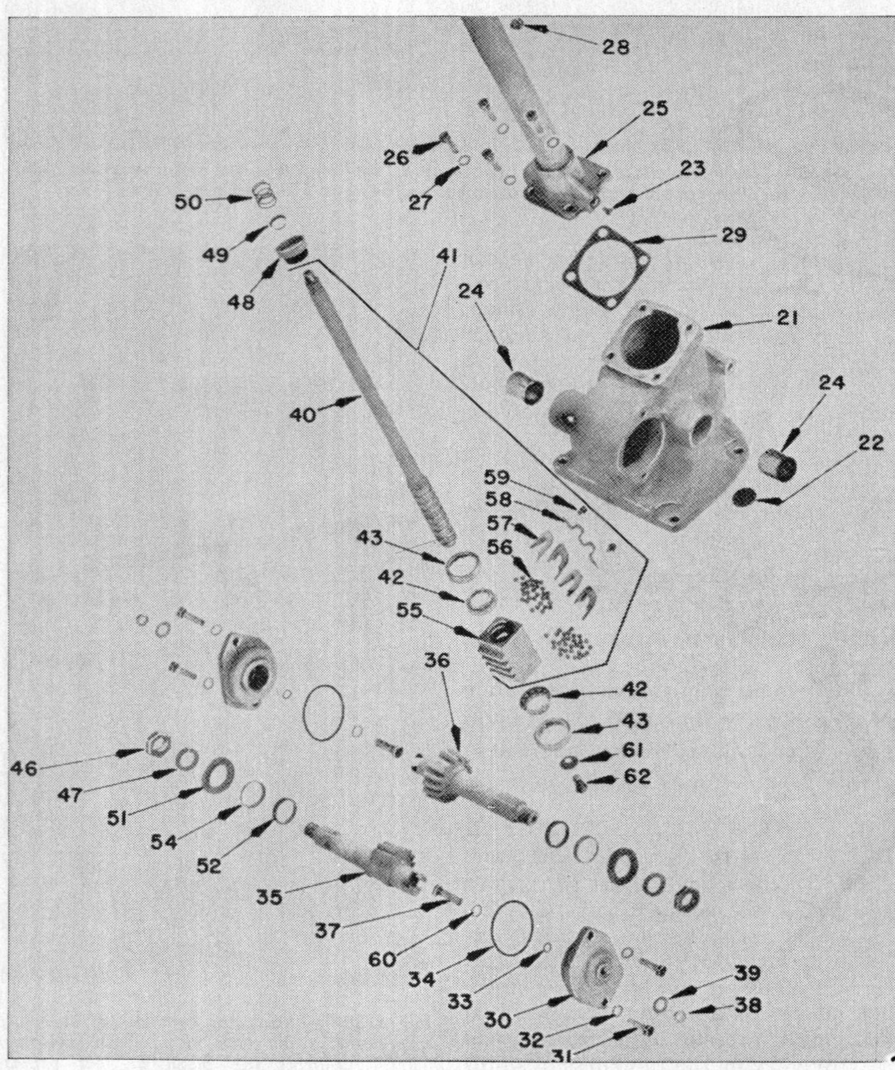

Fig. O655—Details of Saginaw recirculating ball nut type steering gear.

21. Gear housing
22. Expansion plug
23. Oil filler plug
24. Bushings
25. Wormshaft tube and cover
28. Fillister head screw
29. Shim
30. Sector cover
33. Packing
34. "O" ring
35. Sector and shaft, right
36. Sector and shaft, left
37. Lash adjuster
39. Washer
40. Wormshaft
41. Wormshaft and ball nut assembly
42. Bearing cone
43. Bearing cup
48. Upper bearing
49. Spring seat
51. Dust seal
52. Packing
54. Packing retainer
55. Ball nut
60. Shim
61. Retainer
62. Eyelet

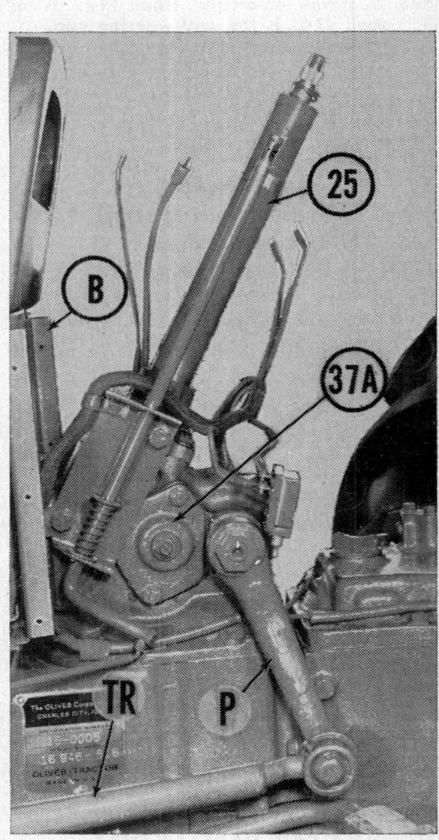

Fig. O656—Ready for removal of steering gear unit, left side view. Lash adjuster in cover (37A) controls mesh adjustment of left sector (pitman) gear with worm (nut) on wormshaft.

B. Battery support
P. Pitman arm
TR. Tie rod (drag link)
25. Wormshaft tube and cover

at the drive unit; two wires at the ammeter; oil gauge tube at the gauge; carburetor choke control wire at the carburetor on HC models only; fuel stop wire at the injection pump on Diesel models only; tail light wire at the fuse block; and governor control ball joint link at the bell crank located on left side of tractor and under the battery support. Remove light switch from the instrument panel. On HC models only, remove the ignition switch from the instrument panel.

Remove governor control hand lever from governor control operating rod. Remove three ¼-inch machine screws attaching instrument panel to hood rear support. Remove six No. 10x24 thread cutting type screws (three from each side) attaching instrument panel to battery support. Remove instrument panel by lifting same up and off of the steering tube.

Remove the sector shaft pitman arms. Remove two cap screws attaching governor control operating rod support to left side of steering gear housing. Remove four cap screws attaching gear unit housing to tractor center housing cover, and lift the gear unit from the tractor.

Fig. O656A—Shims (29) controlling wormshaft end play are accessible by working through large holes (H) in battery support.

Note: An alternate procedure for removing the rear panel assembly is given in the power steering appendix.

GEAR UNIT OVERHAUL

8. Major overhaul of gear unit necessitates the removal of the unit from the tractor as outlined in preceding paragraph 7. To disassemble the unit, proceed as follows: Refer to Fig. O655. Remove cap screws attaching both sector covers (30) to gear housing and withdraw each sector (35 or 36) and cover as an assembly. Separate the sectors from their covers by removing the lock nut from the lash adjusters, and then threading the lash adjusters out of the cover. Remove wormshaft tube and cover assembly (25) from gear housing and withdraw wormshaft and ball nut assembly (41).

8A. Ball nut should move smoothly in grooves of worm and with minimum end play. If worm shows any signs of wear, scoring or other derangement, it is advisable to renew the wormshaft and ball nut as an assembly. However, individual parts which make up the wormshaft and ball nut assembly are available. To disassemble ball nut from wormshaft, proceed as follows: Remove ball retainer clamp (58), retainers (57), and 60 steel balls. Worm ball nut (55) can be removed now from the wormshaft.

Each sector shaft cover (30—Fig. O655) contains a steel backed bronze bushing and similar type bushings are located in the gear housing for the two sectors. Wear in sector cover bushings is corrected by renewal of covers but sector bushings in housing are available separately. Suggested clearance of sector shaft in bushings is 0.0015-0.003.

8B. Select and insert shims (60) between underside of inner head of lash adjusters (37) and "T" slot in sectors to provide zero to 0.002 end play of adjusters in sectors. If the end play exceeds 0.002, it will prevent correct mesh adjustment of the sectors.

8C. To reassemble ball nut, place nut over middle section of worm as shown in Fig. O657. Drop steel balls into one retainer hole of ball nut and rotate wormshaft slowly to carry the balls away from the hole. Continue inserting balls until a total of 17 balls are in each circuit. Next, place 13 balls in each retainer guide while holding the two halves of the guide together; and plug ends of retainer guide with grease to prevent the balls from dropping out. Insert retainer guides in ball nut. Thirty balls are used in each circuit for a total of 60 balls in the ball nut.

8D. After reassembling the steering gear unit, adjust the wormshaft bearings and the sector mesh as outlined in paragraphs 6A and 6B.

Refill gear housing with 1½ pints of SAE 90 oil.

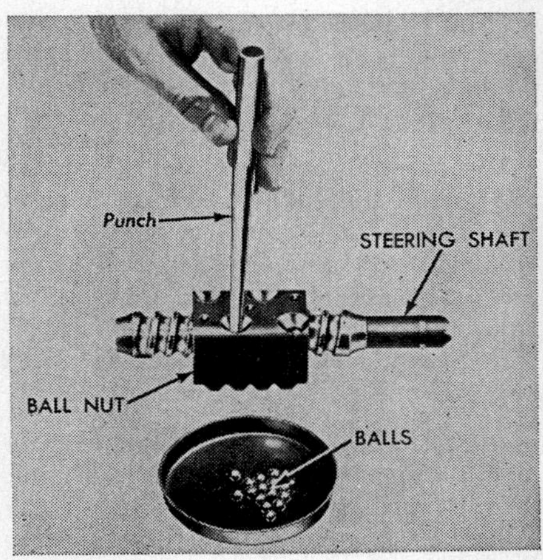

Fig. O657—Aligning ball nut on wormshaft while inserting balls in ball circuit. Insert 17 balls in each circuit, and 13 balls in each ball retainer guide.

ENGINE ASSEMBLY

REMOVE AND REINSTALL

10. To remove engine assembly from tractor chassis, proceed as outlined in paragraphs 10A through 10E. Refer to Figs. O658 and O659.

10A. FUEL TANK. Remove fuel tank as follows: Remove four bolts attaching hood to supports at front and rear. Shut off fuel supply at fuel tank and disconnect fuel line at tank. On diesel models disconnect the excess fuel return line at tank. On series 550 disconnect the oil pressure warning buzzer from its mounting. Remove cap screws attaching fuel tank to hood front and rear supports, and lift fuel tank from tractor.

10B. RADIATOR. Remove radiator as follows: Drain engine cooling system. Remove radiator shell grille pieces. Remove five machine screws and nuts attaching radiator shell to hood front support. Remove two cap screws attaching radiator shell to sides of front main frame (FF), and four cap screws attaching radiator shell to top of front main frame. Lift off radiator shell.

Remove two cap screws (one on each side) attaching radiator support brace to hood front support. Loosen radiator upper and lower hose clamps. Remove two cap screws attaching radiator to front main frame, and lift radiator from tractor.

10C. FRONT MAIN FRAME. Raise and support tractor by blocking-up under tractor center frame (CF). Support tractor front frame (FF). Remove front axle center member pivot pin locking cap screw, and remove the pivot pin. Remove four special bolts attaching engine to front main frame. Remove four cap screws attaching front main frame to tractor center frame, and then remove the front frame by lowering it to the floor.

10D. Remove battery or batteries, battery tray support, and battery radiation shield (RS). Remove link connecting starting motor switch to starter pedal cross shaft and remove starting motor. Disconnect starter pedal at the starter cross shaft rod. Disconnect tachourmeter cable at drive unit, oil pressure gage line at the cylinder block, and generator regulator wiring at the regulator. Remove two wiring harness clips from right side of cylinder block. Remove coolant temperature sending unit from cylinder block.

Disconnect ignition coil primary wire from the coil, and choke wire at the carburetor. Remove governor control rear rod from control bell crank and governor control rod. Disconnect fuel stop wire (FS) at the fuel injection pump, and manifold heater plug cable at the preheater plug.

10E. On all models support engine with a hoist by attaching an eye-hook to the special stud (LS—Fig. O661) which is located between the engine rocker arms cover and the exhaust manifold. Remove cap screws attaching engine to tractor center frame and to center frame cover (CFC).

Move engine forward and away from tractor center frame until engine clutch shaft and power take-off drive shaft are free of the engine clutch, and hoist engine away from tractor.

CYLINDER HEAD

11. **REMOVE HEAD.** Herewith is a procedure for R&R of cylinder head. Remove hood and fuel tank as outlined in paragraph 10A. Refer to Figs. O661 and O662.

Drain engine coolant, and remove engine manifold radiation shield. Remove five machine screws and nuts attaching radiator shell to hood front

Fig. O658—Right hand side of Super 55 Diesel tractor. Fuel tank is located under the hood. Front main frame is (FF), center frame is (CF), center frame cover is (CFC).

Fig. O659—Left hand side of Super 55 Diesel tractor showing fuel stop wire (FS), battery radiation shield (RS) and injection pump and lines. Refer also to Fig. O658.

Fig. O661—Cylinder head screw (OLS) which carries oil supply to rockers should be torqued to 96-100 foot pounds; other head screws should be torqued to 92-96 foot pounds on non-Diesels, 112-117 foot pounds on Diesels.

support. Remove one bolt and nut, from each side, attaching radiator support brace to hood front support. Remove hood front support by removing two bolts attaching same to engine water outlet elbow (thermostat housing). Disconnect radiator upper hose and remove exhaust muffler.

11A. On non-diesel engine models only, remove carburetor from inlet manifold, but do not remove fuel line, air cleaner hose, choke wire and governor linkage from carburetor.

11B. On diesel engine models only, disconnect injection pump to injector nozzle fuel lines at the injector nozzles. Remove fuel return line connecting the injection pump to the No. 1 cylinder injector nozzle. Attach plugs of some kind on the ends of all disconnected fuel lines and injector nozzle openings. Remove hose connecting inlet manifold to air cleaner.

11C. On all engine models, remove rocker arms cover. Remove rocker arms oil line (OL), rocker arms and shaft assembly, and push rods. Remove cylinder head hold down screws and lift off the head.

11D. **REINSTALL HEAD.** Install head gasket so that copper metal side of same contacts the cylinder block. When reinstalling cylinder head, tighten cylinder head hold down screws starting at the center and working toward the ends of the head. Torque tighten all hold down screws except cylinder head oil screw (OLS) to 92-100 ft. lb. torque for non-diesels, and 112-117 ft. lb. torque for diesels. Torque tighten cylinder head oil screw to 96-100 ft. lb.

Adjust all inlet valve tappets to 0.009-0.011 cold and exhaust valve tappets to 0.015-0.017 cold for HC; 0.019-0.021 for diesel.

ROCKER ARMS AND SHAFT

12. To remove valve rocker arms and shaft assembly, proceed as follows: Refer to Figs. O661 and O662. Remove tractor hood and fuel tank as outlined in paragraph 10A. Remove rocker arms cover, and disconnect rocker arms oil line (OL) at the fitting on the rocker arms shaft. Remove nuts retaining rocker arms shaft supports to cylinder head, and lift off rocker arms and shaft assembly.

Note: The early stamped rocker covers have been replaced, on late models, with cast iron covers. When gaskets which have shellac on one side and graphite on the opposite side are used, install same with shellac side toward rocker cover.

12B. Procedure for disassembly of rocker arms and shaft assembly is self-evident. Note that each rocker arm is stamped either "R" or "L" to identify its position on the shaft. Rocker arm bushings are not supplied for service, and bushing wear is corrected by renewal of rocker arm and bushing as an assembly.

Assemble rocker arms to the shaft, as shown in Fig. O661, and install the assembly on the cylinder head so that the oil holes in the shaft face up.

Rocker Arm Shaft Diameter	0.742 –0.743
Reject Shaft If Less Than	0.740
Rocker Arm Bushing Inside Diameter	0.7445–0.7455
Running Clearance	0.0015–0.0035
Renew If Clearance Exceeds	0.005
Shaft Supports, Tightening Torque	25-27 Ft.-Lb.

VALVES, SEATS AND GUIDES

13. Inlet and exhaust valves which seat directly in the cylinder head are not interchangeable. On latest non-diesel engines the exhaust valves are fitted with positive type rotators and seat on inserts in cylinder head.

Inlet valve stems are provided with neoprene oil seals, as shown in Fig. O663, to prevent oil from passing into the combustion chamber via the valve stems. Install new oil seals each time the valves are reseated.

13A. The cast iron, shoulder type inlet and exhaust valve guides are not interchangeable. Inlet guides are shorter in length. Service guides, requiring a press fit for installation, are prefinished and do not require final sizing.

Guide height above machined top surface of cylinder head is 9/16-inch for HC; 43/64-inch for diesel on engines prior to 72 832, or 15/16-inch for HC and diesel on engines after 72 831.

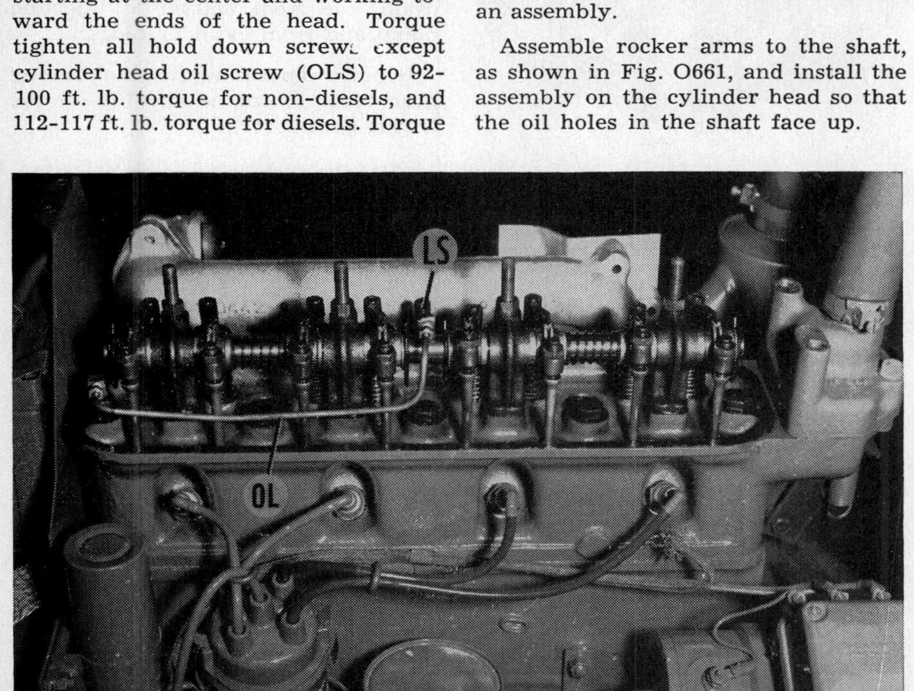

Fig. O662—The offset right and left hand valve rocker arms should be assembled on rocker shaft as shown. Lifting stud is (LS), rocker shaft oil line is (OL).

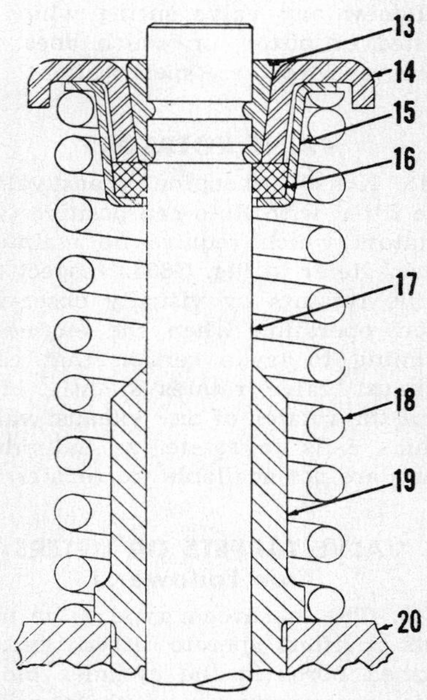

Fig. O663—Details of neoprene gasket and oil guard as installed on inlet valve stems.

13. Spring retainer lock
14. Spring retainer
15. Oil guard
16. Oil guard gasket
17. Inlet valve stem
18. Spring
19. Valve guide
20. Cylinder head

Neither the valves nor the guides for non-diesel engines are interchangeable with those used on diesel engines.

Valve Face Angle....... 44½°
Valve Seat Angle........ 45°
Inlet Valve Tappet
Gap—Cold ....0.009 –0.011
Exhaust Valve Tappet
Gap—Cold (HC) .....0.015 –0.017
(Diesel) ....0.019 –0.021
Inlet Valve Clearance
in Guide 0.0015–0.0035
Renew If Clearance
Exceeds 0.0055
Exhaust Valve Clearance
in Guide 0.0025–0.0045
Renew If Clearance
Exceeds 0.0075
Valve Lift Total.........0.360
Cam Lobe Lift..........0.250

VALVE SPRINGS

14. Inlet and exhaust valve springs used on Oliver engines not equipped with the positive type valve rotators are interchangeable.

Springs with 10½ coils (part No. 1K-198), used on valves without rotators, have a free length of $2\frac{25}{32}$ inches, and should test 44-52 lbs. @ $1\frac{15}{16}$ inches length and 67-77 lbs. @ $1\frac{19}{32}$ inches length.

Valve springs which are used on valves equipped with the positive rotator have a free length of $2\frac{5}{16}$ inches, and should test 43-47 lbs. @ 1 47/64 inches and 78-86 lbs. @ 1 25/64 inches. These springs have white paint on one end.

Renew any valve spring which is rusted or pitted, or which does not conform with these specifications.

VALVE ROTATORS

15. Non-diesel engine exhaust valves are fitted with Roto-cap positive type rotators which require no maintenance. Refer to Fig. O664. Inspect the valve rotators by visually observing their operation when the engine is running to make certain that each exhaust valve rotates slightly. Renew the rotator of any exhaust valve which fails to rotate, as individual parts are not available for repairs.

VALVE TAPPETS OR LIFTERS (Cam Followers)

16. The mushroom type valve tappets or lifters operate directly in unbushed bores in the cylinder block. It is necessary to remove the camshaft as per paragraph 19 before the valve tappets or lifters can be removed.

Valve tappets or lifters which are supplied in the standard size of 0.6240-0.6245 diameter only should have a running clearance of 0.0005-0.002 in a bore diameter of 0.625. Renew cylinder block and/or valve tappets when the clearance exceeds 0.007.

TIMING GEAR COVER

17. To remove timing gear cover, it is necessary to first remove the front main frame (FF—Fig. O653) because of insufficient removal clearance between it and the fan drive pulley.

17A. **FAN DRIVE PULLEY.** To remove the fan drive pulley, proceed as follows: Remove tractor hood. Drain engine coolant. Remove radiator shell grille pieces. Remove five machine screws and nuts attaching radiator shell to hood front support. Remove two cap screws attaching radiator shell to sides of front main frame, and four cap screws attaching radiator shell to top of front main frame. Lift off radiator shell.

Remove two cap screws (one on each side) attaching radiator support brace to hood front support. Loosen radiator upper and lower hose clamps.

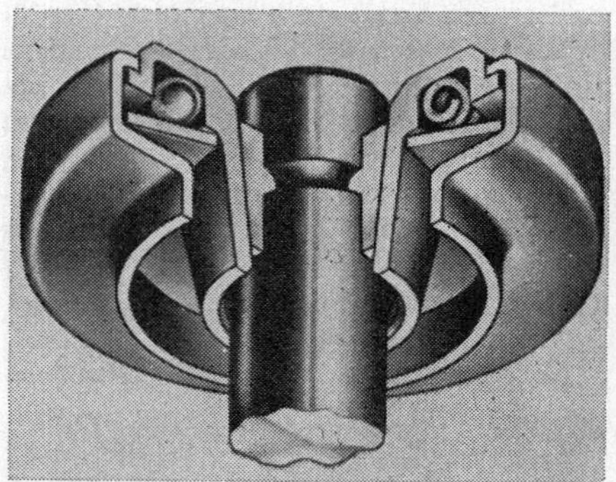

Fig. O664—Rotocap positive type valve rotators as installed on non-Diesel engines. Renew the rotator assembly of any exhaust valve which fails to rotate when engine is running.

Remove two cap screws attaching radiator to front main frame, and lift radiator from tractor.

Support tractor under tractor center frame. Support tractor front frame and remove front axle center member pivot pin locking cap screw, and remove the pivot pin. Remove four special bolts attaching engine to front main frame. Remove four cap screws attaching front main frame to tractor center frame. Remove the front main frame by lowering it to the floor.

Remove starting crank jaw. Use a puller attached to two ⅜-16 cap screws which are threaded into the pulley. The felt dirt seal for the pulley can be renewed at this time.

17B. **TIMING GEAR COVER.** After performing work of removing crankshaft fan drive pulley, as outlined in preceding paragraph 17A, proceed as follows: Remove three cap screws attaching oil pan to timing gear cover and loosen all other oil pan attaching cap screws. Remove water pump.

On non-diesel engines, remove the governor.

On diesel engines, remove the final stage fuel filter, and the outlet fuel line connecting the filter to the injection pump. Remove lubricating oil drain line from bottom of injection pump and from cylinder block.

Remove cap screws attaching timing gear cover to engine and lift off the timing gear cover.

17C. Crankshaft front oil seal (Fig. O665) can be renewed at this time. Soak new oil seal in oil before installing same.

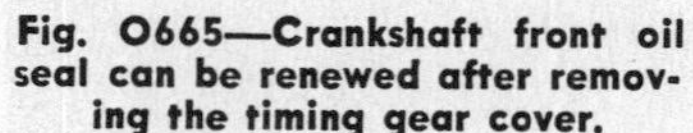

Fig. O665—Crankshaft front oil seal can be renewed after removing the timing gear cover.

30. Seal housing
32. Seal retainer
34. Crankshaft pulley hub
35. Pulley retaining nut
36. Crankshaft
37. Timing gear cover
38. Crankshaft gear
39. No. 1 main bearing cap

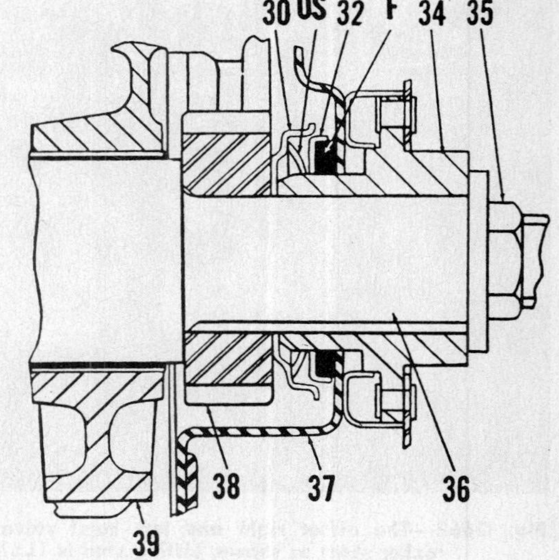

TIMING GEARS

18. Refer to Figs. O667 and O668 for views of the gear trains. Gear train on diesel engines consists of three helical gears and the injection pump gear; on non-diesel engines, two helical gears and a governor gear are used. The cylinder block for all engine models, however, is designed for the installation of an idler gear. On all non-diesel engines, check to make certain that a plug (42-Fig. O667) is installed in what would normally be the idler shaft bushing bore on the diesel engines so as to prevent oil from by-passing.

18A. Remove camshaft gear from camshaft, as shown in Fig. O669, by using a puller attached to two ⅜-16 cap screws which can be threaded into gear. Crankshaft gear can be removed in a similar manner. Avoid pulling the gears with pullers which clamp or pull on the gear teeth.

18B. The camshaft and crankshaft gears are stamped either "S" (standard), "O" (oversize), or "U" (undersize), and the amount of oversize or undersize. Recommended backlash of gears is 0.003-0.005. Renew gears when backlash exceeds 0.007.

18C. When reinstalling the cam gear, remove the oil pan and buckup the camshaft at one of the lobes near front end of shaft with a heavy bar. Heat both the camshaft gear and crankshaft gear in hot oil to facilitate installation.

Reinstall gears with "C" mark on crankshaft gear meshed with a similar mark on camshaft gear as shown in Fig. O667.

18D. The idler gear (48-Fig. O668) is installed only on diesel engines and can be removed after removing the timing gear cover. Running clearance of gear spindle in bushings should not exceed 0.004. When gear is in contact with front face of sleeve, the idler gear face should be flush with crank gear face.

If running clearance is greater than 0.004, or if flush condition is not obtained; renew the sleeve and bushings assembly as neither part is available separately. In rare cases it may be necessary to reface the thrust face of a new sleeve assembly to obtain the flush setting.

Fig. O668—Timing gear installation on Diesel engines. Mesh camshaft gear mark "C" with an identical mark on crankshaft gear. Note call-out "12" indicates Diesel injection pump outlet for No. 1 cylinder.

1. Nozzle return line
10. Timing port
38. Crankshaft gear
44. Camshaft gear
45. Thrust button
47. Injection pump drive gear
48. Idler gear
49. Thrust button
50. Crankshaft engaging ratchet
51. Location of non-diesel engine governor drive gear bushing and sleeve

CAMSHAFT

The front journal on all models rotates in a steel-backed, babbitt lined bushing. The two remaining journals rotate in machined bores in the cylinder block.

19. To remove camshaft, first remove timing gear cover as outlined in paragraphs 17 through 17B. Remove ignition distributor from non-diesel engines, rocker arms and shaft assembly, engine oil pan, and oil pump. On diesel engines, remove primary fuel pump from right side of engine. Remove rocker arm push rods, and block up or support tappets (cam followers). Carefully thread the camshaft and gear forward and out of the cylinder block bores.

19A. Camshaft lobe lift for all models is 0.250. Renew shaft if lift is less than 0.230. Shaft journal sizes are: Front, 1.7495-1.750; second and third, 1.7485-1.7495. Recommended running clearance of number one journal is 0.0015-0.003 with a maximum of 0.005; clearance for all other journals is 0.0025-0.0065 with a maximum permissible clearance of 0.007.

38. Crankshaft gear
40. Governor shaft sleeve
41. Governor shaft bushing
42. Plug (non-diesel)
43. Oil pressure relief valve
44. Camshaft gear
45. Thrust button
46. Flywheel timing port

Fig. O667—Crankshaft and camshaft gear installation on non-Diesel engines. Mesh camshaft gear mark "C" with an identical mark on crankshaft gear. Diesel engines are similar. Plug "42" is installed on all non-Diesel engines.

Fig. O669—Removing camshaft gear on non-Diesel engines by using a puller attached to two ⅜-16 bolts which can be threaded into gear. Diesel engine camshaft gear removal is similar.

When the running clearance in unbushed bores exceeds 0.007, correction can be made by renewing the camshaft and/or the cylinder block, or by reboring the bores in the cylinder block to a diameter of 1.8745-1.8755 inches and installing bushings. Presized service bushings, as supplied for number one journal, can be used and should be installed using a close fitting piloted driver.

Camshaft end play is controlled by a spring loaded thrust button. Thrust button spring has a free length of 1$\frac{3}{16}$ inches and should have a pressure of 15.5-18.5 lbs. when compressed to a height of $\frac{25}{32}$ inch.

If for any reason the camshaft gear was removed from the shaft, the gear should be pressed on the shaft before installing the shaft in the engine.

ROD AND PISTON ASSEMBLIES

20. Piston and connecting rod assemblies are removed from above after removing the cylinder head as per paragraphs 11 through 11C and the oil pan. Connecting rods are offset. Numbers 1 and 3 piston and rod assemblies should be installed with the long part of the rod journal bearing facing toward the flywheel. Numbers 2 and 4 piston and rod assemblies should be installed with the long part of the rod journal bearing toward timing gears.

Rod bolts should be torqued to 46-50 foot pounds.

PISTONS, SLEEVES AND RINGS

21. **PISTONS.** Pistons of aluminum alloy, are cam ground. Bare pistons are not catalogued, but pistons with rings, piston pin, piston pin retainers and sleeve are available as a set for one or more cylinders. Desired piston skirt clearance varies as follows: If cast-in number on piston is 180704, desired clearance is 3-6 lbs. pull using a 0.0015x½-inch feeler. On all other engines, it is 3-6 lbs. pull using a 0.002x ½-inch feeler. Wear limit of pistons and sleeves is when a 0.006x½-inch feeler gage requires less than a 5-10 lbs. pull on a spring scale to withdraw it.

22. **SLEEVES.** The wet type cylinder sleeves should be renewed when any of the following conditions exist: Taper, 0.008; out-of-round, 0.002; wear, 0.010.

Before installing a new sleeve, clean the cylinder block sealing surfaces; then, install the sleeve without any sealing rings so as to check the relationship of the sleeve flange to the cylinder block.

The top surface of the sleeve flange should extend 0.001-0.004 above the top surface of the cylinder block. If standout (protrusion) is in excess of 0.004, check for foreign material under sleeve flange. Excessive standout will cause water leakage at cylinder head gasket.

After checking the sleeve standout, install the two neoprene sealing rings on the sleeve. Use a white lead paste on the sealing rings to facilitate installation of the sleeve and to help seal the sealing rings.

23. **RINGS.** There are three compression rings and one oil control ring per piston. Rings are supplied in standard size only. The three compression rings are taper faced and should be installed so that the largest diameter of the ring faces toward the bottom of the piston.

Recommended end gap for all compression rings is: 0.010-0.020 with a reject value of 0.045; end gap for the oil control ring is 0.010-0.018 with a reject value of 0.045 for service. Recommended side clearance of all rings in grooves is 0.0015-0.003 with a reject value of 0.006.

PISTON PINS AND BUSHINGS

24. The 1.2494-1.2495 diameter full floating type piston pin is retained in piston bosses by snap rings. Piston pins are available in the standard size and oversizes of 0.005 and 0.010.

The split type graphite bronze piston pin bushings should be installed in the rod so that bushing outer edge is flush with outer edge of rod bore and the split side is at the top of the rod. Two bushings are installed in each rod. Note: Late type steel-backed bushings are interchangeable with the early bronze bushings providing they are installed in pairs.

Bushings should be sized after installation to provide the piston pin with a .0005-.001 clearance, and the piston bosses sized to provide the piston pin with a .0002-.0004 clearance.

CONNECTING RODS AND BEARINGS

25. Rod bearings are renewable from below without removing rods from engine. Diesel and HC connecting rod forgings are interchangeable. All engines after serial 961990 have flame hardened crankshafts riding in precision type shells micro lined with lead-tin and an intermediate layer of copper. These copper lead bearings should be used only with hardened shafts which have the number 181411 stamped on the shaft.

Bearings are supplied in undersizes of 0.003 and 0.020 for hardened shafts; 0.003, 0.030 and 0.033 for unhardened shafts.

Check crankpins and bearings for wear, scoring and out-of-round.

Crankpin Diameter........2.249-2.250
Running Clearance....... .0005-.0015
Renew if Clearance Exceeds.... .0025
Side Clearance........... .0075-.0135
Rod Length, C to C.............6.750
Rod Bolt Torque, Ft. Lbs........46-50

CRANKSHAFT AND BEARINGS

26. All engines after serial 961990 have flame (Tocco) hardened crankshaft journals and pins and ride in precision type bearing shells micro lined with lead-tin and an intermediate layer of copper. These shafts can be identified by the number 181411 stamped on the shaft.

The HC engines prior to above serial number have plain (unhardened) crankshafts and only the old style precision lead bearing shells should be used with these shafts. Unhardened shafts can be identified by the number 180211 stamped on the shaft.

Bearings are supplied in undersizes of 0.003 and 0.020 for hardened shafts; 0.003, 0.030 and 0.033 for unhardened shafts.

26A. To remove crankshaft from engine, proceed as follows: Remove engine from tractor as outlined in paragraphs 10A through 10E. Remove oil pan, timing gear case cover, engine clutch, flywheel, and main bearing caps. On diesel engines only, remove front plate from cylinder block. Unbolt connecting rod bearing caps and lift out the crankshaft.

When reinstalling crankshaft, mesh "C" mark on crankshaft gear with a similar mark on camshaft gear.

27. All main bearings shells can be renewed from below without removing the crankshaft.

Check main bearings and crankshaft journals for wear, scoring, end play, and out-of-round, using the tabulated data in paragraph 27A.

27A. Crankshaft and main bearing data:

Main Journal Diameter....2.249-2.250
Permissible Out-of-Round ..... .0005
Crankpin Diameter.......2.249-2.250
Journal Running
Clearance .... .0005-.0035
Renew if Clearance
Exceeds .... .005
Crankpin Running
Clearance .... .0005-.0015
Renew if Clearance
Exceeds .... .0025
End Play .... .004-.009
End Play Control.........No. 2 Brg.
Main Bearing Cap Screw
Torque—Ft.-Lbs. 88-92
Rod Bolt Torque—Ft.-Lbs. 46-50

CRANKSHAFT OIL SEALS

29. **FRONT OIL SEAL.** Crankshaft front oil seal Fig. O665 is contained in the timing gear cover. It can be renewed after removing the timing gear cover as outlined in paragraphs 17A through 17C.

30. **REAR OIL SEAL.** Crankshaft one piece rear oil seal of treated cork can be renewed after performing a tractor split as outlined in paragraph 81A, and removing the engine clutch and flywheel. Refer to Fig. O672.

Soak new seal in oil before installing same.

FLYWHEEL AND GEAR

31. **FLYWHEEL.** Flywheel can be removed after splitting the tractor as in paragraph 81A and removing the clutch.

Flywheel runout, checked at rear outer face of wheel, should not exceed .005 total indicator reading. Flywheel retaining cap screws should be torqued to 67-69 foot pounds.
same to 450 deg. F. and install gear with beveled end of teeth facing the engine.

32. **TIMING MARKS.** Flywheel timing marks are viewed through an inspection port, located on the right side of the engine. Non-diesel engine
To install a new ring gear, heat

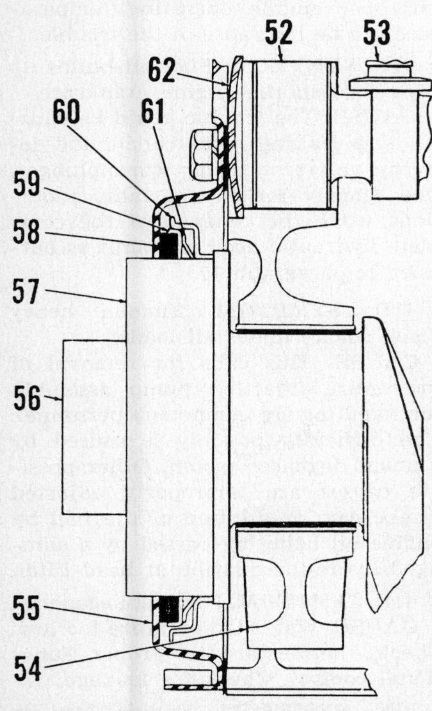

Fig. O672—Cross-sectional view showing the crankshaft rear oil seal which is made of specially treated cork.

52. Camshaft
53. Cam follower
54. Rear main bearing cap
55. Oil retainer
56. Crankshaft
58. Cork seal
59. Seal retainer
60. Seal assembly retainer
61. Spring
62. Expansion plug

flywheels are stamped as follows: TDC indicating top center position of number one piston; INT indicating inlet valve opens; and IGN-HC indicating fully advanced ignition timing (28 degrees before TDC).

32A. Diesel engine flywheels are stamped with the fuel injection pump timing mark "FP" in addition to the TDC and INT marks used on the flywheels of non-Diesel engines. Early

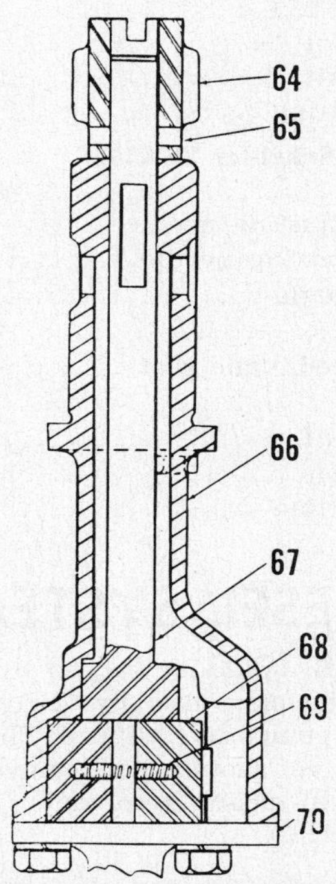

Fig. O673—Vane type engine oil pump. Install vanes so that the flat sides will face in the direction of rotation which is counter-clockwise when viewed from the vane end.

64. Drive gear
65. Retaining pin
66. Pump body
67. Drive shaft
68. Vanes
69. Spring
70. Cover

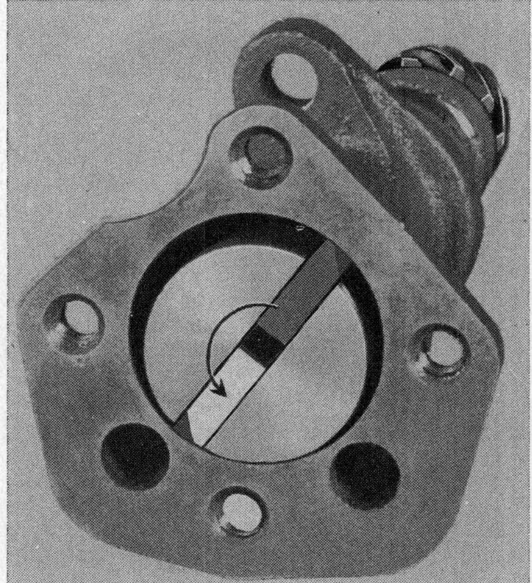

Fig. O674—Oil pump vanes should be installed so that the flat sides will face in the direction of rotation (counter-clockwise).

production Diesel engines were equipped with No. MS135A flywheel which has the "FP" mark located 23 degrees or 2½ inches before TDC. Later production tractors after No. 8391500 are equiped with No. LS135C flywheel on which the "FP" mark is located 26 degrees before TDC.

The 26 degree timing is recom mended for Super 55 Diesels therefore early flywheels should be checked for mark location. If "FP" mark is $2\frac{27}{32}$ inches ahead of TDC it is OK for 26 degree timing. If "FP" is 2½ inches ahead of TDC mark deface it and make a new one at the $2\frac{27}{32}$-inch location or $\frac{11}{32}$-inch away from the original "FP" mark.

OIL PUMP AND RELIEF VALVE

33A. **PUMP.** The vane pump, Fig. O673, which is driven by the camshaft, is mounted on the underside of the cylinder block. Pump removal requires removal of oil pan and one cap screw attaching the pump body to the cylinder block. On non-Diesel engines, it will be necessary to retime the ignition unit whenever the oil pump is removed, as the ignition unit drive is supplied through the oil pump drive shaft.

33B. Disassembly is self-evident. Vanes should be installed to oil pump drive shaft so that flat sides of vanes will be facing the direction of normal rotation when viewed from the vane end, as shown in Fig. O674. The pump and cover are assembled without a gasket. If the pump becomes worn, do not attempt to repair it; renew the entire unit.

33C. Before installing the oil pump to non-Diesel engines, rotate the engine crankshaft until the number one piston is on compression stroke and the flywheel mark "TDC" is indexed at the inspection port. Install the oil pump so that the narrow side of the pump shaft, as divided by the slot (ignition unit drive slot) is on the crankshaft side and parallel to the crankshaft. Refer to Fig. O675 which

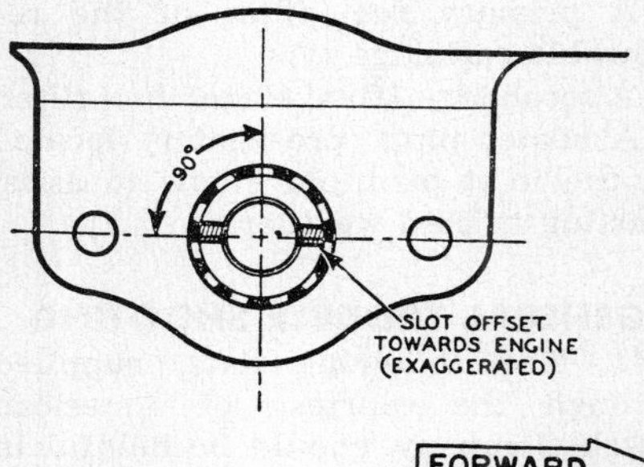

Fig. O675—Correct position of the ignition unit drive slot when viewed from the ignition unit mounting pad surface.

shows the correct position of the ignition unit drive slot when viewed through the ignition unit shaft hole in the cylinder block. If the drive slot is not in the position as shown, remove the oil pump and remesh the pump drive gear. Reinstall the ignition unit and check the timing.

33D. **RELIEF VALVE.** The non-adjustable piston type oil pressure relief valve is located externally on right side of engine in vicinity of timing gear cover gasket surface. Recommended operating pressure is 15 psi at a crankshaft speed of 1600 rpm.

Early type relief valve spring has a free length of $1\frac{9}{16}$ inches and should test 2¼-2¾ lbs. at a length of $1\frac{1}{16}$ inches.

Late type relief valve spring has a free length of 2 inches and should test 5½-C½ lbs. at a length of 1-inch.

CARBURETOR

34. The following carburetors have been used; Carter UT2257S; Marvel-Schebler TSX603 and TSX775 and Zenith 12427.

Calibration and manufacturers repair part numbers are as follows:

Marvel-Schebler TSX603

Part	Number
Idle jet	49-345
Main adjusting needle	43-627
Idle adjusting needle	43-33
Main nozzle	47-589
Venturi	46-A38
Float needle and seat	233-536
Float	30-600
Gasket kit	16-592
Repair kit	286-1067
Float setting	*¼-inch

Marvel-Schebler TSX775

Part	Number
Idle jet	49-101-L
Main adjusting needle	43-277
Idle adjusting needle	43-33
Main nozzle	A7-A24
Venturi	46-A125
Float needle and seat	233-543
Float	30-739
Gasket kit	16-654
Repair kit	286-1252
Float setting	*¼-inch

Carter UT2257S

Part	Number
Idle jet	11B-236
Main adjusting needle (jet)	159-165S
Idle adjusting needle	30A-37
Main nozzle	12-439S
Venturi	58-91
Float needle and seat	25-233S
Float	21-121S
Gasket kit	259
Repair kit	1808
Float setting	*17/64-inch

Zenith 12427

Part	Number
Idle jet	C55-22-10
Main adjusting needle	C71-54
Idle adjusting needle	C46-25
Main nozzle	C66-109-55
Venturi	B38-73-23
Float needle and seat	C81-1-40
Float	C85-115
Gasket kit	C181-325
Repair kit	K12427
Float setting	*$1\frac{5}{32}$-inch

*Measured from gasket surface of throttle body to nearest face of float on all models except Zenith which is measured from farthest face of float to gasket surface of throttle body.

DIESEL SYSTEM

40. Main components of the Diesel engine fuel system are: An American Bosch single plunger, constant stroke, sleeve control, type PSB fuel injection pump. Injection pump is driven at crankshaft speed by idler gear which meshes with the crankshaft gear.

American Bosch or British CAV (Lucas) closed pintle type fuel injectors. These are interchangeable as assemblies.

Pre-combustion chambers (energy cells) located in the combustion chambers of the cylinder head.

A flyweight type governor which is used to control the fuel delivery as a function of speed control is an integral part of the injection pump.

A camshaft operated automotive, diaphragm type primary fuel supply pump is used to furnish fuel to the inlet side of the injection pump.

A primary fuel filter of the renewable cartridge type.

A secondary (final stage) fuel filter.

A heater plug (pre-heater) located in the inlet manifold elbow to assist starting in cold weather.

GENERAL TROUBLE SHOOTING

41. The following data, supplied through the courtesy of American Bosch Company, should be helpful in shooting trouble on the Oliver Super 55 and 550 Diesel tractor.

41A. **SYMPTOM.** Engine does not idle well; erratic fluctuations.

CAUSE. Could be caused by faulty fuel injectors, also by dirty overflow on pumps so equipped. The overflow valve should be removed and washed in cleaning solvent.

Fig. O676—Showing bleed screw (BS2) on final stage filter. Refer to paragraph 42 for bleeding procedure.

41B. **SYMPTOM.** Intermittent or continuous puffs of black smoke from exhaust.

CAUSE. Faulty fuel injector, also improper engine operating temperature can be the cause of the trouble.

41C. **SYMPTOM.** Fuel oil builds up (dilution) in the engine crankcase.

CAUSE. The trouble could be caused by a leaking gasket under the delivery valve, or badly worn plunger. The remedy for any of these conditions would be renewal of the complete hydraulic head as a unit as outlined in paragraph 45.

41D. **SYMPTOM.** Sudden heavy black smoke under all loads.

CAUSE. This calls for removal of the entire injection pump assembly for handling by competent personnel. The difficulty possibly is caused by a stuck displacer piston. Other possible causes are improperly adjusted smoke cam or dilution of the fuel by engine oil being by-passed by a damaged hydraulic distributor head filter.

41E. **SYMPTOM.** Poor fuel economy.

CAUSE. Water temperature too low. Check thermostat for proper functional control. Check for leakage.

41F. **SYMPTOM.** Engine low in power.

CAUSE. Filter between supply pump and injection pump may be clogged, or fuel supply pump is faulty. Due to type of fuel used, it may be necessary to advance the timing. Under no circumstances should the timing be advanced more than 4 degrees.

41G. **SYMPTOM.** Engine rpm too low at full throttle position.

CAUSE. Could be caused by improper setting of the throttle linkage. Remove pump control lever cover and check if full travel is obtained at full load position of throttle control lever.

BLEEDING THE SYSTEM

42. Refer to Figs. O676 and O677. The diesel fuel system should be bled or purged whenever the system has been disconnected or fuel tank emptied. To bleed the low pressure side of the system (fuel tank to injection pump), proceed as follows: Loosen bleed screws (BS1) and (BS2) located on the primary fuel filter and the final stage filter. Operate priming lever (PH) on fuel supply pump with full strokes until clear fuel (free of air bubbles) flows past fuel filter bleed screw on the primary filter; then close the bleed screw. Repeat this procedure on the secondary filter.

To bleed the high pressure system, loosen the fuel line connections at the fuel injectors; then using the starting motor, rotate engine until clear fuel (free of air bubbles) flows past the fuel injector connections.

INJECTION PUMP

42A. The Oliver and Bosch identifications of the pumps that have been used on Super 55 and 550 are as follows: Bosch PSB4A70Y-3797 or A1 is Oliver 1LS1701C; Bosch PSB4A70V-S4527-A1 is Oliver 101406AS; Bosch PSB4A70VS4801A is Oliver 101406-ASA. Pumps are single plunger, sleeve control type and are driven at crankshaft speed.

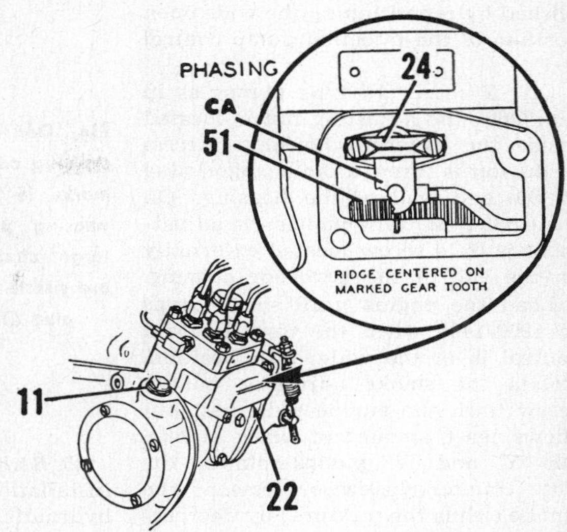

Fig. O678—View of injection pump control unit assembly (24) as seen with timing window (22) removed. Refer to text for checking procedure.

43. **TROUBLE-SHOOTING.** The following data, paragraphs 43A and 43B, should be helpful in shooting trouble on the injection pump.

43A. FAULTY ACCELERATION. If the engine fails to accelerate or to respond to the throttle, the trouble may be caused by excess friction in the pump control unit assembly (24—Fig. O678). To check for excess friction in the pump control unit assembly, proceed as follows: With engine running ,remove timing window (22) from side of pump and observe action of the pump control arm (CA) which is connected to the governor control rod.

If arm is sticking or moving erratically, stop the engine and disconnect the governor control rod. When arm is manually moved to either extreme of its movement, it should drop to neutral by its own weight. If movement is sticky, remove the pump control unit assembly by removing two (usually wired) attaching screws. Wash the control unit in an approved solvent. If the treatment does not produce free movement, disassemble the unit and clean the bearing surfaces by lapping with mutton tallow. Make sure that pump is equipped with latest control unit which may be identified by its use of hexagon nut to attach the control lever to the shaft.

If control unit arm swings freely when it is disconnected from the governor control rod, check the governor rod for full travel and smooth operation. If binding is encountered, remove governor housing from end of injection pump and clean or renew the parts as required to remove wear or eliminate sticking.

43B. HARD STARTING, POOR PERFORMANCE. Before condemning the pump assembly as the cause of these troubles, check condition of pump plunger as follows: Remove hydraulic head assembly as outlined in paragraph 45. Unload plunger spring (57—Fig. O684) and remove split cone locks (55). Pry off gear retainer cover (59) and remove the plunger and control sleeve. Plunger, control sleeve and hydraulic head are matched parts.

Inspect these parts for scuff marks, scratches and dull appearance of the lapped surfaces. A dull appearance indicates considerable wear. If any of these conditions exist, install a new hydraulic head assembly as per paragraph 45.

Install control sleeve with slot facing towards nameplate side of pump head; then carefully install the plunger.

44A. **SMOKE ADJUSTMENT.** This setting is considered as an emergency measure and when needed is accom-

Fig. O677 — Showing bleed screw (BS1) on first stage (primary) filter. Priming handle on fuel pump is (PH). Refer to paragraph 42 for bleeding procedure.

plished by repositioning the wide open position of the injection pump control rod.

On "Y" and "V" type pumps as in Fig. O689 the adjusting nut is located inside the governor housing. Access to the nut is through the hexagon plug in the end wall of the housing. On the earlier "Z" type pumps the adjustment is by a screw located externally on the end of the governor housing.

Load the engine until speed drops to 1200-1400 when the throttle hand control is in the wide open position. Rotate the smoke adjusting nut or screw until the engine pulls best but shows least amount of black smoke. On "Y" and "V" pumps rotating the nut counter-clockwise lessens the smoke (leans the mixture) by decreasing the amount of fuel delivered. On "Z" type pumps the effect is reversed and rotating the screw counter-clockwise (richens the mixture) increases the amount of smoke.

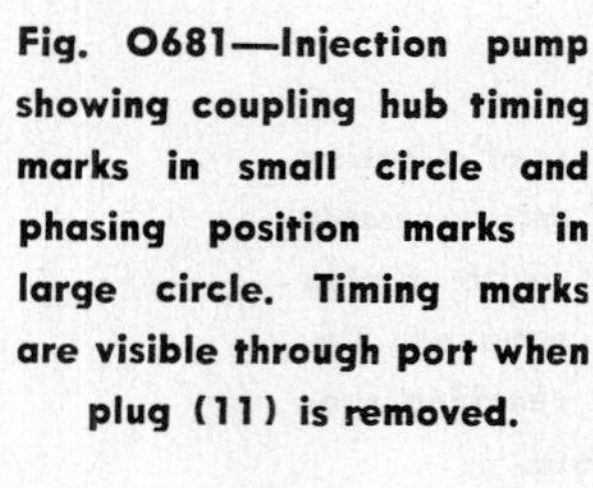

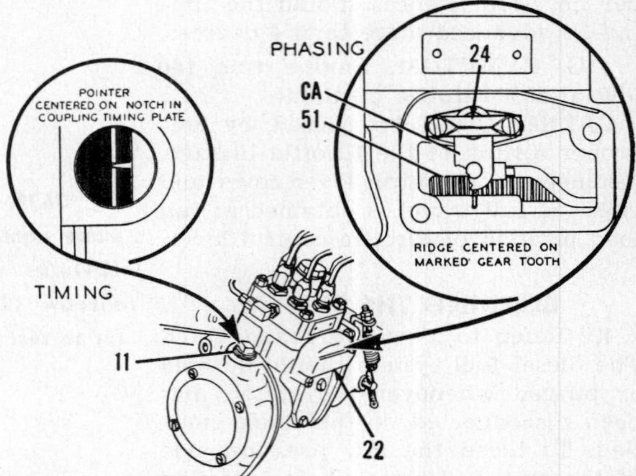

Fig. O681—Injection pump showing coupling hub timing marks in small circle and phasing position marks in large circle. Timing marks are visible through port when plug (11) is removed.

45. **R&R HYDRAULIC HEAD UNIT.** Installation of a new or exchange hydraulic head unit is sometimes the indicated remedy when injection pump trouble is encountered. To remove the hydraulic head, proceed as follows: Remove pump timing window (22—Fig. O681) and rotate the engine until the marked tooth of plunger drive gear (51) is approximately in register either with stamped arrow head mark or stamped "O" mark. Remove pump control unit (24), and lubricating oil filter (38—Fig. O679). Remove fuel lines, and nuts

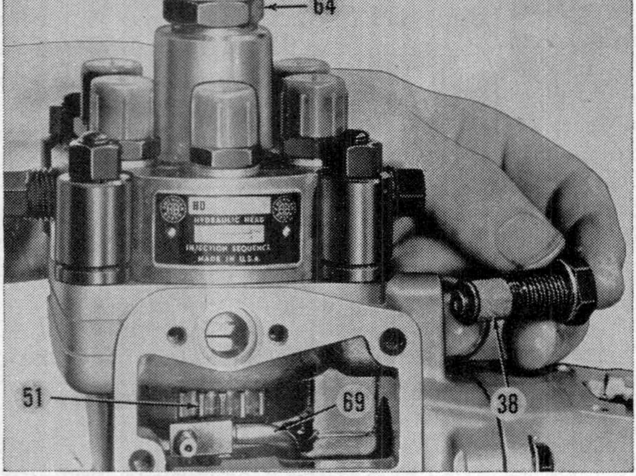

Fig. O679—Removing lubricating oil filter (38) preparatory to removing the injection pump hydraulic head assembly as shown in Fig. O680. Refer to paragraph 45 for procedure.

Fig. 0680 — Removing hydraulic head unit from Bosch PSB type injection pump. Marked tooth of gear (51) must be registered with index mark on window ledge before gears will unmesh. Refer to paragraph 45 for procedure.

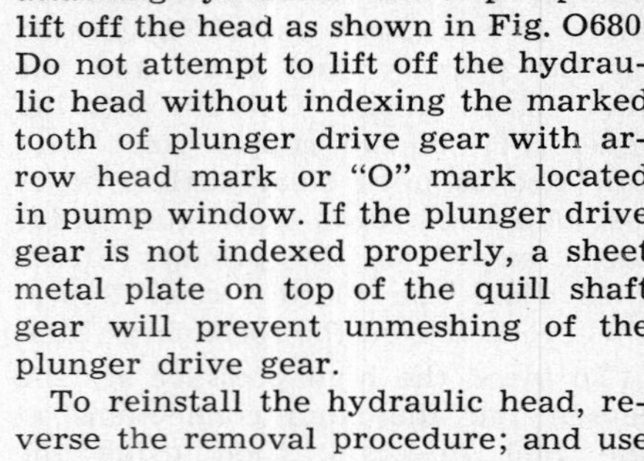

attaching hydraulic head to pump and lift off the head as shown in Fig. O680. Do not attempt to lift off the hydraulic head without indexing the marked tooth of plunger drive gear with arrow head mark or "O" mark located in pump window. If the plunger drive gear is not indexed properly, a sheet metal plate on top of the quill shaft gear will prevent unmeshing of the plunger drive gear.

To reinstall the hydraulic head, reverse the removal procedure; and use new "O" rings.

46. **TIMING PUMP TO ENGINE.** Injection pump should be timed to engine so that closing of pump plunger port for number one outlet occurs when engine flywheel mark "FP" is indexed with flywheel inspection port notch. When it is known that the pump internal timing is OK, the pump can be timed to the engine as follows:

46A. Remove inspection port plug (11—Fig. O681) located in the timing gear cover directly above the pump drive gear. Also, remove the timing window cover (22) from side of pump.

Rotate engine crankshaft until number one cylinder is coming up on compression stroke; then slowly, until flywheel mark "FP" (See note below) is aligned with flywheel housing inspection port notch.

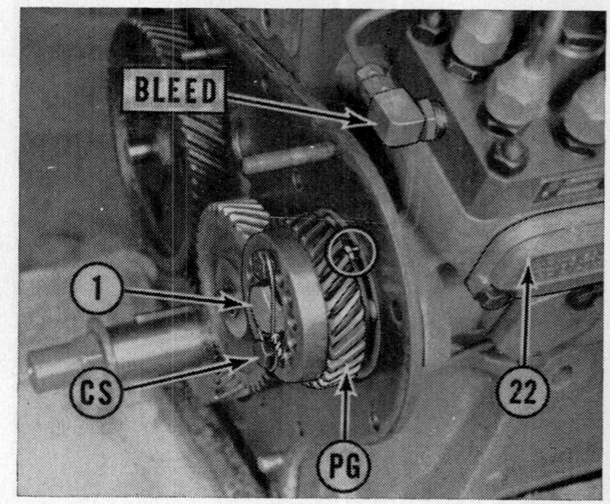

Fig. O682—View of injection pump and drive with timing gear cover removed. Position of pump gear (PG) on pump shaft may be changed without removing the timing gear cover after removing small auxiliary cover from timing gear case cover. Refer to paragraph 46B for procedure. Pump timing pointer is shown in circle. Note bleed line.

NOTE:

On early production engines, "FP" mark is 23 degrees or 2½ inches before TC. After serial 8391500 the "FP" mark is 26 degrees or 2 27/32 inches before TC. All engines should be timed 26 degrees before TC and old flywheels should be re-marked accordingly as outlined in paragraph 32.

NOTE: Compression stroke of No. 1 cylinder can be determined either by removing the pipe plug and cap from the No. 1 cylinder energy cell, or by removing the valve rocker arms cover and observing the closing of the No. 1 cylinder inlet valve.

At this time, the line mark on the injection pump coupling hub should be in register within $\frac{1}{32}$-inch with the timing pointer which extends from the front face of the pump. Refer to left circle in Fig. O681. Both the line mark and timing pointer are viewed through the inspection port plug opening (11). If injection pump coupling hub mark is not in register with the timing pointer, as stated, it will be necessary to remesh the pump gear as outlined in paragraph 46B.

46B. REMESHING PUMP GEAR. To remesh pump gear for retiming

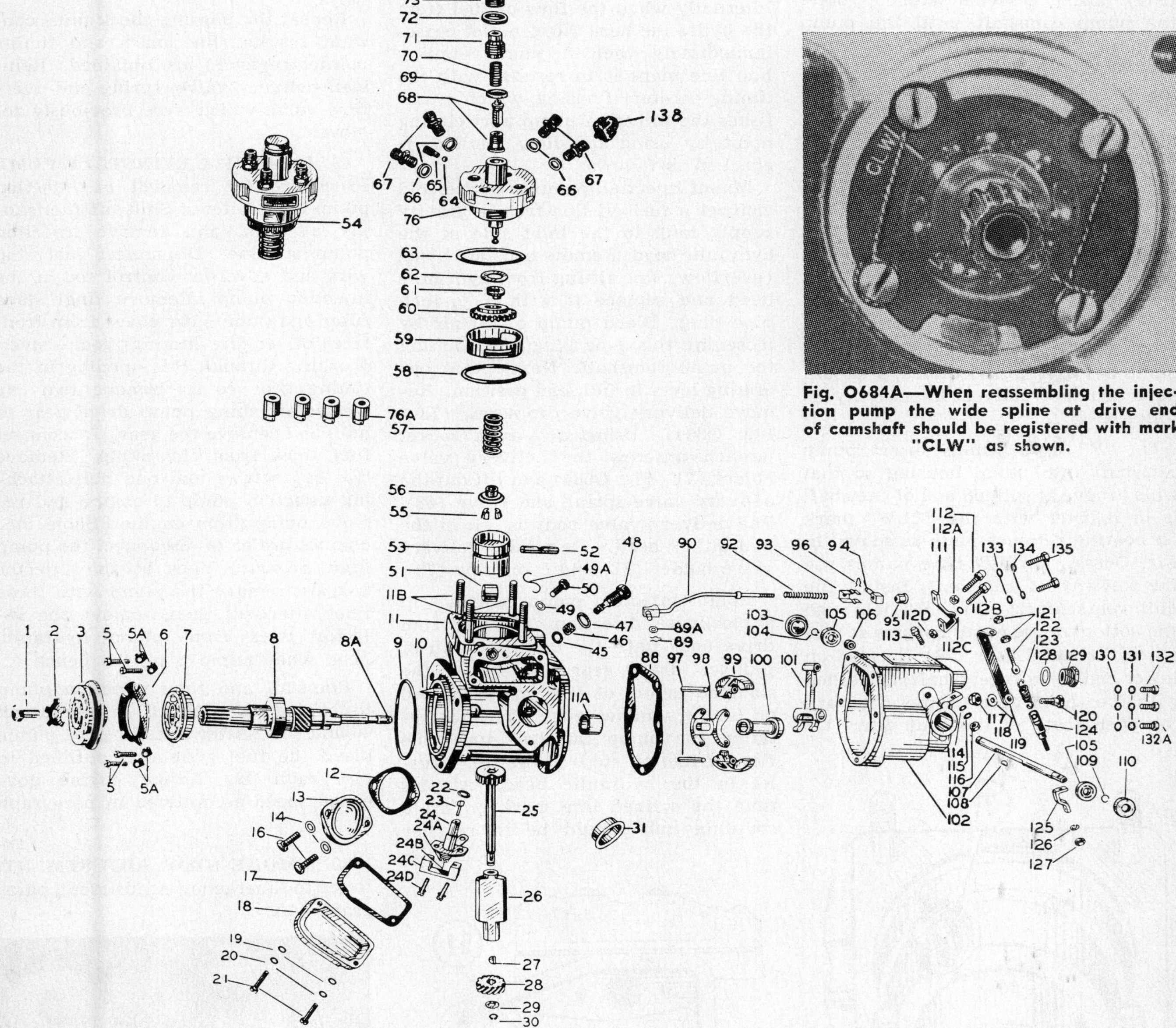

Fig. O684A—When reassembling the injection pump the wide spline at drive end of camshaft should be registered with mark "CLW" as shown.

Fig. O684—Disassembled view of American Bosch PSB4A70Y3797 injection pump used on early Super 55 Diesel tractor. Latest pumps are similar except that camshaft (8) and hub (3) are tapered instead of splined.

3. Drive hub
6. Retainer
7. Camshaft ball bearing
8. Camshaft
9. Rubber ring
10. Timing pointer
11. Pump housing
11A. Camshaft bushing
18. Timing window cover
24. Pump control unit
24C. Retainer for item 24
25. Quill shaft and gear
26. Quill shaft bushing
28. Camshaft driven gear
31. Closing plug
46. Filter screw
51. Tappet roller
53. Tappet guide
54. Hydraulic head assembly
55. Plunger lock
56. Spring lower seat
57. Plunger spring
59. Gear retainer cover
60. Plunger drive gear
61. Plunger guide
64. Screw sealing ball
67. Discharge fitting
68. Delivery valve
70. Delivery valve spring
71. Delivery valve holder
76. Hydraulic head
90. Control rod
92. Control rod spring
93. Retainer clip
94. Control rod pin
95. Load adjusting nut
98. Governor weight spider
99. Governor weight
100. Sleeve and brg.
101. Fulcrum lever
102. Governor housing
105. Bearing for shaft 107
107. Fulcrum lever shaft
110. Seal for shaft 107
111. Stop plate
119. Operating lever
122. Spring adjusting screw
123. Extension spring for 119
124. Shut off screw for 119
125. Shut off plate

pump to engine, proceed as follows: First position engine crankshaft so that flywheel mark "FP" is indexed with flywheel housing inspection port notch when No. 1 cylinder is on compression stroke. Remove secondary or final stage fuel filter, and pump gear cover which are located on front face of engine timing gear cover. Working through pump gear cover opening, remove two cap screws (CS—Fig. O682) retaining pump drive gear (PG). Using a socket wrench, rotate the pump camshaft until the pump coupling hub line mark registers with the timing pointer; then reinstall pump drive gear retaining cap screws. Several trials by remeshing the pump drive gear with the idler gear may be necessary before finding the holes which will admit the cap screws without throwing the hub line mark out of register with the timing pointer.

Recheck injection pump to engine timing and lock the pump drive gear cap screws with wire.

47. **PUMP INTERNAL TIMING.** If pump has been disassembled, it should be timed internally at reassembly as follows:

On "Y" type pumps insert pump camshaft into pump housing so that wide groove at splined end of camshaft is in register with the "CLW" mark on bearing retainer plate as shown in Fig. O684A. On "V" type pumps use the keyway of camshaft. Install the quill gearshaft (25—Fig. O684) through the bottom of the pump housing so that when the spiral gear (28) (located on lower end of quill gearshaft) is meshed with the spiral gear on the camshaft, the open tooth of the spur gear (located at upper end of quill gearshaft) will be in register with the drill mark which is located on the counterbore of the pump housing as shown in Fig. O685. Now, refer to Fig. O686 and install the hydraulic head assembly so that the line marked tooth of the plunger drive gear is in register with the stamped arrow or "O" mark which is located on the timing window ledge.

48. **PUMP INTERNAL PHASING.** Injection pump is correctly phased internally when the flow of fuel from the hydraulic head No. 1 outlet ceases immediately when the pump coupling hub line mark is in register with the timing pointer. Phasing, which establishes the injection pump port closing point by using the flow method, is checked as follows:

Mount injection pump in a vise, and connect a fuel oil line from a gravity supply tank to the inlet side of the hydraulic head. Remove constant bleed (overflow) line fitting from hydraulic head and replace it with a ¼-inch pipe plug. Bleed pump of all air by loosening this pipe plug and rotating the pump camshaft. Next, place operating lever in full load position. Remove delivery valve cap screw (73—Fig. O684). Using a $\frac{7}{16}$-inch socket wrench unscrew the delivery valve holder (71—Fig. O684) and lift out the delivery valve spring and valve (68). The delivery valve body is left in the hydraulic head. Reinstall delivery valve holder (71) and cap screw (73).

Rotate injection pump camshaft in a clockwise direction (viewed from drive end) until the marked tooth of plunger drive gear approaches the stamped arrow or "O" mark located on timing window ledge. Continue rotating the pump camshaft until the flow of fuel oil stops at the No. 1 outlet in the hydraulic head. At this time the scribed line mark on drive coupling hub should be in register within $\frac{1}{32}$-inch with the timing pointer, and the marked plunger gear tooth will be about one tooth to the right of the stamped arrow or "O" mark located on timing window ledge.

If the drive coupling hub line mark and timing gear pointer are out of register more than ⅜-inch, the pump is assembled incorrectly. If the out-of-register is ⅛-inch or less, remove the old line mark and affix a new line mark on the drive coupling hub.

Repeat the phasing check until constant results (line mark and timing pointer register) are obtained. Reinstall delivery valve, spring and overflow valve which were previously removed.

49. **REMOVE & REINSTALL PUMP** Procedure for removal of injection pump is as follows: Shut-off fuel supply at tank, and remove injection pump oil line. Disconnect fuel stop wire and governor control rod at the injection pump. Remove final stage filter and pump gear cover from front face of engine timing gear cover. Working through this opening in the timing gear cover, remove two cap screws attaching pump drive gear to hub, and remove the gear. Disconnect fuel lines from the pump. Remove two cap screws and one bolt attaching injection pump to engine and remove pump from engine. Some mechanics prefer to disconnect the pump high pressure lines at the injector nozzles; remove the pump with these lines attached; then, remove the injector lines from pump hydraulic head when pump is on the bench.

Reinstall and time injection pump to engine as outlined in paragraph 46 and 46B. After installing the pump, bleed the fuel system as outlined in paragraph 42. Adjust engine governed speed as outlined in paragraph 52.

50. **SMOKE STOP ADJUSTMENT.** Refer to Emergency Adjustment, paragraph 44A.

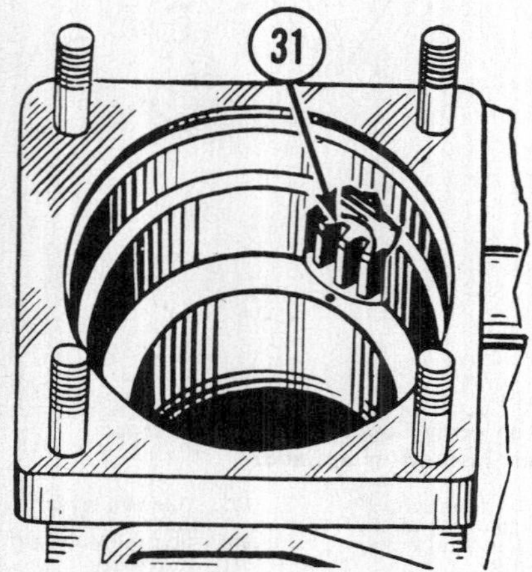

Fig. O685 — When assembling injection pump, the center open tooth of gear at top of quill shaft (31) should be registered with mark on body as shown, when pump camshaft is positioned as shown in Fig. O684A.

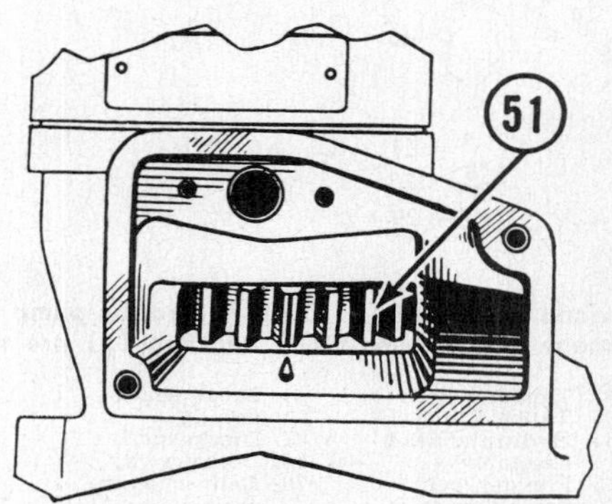

Fig. O686—Pump is phased to fire number one cylinder when the marked tooth of plunger drive gear (51) is approximately in register with the "O" mark or arrow on timing window ledge as shown.

Fig. O688—When pin "P" is removed from control arm, the control arm shaft should rotate freely in shaft bearings. This is important.

INJECTION PUMP GOVERNOR

The diesel engine governor (flyweight type) is an integral part of the injection pump.

51. **LINKAGE FREEDOM.** The necessity for free movement of the governor linkage extends to the pump delivery control unit, located on the injection pump under the timing window. Refer to Figs. O678 and O688. To check freedom of the pump control unit, remove pin (P) from control unit arm. If arm shaft is tight in its sleeve, remove the pump control unit and free-up or renew the unit as outlined in paragraph 43A.

52. **GOVERNOR ADJUSTMENT.** Engine idle no-load speed of 675 crankshaft rpm (221-238 PTO rpm or 429-462 belt pulley rpm) is adjusted by means of screw (101—Fig. O689) located on the right side of the injection pump housing. Rotate the adjusting screw in or out to decrease or increase the engine idle speed.

The high idle engine speed of 2200 crankshaft rpm (749 PTO rpm) is adjusted by varying the tension of governor spring (91) with adjusting nuts (90).

FUEL INJECTORS

Users of the I&T SHOP SERVICE are warned in the individual manuals to play safe when servicing fuel injectors. The extreme pressure of the injector nozzle spray is dangerous and can cause the fuel to penetrate the human flesh. Avoid this source of danger when checking the injector nozzles by directing the spray away from your person.

55. **LOCATING A FAULTY INJECTOR.** If one engine cylinder is misfiring, it is reasonable to suspect a faulty fuel injector. Generally, a faulty injector can be located by loosening the high pressure line fitting on each injector in turn; thereby preventing fuel from entering the combustion chamber. As in checking spark plugs in a spark ignition engine, the faulty injector is the one which, when its fuel line is loosened, least affects the running of the engine. Remove the suspected injector from the engine as outlined in paragraph 56; then reconnect the fuel line to the injector. With the discharge end of the injector directed where it will do no harm, crank the engine and observe the spray pattern as shown in Fig. O690.

If the spray pattern is ragged, it is likely that the injector nozzle is the cause of the misfiring. To check the diagnosis install a new or rebuilt injector or an injector from a cylinder which is firing regularly. If the cylinder fires regularly with the other injector, the condemned injector should be serviced as outlined in paragraphs 57 through 57E.

56. **R&R INJECTOR UNITS.** Before loosening any lines, wash all connections with fuel oil. Injector removal procedure which varies according to the cylinder location, is as follows:

Number 1 injector unit can be removed after disconnecting all injector nozzle fuel lines and lifting the fuel line harness assembly off the engine.

Number 2 injector unit can be removed after removing the two clamps from the fuel line harness and springing slightly the number 3 injector nozzle high pressure fuel line for removal clearance.

Number 3 injector unit can be removed after removing the number 1 injector nozzle high pressure fuel line and removing two clamps from the fuel line harness.

Number 4 injector unit can be removed after disconnecting the leak-off line and the high pressure fuel line.

After disconnecting the high pressure and leak-off fuel lines, cover open ends of lines and pump with composition caps to prevent entrance of dirt. Remove injector unit holder body stud nuts and carefully withdraw the injector unit from cylinder head.

56A. Thoroughly clean the injector recess in the cylinder head before reinstalling the injector. It is important

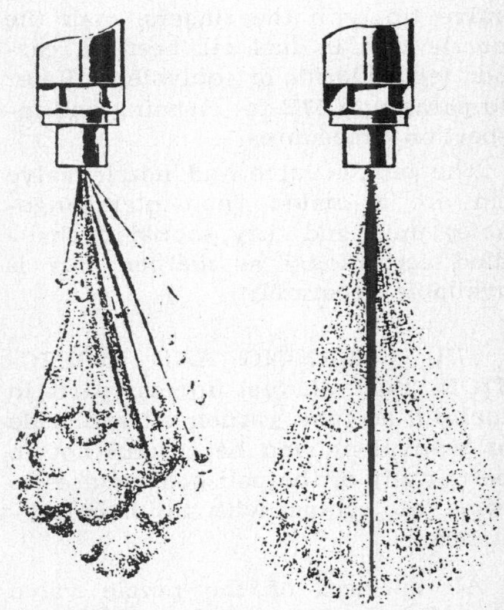

Fig. O690 — Spray patterns of a standard pintle type nozzle. Left: A poor spray pattern. Right: Ideal spray pattern.

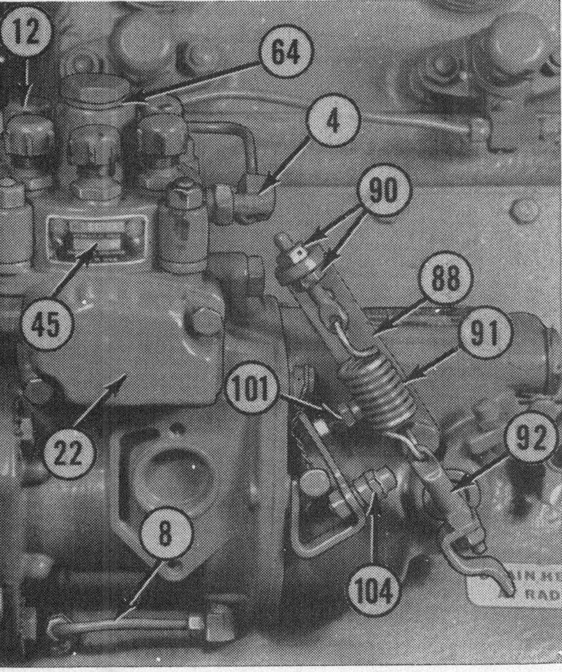

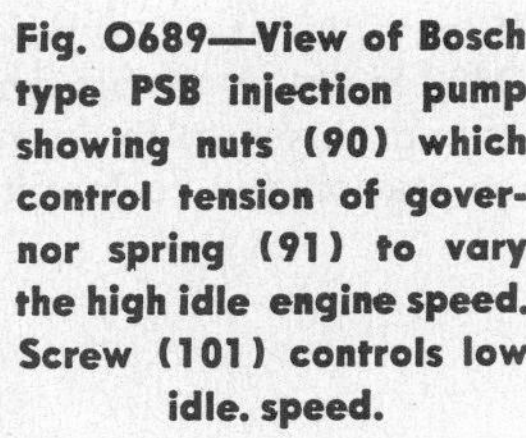

Fig. O689—View of Bosch type PSB injection pump showing nuts (90) which control tension of governor spring (91) to vary the high idle engine speed. Screw (101) controls low idle. speed.

8. Oil line (lubricating)
12. Outlets to nozzles
22. Timing window cover
45. Hydraulic head
64. Delivery valve cap
88. Operating lever
101. Low idle screw
104. Shut-off spacer

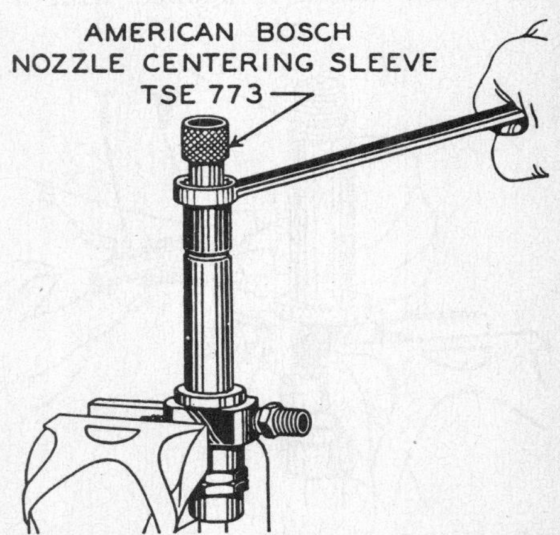

Fig. O691—Using a Bosch tool No. TSE-773 to center the nozzle tip in the cap nut.

that the seating surfaces of the recess be free of even the smallest particle of carbon which could cause the injector to be cocked and result in blow-by of hot gases. No hard or sharp tools should be used for cleaning. Bosch recommends the use of a wooden dowel or brass bar stock which can be shaped for effective cleaning. Do not re-use the copper ring gasket; install a new one. Tighten the nozzle holder stud nuts to 14-16 foot pounds torque.

57. **SERVICING FUEL INJECTORS.** Hard or sharp tools, emery cloth, crocus cloth, grinding compounds or abrasives of any kind should **NEVER** be used in the cleaning of fuel injectors.

57A. DISASSEMBLY OF INJECTOR NOZZLE. Carefully clamp the nozzle holder body in a vise, and remove the nozzle cap nut as shown in Fig. O691. Remove the injector nozzle unit consisting of the valve (V—Fig. O692) and the tip (T). If nozzle valve cannot be readily withdrawn from nozzle valve tip with the fingers, soak the nozzle unit in fuel oil, acetone, carbon tetrachloride or equivalent. Refer to paragraph 57B for cleaning and inspection procedures.

The nozzle valve and nozzle valve tip are a mated (non-interchangeable) unit, and they should be handled accordingly as neither part is available separately.

57B. CLEANING AND INSPECTION. Soak all fuel injector parts in fuel oil, acetone ,carbon tetrachloride or equivalent, and be careful not to permit any of the polished surfaces to come into contact with any hard substance.

All surfaces of the nozzle valve should be mirror bright except the contact line of the beveled seating surface. Polish the valve with mutton tallow which is applied with a soft cloth or felt pad. The valve may be held by its stem in a revolving chuck during the polishing operation. A piece of soft wood well soaked in oil, or a brass wire brush will be helpful in removing carbon from the valve. If the valve shows any dull spots on its sliding surfaces, or if a magnifying glass inspection shows any nicks, scratches or etching, discard the valve and the nozzle valve tip or send them to an authorized service station for possible overhaul.

The inside surface of the valve tip can be cleaned with a piece of soft wood which is formed to a point that corresponds to the angle of the valve seat. The wood should be well soaked in oil. Some Bosch mechanics use an ignition distributor felt oiling wick instead of the soft wood for cleaning the seat in the valve tip. A Delco-Remy distributor oiling wick, part DR804076, is suitable for this purpose. Shape the end of the wick to conform to the seat angle, and coat the formed end with tallow for polishing.

The orifice at the end of the valve tip can be cleaned with a wood splinter. Outer surfaces of the nozzle should be cleaned with a brass wire brush and a soft cloth soaked in carbon solvent. If there is any erosion or loss of metal at or adjacent to the valve tip orifice, discard both the valve tip and valve unit.

57C. Clean the exterior of the nozzle holder, except the lapped sealing surface, with a brass wire brush and carbon solvent.

The sealing surface of the valve tip, where it contacts the nozzle holder body, must be flat and shiny. Remove any discoloration on these lapped surfaces by re-lapping with fine compound on a lapping plate using a figure 8 motion as shown in Fig. O693. If equipment of type shown in Fig. O693 is not available send nozzle holder and valve tip to an authorized service station for re-lapping.

Clean deposits from nozzle holder body cap nut with a brass wire brush and carbon solvent. Shallow irregularities which would provide a possible leakage path across the gasket contacting surface of cap nut can be removed by reduction lapping, using emery cloth laid on a flat surface. If the irregularities cannot be corrected in a reasonably short time, renew the cap nut.

57D. DISASSEMBLY OF NOZZLE HOLDER BODY. If the shop is not equipped with fuel injector tester, disassembly of the nozzle holder body should not be attempted. If a tester is available, proceed as follows: Remove the protection cap, spring retaining cap nut, pressure adjusting spring, spring upper seat, and lower spring seat and spindle.

Carefully check the lower spring seat and spindle. Renew the spindle if it is bent, or if its end which contacts the nozzle valve is cracked or otherwise damaged, or if the spring seat shows any signs of wear.

Carefully check the pressure adjusting spring. Renew the spring if it is rusted, pitted or shows any surface cracks.

57E. REASSEMBLY OF INJECTOR. After all of the fuel injector parts have been cleaned and checked, lay all of them in a pan containing clean fuel oil. Withdraw the parts as needed and assemble them while they are still wet. Observe the following points before and during assembly:

Nozzle valve should have a minimum clearance free fit in the valve tip. If valve is raised approximately ⅜ of an inch off its seat it should be free enough to slide down to its seat without aid when the assembly is held at a 45 degree angle.

If parts have been properly cleaned and polished, all sliding and sealing surfaces should have a mirror finish.

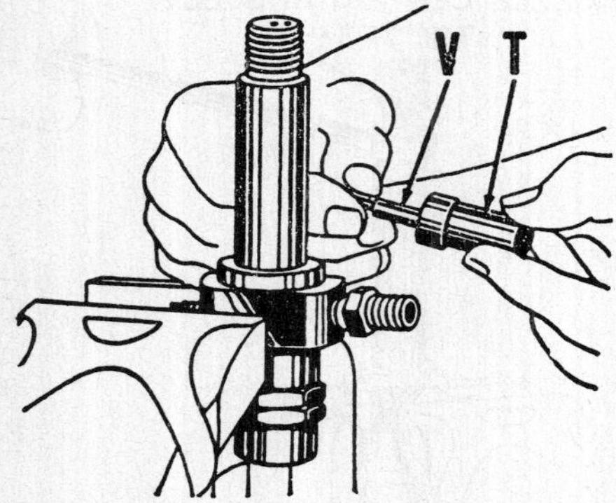

Fig. O692—Removing Oliver Diesel engine injector nozzle tip (T) and the valve (V).

Fig. O693—When cleaning lapped flat surfaces with tallow on lapping plate, use the "figure eight" motion as shown.

Renew any part which does not meet this condition, or send the mating pieces to an authorized service station for possible reconditioning.

It is desirable that the nozzle valve tip be perfectly centered in the nozzle valve tip cap nut. A centering sleeve, American Bosch tool TSE773, is available and should be used for this purpose as shown in Fig. O691. Nozzle valve tip cap nut should be torqued to 50-55 foot pounds.

Adjust the fuel injector opening pressure and check the condition of the fuel injector as outlined in paragraphs 58A through 58C.

58. **TEST AND ADJUST FUEL INJECTORS.** The job of testing and adjusting the fuel injector requires the use of an approved nozzle tester. The fuel injector should be tested for leakage, spray pattern and opening pressure.

58A. OPENING PRESSURE. Recommended fuel injector opening pressure of 1750 psi can be obtained by rotating the pressure adjusting screw. If a new pressure adjusting spring has been installed, adjust the opening pressure to 1770 psi or 20 psi higher to compensate for subsequent spring set.

While operating the injector tester handle, observe the gage pressure when the valve opens. The gage pressure drop when the valve opens should not exceed 300 psi. A greater pressure drop usually indicates a sticking valve.

58B. LEAKAGE. To check the fuel injector for valve leakage actuate the tester handle slowly, and as the gage needle approaches 1750 psi observe the valve tip orifice for drops of fuel. If drops of fuel are observed at pressures less than 1750 psi, the nozzle valve is not seating properly.

A slight overflow of fuel at the leak off port is normal. If the overflow is greater than that of a new fuel injector it indicates excessive wear between the cylindrical surfaces of the nozzle valve and the valve tip.

58C. SPRAY PATTERN. Operate the tester handle slowly until nozzle valve opens, then stroke the lever with light quick strokes at the rate of approximately 100 strokes per minute. The correct spray pattern is shown in Fig. O690. If the spray pattern is ragged disassemble the nozzle valve tip and carefully check and recondition the parts as required.

DIESEL ENERGY CELLS

60. These assemblies are mounted directly opposite from the fuel injectors in the cylinder head, as shown in Fig. O695. Oliver Corp. catalogs the cell body with orifice insert and cell cap separately. It is suggested, however, that both of these parts be renewed if either one requires renewal.

In almost every instance where a carbon-fouled or burned energy cell is encountered, the cause is traceable either to a malfunctioning injector, incorrect fuel or incorrect installation of the energy cell. Manifestations of a fouled or burned unit are misfiring, exhaust smoke, loss of power or pronounced detonation (knock).

60A. **REMOVAL.** Any energy cell can be removed without removing any engine component. To remove an energy cell, first remove the threaded cell holder plug (97) and take out the cell holder spacer or retainer (98). With a pair of thin nosed pliers, remove cell cap (99). To remove the cell body (100) screw a $\frac{15}{16}$-20 NF bolt into the threaded end of the cell body. A nut and collar on the bolt will make it function as a puller. If no puller is available, remove the fuel injector and use a brass rod to drift the cell body out of the cylinder head.

60B. **CLEANING.** Clean all carbon from front and rear crater of cell body using a brass scraper or a shaped

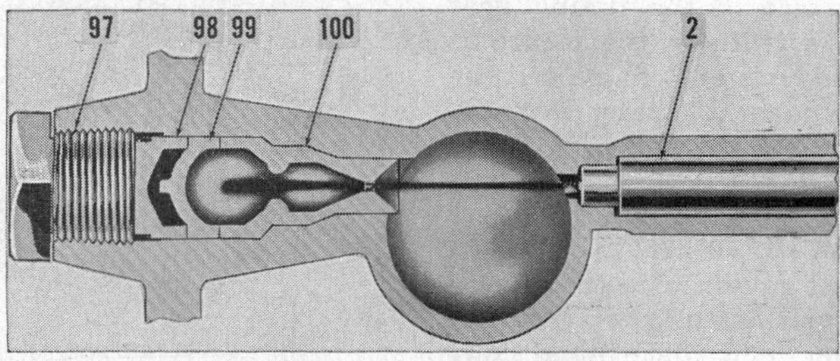

Fig. O695—Energy cells (pre-combustion chambers) one for each cylinder, are located in cylinder head facing the injectors from opposite side of head.

2. Nozzle 97. Retaining plug 98. Retainer 99. Cell cap 100. Cell body

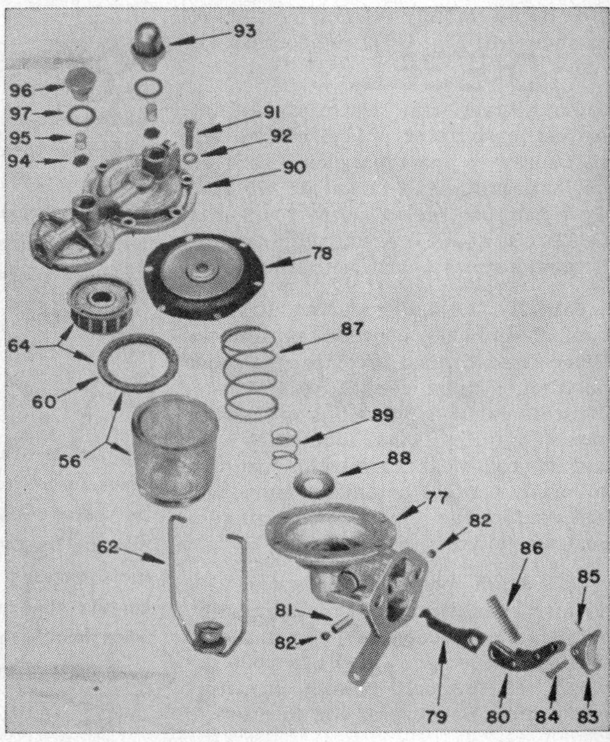

Fig. O695A—Components of AC automotive type fuel supply pump used on Diesels.

56. Bowl and gasket
62. Bale
64. Strainer and gasket
77. Body and primer
78. Diaphragm assembly
79. Link
80. Rocker arm
83. Rocker arm shoe
86. Rocker arm spring
87. Diaphragm spring
89. Diaphragm spring
90. Top cover
93. Air dome
94. Pump valve
95. Valve spring

piece of hard wood. Clean the exterior of the energy cell with a brass wire brush, and soak the parts in carbon solvent. Reject any part which shows signs of leakage or burning. If parts are not burned re-lap sealing surfaces (99 and 100) by using a figure 8 motion on a lapping plate coated with fine lapping compound.

STARTING AID (PREHEATER)

61. Diesel engines are equipped with separate inlet and exhaust manifolds. To aid in starting in cold weather a heating coil located in the inlet manifold elbow is used to preheat the air.

Operation of the preheater can be checked by depressing the control switch (which is located on the instrument panel) for approximately 30 seconds; then place your hand on the base of the preheater to see if it is warm.

PRIMARY FUEL SUPPLY PUMP

62. Diesel engines are equipped with an AC automotive, diaphragm type primary fuel supply pump, Fig. O695A, mounted on the right side of the engine. The pump which is equipped with a hand operated priming lever for use in bleeding the low pressure side of the fuel system is actuated by the engine camshaft. A satisfactory pump will show a 11-15 psi gage reading when checked at the outlet side.

62A. Removal and disassembly of the fuel supply pump is self-evident after an examination of the unit. To eliminate the possibility of overstretching the fabric material, when installing a new diaphragm, place the priming lever in a perpendicular (parallel with pump mounting pad surface) position, when tightening the cover retaining screws.

NON-DIESEL GOVERNOR

On non-diesel engines the engine speed governor, Fig. O696, is mounted on the front face of the timing gear cover and is driven by the camshaft gear. For adjustment data on the diesel engine governor, refer to paragraph 52.

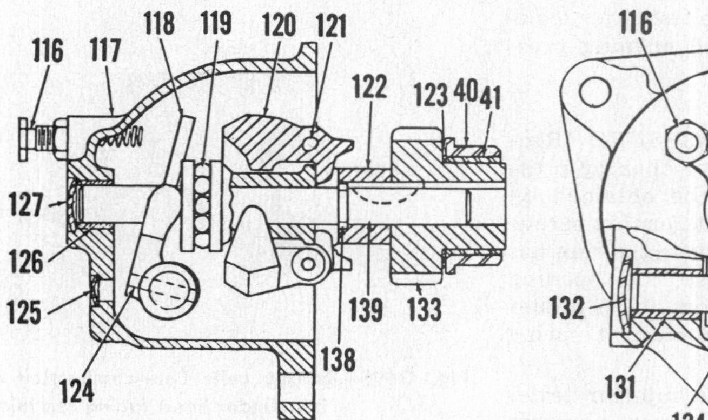

Fig. O696—Sectional view of speed governor assembly used on non-Diesel engines. Sleeve (40) and bushing (41) are located in the front face of the cylinder block.

ADJUSTMENT

65. To adjust governor to carburetor linkage, proceed as follows: Remove the air cleaner, and place hand control lever in the full speed position. Disconnect carburetor to governor rod from governor operating lever. Place carburetor throttle valve in wide-open position and the governor unit lever in full forward position. Adjust length of governor to carburetor rod to provide approxmately $\frac{1}{16}$-inch over-travel as shown in Fig. O697. Reconnect the rod.

65A. Start and warm-up engine. Adjust carburetor mixture and idle stop screw to provide a closed throttle, no-load crankshaft speed of 375 rpm. Turn bumper spring screw (116—Fig. O697) out about 4 revolutions or until it breaks contact with actuating lever.

65B. To adjust governor to hand control linkage, proceed as follows: Place hand control lever in full speed position. Adjust length of governor control rear rod by means of the two nuts (GN—Fig. O653) located at front end of rod near carburetor, so that governor control actuating lever will just contact the governor control support as shown in Fig. O698.

65C. High speed adjustment of 2200 no-load crankshaft rpm is obtained by varying the tension of governor control spring shown in Fig. O699 by means of the self-locking adjusting screw nut. Because of the location of

Fig. O697 — Carburetor to governor rod length should be adjusted to provide 1/16-inch over-travel (should be 1/16 too long) to assure full opening of throttle valve. Refer to paragraph 65 for procedure.

Fig. O698—Length of governor control rear rod should be adjusted by means of nuts at its forward end so that actuating lever will just contact the control support when hand control is in full speed position.

Fig. O699—High idle governed speed is controlled on non-Diesel engines by the self locking adjusting nut called "adjusting screw lock nut" above. Refer to paragraph 65C for procedure.

the governor spring adjusting screw and nut, it is advisable to stop the engine when changing the adjustment. Rotate the adjusting screw nut clockwise to increase the speed; counterclockwise to decrease the speed.

65D. After adjusting the high idle no-load speed, rotate bumper spring screw inward to eliminate any tendency for engine to surge but not enough to increase the engine speed. If surging persists make sure that bumper spring is of correct $\frac{7}{16}$-$\frac{1}{2}$-inch length.

A tendency for the hand throttle to creep can be corrected by putting more friction on the control linkage bell crank assembly. Bell crank is pivoted on a $\frac{3}{8}$-inch cap screw with interposed discs of friction material.

R&R AND OVERHAUL

66. To remove the governor from the engine, proceed as follows: Remove air cleaner assembly. Unhook governor control spring from governor operating lever. Disconnect governor to carburetor rod at the governor. Remove 3 cap screws attaching governor to front face of engine timing gear cover, and carefully withdraw the governor unit so as to prevent the loss of the drive gear bronze thrust washer.

66A. Governor drive gear (133—Fig. O696) is a press fit and is keyed to the governor shaft (139). Gear can be removed with the use of a suitable puller. Governor weight carrier (122) is a press fit and keyed to the shaft, and it should be removed by pressing on the gear end of the shaft. Disassembly of the weight unit is self-evident after an examination. Governor shaft bushing (126), located in governor housing, requires final sizing (0.4375-0.4385 inside diameter) after installation to provide a diametral clearance of 0.0015-0.002.

Governor gear hub rotates in a babbitt bushing (41—Fig. O700) and sleeve assembly which is pressed into the front face of the cylinder block. To renew the bushing and sleeve assembly, it will be necessary to remove the timing gear cover, and use a suitable puller to remove same as shown in Fig. O700. Renew bushing when running clearance exceeds 0.007. Pre-sized service bushings have a bore diameter of 1.002, and should be installed with a piloted drift. Note: Governors used on tractors after serial number 72 831 differ somewhat in detail but these changes are obvious upon examination.

Fig. O700—Using a puller to remove the governor gear sleeve (40) and bushing (41) assembly. Diesel engine idler gear shaft bore (42) is plugged on non-Diesel engines.

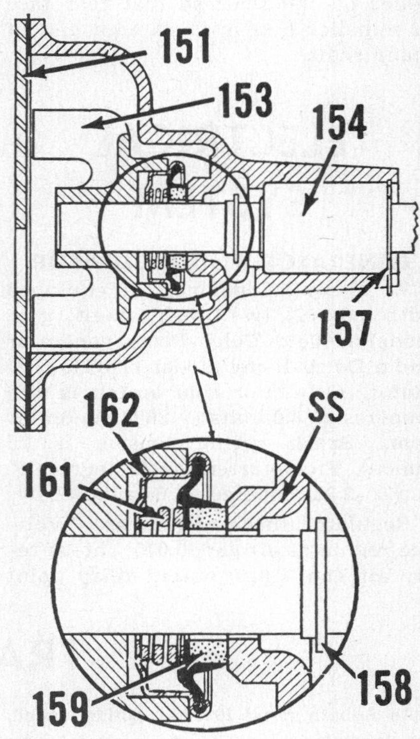

Fig. O701—Sectional view of coolant pump. Refer to paragraph 73A for overhaul procedure.

COOLING SYSTEM

RADIATOR

70. A pressure cap is used on later radiators. To remove the radiator, proceed as follows: Drain engine coolant and remove tractor hood. Remove five machine screws attaching radiator shell to hood front support. Remove the grille pieces and two cap screws attaching radiator shell to sides of tractor front frame. Remove four cap screws attaching shell to top of tractor front frame, and lift off the shell. Unbolt radiator support braces from radiator. Disconnect radiator hoses. Unbolt radiator support from tractor front frame and lift off the radiator .

THERMOSTAT

71. Thermostat, which is rated at 175 degrees F., is located in engine water outlet elbow, located on front upper surface of cylinder head. To renew the thermostat, remove hood, fuel tank, hood front support and engine water outlet elbow.

FAN BELT AND BLADES

72. Fan belt can be renewed without removing the radiator or fan blades. Fan blades assembly can be renewed without removing the radiator or water pump.

WATER PUMP

73. To remove water pump, proceed as follows: Drain engine coolant and loosen pump to radiator hose clamps. Remove fan blades, fan spacer, and pulley. Remove fan belt and four cap screws attaching pump to cylinder

block and lift off pump from right hand side of tractor.

73A. To disassemble pump, proceed as follows: Remove pump cover (151—Fig. O701). Remove snap ring (157) from front of water pump housing. Press shaft and bearing assembly (154) toward fan blades end of pump and out of impeller (153) and pump housing. Pump seal assembly can be renewed at this time. When renewing seal, be sure seal is the type with exposed spring.

Check condition of seal seat (SS) which is integral with pump body. If seat is rough or scored, renew pump body or resurface the seal seat. The shaft and prelubricated bearings are serviced only as an assembly.

When reassembling pump, press impeller on the shaft so that rear face of impeller hub is flush with end of pump shaft.

ELECTRICAL SYSTEM

GENERATOR AND REGULATOR

75. Early HC engines are equipped with a 6-volt, two brush (shunt) type model 1100029 Delco-Remy generator and a Delco-Remy model 1118790 regulator. Generator cold output is 35.0 amperes @ 8.0 volts @ 2650 generator rpm. Brush spring tension is 28 ounces. Field current @ 80 degrees F. is 1.85-2.03 amperes @ 6 volts.

Regulator specifications are: Voltage regulator air gap, 0.075; cutout relay air gap, 0.020; cutout relay point opening, 0.020; cutout relay closing voltage, 5.9-7.0 volts; and voltage regulator setting, 6.8-7.4 volts.

75A. Later HC and all diesel engines have a 12-volt shunt type model 1100-314 or 1100322 Delco-Remy generator and a Delco-Remy model 1118791 regulator. Generator cold output is 20.0 amperes @ 14.0 volts @ 2300 generator rpm. Brush spring tension is 28 ounces. Field current @ 80 degrees F. is 1.58-1.67 amperes @ 12 volts.

Regulator specifications are: Voltage regulator air gap, 0.075; cutout relay air gap, 0.020; cutout relay point opening, 0.020; cutout relay closing voltage 11.8-14.0 volts; and voltage regulator setting, 14.0-15.0 volts.

STARTING MOTOR

76. Early HC engines are equipped with a Delco-Remy 1107147, 6 volt starting motor fitted with Bendix drive. Tested at no load, current draw should be 70.0 amperes @ 5.65 volts @ 5500 rpm. Lock torque minimum should be 11.0 pounds feet @ 550 amperes @ 3.25 volts. Brush spring tension is 24 ounces.

76A. Late HC and all diesel engines have a Delco-Remy 1108610, 1108632, 1108648 or 1107682, 12-volt starting motor fitted with an over-running clutch. On the 1108610 and 1108632, current should be 75.0 amperes @ 11.25 volts @ 6000 rpm. Lock torque minimum should be 29 pounds-feet @ 615 amperes @ 5.85 volts. Brush spring tension is 24 ounces. The 1107682 draws 112 amperes @ 10.6 volts @ 3240 rpm. On lock test the draw is 320-385 amperes @ 3.5 volts. Brush tension is 35 ounces. On the 1108648 brush tension is 24 ounces. Current draw is 40-70 amperes @ 11.8 volts and 6800-9200 rpm. On lock test the draw is 615 amperes @ 89 pounds feet.

IGNITION SYSTEM

77. Non-diesel engines are equipped with a 6- or 12-volt battery ignition system using a Model 1112555 Delco-Remy distributor.

Distributor specifications are as follows: Breaker point gap, .022; cam angle 25-34 degrees; and rotation, counter-clockwise. Automatic advance specifications in distributor degrees and distributor rpm are: 0-2 degrees @ 200 rpm; 6-8 degrees @ 600 rpm; and 13-15 degrees @ 1075 rpm. Condenser capacity is 0.18-0.23 mfd. Breaker spring tension 17-21 ounces.

77A. **TIMING TO ENGINE.** Firing order of engine is 1-2-4-3. To time distributor to engine, proceed as follows: Adjust breaker contacts to .022.

Crank engine until No. 1 piston is on compression stroke; then, rotate slowly until flywheel mark "TDC" is indexed with flywheel housing inspection port notch. Install distributor to engine with distributor rotor positioned to fire No. 1 cylinder. Rotate distributor body until contacts just start to open; then temporarily tighten the distributor body clamps.

Run engine @ 2200 rpm (high idle) at which time a timing light should show spark occurring when flywheel mark IGN-HC (28 degrees before TDC) registers with notch at inspection port. Rotate distributor body to obtain registration; then tighten clamp bolt securely.

TRACTION (ENGINE) CLUTCH

An Auburn model 100083-1, spring loaded, dry, 9¼-inch single disc clutch is used for the traction (engine) clutch on tractors prior to serial 6000-48652. On later tractors, Auburn cover assembly 100167 (Oliver 100691AS) having six instead of three pressure springs is used. Refer to paragraphs 80 through 81D for adjustment, R&R and overhaul data applying to the traction clutch.

A 5-inch diameter Twin Disc model SP-505, over-center, dry multiple disc clutch is used in the power take-off drive. Refer to paragraphs 130 through 133 for adjustment, R&R, and overhaul data applying to the PTO clutch.

ADJUSTMENT

80. Refer to Fig. O702. Recommended free travel of clutch pedal is ½ to ¾-inch when measured from pedal stop (S) to top of pedal (36). To adjust the pedal free travel, rotate nut (not shown) on lower end of clutch lever connecting link (43).

R&R AND OVERHAUL

81. Traction clutch or power take-off clutch can be removed after performing a tractor split (consisting of separating engine and tractor center frame cover from tractor center frame).

Fig. O702—Main (traction) clutch pedal free travel of ½-¾-inch is obtained by turning the nut at lower end of link (43).

81A. **TRACTOR SPLIT.** Refer to Fig. O703. To perform a tractor split, proceed as follows: Block-up under tractor center frame in a manner so as not to raise tractor wheels from the floor. Support front end of tractor with hoist by placing a chain or rope sling around the tractor hood, engine and rear portion of tractor front main frame.

Unbolt radius rods from front axle center member. Remove four cap screws attaching center frame cover to transmission housing. Remove six cap screws attaching center frame cover to tractor center frame. Disconnect fender light wire from fuse block under the panel. Remove two bolts (one from each side) attaching tractor center frame to engine rear plate. Remove four cap screws (two from each side) attaching tractor front main frame to tractor center frame. Complete the tractor split by moving the engine part of tractor away from the tractor center frame and transmission assembly.

Fig. O703—"Splitting" the tractor for access to main traction clutch. Center frame is (CF), center frame cover is (CFC) and main front frame is (FF). Refer to paragraph 81A for procedure.

81B. **R&R AND OVERHAUL.** Refer to Figs. O704 and O705. After performing a tractor split correlation mark clutch cover and flywheel; then, unbolt clutch from flywheel.

Check the release bearing and clutch shaft pilot bearing for wear and damage, and renew if necessary. Check also the clutch release shaft, located in the tractor center frame, in its unbushed bores. Reinstall clutch lined plate with long hub of it facing flywheel.

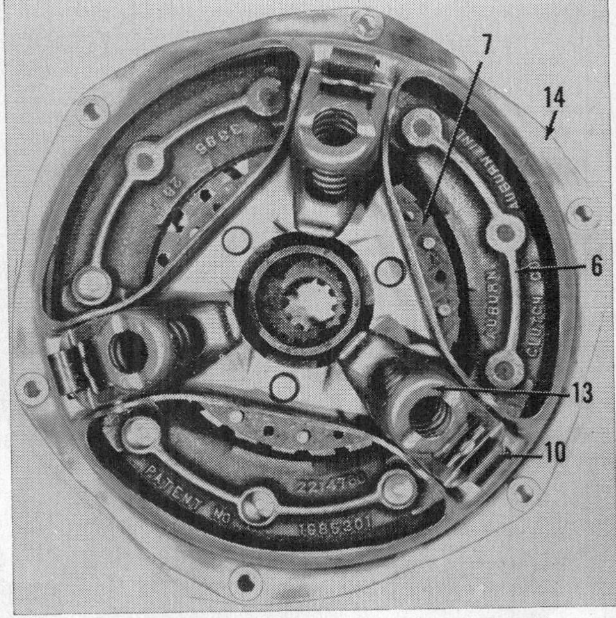

Fig. O704—Rear view of main (early model) clutch. Refer to paragraph 81B for removal procedure.

6. Pressure plate
7. Driven (lined) plate
10. Lever return clip
13. Release lever
14. Bracket and drive hub (cover)

81C. The following specifications apply to the model 100083-1 clutch equipped with a 3-spring 100084-1 cover assembly. Release lever height should be $1\frac{15}{16}$ inches when lever height is checked with key stock of .285-inch thickness inserted in place of the lined plate. Pressure springs, which are painted with a yellow stripe for identification have a free length of 2.518 inches, should test 279 to 309 pounds @ a length of 1.812 inches.

The 6-spring cover assembly No. 100167 lever settings are same as on the 3-spring model. Pressure springs should test 173-191 pounds @ $1\frac{9}{16}$ inches.

81D. Clutch shaft, which passes through the hollow pto drive pinion shaft, is also the transmission input shaft. For procedure to remove or renew the clutch shaft, PTO drive pinion shaft and/or seals and bearings for either shaft, refer to paragraph 89.

Fig. O705—Rear view of flywheel with clutch removed. Clutch pilot bearing is (PB) and crankshaft rear oil seal retainer is (SR).

TRANSMISSION

84. The six forward and two reverse speed transmission is of the helical gear, constant mesh type. The transmission shafts, bevel ring gear, and final drives (consisting of spur type bull pinions and bull gears, and not the rear wheel axle shafts) are contained in the transmission housing (tractor rear frame).

Refer to Fig. O706. In this manual the transmission is considered as comprising the top cover (which also contains the main hydraulic lift system pump, reservoir, work cylinder, and lift arms and cross shaft), shifter rails and forks, engine clutch (transmission input) shaft, bevel pinion shaft and gears, countershaft and gears, and reverse idler gear and shaft. The pto drive shaft and pinion (35) which is installed on all models, including models without pto, is considered also as part of the transmission.

BASIC PROCEDURE

85. Although most transmission repair jobs involve overhaul of the complete unit, there are infrequent instances where the failed or worn part can be repaired without completely disassembling the transmission. In effecting such localized repairs time will be saved by observing the following procedures:

85A. Shifter Rails And Forks. The three shifter rails and the detent (poppet) block can be serviced after removing the transmission cover (hydraulic housing) and unbolting the forks from the rails.

Right shifter fork can be removed after unbolting the fork from the rail.

Left shifter fork can be removed after removing the shifter rails.

The center shifter fork can be removed on current production models or models equipped with a fork which has been modified as shown in Fig. O710 after unbolting the fork from the rail. On early production models or models equipped with a fork which hasn't been modified as shown in Fig. O710 the center fork can be removed only after removing the transmission input double gear as outlined in paragraph 85D.

85B. PTO Drive Shaft And Pinion. This shaft (35—Fig. O706) is installed in all models. It and the oil seal for same can be renewed after performing a tractor split, removing pto clutch and spider, and removing engine clutch release bearing tube carrier.

Fig. O707—"Splitting" the tractor for access to clutches and flywheel. Center frame is (CF), center frame cover (CFC), front frame is (FF).

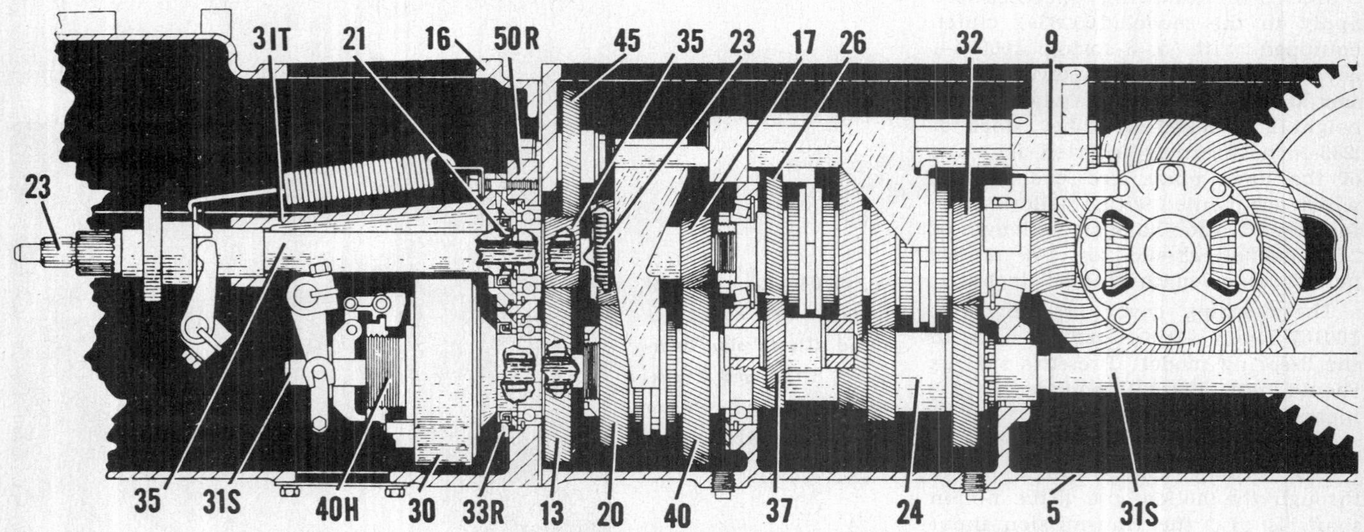

Fig. O706—Drive line from clutch shaft rearward. All parts shown are contained in the tractor center frame (16) and the transmission housing or rear frame (5).

9. Bevel pinion shaft
13. PTO drive gear
17. Input double gear
20. High range drive gear
23. Engine clutch shaft
24. Countershaft
26. Reverse gear
30. PTO clutch spider
31S. PTO drive shaft
31T. Bearing tube carrier
32. Output gear (25 teeth)
33R. Retainer
35. PTO drive pinion shaft
37. Reverse idler gear
40. Low range drive gear
40H. PTO clutch hub
45. Pump idler gear
50R. PTO pinion retainer

Fig. O707A—Top view open portion of center frame after tractor split has been performed. For procedure covering removal of these parts, refer to paragraphs 132 through 136.

AS. Allen screw
GT. Grease tube
GTN. Grease tube nut
23. Engine clutch shaft
24. Engine clutch release shaft
25. Engine clutch release fork
29. Release bearing sleeve
31. Bearing tube carrier
35. PTO drive pinion
41. Pedal return spring
50. Bearing retainer
58. Release bearing
62. PTO clutch fork

85C. Engine Clutch Shaft. Engine clutch shaft lip type oil seal, located in bore of hollow pto drive shaft and pinion, or the engine clutch shaft (23) can be renewed after removing the pto drive shaft and pinion.

85D. Transmission Input Double Gear. The transmission input double gear (17) which is located on forward end of bevel pinion shaft, can be removed after performing a tractor split, detaching tractor center frame from transmission housing, removing transmission cover, removing hydraulic pump drive idler gear, and unbolting center shifter fork from shifter rail.

Fig. O707B—Front view of center frame and transmission (rear frame) housing after pto clutch plates have been removed.

23. Engine clutch shaft
28. Clutch spider nut
30. PTO clutch spider
31S. PTO drive shaft
35. PTO drive pinion shaft

85E. Countershaft High And Low Range Gears. These gears (20 and 40) which are located on forward end of transmission countershaft, can be removed after removing the transmission input double gear, and pto drive shaft.

85F. Bevel Pinion Shaft. The bevel pinion shaft (9) with gears assembled to shaft, can be removed after removing the countershaft high and low range gears. It is not necessary to remove the bevel ring gear and differential assembly.

85G. Countershaft. The countershaft (24) can be removed after removing the bevel pinion shaft.

85H. Reverse Idler. The reverse idler gear (37) and shaft can be removed after removing the bevel pinion shaft, and countershaft bearing retainer plate.

R&R AND OVERHAUL

Data on removing and overhauling the various components which make-up the transmission are as follows:

86. **TRACTOR SPLIT.** Refer to Fig. O707. To perform a tractor split (detaching engine and tractor center frame cover from tractor center frame and transmission housing) proceed as follows: Block-up under tractor transmission housing in a manner so as not to raise tractor wheels from the floor. Support front end of tractor with a hoist by placing a chain or rope sling around the tractor hood, engine and rear portion of tractor front main frame.

Unbolt radius rods from axle center member. Disconnect light wire from fuse block and remove wire from clips located on steering gear housing. Remove dust shield from lower side of tractor front and center frames. Remove six cap screws attaching center frame cover to tractor center frame. Remove two bolts (one from each side) attaching tractor center frame to engine rear plate. Remove four cap screws (two from each side) attaching tractor front main frame to tractor center frame. Complete the tractor split by moving engine part of tractor away from tractor center frame and transmission assembly.

87. **R&R TRACTOR CENTER FRAME.** To remove tractor center frame or detach tractor center frame from transmission, proceed as follows: First, perform a tractor split as outlined in paragraph 86 then remove in order, the pto clutch plates, pto clutch spider (drum), clutch release bearing tube carrier and pto drive pinion shaft as outlined in paragraphs 87A through 87C. The center frame may be removed without removing clutch spider (drum) and pto pinion shaft separately, but such a procedure involves the likelihood of damaging the oil seal for the engine clutch shaft. This seal is located in the hollow pto drive pinion shaft and must pass over the splines at front end of engine clutch shaft. The weight of the center frame is such that most mechanics will prefer to remove the release bearing tube carrier, pto clutch drum and the pto drive shaft pinion separately, from the center frame.

87A. REMOVE PTO CLUTCH (PLATES). Refer to Fig. O707A. Remove engine clutch release fork (25) and release shaft (24). Remove pto clutch actuating fork and release shaft (62). From right side of tractor center frame remove nut (GTN) which retains shift collar oil tube (GT) to center frame. Remove outer Allen set screw (AS) attaching clutch hub to pto drive shaft and loosen the inner Allen screw. Two screws should always be used at this location. Refer to Caution paragraph 138. Pull the internal parts of pto clutch assembly (hub, driving and driven plates, sliding sleeve and collar) off of front end of pto drive shaft.

87B. REMOVE PTO CLUTCH (DRUM) SPIDER. After removing pto clutch plates assembly, remove three nuts attaching engine clutch release bearing tube carrier (31) to center frame and lift off the tube. Unstake and remove the pto clutch spider retaining spanner type nut (28—Fig. O707B) which has a right hand thread. An Owatonna ED3434 spanner wrench or equivalent will facilitate removal of this nut. Pry off clutch spider from splined hub of pto drive gear. An "O" ring seal for inside diameter of spider hub can be renewed at this time. The lip type oil seal for outside diameter of spider hub can be renewed at this time after removing the retainer (33—Fig. O707C). When reassembling tighten the spanner nut (28—Fig. O707B) to 150 foot pounds then lock same by staking it in three places.

87C. PTO DRIVE PINION SHAFT. Refer to Fig. O707D. Pry off the pto drive pinion shaft bearing retainer (50). Using a punch inserted in pinion shaft hole (H) bump pinion shaft and bearing out of center frame wall. Wind a layer of friction or adhesive tape over splines of engine clutch shaft (23) then carefully slide the pinion shaft off the engine clutch shaft to prevent damaging the oil seal located on the inside of the hollow pto pinion shaft. Be equally careful when reassembling the pto pinion shaft.

The pto drive pinion shaft contains both a needle bearing and lip type oil seal for the engine clutch shaft. To renew the lip type seal, it will be necessary to remove the needle bearing. Install the needle bearing so that approximately $\frac{1}{16}$-inch of space exists between the seal and bearing.

87D. CENTER FRAME. Now remove the cap screws attaching center frame to front of transmission (rear frame) and lift or hoist the center frame off the transmission. Refer to Fig. O707E for view after center frame and attached pto drive gear has been removed.

88. **OVERHAUL PTO CLUTCH.** Refer to paragraph 133 for data on overhauling the PTO clutch.

89. **ENGINE CLUTCH (TRANSMISSION INPUT) SHAFT.** To remove engine clutch shaft (23—Fig. O707E) first perform a tractor split as outlined in paragraph 86. Remove tractor center frame from transmission housing as outlined in paragraphs 87 through 87D. Remove Truarc snap ring (63) which retains the engine clutch shaft to the transmission input double gear, and withdraw the engine clutch shaft.

90. **TRANSMISSION TOP COVER.** To remove transmission housing top cover which contains hydraulic lift system pump, reservoir, work cylinder, and lift arms and shaft, proceed as follows: Remove seat pan by sliding it off of seat spring assembly. Remove gear shift lever. Disconnect lifting link assemblies from hydraulic lift arms. Remove cap screws attaching top cover to transmission housing. Attach a chain or rope sling on cover, as shown in Fig. O708; and lift cover off housing.

Fig. O707D — Showing hollow PTO clutch drive pinion shaft (35) and solid engine clutch shaft (23) after release bearing tube has been removed. See (31) in Fig. O707A. Procedure for removal of drive pinion shaft is in paragraph 135.

Fig. O707C—Showing splined hub of PTO drive gear (13) from which PTO clutch spider (drum) has been removed. Item (10) is inspection plate, (23) is engine clutch shaft. An oil seal is housed in retainer (33).

Fig. O707E—Front view of transmission with center frame and top (hydraulic housing) cover removed.

17. Input double gear
20. High range drive gear
23. Engine clutch shaft
31. PTO drive shaft
63. Snap ring

Fig. O708—Removing transmission top cover which is called the hydraulic housing. This contains the pump, reservoir, work cylinder and lift arms and shaft of the hydraulic lift system.

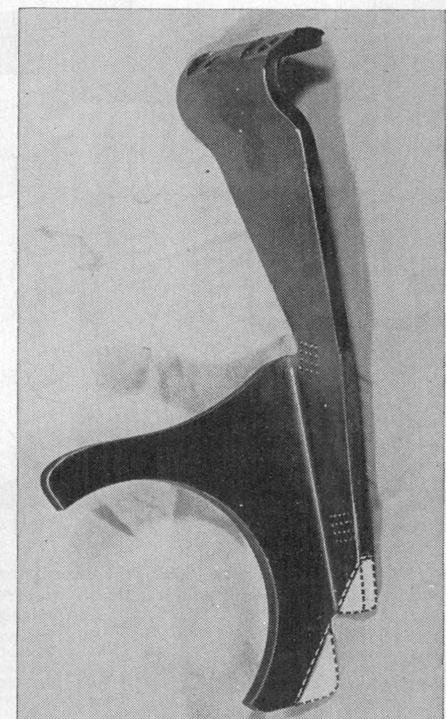

Fig. O710—If transmission center shifter fork is relieved (by grinding) at area indicated by broken line it can be removed without disturbing the input double gear. Refer to paragraph 91.

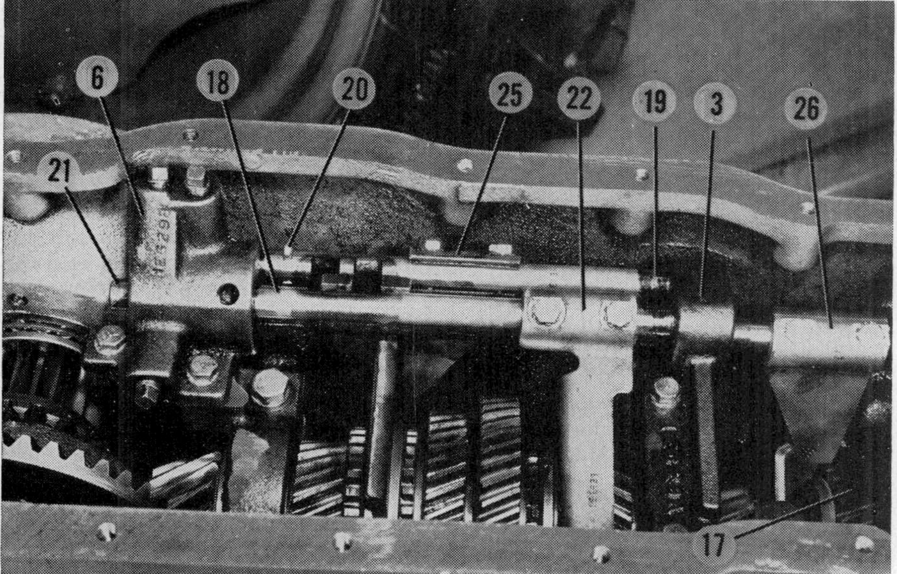

Fig. O709—Transmission shifter forks and rails as seen with transmission top cover removed.

3. Front shifter rod bracket
6. Rear shifter rod bracket
17. Input double gear
18. Right shifter rod
19. Left shifter rod
20. Cotter pin
21. Center shifter rod
22. Right shifter fork
25. Left shifter fork
26. Center shifter fork

91. **SHIFTER RAILS AND FORKS.** Refer to Fig. O709. The shifter rails, detent poppet block (6) also known as rear shift rod bracket, right shifter fork (22), and left shifter fork (25) can be removed after removing the transmission housing top cover.

On current production models or where the center shifter fork (26) has been modified as shown by the broken line area in Fig. O710, the center fork can be removed after unbolting the fork from its rail. On early production units equipped with a fork which has not been modified, it will be necessary to remove the transmission input double gear, as outlined in paragraph 92, in order to remove the center fork.

91A. Early type center shifter forks can be modified by grinding off the area which is indicated by the broken line in Fig. O710.

92. **TRANSMISSION INPUT DOUBLE GEAR.** Refer to Fig. O711. To remove transmission input double gear which is located on forward end of bevel pinion shaft proceed as follows: First remove transmission housing top cover as outlined in paragraph 90. Perform a tractor split as outlined in paragraph 86. Detach tractor center frame as outlined in paragraphs 87 through 87D. Remove engine clutch shaft, if not already done in detaching the tractor center frame, by removing

CSF. Center shifter fork
9. Bevel pinion shaft
17. Input double gear
20. High range drive gear
24. Countershaft
45. Hydraulic pump idler
47. Pump idler shaft
49. Nut
50. Clamping nut
62. Snap ring, small
76. Ball bearing

Fig. O711—Front of transmission after center frame, engine clutch shaft and pto drive shaft have been removed. At this time the counter shaft high and low range drive gears can be removed without disturbing the counter shaft. Refer to paragraph 94 for procedure.

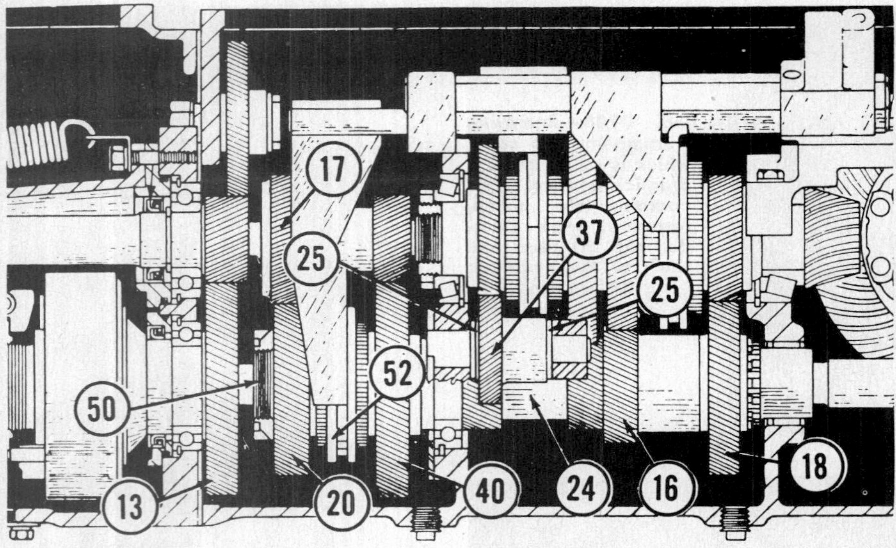

Fig. O711A—Sectional view of transmission showing countershaft high and low range gears (20) and (40) which can be removed without removing the countershaft. Refer to paragraph 94.

13. PTO drive gear
16. Countershaft gear (24 teeth)
18. Countershaft gear (38 teeth)
20. High range gear (37 teeth)
24. Countershaft
25. Idler gear washer
37. Reverse idler
40. Low range gear (47 teeth)
50. Clamping nut
52. Sliding coupling (25 teeth)

the Truarc snap ring which retains the clutch shaft to the transmission input double gear.

Unbolt center shifter fork (CSF) from shifter rail. Remove hydraulic pump idler gear shaft nut (49). Remove idler gear and shaft by bumping shaft rearward and out of transmission housing front wall. Remove small Truarc snap ring (62) from front end of bevel pinion shaft. Working through top cover opening in transmission housing bump input double gear (17) off bevel pinion shaft. Input double gear ball bearing and needle roller bearing can be renewed at this time.

Note: On late model tractors the needle bearing is held in position in bore of input double gear by a snap ring.

93. **PTO DRIVE GEAR.** To service the pto drive gear (13—Fig. O711A) it will be necessary to perform a tractor split, remove the pto clutch and spider, and detach the tractor center frame from the transmission housing. For complete overhaul data applying to this gear refer to paragraph 133B in the POWER TAKE-OFF section.

94. **COUNTERSHAFT HIGH & LOW RANGE GEARS.** Refer to Figs. O711 and O711A. To remove countershaft high and low range gears (20 & 40) which are located on forward end of countershaft (24), proceed as follows: First perform a tractor split as per paragraph 86, and detach the tractor center frame from the transmission housing as per paragraphs 87 through 87D. Remove transmission input double gear (17) as outlined in paragraph 92. Remove power take-off drive shaft, as outlined in paragraph 138.

Unstake countershaft clamping spanner nut (50), which has a left hand thread. Using a spanner wrench (OTC ED3434 tool or equivalent), remove the nut. Lift off high range gear (20), mounting sleeve, coupling (52), collar, low range gear (40), and sleeve from countershaft.

94A. High and low range gears are fitted with bronze bushings which are not available for service. Inside diameter of new bushings is 2.250-2.251 inches.

94B. Reinstall the high and low range gears on countershaft as shown in Fig. O721. Tighten countershaft clamping spanner nut to a minimum torque of 150 ft.-lbs.; then lock the spanner nut by staking it, in three different places, between the splines.

95. **BEVEL PINION SHAFT.** To remove bevel pinion shaft, proceed as follows: First, perform a tractor split as outlined in paragraph 86. Detach tractor center frame from transmission housing as outlined in paragraphs 87 through 87D. Remove transmission housing top cover as outlined in paragraph 90. Remove shifter rails and forks as outlined in paragraph 91. Remove transmision input double gear, located on forward end of bevel pinion shaft, as outlined in paragraph 92. Remove countershaft high and low range gears as outlined in paragraph 94. It is not necessary to remove the bevel ring gear and differential assembly at this time.

95A. Remove bevel pinion shaft bearings adjusting spanner nut (8—Figs. O718 and O719A) which has a left-hand thread. This is the larger of the two nuts on front end of shaft. An Owatonna ED3434 spanner wrench will facilitate the removal of this nut. Do not remove the smaller clamping nut from the shaft at this time.

Remove bevel pinion shaft bearing cap (BC—Fig. O714). Make up a pry bar from ⅛-inch flat stock as shown in Fig. O715. Raise front end of shaft slightly, engage bar between bearing cone and flange of bearing sleeve and pry cone off shaft.

Remove cap screws (CS—Fig. O723) and right hand brake cover and pressure plate as a unit. Remove right hand differential bearing cage (28—Fig. O724) by engaging a puller into the slots provided in the cage flange. Move bevel ring gear to right as far as possible then lift bevel pinion shaft and gears assembly from the transmission.

To remove the bevel ring gear and differential unit at this time refer to paragraphs 100B and 100C.

Fig. O714—Using the pry bar (SST) shown in Fig. O715 to remove roller bearing cone from front end of bevel pinion shaft. Refer to paragraph 95.

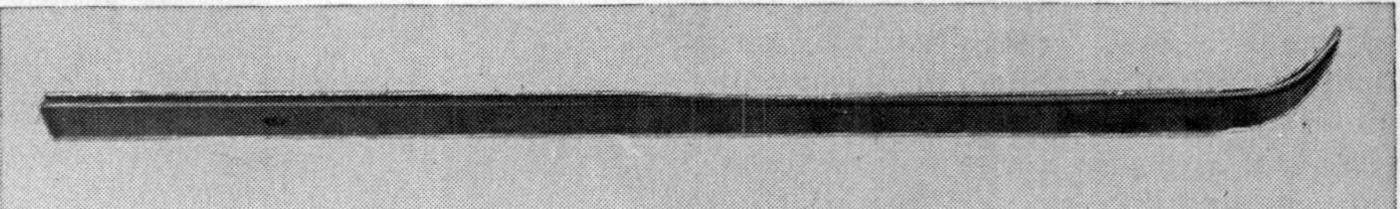

Fig. 0715—Pry bar for removing roller bearing cone from front end of bevel pinion shaft is shop made from ⅛-inch stock, ¾ wide by 11 inches long.

95B. Disassemble the bevel pinion shaft and gears assembly by unstaking and removing the left-hand threaded clamping nut (15—Fig. O717). An Owatonna ED3433 spanner wrench will facilitate the removal of this nut.

Gears (21, 26, 32 and 39) are fitted with steel backed, bronze bushings which are not supplied for service. Renew the gears and/or gear mounting sleeve if running clearance exceeds .006. Inside diameter of new bushings is 2.125-2.126 inches. It is advisable, but not necessary, to renew the bevel ring gear and pinion as a matched set. Pinion and ring gear are available separately and also as a matched set.

When reassembling the bevel pinion shaft, tighten the internally and externally threaded clamping nut (15) to a minimum torque of 150 ft.-lbs.; then lock by staking it in two places.

Reinstall bevel pinion shaft and gears assembly by reversing the removal procedure. Make sure that snap ring (61—Fig. O717) at rear bearing is entered into groove in transmission case. Tighten pinion shaft bearing cap retaining screws to a torque of 75-80 ft.-lbs. Before reinstalling the right hand differential bearing cage adjust the pinion shaft bearings as outlined in paragraph 95C.

95C. BEARING ADJUSTMENT. Bevel pinion mesh (fore and aft position) is fixed and non-adjustable. Adjust bearings, with spanner nut (8—Fig. O718), to a slight pre-load, or when 6-8 spring scale pounds pull measured at the teeth of gear (39) is required to rotate the shaft when bevel pinion and ring gear teeth are unmeshed.

To adjust the bearings when bevel pinion and ring gear teeth are meshed, proceed as follows: Carefully tighten the spanner nut until pinion shaft has zero end play; then, rotate the spanner nut approximately one-twelfth of a full turn tighter.

95D. BEVEL GEARS BACKLASH ADJUSTMENT. First, adjust differential shaft bearings with shims inserted under differential bearing cages to provide zero to 0.003 end play; then, adjust backlash of bevel gears to the amount stamped on outer edge of ring gear by removing shims from under one bearing carrier and installing the same shims under the opposite bearing cage. To reduce backlash, remove a shim from left hand cage and install it under the right cage.

Fig. O716—Preparing to lift bevel pinion and gears as an assembly from the transmission. As shown here, the bearings adjusting nut (8—Fig. O718) has already been removed from pinion shaft. Refer to paragraph 95A.

9. Bevel pinion and shaft
15. Clamping nut
21. Output gear (39 teeth)
26. Reverse gear
32. Output gear (25 teeth)
39. Output gear (45 teeth)
51. Drive collar
52. Coupling (25 teeth)
56. Coupling (31 teeth)
57. Coupling adapter
61. Snap ring
75. Rear roller bearing

Fig. O717—Bevel pinion shaft and gears assembly. Function of nut (15) is to clamp gears etc. to shaft. Another nut (8—Fig. O718) at same end of shaft (not shown here) adjusts, the tapered roller bearings. Refer to Fig. O716 for parts legend and to paragraph 95 for service procedure.

Fig. O718—Taper roller bearings supporting bevel pinion shaft should be adjusted to a slight preload by means of the larger diameter spanner nut (8) on shaft. Refer to paragraph 95C.

Fig. O719—Top view of countershaft assembly after bevel pinion shaft has been removed. Refer to paragraph 94 for procedure covering removal of gears (20 and 40) and to paragraph 96 for removal of remainder of countershaft unit (24).

16. Countershaft gear
20. High range drive gear
24. Countershaft (24 teeth)
25. Idler gear washer
34. Gear mounting sleeve
37. Reverse idler
40. Low range drive gear
41. Bearing retainer plate
50. Clamping nut
52. Gear coupling (25 teeth)

96. **COUNTERSHAFT.** This shaft (24—Fig. O719) can be removed after removing the countershaft high and low range drive gears (20 and 40) as per paragraph 94 and the bevel pinion shaft as per paragraph 95. Remove bearing retainer (41). Place a punch on rear face of countershaft ball bearing outer race, and bump the bearing and countershaft forward until bearing is out of its bore in transmission housing. Remove snap ring which retains the two rearmost gears (16 and 18) on the countershaft; then withdraw the countershaft forward and out of these two gears.

96A. Ball bearing (72—Fig. O720) located on forward end of countershaft, and the needle roller bearing (73—Fig. O719A) located in transmission housing wall, can be renewed at this time. Install this needle roller bearing so that it is centered in its bore in the transmission wall. The bore of the countershaft also contains two needle roller bearings which support the pto drive shaft. When installing these needle roller bearings, press on the end of the bearing which is stamped with the manufacturers name.

96B. Reinstall the countershaft and gears assembly be reversing the removal procedure.

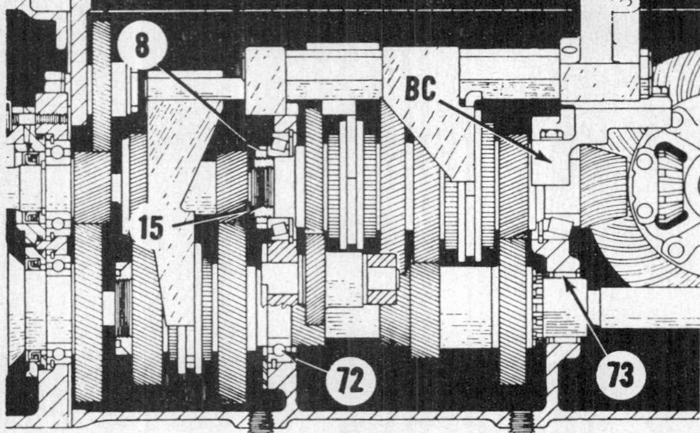

Fig. O719A — Side section through a portion of transmission housing and tractor center frame. Nut (8) controls adjustment of bevel pinion shaft bearings. Nut (15) clamps the gears, sleeves on to the bevel pinion shaft.

97. **REVERSE IDLER GEAR.** Refer to Figs. O719 and O720. To remove the reverse idler gear (37), first remove the bevel pinion shaft and gears as outlined in paragraphs 95 and 95A; then proceed as follows: Remove countershaft bearing retainer (41) and withdraw reverse idler gear shaft (64).

Reverse idler gear rotates on two steel backed bronze bushings which are not supplied for service. Renew gear and bushings when running clearance exceeds .006. Inside diameter of new idler gear bushings is 1.001-1.002. Diameter of a new idler gear shaft is .9990-.9995.

Fig. O720—Front view of hollow countershaft and front bearing as seen with bevel pinion shaft and countershaft high and low range gears removed.

24. Countershaft
64. Reverse idler shaft
72. Front bearing
78. Needle bearing for PTO drive shaft

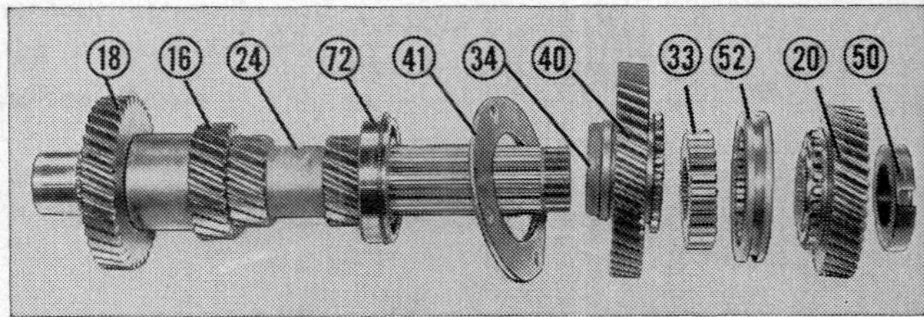

Fig. O721—Components of transmission countershaft assembly.

16. Gear (24 teeth)
20. High range drive gear
24. Countershaft
33. Drive collar (22 splines)
34. Gear mounting sleeve
40. Low range drive gear
50. Clamping nut
52. Sliding coupling
72. Front ball bearing

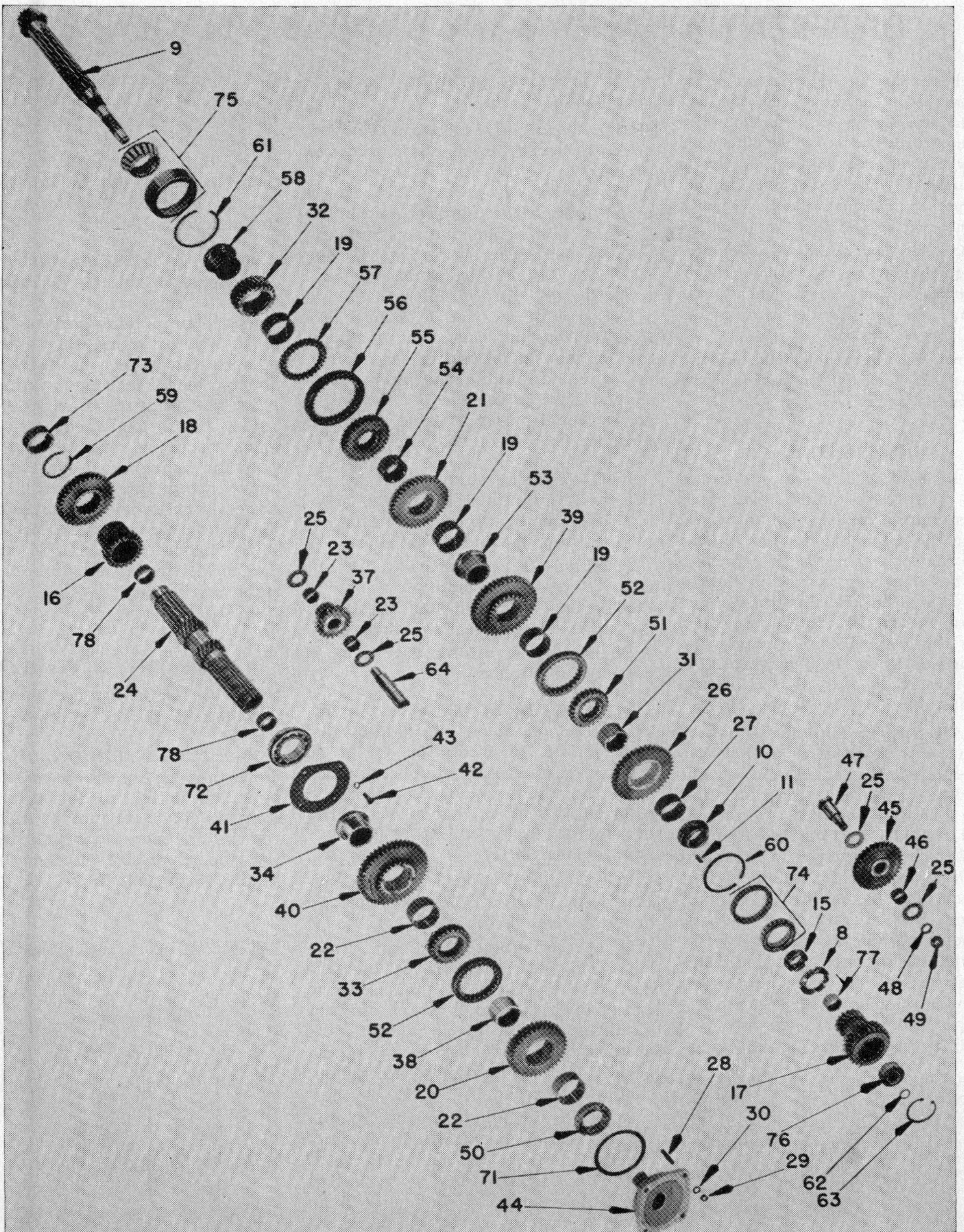

Fig. O722—Components of transmission portion of rear frame but not including pto drive elements which are shown in Fig. O746. On late models, bearing (77) is retained in gear (17) by a snap ring.

8. Bearing adjusting nut
9. Bevel pinion and shaft
10. Front bearing sleeve
15. Shaft clamping nut
16. Countershaft gear (24 teeth)
17. Input double gear
18. Countershaft gear (38 teeth)
19. Bushing (not supplied)
20. High range drive gear
21. Output gear
22. Bushing (not supplied)
24. Countershaft
26. Reverse gear
31. Reverse gear mounting sleeve
32. Output gear (25 teeth)
33. Drive collar
34. Mounting sleeve
37. Reverse idler
38. Mounting sleeve
39. Output gear
40. Low range drive gear
41. Bearing retainer plate
44. Countershaft pilot support
45. Hydraulic pump idler gear
46. Bushing (not supplied)
47. Hydraulic pump idler shahft
50. Clamping nut
51. Drive collar
52. Sliding coupling
53. Gear mounting sleeve
54. Gear mounting sleeve
55. Drive collar (31 teeth)
56. Sliding coupling
57. Gear coupling adapter
58. Gear mounting sleeve
59. Countershaft snap ring
60. Snap ring 3 45/64" O. D.
61. Snap ring 3 45/64" square end
62. Snap ring 25/32" internal
63. Snap ring 2½" O. D.
64. Reverse idler shaft
72. Countershaft front bearing 43210
73. Countershaft rear needle bearing
74. Pinion front roller bearing
75. Pinion rear roller bearing
76. Double gear ball bearing
77. Double gear needle bearing
78. PTO drive shaft needle bearing

DIFFERENTIAL AND MAIN DRIVE BEVEL GEARS

Differential unit which is of the two pinion open case type is mounted in the after section of the transmission housing. Main drive bevel ring gear is attached to the differential spider with rivets and on tractors prior to serial number 66 750-000 (approx.), is available either separately from the bevel pinion or in a matched set with the bevel pinion.

On all tractors after the above serial number, the ring gear is secured to the differential by a cold-riveting process and only complete ring gear and differential assemblies are available for service.

Also mounted on the differential are two spur type bull pinions which mesh with the final drive bull gears.

DIFFERENTIAL

100. **REMOVE.** Procedure for removing differential and bevel ring gear assembly varies depending on whether the main drive bevel pinion is to be removed or not to be removed. If the bevel pinion is not to be removed, it will be necessary to remove both bull gears. In either case, first remove the transmission housing top cover as outlined in paragraph 90; then proceed as outlined either in paragraph 100B, or 100A and 100B.

100A. If the bevel pinion is not to be removed, proceed as follows to remove both bull gears: Remove right rear wheel and tire assembly and bearing cage (7—Fig. O728) from right rear wheel axle shaft carrier. Remove snap ring from inner end of right rear wheel axle shaft. Support right bull gear, and pry axle shaft out of bull gear. Lift out bull gear. Remove left wheel and axle bearing cap without disturbing the pto drive shaft. Remove snap ring from inner end of left rear wheel axle shaft. Support the left bull gear, and pry bull gear off axle shaft. Lift out bull gear. Continue removal procedure as outlined in paragraph 100B.

Fig. O723—Showing location of one brake cover (23) which must be removed for access to one bull pinion.

100B. With both bull gears removed, or with bevel pinion shaft removed proceed as follows: Remove both brake covers (23—Fig. O723), brake discs, brake actuating discs, and brake pressure plates. Remove five nuts attaching each differential bearing cage (28—Fig. O724) to transmission housing. Support differential unit. Using a puller with reaction legs, pull differential bearing cages out of transmission housing. Bearing cage flange is provided with two rectangular shaped slots to receive the jaws of a conventional puller. Extract both bull pinions and differential side gears.

100C. If bevel pinion shaft is out, lift out differential assembly by moving the assembly toward the left and raising the right end of the shaft.

If both bull gears are out, lift out the differential assembly by moving the assembly toward the right and raising the left end of the shaft.

Reinstall the differential unit as outlined in paragraph 102.

101. **OVERHAUL.** Observe the following points when overhauling the differential: Refer to Fig. O725. To remove differential pinions, extract snap rings (1) which retain pins in differential spider; then use a puller in threaded hole to extract pin (17) from spider (16).

In some installations it may be necessary to de-rivet the ring gear in order to remove the pins (17).

When reriveting the main drive bevel ring gear, temporarily bolt the gear to the spider. Use cold rivets to attach the ring gear being careful not to distort the gear or spider in the process. Check trueness of ring gear back face after riveting. Total runout should not exceed .004.

Before installing the overhauled differential check backlash between differential pinions (8) and side gears (9). This can be done by mounting the bull pinion and side gear in an upright position. Lower differential assembly into operating position on side gear, and measure the backlash. Measure the opposite side in a similar manner. If backlash is less than .006 on each side, make up and install a shim washer to the spider shaft to obtain .006 minimum backlash.

102. **REINSTALL.** When reinstalling the differential, observe the following points: Assemble new oil seal (27) in differential bearing cage with lip of seal facing inward. Sealing effect between bearing cages and transmission housing is obtained with "O" ring (32). Install differential spider thrust bearing (68) in bull pinion so that inner race of bearing contacts the differential spider.

102A. Adjust differential bearings by means of shims (7) located between bearing cage and transmission housing, to obtain zero to .003 end play. When making this adjustment, be sure that bevel ring gear teeth are not bottomed in bevel pinion teeth. After adjusting the bearings end play, adjust bevel gears backlash to the amount stamped on outer edge of ring gear by removing shims from under one bearing cage and installing the same shims under the opposite bearing cage. To reduce backlash, remove shims from under left bearing cage and install the same shims under the right bearing cage.

MAIN DRIVE BEVEL GEARS

Main drive bevel ring gear is attached to the differential spider with rivets.

104. **BEVEL PINION.** The mesh (fore and aft) position of the bevel pinion, which is also a transmission shaft, is fixed and not adjustable. To remove or renew the bevel pinion follow the procedure outlined in paragraphs 95 through 95D.

Fig. O724—Right side of tractor showing right bull pinion shaft (2) as seen after brake has been removed.

2. Bull pinion
3. Expansion plug
27. Oil seal
28. Differential bearing cage

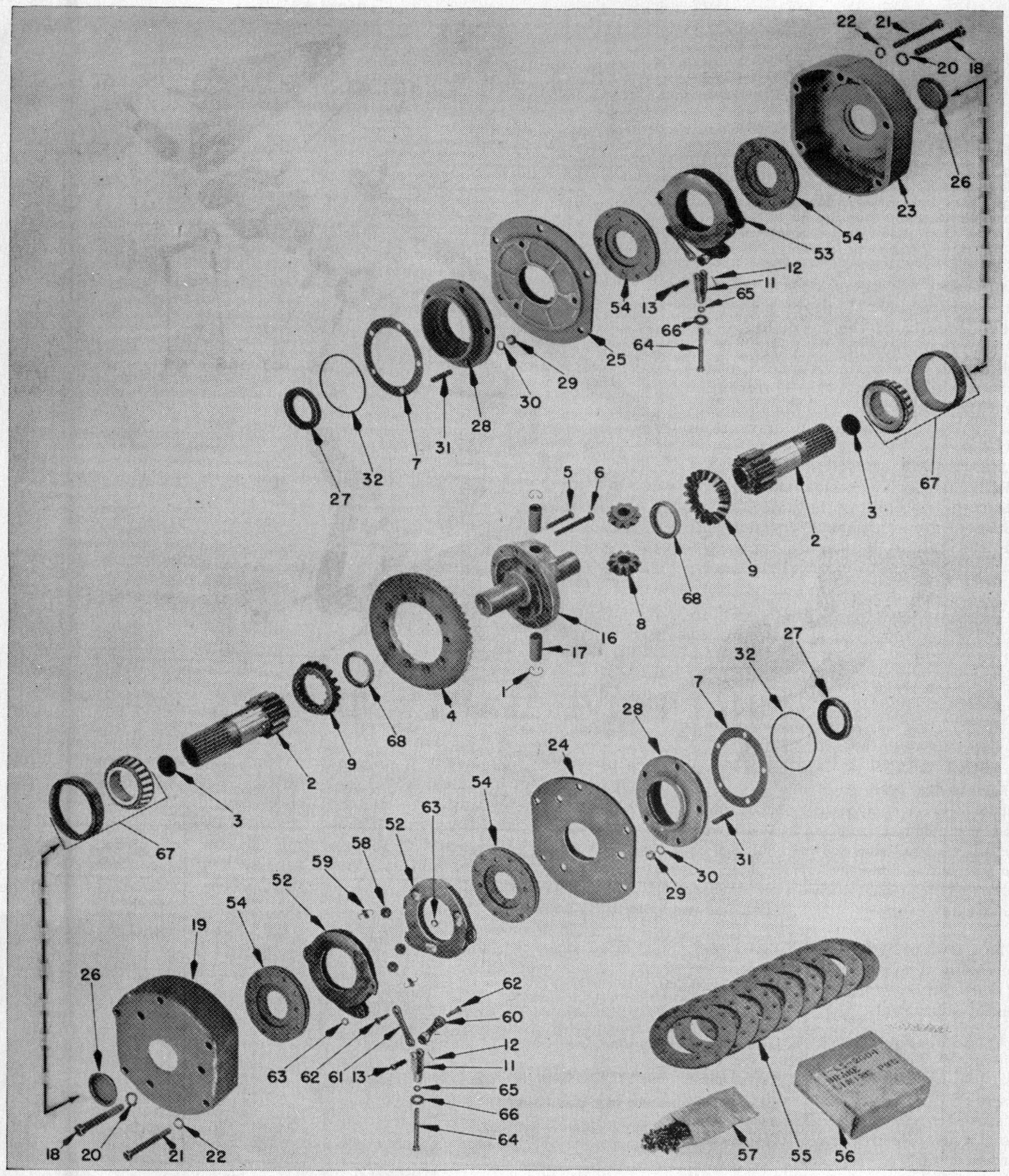

Fig. O725—Differential, bull pinions, bevel ring gear and brakes and their components.

1. Pin retaining ring
2. Bull pinion
3. Expansion plug
4. Bevel drive gear
7. Bearing cage shim
8. Differential pinion
9. Differential (side) gear
11. Brake rod end
16. Shaft & spider
17. Differential pinion pin
19. Brake cover
23. Brake cover
24. Brake pressure plate
25. Brake pressure plate
26. Cover cap
27. Bearing oil seal
28. Differential bearing cage
32. "O" ring
52. Activating disc
53. Activating disc assembly
54. Middle disc and lining
55. Lining
58. Ball ⅞"
60. Activating yoke link
61. Actuating plain link
67. Differential roller bearing
68. Differential thrust bearing

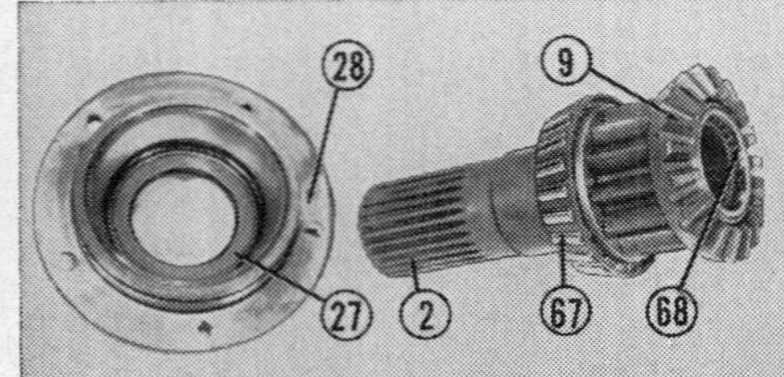

Fig. O726—Right view shows assembly of bull pinion (2), side gear (9), roller bearing (67) and thrust bearing (68). Left view is differential bearing cage (28) with oil seal (27). Refer to paragraph 110 for removal procedure.

105. **BEVEL RING GEAR.** To remove or renew the bevel ring gear independently of the bevel pinion or in conjunction with the bevel pinion follow the procedure outlined in paragraphs 100 through 102A. Differential bearings and bevel gears backlash adjustment is outlined in paragraph 102A.

106. **BEVEL PINION AND RING GEAR.** To renew the bevel pinion and ring gear, follow the procedure as outlined in paragraphs 95 through 95D for the pinion, and paragraphs 100 through 102A for the bevel ring gear.

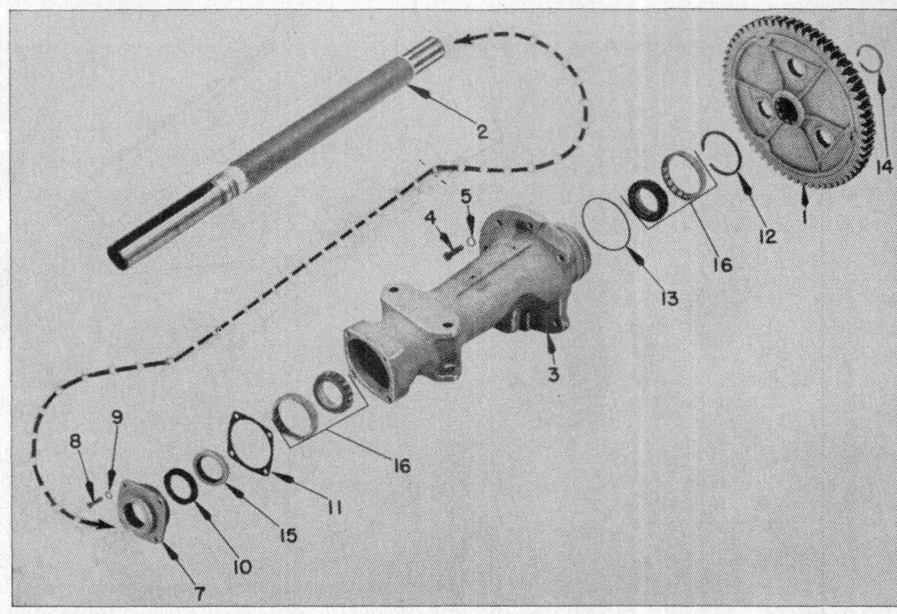

Fig. O726A—Rear axle carrier, wheel axle shaft bull gear and attendant parts.

1. Bull gear
2. Rear wheel axle
3. Axle carrier, right
7. Axle bearing cage
10. Felt washer
11. Bearing cage shim
12. Snap ring 4 1/16" O. D.
13. "O" ring
14. Axle snap ring 2 3/32"
15. Axle oil seal
16. Axle roller bearing

FINAL DRIVE GEARS AND AXLES

BULL GEARS AND PINIONS

Refer to Fig. O727. Final drive gears, consisting of two spur type bull pinions which mesh with two bull gears, are located in after section of transmission housing. Bull pinions are mounted on differential spider shaft, and bull gears are splined to inner ends of wheel axle shafts.

110. **BULL PINION & BEARINGS.** Herewith is a procedure for renewal of bull pinion and bearings, and bull pinion shaft seals.

110A. First remove transmisssion housing top cover as follows: Remove seat pan by sliding it off of seat spring assembly. Remove gear shift lever. Disconnect lifting link assemblies from hydraulic lift arms. Remove cap screws attaching top cover to transmission housing, and lift cover off transmission housing.

110B. Remove brake cover (23) brake discs, brake actuating disc and brake pressure plate. Remove five nuts attaching differential bearing cage (28—Fig. O724) to transmission housing. Support differential unit and remove cage. Bearing cage flange is provided with two rectangular shaped slots to receive the jaws of a conventional puller. Bull pinion shaft oil seal (27), located in bearing cage, can be renewed at this time. Extract bull pinion. Differential side gear will remain in transmission housing unless other bull pinion is removed.

110C. Refer to Fig. O726. Install new oil seal (27) in differential bearing cage with lip of seal facing inward. Install differential spider thrust bearing (68) in bull pinion so that inner race of bearing contacts the differential spider. Ordinarily, the renewal

Fig. O727—Gear arrangement in transmission (rear main frame) housing.

1. Bull gear
2. Bull pinion
23. Brake cover

of pinion or bearings should not change the bearing adjustment or the backlash adjustment of the main drive bevel gears enough to warrant readjustment, providing the original shims are reinstalled. It will be advisable, however, to check need for adjustment and the procedure for adjusting by referring to paragraph 102A.

Fig. O728—Axle carrier as seen when wheel is off. Item (2) is axle shaft. Item (7) is axle bearing cage.

111. **BULL GEAR.** To remove or renew either bull gear, proceed as follows: First, remove transmission housing top cover as outlined in paragraph 110A.

111A. Remove tire, rim and disc assembly from rear wheel hub. Refer to Fig. O728. Unbolt bearing cage (7) from outer end of wheel axle shaft carrier. Extract snap ring used to anchor bull gear to inner end of axle shaft. Support bull gear, and pry axle shaft out of bull gear.

111B. If tractor has no pto the other bull gear can be removed at this time by extracting snap ring from gear end of shaft and prying the gear off the shaft.

If tractor has a pto the second bull gear can be removed either by removing the pto drive shaft as in paragraph 138 and doing the work listed in paragraph 111B or preferably by removing the other rear wheel and proceeding as in paragraph 111A.

111C. Reinstall bull gear so long part of bull gear hub faces inward. Coat rear axle shaft bearing adjusting shims with castor oil to obtain sealing. Check bearing adjustment and adjust if necessary, as per paragraph 115B.

REAR WHEEL AXLE SHAFTS

114. **AXLE SHAFT OIL SEAL.** Refer to Fig. O726A. Wheel axle shaft oil sealing is obtained by a felt washer (10), and lip type seal (15) of neoprene. Seal assembly, located in wheel axle shaft bearing cage, can be renewed as follows: Unbolt rear wheel hub from axle shaft and remove the tire, rim, disc, and hub assembly. Unbolt bearing cage from outer end of wheel axle shaft carrier, and remove the bearing cage.

Install lip seal with lip facing the bull gears.

115. **WHEEL AXLE SHAFT AND BEARINGS.** To renew one wheel axle shaft follow procedure for removal of the first bull gear as outlined in paragraphs 111 and 111A. Renew the other axle shaft in a similar manner, but do not remove the pto drive shaft.

115A. Inner bearing cup can be removed from axle carrier after extracting a snap ring. Install bearing cones so that their tapers face in opposite directions as shown in Fig. O729. Install lip type oil seal in bearing cage so that lip faces the bull gear. Coat bearing adjusting shims with castor oil to obtain sealing.

115B. Axle Shaft Bearings Adjustment. Adjust axle shaft bearings to zero-.002 preload by means of shims (11—Fig. O729), located between bearing cage and outer end of wheel axle shaft carrier. If bull gear or pinion is out of transmission, adjust bearings to 5-7 pounds NET pull on spring scale (attached in one of the wheel hub cap screw holes) to rotate axle shaft. Net pull means seal drag in pounds on scale, plus 6 pounds.

Shims are available in thicknesses of 0.004, 0.007 and 0.015. Coat shims with castor oil to obtain sealing.

116. **AXLE CARRIERS.** To remove the axle carrier proceed as follows: Remove transmission housing top cover as per paragraph 110A. Remove tire, rim, and disc assembly from rear wheel hub. Remove fender from carrier. Remove nut retaining drawbar lower link to link support stud. Remove six cap screws attaching axle carrier to transmission housing. Extract snap ring retaining bull gear on inner end of axle shaft. Support axle carrier, and pry axle shaft out of bull gear. The "O" ring gasket, providing a seal between axle carrier and transmission housing, can be renewed at this time. Remove axle shaft from carrier by unbolting bearing cage from outer end of axle carrier.

Refer to paragraphs 115A and 115B for overhaul and bearing adjusting procedures.

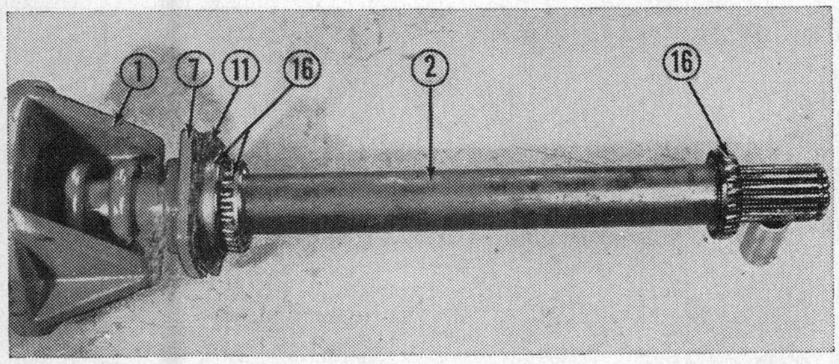

Fig. O729—Rear wheel axle shaft and bearings assembly with wheel hub (1). Note tapers of bearings (16) face in opposed direction.

2. Wheel axle shaft
7. Axle bearing cage
11. Shims
16. Bearing cup and cone

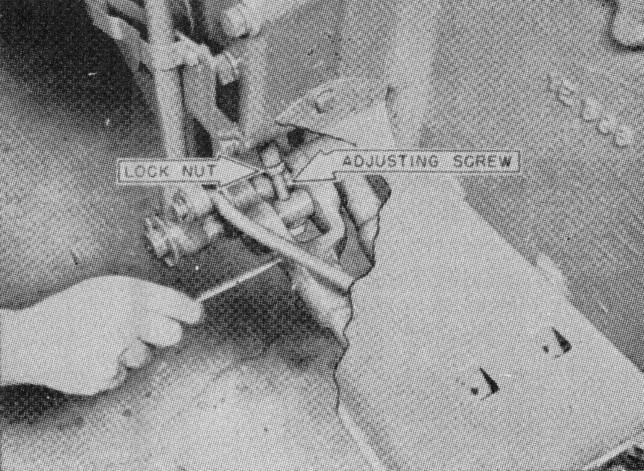

Fig. O732—To "take up" on brake, loosen lock nut and rotate adjusting screw clockwise.

BRAKES

Double disc brakes are mounted on outer ends of bull pinion shafts.

Effective at tractor serial number 72 832-500, increased ratio brake actuating assemblies (53—Fig. O725) have been installed. These brakes are identified by a "5" stamped on the actuating lug. Do not intermix the early and late type assemblies. When installing the late type brakes in tractors prior to the above serial number, it will be necessary to grind out the inside of existing brake covers (19 and 23) to provide operating clearance or install new type brake covers.

120. **ADJUSTMENT.** Refer to Fig. O732. To "take-up" on the brakes, loosen the brake adjusting screw locknut; then rotate adjusting screw in a clockwise direction. Do the same to the other brake. Equalize the brakes by backing off the adjustment on the tight brake.

121. **DISCS, DRUMS AND SEALS.** Refer to Fig. O723. To remove lined discs, first disconnect brake linkage by removing adjusting screw from rod end of actuating yoke. Remove one cap screw attaching platform to brake cover. For right brake, remove pto clutch control lever (68). Remove five cap screws (CS) and lift brake cover and pressure plate off tractor. Remove three 5/8-inch cap screws (CSL) which position the brake actuating disc. Both lined discs, actuating disc, and plate shown in Fig. O733 can now be removed from cover.

Fig. O733—Right brake assembly as seen with brake cover (23—Fig. O723) removed. Item (48) is brake cross shaft.

The presence of oil in the brake compartment is corrected by renewal of differential bearing cage "O" ring seal or bull pinion lip type oil seal. To renew either seal, remove the differential bearing cage as outlined in paragraph 110B.

122. **BRAKE PEDAL CROSS SHAFT.** The brake pedal cross shaft (48—Fig. O733) operates in machined bores in the transmission housing. The cross shaft is equipped with lip type oil seals, one at each end, which are pressed into the transmission housing.

Either oil seal can be renewed without removing the cross shaft. To renew the seal on the right side, remove the cross shaft support and both brake pedals. To renew the left side seal, remove the cross support and brake actuating arm.

To remove the pedal cross shaft, it will be necessary to remove a rear wheel tire, rim and disc assembly.

BELT PULLEY UNIT

The belt pulley unit Fig. O735, which is supplied as extra equipment is driven by the pto drive shaft. The pulley unit may be mounted and operated either in a right or a left horizontal position, or in a down vertical position. Belt pulley revolves at 1451 rpm when the engine crankshaft speed is 2200 rpm. To operate the belt pulley engage the power take-off clutch.

Belt pulley is lubricated with one pint of seasonal grade transmission lubricant which is either SAE 140 for temperatures above 32 degrees F., or SAE 90 for temperatures below 32 degrees F.

R&R AND OVERHAUL

123. Removal of belt pulley unit requires removal of four nuts which attach unit to rear face of transmission housing.

124. Refer to Fig. O735. To renew the pulley housing oil seal (16), it will be necessary to disassemble the complete unit. Pulley base oil seal (22) can be renewed after detaching pulley housing base (17) from pulley housing (11).

124A. Overhaul procedure is as follows: Drain lubricant and remove pulley. Remove pulley housing base (17) from pulley housing (11), and lift out pulley drive gear and shaft (20). Remove castellated nut (5) from inner end of pulley shaft, and lift out pulley shaft.

124B. According to Oliver's parts catalog, the pulley drive pinion and pulley drive gear are supplied separately or in a matched set. Mesh position of the pulley drive pinion is fixed and non-adjustable. Backlash of these gears is adjusted with shims (1 & 2) which are located under the pulley drive gear bearing carrier and pulley housing base respectively.

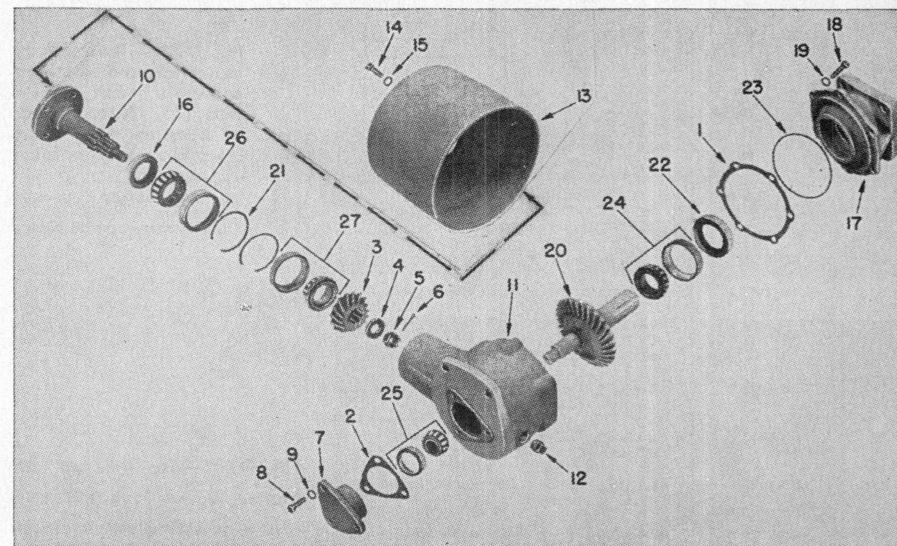

Fig. O735—Components of belt pulley assembly. Mesh position of bevel gears is not adjustable but backlash is controlled by shims (1) and (2).

1. Pulley housing shim
2. Bearing carrier shim
3. Pulley drive pinion
4. Washer
7. Drive gear bearing carrier
10. Pulley shaft
11. Pulley housing
13. Pulley
16. Housing oil seal
17. Housing base
20. Pulley drive gear
21. Snap ring
22. Base oil seal
23. "O" ring
24. Front bearing for 20
25. Rear bearing for 20
26. Outer bearing for 10
27. Inner bearing for 10

Fig. O735A—Belt pulley installation.

Fig. O735B—View showing method of holding pulley shaft while tightening nut to preload shaft. Be sure to check and record the amount of seal drag before tightening nut. Tighten nut until pre-load is 2-4 pounds greater than the oil seal drag.

Install both oil seals with their lips facing the bevel gears. When re-installing pulley shaft in pulley housing, insert a suitable sized split collar ¼-inch thick, or two pieces of ¼-inch key stock between inner face of pulley shaft flange and outer face of oil seal (16). This will force the oil seal squarely into position and flush with end of pulley housing.

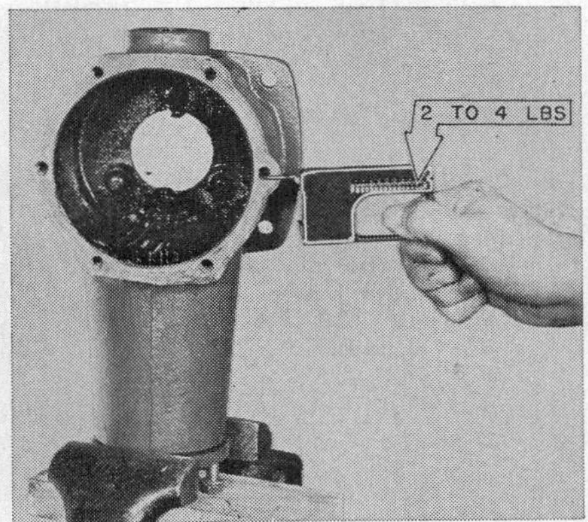

Fig. O736A — With flange of pulley shaft clamped in a vise and a spring scale attached as shown, check the pre-load. Preload should be 2-4 lbs. greater than seal drag. Refer to text.

Adjust pulley shaft bearings to a just perceptible preload by means of castellated nut (5). With nut loosely installed, rotate shaft and record the amount of oil seal drag. Then tighten nut until preload is 2-4 pounds more than the seal drag when measured with a spring scale as shown in Fig. O736A.

Adjust pulley drive gear shaft bearings to .001-.005 end play with shims (1 & 2—Fig. O635) interposed at bearing carrier (7) and pulley base (17). Shims are supplied in thicknesses of 0.004, 0.007, and 0.015.

After adjusting the pulley drive gear shaft bearings, adjust bevel gears backlash to the amount stamped on outer edge of pulley drive gear by removing shims (1 or 2) from one location and inserting the same thickness at the opposite location. To reduce backlash remove a shim or shims (2) from bearing carrier side and add the same thickness of shims (1) on the housing base side.

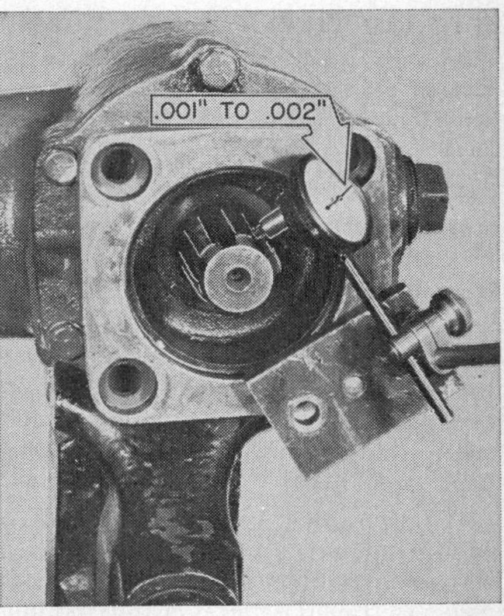

Fig. O736B—View showing method of mounting a dial indicator to check gear backlash. Refer to text for procedure and specifications.

Backlash value stamped on drive gear applies to measurements at the gear teeth. Teeth are somewhat inaccessible and if backlash is measured with a dial indicator at splines of drive pinion it should be one-third of the amount etched on drive gear.

POWER TAKE-OFF UNIT

The continuous (independent) type pto, which is supplied as extra equipment, is equipped with an output shaft which is reversible so as to provide a spline size either of 1⅛ inches or 1⅜ inches. This pto output shaft is spline coupled to the pto drive shaft. PTO shaft revolves at 749 rpm when engine crankshaft speed is 2200 rpm.

Power flow to and through the continuous pto is as follows: Refer to Fig. O738. The hollow pto drive pinion shaft (35) which fits over the engine clutch shaft is splined into the engine clutch cover. The integral pinion, located at the other end of this shaft, meshes with gear (13) which is splined to the spider of the pto clutch (30). From the multiple disc, over-center type pto clutch (30), the power is transmitted to a long drive shaft (31) which passes through the hollow transmission countershaft and terminates aft of the rear wall of transmission (rear frame) as shown.

ADJUST PTO CLUTCH

130. If multiple disc clutch slips under load, it should be tightened as follows: Refer to Fig. O739. Remove rectangular shaped hand hole cover from bottom of tractor center frame. Working through cover opening, depress the clutch adjusting ring lock pin and rotate the adjusting ring clockwise (when facing clutch sleeve collar end) one notch at a time or until 32 pounds pull, measured two feet above the pivot of hand lever, is required to engage the clutch.

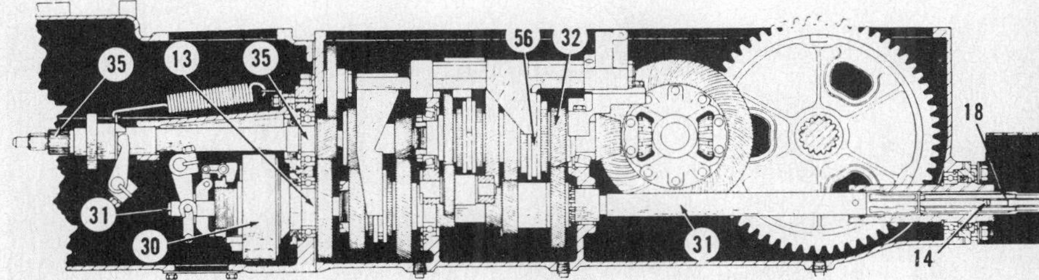

Fig. O738—Power flow through pto system. The pto drive pinion (35) is attached to, and rotates with, the flywheel. Teeth on rear of drive pinion are in constant mesh with gear (13) which is connected to and disconnected from the long pto drive shaft (31) at will by means of the pto clutch (30).

PTO CLUTCH PLATES AND SPIDER

Data on removing and overhauling the various pto clutch components which are housed in the tractor center frame are outlined in the following paragraphs.

131. **TRACTOR SPLIT.** To perform a tractor split, as shown in Fig. O740, proceed as follows: Block-up under tractor center frame in a manner so as not to raise tractor wheels from the floor. Support front end of tractor with a hoist by placing a chain or rope sling around the tractor hood, engine and rear portion of tractor front main frame.

Unbolt radius rods from front axle center member. Remove four cap screws attaching center frame (CF) cover to transmission housing. Remove six cap screws attaching center frame cover (CFC) to tractor center frame. Disconnect fender light wire from fuse block under the panel. Remove two bolts (one from each side) attaching tractor center frame to engine rear plate. Remove four cap screws (two from each side) attaching tractor front main frame (FF) to tractor center frame. Complete the tractor split by moving engine part of tractor away from tractor center frame and transmission assembly.

132. **R&R CLUTCH.** To remove either the pto clutch assembly less spider as shown in Fig. O744, or clutch spider (drum) Fig. O742 or clutch actuating fork and shaft Fig. O741 first perform a tractor split as outlined in paragraph 131. Refer to Fig. O741. Remove engine clutch release fork (25) and shaft (24). Remove pto clutch actuating fork and shaft (62). From right side of tractor center frame, remove nut (GTN) retaining pto shift collar oil tube (GT) to center frame. Loosen Allen set screw (AS) retaining clutch hub to pto drive shaft. Refer to Caution in paragraph 138. Pull internal parts of pto clutch assembly (hub, driving and driven plates, floating plates, sliding sleeve and collar) off of pto drive shaft. Overhaul data for the clutch assembly is outlined in paragraph 133.

Fig. O739 — Access to pto clutch adjusting ring is obtained by removal of an inspection plate on bottom surface of tractor center frame. Refer to paragraph 130 for procedure.

132A. Clutch Spider. To remove pto clutch spider (drum), at this time, proceed as follows: Refer to Fig. O742. After removing pto clutch, remove three nuts attaching engine clutch release bearing tube (31) to tractor center frame, and lift off the tube. Unstake and remove pto clutch spider retaining spanner type nut (28) which has a right hand thread. An Owatonna ED3434 spanner wrench or equivalent will facilitate the removal of this nut. Pry off pto clutch spider from splined hub of pto drive gear.

An "O" ring seal for inside diameter of spider hub can be renewed at this time. Lip type oil seal for outside diameter of spider hub can be renewed after removing drive sleeve bearing retainer (33—Fig. O743A).

Tighten the clutch spider retaining spanner type nut to a minimum torque of 150 ft.-lbs.; then lock the spanner nut by staking it in three places.

Fig. O740—"Splitting" the tractor for access to clutches. Refer to paragraph 131 for procedure.

133. **OVERHAUL PTO CLUTCH.** Observe the following points when overhauling the already removed pto clutch and clutch spider. To disassemble the removed clutch, Fig. O744, depress clutch adjusting ring lock pin (59); and rotate the adjusting ring (53) in a counter-clockwise direction off clutch hub (40). The floating plate (43), and driving and driven plates, can now be lifted off clutch hub. Clutch contains five friction type driving plates and four steel type driven plates.

Clutch hub (40—Fig. O745) contains two lip type oil seals (37) which are installed with their lips facing each other.

Carefully inspect all parts for wear, scoring, and damage; and renew if necessary. Use an oil stone to smooth rough and ridged splines.

Use sealing compound on the metal part of all lip type oil seals. Use Lubriplate grease on all "O" ring seals.

Fig. O741—View of engine clutch and pto clutch elements seen after tractor split has been performed. For procedures covering removal of these parts, refer to paragraphs 132 through 136.

AS. Allen screw
GT. Grease tube
GTN. Grease tube nut
23. Engine clutch shaft
24. Engine clutch release shaft
25. Engine clutch release fork
29. Release bearing sleeve
31. Bearing tube carrier
35. PTO drive pinion
41. Pedal return spring
50. Bearing retainer
58. Release bearing
62. PTO clutch fork

Fig. O743A — Looking rearward after pto clutch, clutch spider and drive pinion have been removed from center frame.

CF. Center frame
CS. Cap screws
10. Inspection plate
13. PTO drive gear hub
23. Engine clutch shaft
33. Bearing retainer

PTO DRIVE GEAR

133B. **PTO DRIVE GEAR.** To remove pto drive gear (13—Fig. O738 and O743), first remove the pto clutch spider (drum) as outlined in paragraphs 132 and 132A. After removing pto clutch spider, pry off pto pinion shaft bearing retainer (50). Using a punch inserted in the $\frac{5}{16}$-inch diameter hole (H) located in pinion shaft (35), bump pinion shaft and bearing out of tractor center frame wall. Carefully slide pinion shaft off engine clutch shaft to prevent damaging lip type seal, located in inside diameter of pinion shaft. Refer to Fig. O743A and remove cap screws (CS) and nuts attaching tractor center frame (CF) to transmission housing and lift off center frame.

Remove pto drive gear bearing retainer (33). Press or pull pto drive gear (13—Fig. O743B) out of its bearing and tractor center frame (TCF).

Fig. O743B — Rear face of tractor center frame (TCF) with pto drive gear (13) mounted therein. Procedure for R&R of gear is in paragraph 133B.

Fig. O742 — Frontal view of center frame and transmission housing (rear frame) after pto clutch plates have been removed.

23. Engine clutch shaft
28. Clutch spider nut
30. PTO clutch spider
31. Bearing tube carrier
31S. PTO drive shaft
35. PTO drive pinion

Fig. O743—View of pto drive pinion shaft (35) after pto clutch and release bearing tube carrier (31—Fig. O742) have been removed. Procedure for removal of pto drive pinion shaft is in paragraph 135.

H. Pinion shaft hole
13. PTO drive gear
16. Center frame
33. Bearing retainer for 13
34. Pinion shaft oil seal
35. PTO pinion shaft
50. Bearing retainer for 35

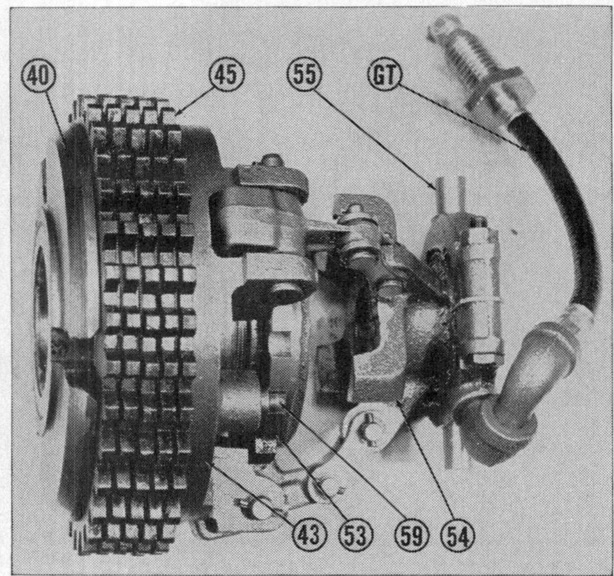

Fig. O744—Pto clutch assembly minus the spider or drum (30) shown in Fig. O742. Procedure for removal of clutch is in paragraph 132.

GT. Grease tube
40. PTO clutch hub
43. Floating plate
45. Driving plates
53. Clutch adjusting ring
54. Sliding sleeve
55. Release collar
59. Lock pin

Fig. O745A—Rear of tractor showing pto output shaft (18), hub (10) and bearing retainer (20).

Fig. O745—Some components of pto clutch assembly shown in Fig. O744.

37. Oil seal
40. PTO clutch hub
41. Hub drive key
43. Floating plate
44. Clutch driven plate
45. Clutch driving plate

133C. Inside diameter of drive gear hub contains two needle roller bearings (78) for support of pto drive shaft. Install bearings by pressing on end of bearing which is marked with the manufacturer's name. Install one bearing in gear end of hub $\frac{3}{32}$-inch below surface; and install other bearing $\frac{9}{16}$-inch below the end of the hub.

PTO DRIVE PINION SHAFT

The pto drive pinion shaft transmits power from the engine clutch cover to the pto drive gear (clutch).

135. **R&R AND OVERHAUL PTO PINION SHAFT.** To remove the pto pinion shaft (35—Fig. O741) either for removal of the engine clutch shaft, renewal of the pinion shaft, renewal of lip type seal located in inside diameter of pto shaft, or renewal of oil seal located in pinion shaft bearing retainer, first perform a tractor split as outlined in paragraph 131. Remove pto clutch and clutch spider as outlined in paragraphs 132 and 132A.

Pry off pto pinion shaft bearing retainer (50—Fig. O743). Using a punch inserted in the $\frac{5}{16}$-inch hole (H) located in pinion shaft, bump pinion shaft and bearing out of tractor center frame wall. Carefully slide pinion shaft off engine clutch shaft to prevent damaging lip type seal, located in inside diameter of pinion shaft.

136. Install needle bearing in inside diameter of pinion shaft by pressing on end of bearing which is marked with the manufacturer's name. Install this bearing within $\frac{1}{16}$-inch of the lip type oil seal, located in inside diameter of pinion shaft.

PTO DRIVE SHAFT

Refer to Fig. O738. The pto drive shaft (31) transmits power from the pto clutch to the pto (output) shaft. The pto (output) shaft (18) which is reversible to provide a spline size either of 1⅜ inches or 1⅛ inches can be removed after removing bolt (14).

138. **R&R PTO DRIVE SHAFT.** To remove pto drive shaft, proceed as follows: Remove rectangular-shaped hand hole plate from lower side of tractor center frame. Working through cover opening, loosen Allen set screw

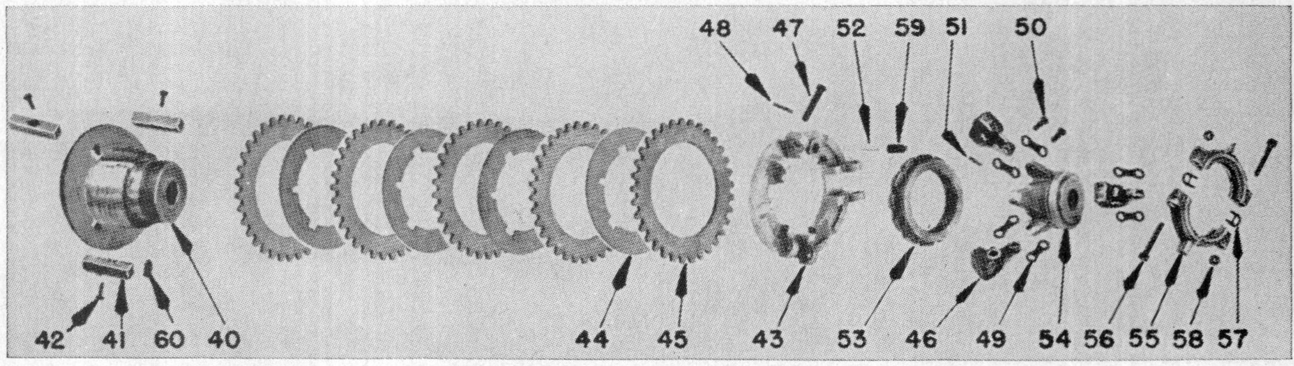

Fig. O745B—Components of pto clutch and hub assembly shown at (39) in Fig. O746.

40. PTO clutch hub
41. Drive key
44. Driven plate
45. Driving plate
46. Clutch lever
47. Lever pin
48. Roll pin
49. Lever link
51. Roll pin
52. Lock pin spring
53. Adjusting ring
54. Sliding sleeve
55. Sleeve (release) collar
57. Collar shim
58. Self-locking nut
59. Clutch lock
60. Self-locking set screw

(AS—Fig. O741) retaining pto clutch hub to pto drive shaft.

CAUTION: If hub has only one Allen screw discard it and install in its place one Oliver 1E-1677A dog point screw 3/8x1/2 also one 1E-1677B hollow lock screw 3/8x3/16-inch.

From rear face of transmission housing, remove pto shaft shield and pto (output) shaft retaining bolt (14—Fig. O745A). Extract pto (output) shaft (18), and remove bearing retainer (20). Lip type oil seal (32—Fig. O746), located in the bearing retainer, can be renewed at this time. Pull pto drive shaft (31) out of transmission housing.

CAUTION: When reinstalling shaft be careful not to push the pto clutch parts out of the clutch spider. Clutch assembly can be held in place by hand through the inspection port in bottom of center frame.

139. **OVERHAUL PTO DRIVE SHAFT.** The pto drive shaft rotates in five anti-friction type bearings which are located as follows: Two needle roller bearings (78) in bore of pto drive gear hub, two needle roller bearings (78—Fig. O722) in bore of transmission countershaft, and one ball bearing on pto hub.

To renew needle bearings in bore of pto drive gear, first remove drive gear as in paragraphs 133B and 133C. Needle bearing in front end of countershaft can be renewed at this time. Rear bearing in countershaft can be removed from opposite end of shaft by using an inertia type puller and installed by using a suitable Owatonna driving collar.

An "O" ring (12—Fig. O746) located between pto drive shaft and pto hub can be renewed after removing a 1/4-inch Roll pin (11).

Install oil seal (32) with lip facing the transmission gears.

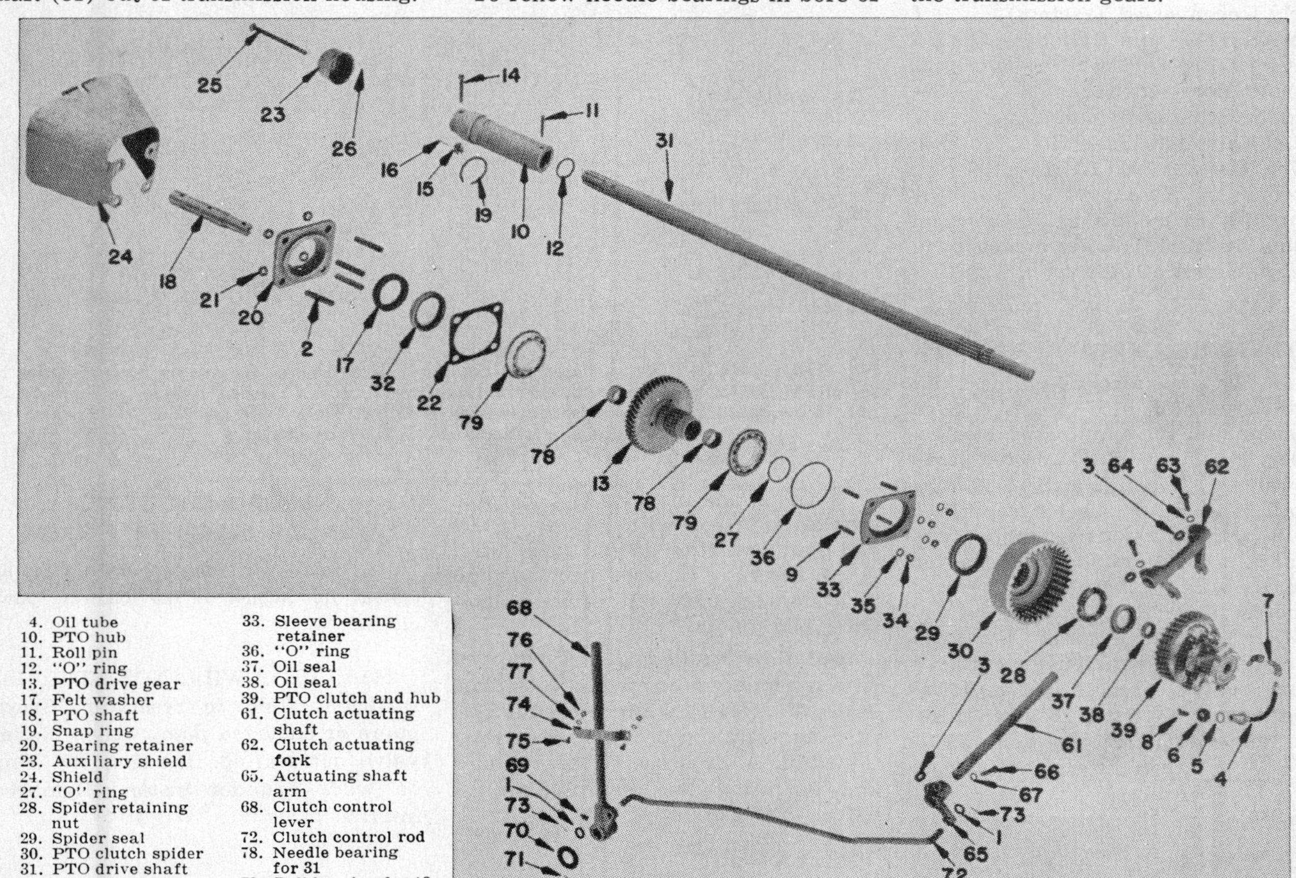

Fig. O746—Gear (13) drives all of the pto system but is in turn driven by the hollow pto pinion shaft (35—Fig. O743) which surrounds the engine clutch shaft and is splined into the engine clutch cover.

HYDRAULIC LIFT SYSTEM

Two types of systems have been used. Both types include a constant running Vickers vane type pump, a control valve, a single-acting cylinder, a reservoir, filter, lift arms and shaft, and three-point hitch linkage. All of the aforementioned components except the three-point hitch linkage are located on the hydraulic housing which is also the transmission housing top cover. A control valve which is mounted externally on the hydraulic housing is available as extra equipment for remote cylinder operation.

On series 550 tractors at serial number 64 218-500, the relief valve setting was raised from 1500 psi to 1700 psi. This increase in relief valve opening pressure was accomplished by changing the hydrostat valve (control valve plunger) and modifying the pump and outlet adapter.

Hydraulic systems on tractors prior to serial number 64 218-500 may be converted from 1500 psi to 1700 psi operating pressure by installing the heavy duty pump and the control valve (or valves), or in some cases, by changing the hydrostat valves. Service procedures remain the same.

142. In the first type system draft control and position control of implements were obtained by a control lever plus a separate selector lever. This first system called the "manual selector" type was superseded by the "double feed-back" system which does away with the separate selector lever. The pump, internal valve assembly, piston, draft control shaft, external lever, cross shaft, lift arms and three-point

hitch of the manual selector system are interchangeable with those of the double feed-back system.

The double feed-back system became effective in production at tractor serial 46001-500. Field changeover kits are available for converting the manual selector systems to the double feed-back type.

SERVICE SPECIFICATIONS

143. Specifications apply to both the manual selector and double feed-back systems:

Reservoir—Quarts 9
(Early Unconverted Tractors).......6
External Valve and Cylinder—Qts...2
Delivery GPM @ 2000 RPM—Qts. ..22
Cylinder Bore—Inches 3⅛
Cylinder Stroke—Inches 4⅞
*Relief Valve Setting—PSI.......1500
Lifting Time @ 2000 RPM—
Seconds 2½

*Applies to models prior to serial number 64 218-500. Later models have relief valve set at 1700 psi.

HYDRAULIC OPERATING FLUID

144. Recommended operating fluid for the hydraulic lift system is SAE 5W engine oil for temperatures below 32 deg. F., and SAE 30 for temperatures above 32 deg. F., or SAE 10W-30 for all seasons. Reservoir capacities are: 9 qts, for systems without external or remote cylinders, 11 qts. for systems with external or remote cylinders.

144A. To drain a system which is not equipped with a remote cylinder control valve proceed as follows: Place three point hitch in down position. Remove ¼-inch pipe plug (PP—Fig. O747) from right side of hydraulic housing, and install a ¼-inch pipe nipple, equipped with a hose. Start and operate engine at idle speed until all oil has been pumped from system; then stop the engine to prevent damaging the pump.

144B. If hydraulic system is equipped with a remote cylinder control valve, drain the system while the engine is running at idle speed by disconnecting the hose or hoses at the cylinder and operating the external control valve.

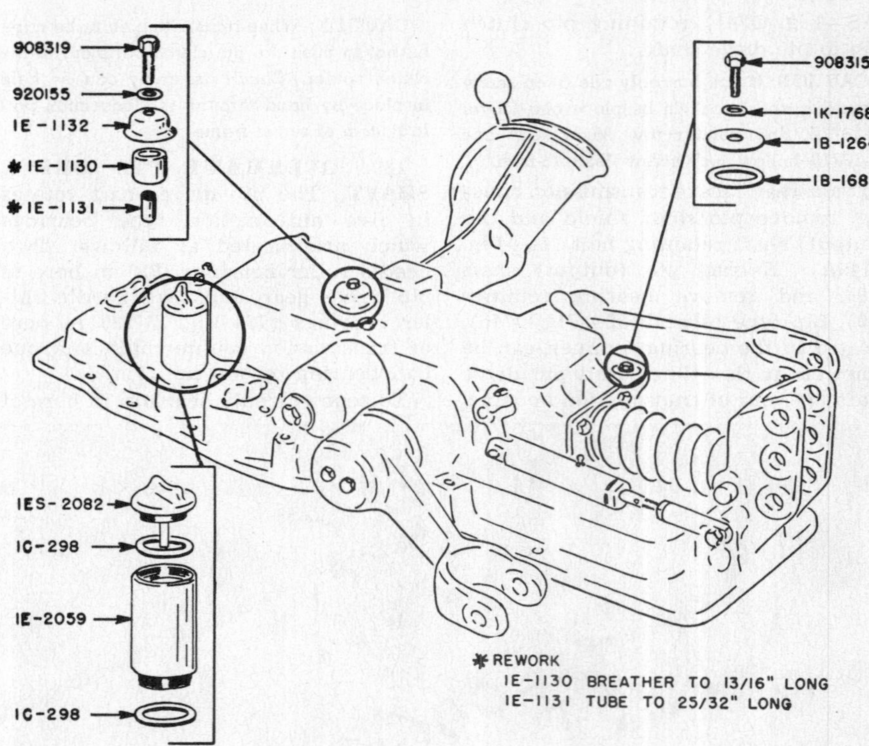

Fig. O747A—To provide sufficient oil capacity for hydraulic systems on tractors prior to serial 23131-500 equipped with external control valve and remote cylinder, a Field Package is available. Shown are the elements of Oliver Field Package 1ESR-2059 which includes a standpipe 1E-2059 as shown.

TROUBLE-SHOOTING MANUAL SELECTOR SYSTEM

145. Causes for faulty operation of the lift system are outlined in the following paragraphs.

145A. UNIT WILL NOT LIFT. Insufficient fluid in reservoir. Worn pump or damaged pump drive. Relief valve not seating. Leaking "O" ring on work cylinder piston. Damaged internal parts.

Fig. O747—Hydraulic systems having no remote cylinder are drained at pipe plug (PP). Draining of system with remote cylinder is as in paragraph 144B.

Fig. O750—External parts of manual selector hydraulic lift system.

L. Hitch link, lower
LL. Hitch lifting link
LS. Lift link clevis
PTO. Clutch control for PTO
16. Hydraulic housing
93. Lift arm
101. Lift rocker
118. Control lever
134. Draft control spring
140. Spring yoke pin
152. Selector control lever

145B. WILL NOT LIFT IN DRAFT CONTROL. Wrong draft control adjustment. Readjust as in paragraphs 148 and 150. On tractors prior to serial 18135-500 look for worn internal control shaft, or adjusting plunger or external link spring that has taken a permanent set.

145C. ERRATIC DRAFT CONTROL. Same as 145B but also check for worn or binding hydraulic rocker, sticky control valve.

145D. EXCESSIVE SETTLING OR CYCLING. Leakage in system usually at cylinder or between spool valve and bore or at ball check valve.

145E. RELIEF VALVE BLOWS AFTER LIFTING. Pilot valve spool not returning to neutral, worn cross shaft bushings or draft control shaft.

145F. PUMP NOISY IN OPERATION. Air leak on inlet side of pump. If noise occurs only when external or remote cylinder is operated, it indicates insufficient fluid in reservoir. If reservoir fluid level is O.K., trouble could be caused by insufficient fluid capacity in reservoir. Increase reservoir capacity by installing stand pipe kit (Oliver Field Package No. 1E SR-2059) as shown in Fig. O747A.

Other causes of noisy pump operation are: (a) Pump vanes sticking, (b) relief valve chattering, (c) poor quality of oil causing air bubbles, or (d) oil too heavy.

BLEEDING LIFT SYSTEM

146. Systems equipped with an external control valve will require bleeding whenever the system is drained or the hoses at the cylinder are disconnected. To bleed the system, proceed as follows: With engine operating at idle speed, loosen the connection of one hose at the cylinder. Hold the control valve lever in either extreme position until there is a continuous flow of oil without air bubbles at the loosened connection. Repeat this operation for the other hose if system is equipped with a double-acting cylinder; then recheck oil supply in the reservoir.

MAIN SYSTEM LINKAGE ADJUSTMENTS

MANUAL SELECTOR SYSTEM

There are three adjustments on this early production system as follows: Position Control, Draft Control and Draft External Linkage. The draft control external linkage adjustment is made without disturbing the hydraulic housing, the other two adjustments require removal of the hydraulic housing which is the transmission top cover.

148. **DRAFT CONTROL EXTERNAL LINKAGE.** To determine if the link adjusting plunger (149—Fig. O751) is adjusted correctly, proceed as follows: Place selector lever (152—Fig. O750) in draft control (straight-up) position. With engine operating at 2000 rpm, slowly move control lever (118) from DOWN position to UP position. Lift arms should start to raise when control lever reaches a point 2½ inches (not 1¼-1½) from extreme UP position. If control lever is not so positioned at start of raise refer to Fig. O751 and loosen locknut (150) on threaded portion of plunger and rotate the adjusting plunger (149) until these limits are obtained. To rotate the adjusting plunger insert a small diameter punch in drilled hole of same.

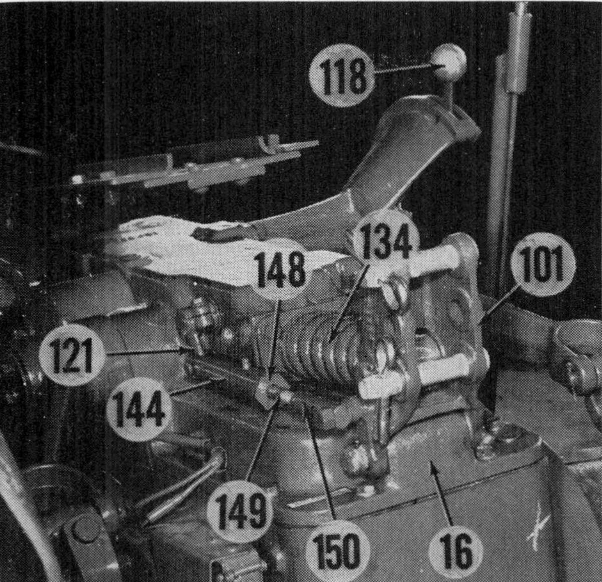

Fig. O751—Showing rear and left side of hydraulic housing (16) and draft control external linkage used on manual selector system.

16. Hydraulic housing
101. Lift rocker
118. Control lever
121. Draft control external lever
134. Draft control spring
144. External link front end
148. External link nut
149. External link adjusting plunger
150. Lock nut

149. **POSITION CONTROL.** To check the position control adjustment when the hydraulic housing (transmission top cover) is installed, proceed as follows: With engine operating and lift arms raised and selector lever (152—Fig. O750) in the position control (down) location, it should be possible to manually raise the lift arms approximately one inch higher. If lift arms behave otherwise than described, move the control lever (engine running) to down position. While lift arms are in down position move selector lever (152) from draft control to position control. When changing from draft control to position control the lift arms should not move. If there is any movement at the lift arms, the position control is in need of adjustment as outlined in paragraph 149A.

149A. To adjust the position control linkage, which should be adjusted before adjusting the internal draft control linkage, proceed as follows. Drain lift system reservoir as per paragraph 144A. Remove hydraulic housing from transmission housing, and reservoir oil pan from hydraulic housing. View of hydraulic housing with reservoir oil pan removed is shown in Fig. O759.

Place selector lever (152—Fig. O750) in position control (down) and control lever (118) in full raise (up) position. Refer to Fig. O752, and insert a ⅜-inch thick spacer between hydraulic housing and stop on rocker arm as shown. Hold rocker arm against this spacer, and check location of machined shoulder on pilot valve spool. The machined shoulder on pilot valve spool should be flush with end of valve body. If shoulder is not flush, disconnect adjusting link (A) from end of cam follower and rotate the clevis end until flush condition of pilot valve spool is obtained.

150. **DRAFT CONTROL INTERNAL LINKAGE.** To adjust the draft control internal linkage, which should be adjusted after adjusting the position control linkage, proceed as follows: Drain lift system reservoir as outlined in paragraph 144A. Remove hydraulic housing from transmission housing, and reservoir oil pan from hydraulic housing. Disconnect draft control external linkage from the lift rocker.

Place selector lever (152—Fig. O750) in draft control (up) position and control lever (118) 2½ inches from extreme UP position. The former factory recommended dimension of ½ to ¾ or 1¼ to 1½ inches for control

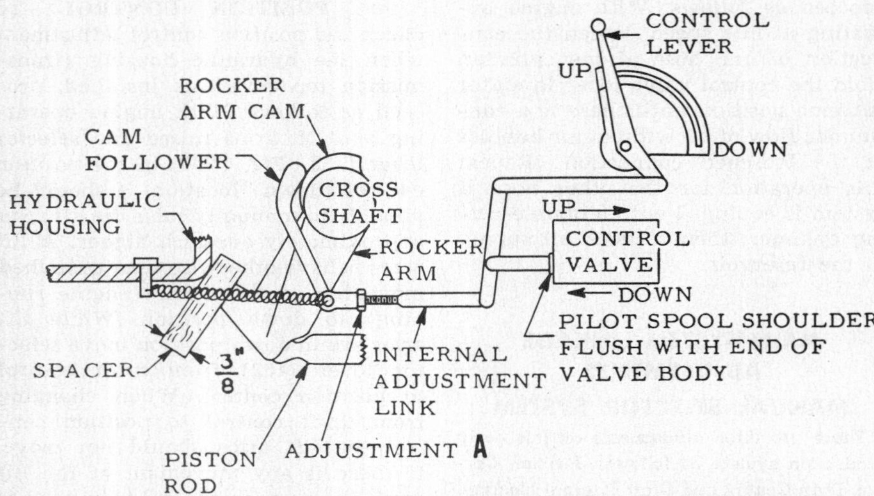

Fig. O752—Manual selector system simplified diagram of set up for making position control internal adjustment of hydraulic system. Refer to paragraphs 149 and 149A for checking and adjusting procedure.

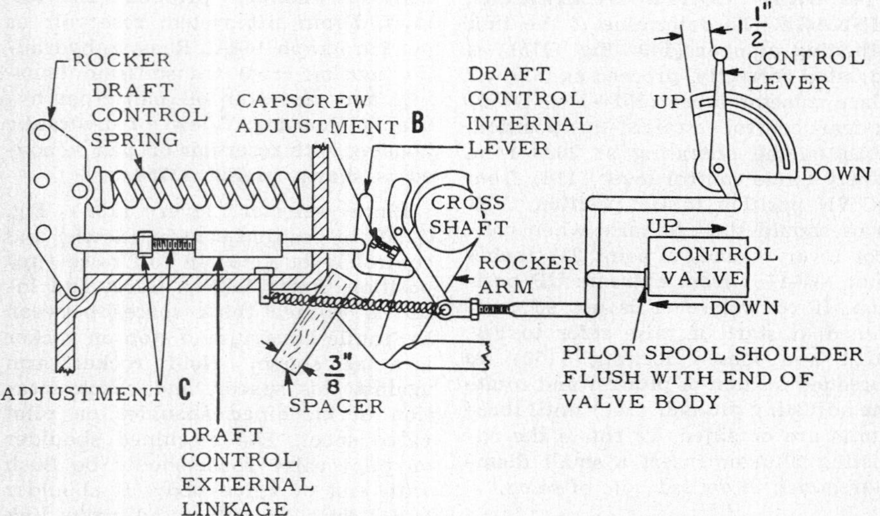

Fig. O753—Manual selector system simplified diagram of set up for making draft control internal adjustment. Refer to paragraphs 150 and 150A for checking and adjusting procedure.

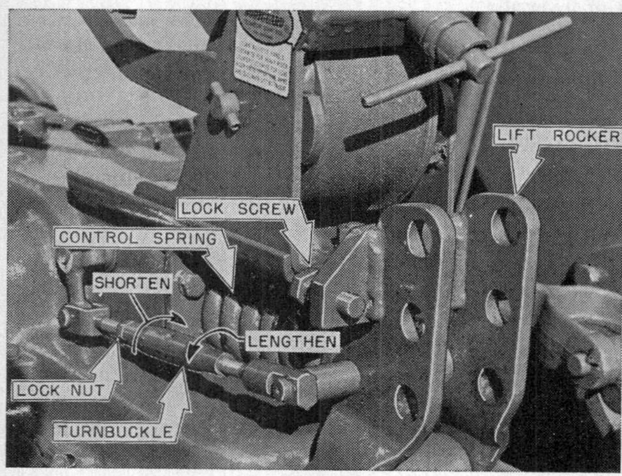

Fig. O754—The ½- to 1-inch free travel of lift arm is controlled by adjusting turnbuckle as shown above. This is the only adjustment on the double feed-back system.

lever position has been cancelled. Refer to Fig. O753, and insert a ⅜-inch thick spacer between hydraulic housing and stop on rocker arm as shown. Hold rocker arm against this spacer, and push draft control internal lever forward until adjusting cap screw in draft control lever touches the rocker arm. At this time, the machined shoulder on pilot valve spool should be flush with end of valve body. If machined shoulder is not flush, rotate the adjusting cap screw (B) until flush condition of pilot valve spool is obtained.

Recheck the adjustment, and reconnect the valve spool return spring.

DOUBLE FEED-BACK SYSTEM

These systems can be distinguished from the manual selector system by the fact they have only a single control lever instead of a control lever and selector lever. The single adjustment can be made without disturbing the hydraulic housing.

150A. CONTROL ADJUSTMENT. With engine operating at idle speed and the three-point hitch freed of any implement load, move the control lever to the full raised position at the top of the quadrant. The system is correctly adjusted if the lift arms can now be raised an additional ½- to 1-inch by hand.

To obtain the ½- to 1-inch distance rotate the turnbuckle (Fig. O754) clockwise (viewed from the rear) to increase, counter-clockwise to decrease the distance. Recheck and readjust if necessary.

TESTING SYSTEM PRESSURE

151. SYSTEMS WITHOUT EXTERNAL CONTROL VALVE. Connect a high reading pressure gauge after removing plug (45—Fig. O763) from hydraulic passage cover (42). Relief valve unseats at a pressure of 1500 psi on early models, or 1700 psi on late models.

151A. SYSTEMS WITH EXTERNAL CONTROL VALVE. A check of the relief valve unloading pressure (located in body of external control valve) can be made by connecting a high reading pressure gauge into one of the valve to cylinder hoses. Relief valve should unseat at 1500 psi on early models, or 1700 psi on late models.

MAIN SYSTEM R&R AND OVERHAUL

This manual selector system differs from the double feed-back system mainly in the details of the internal and external control linkage. Servicing information contained in paragraphs 152 through 158 applies therefore to both systems. The accompanying illustrations will show the details of the two linkage systems.

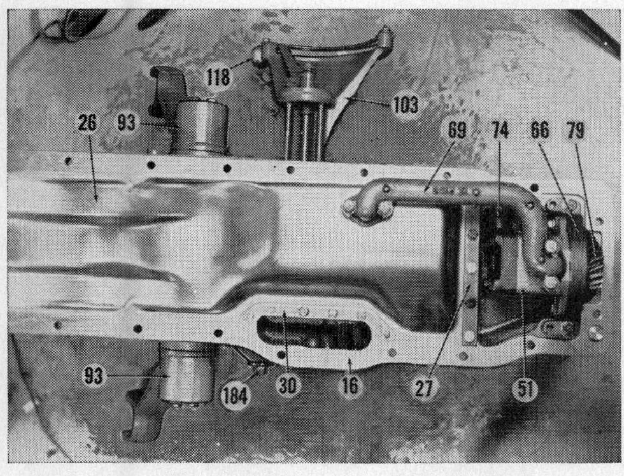

Fig. O755 — Underside view of removed hydraulic housing with attached Vickers pump.

16. Hydraulic housing
26. Reservoir oil pan
27. Pan reinforcement, front
30. Pan reinforcement, left
51. Pump
66. Pump bracket
69. Inlet adapter
74. Outlet adapter
79. Pump gear
93. Lift arm
103. Control quadrant
118. Control lever
184. Filter body

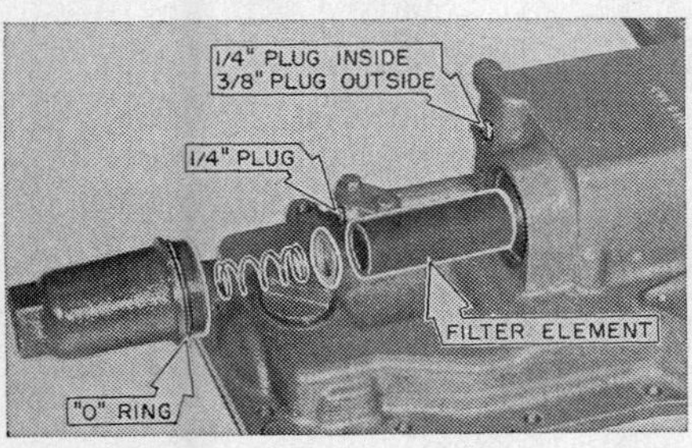

Fig. O755B—View of the full flow oil filter exploded from housing. Unit is equipped with a spring loaded by-pass valve to prevent oil starvation should filter element become plugged.

152. **R&R HYDRAULIC HOUSING.** To remove hydraulic housing (transmission housing top cover), proceed as follows: First drain hydraulic system as outlined in paragraph 144A. Remove seat pan by sliding it off of seat spring assembly. Remove transmission gear shift lever. Disconnect lifting link assemblies from lift arms. Remove cap screws attaching top cover to transmission housing, and lift cover off transmission housing.

153. **PUMP.** The Vickers vane type pump which is mounted on the underside of the hydraulic housing receives its drive through an idler gear which meshes with the pto shaft pinion. Hydraulic system relief valve which is non-adjustable is located in control valve body.

To remove the pump, first drain the system as outlined in paragraph 144A, and remove the hydraulic housing as outlined in paragraph 152. Refer to Fig. O755, and remove inlet adapter (69). Remove two cap screws attaching outlet adapter (74) to hydraulic housing. Remove cap screws attaching pump bracket (66) to hydraulic housing, and lift off pump.

To remove the pump drive idler gear, it will be necessary either to perform a tractor split or to remove the steering gear unit as outlined in paragraph 7. Idler rotates on a bronze bushing which is not supplied separately for service.

153A. To disassemble a removed pump unit refer to Fig. O756 and proceed as follows: Remove cap screws (2) attaching pump cover (1) and lift off cover. Lift off compression spring (3), and pressure plate (4). Before removing rotor (7) carefully note position of embossed arrow on outer surface of ring (6) in relation to pump body. Remove ring (6); then pull rotor (7) and vanes (8) from splined end of shaft (13).

To inspect pump shaft bearings (16 and 17) or renew seal (12), unbolt pump bracket from pump body and remove snap ring (15). Pull shaft and bearings (17) as a unit from pump body. Bearing (16) will remain in pump body. To remove inner bearing

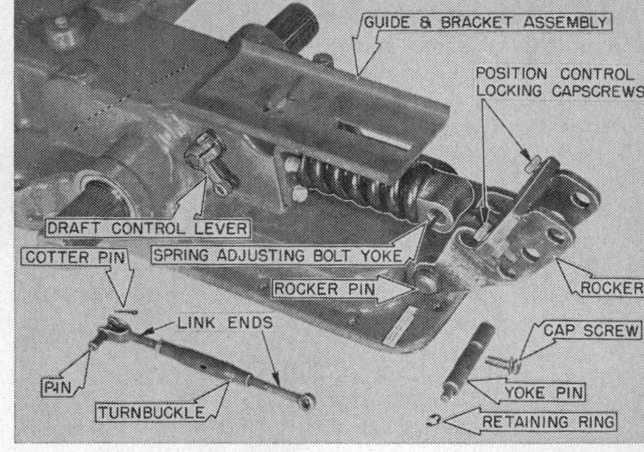

Fig. O756A—View showing hydraulic housing of double feed-back system with external linkage removed.

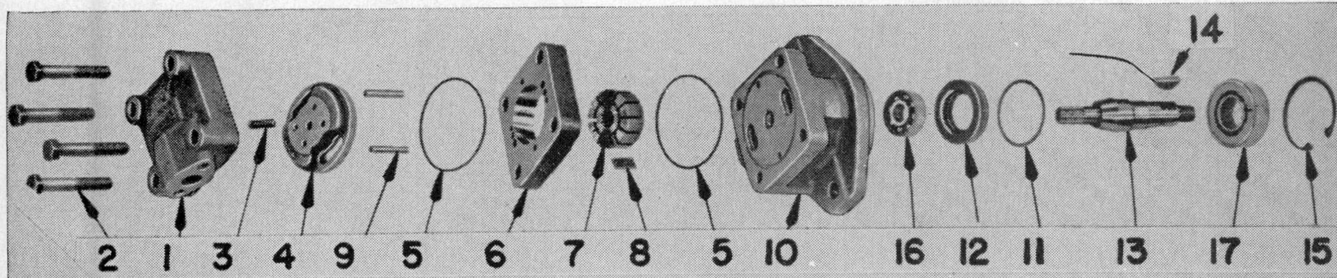

Fig. O756—Components of Vickers vane type hydraulic pump.

1. Cover
2. Cover screws
3. Compression spring
4. Pressure plate
5. Sealing ring
6. Pump ring
7. Rotor assembly
8. Rotor vane
9. Dowel pin
10. Pump body
11. Bearing spacer
12. Shaft seal
13. Pump shaft
15. Snap ring
16. Inner ball bearing
17. Outer ball bearing

(16) from pump body first extract seal (12) and spacer (11). Using a drift bump bearing (16) out of pump body.

153B. Thoroughly wash all parts and inspect their rubbing surfaces for wear, scoring and pitting. All vanes should slide freely in their rotor slots.

153C. Observe the following points when reassembling the pump. Carefully oil all parts with engine oil prior to reassembly. Refer to Fig. O757 and install new shaft seal into housing counterbore keeping the installing sleeve (IS) in place. Assemble shaft through seal being careful not to push out the installing sleeve until lip of seal is riding on polished surface of shaft.

Assemble vanes to rotor and ring with rounded ends of vanes contacting the race of the ring.

Tighten pump cover screws (2—Fig. O756) to a torque of 65-75 Ft. Lbs.

154. **PUMP TEST SPECIFICATIONS.** Test specifications for Vickers pump are: Pump delivery is 5.0 gallons per minute at 2000 pump rpm at a gauge pressure of 1500-1700 psi. Hydraulic system pressure relief valve which is non-adjustable is located in control valve body.

155. **INTERNAL CONTROL VALVE.** The internal valve body mounted on the hydraulic cylinder contains a three spool open center pilot valve, a three spool directional valve, a spool type hydrostat plunger and a ball type check valve as shown in Fig. O760.

To remove the valve body assembly, first drain the system as outlined in paragraph 144A; then remove the hydraulic housing as outlined in paragraph 152. Unbolt hydraulic inlet adaptor from pump and reservoir pan, and reservoir oil pan from hydraulic housing. Refer to Fig. O759 and remove clevis pin or retainer attaching connecting rod (89) to rocker arm (94). Detach linkage adjusting rod (164) from valve actuating lever (166). Remove four cap screws attaching cylinder to hydraulic housing, and lift off cylinder and valve body assembly. Remove four cap screws attaching valve body to cylinder. Note five sealing "O" rings.

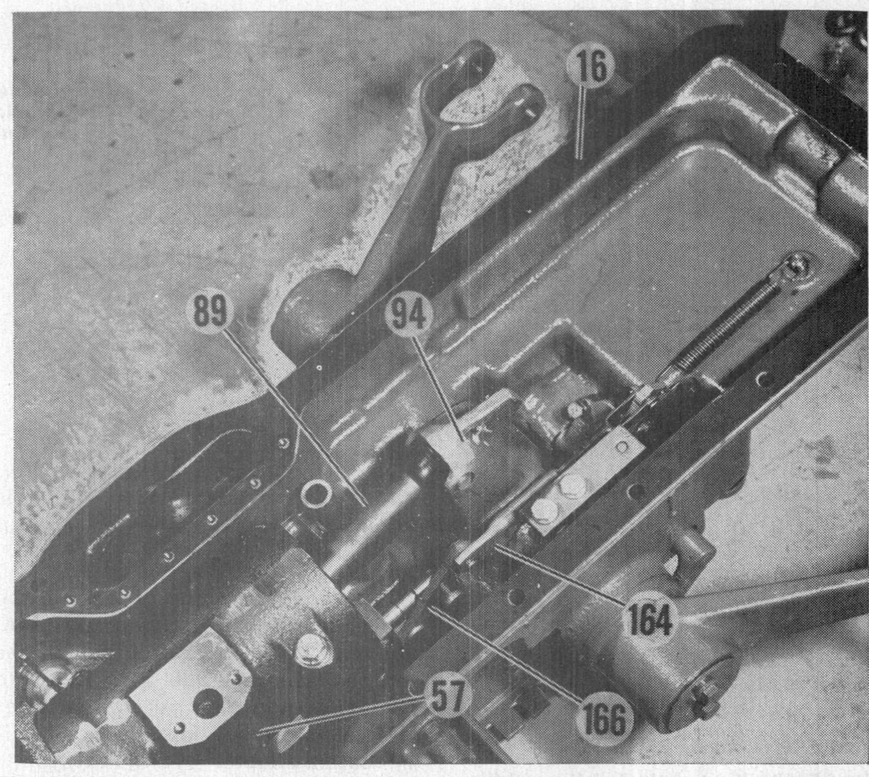

Fig. O759—Manual selector system. Underside view of removed hydraulic housing with reservoir oil pan removed.

16. Hydraulic housing
57. Hydraulic cylinder
89. Piston connecting rod
94. Hydraulic rocker arm
164. Hydraulic linkage adjusting rod
166. Hydraulic valve actuating lever

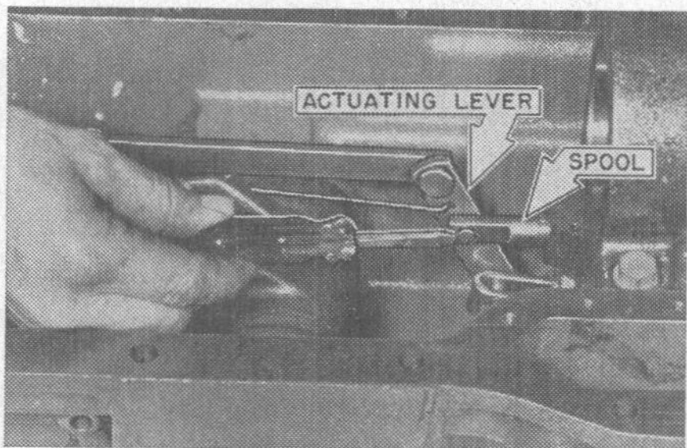

Fig. O759A — Disengage valve spool end from actuating lever by removing retaining ring as shown.

Fig. O757 — Details of shaft seal used on Vickers hydraulic pump. Seal installing sleeve (1S) is furnished with each new seal and must be removed after seal is installed.

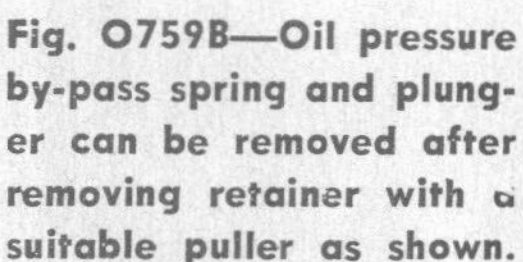

Fig. O759B—Oil pressure by-pass spring and plunger can be removed after removing retainer with a suitable puller as shown.

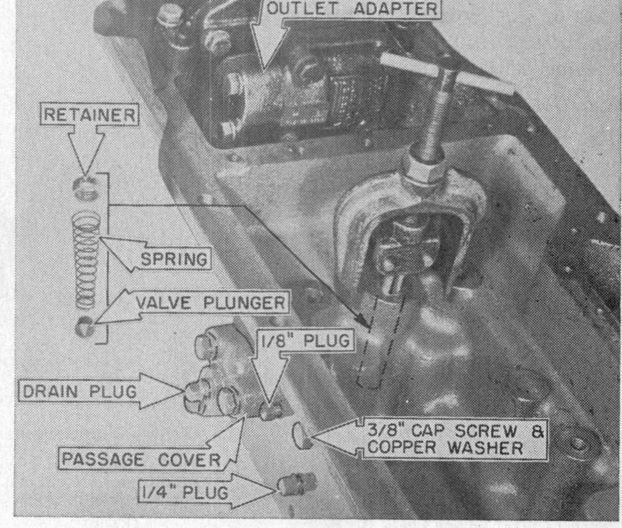

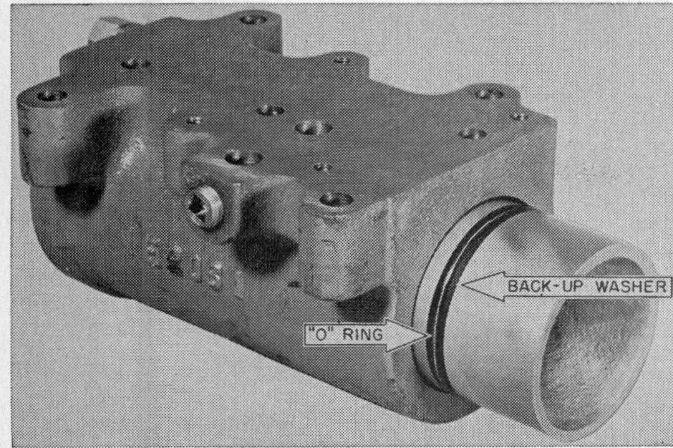

Fig. O759C—View showing piston partially removed from cylinder. Note position of "O" ring and back-up washer.

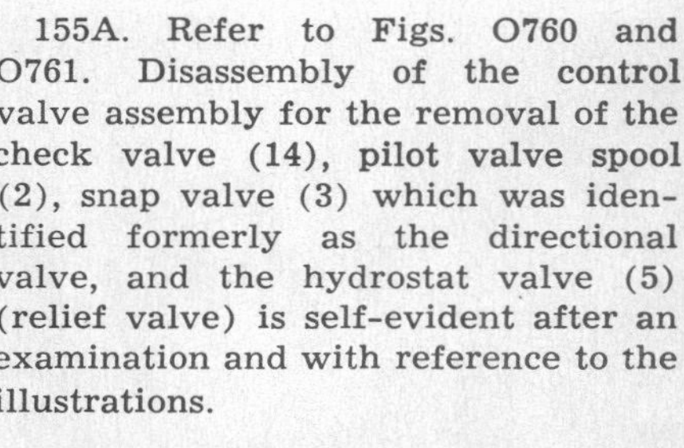

155A. Refer to Figs. O760 and O761. Disassembly of the control valve assembly for the removal of the check valve (14), pilot valve spool (2), snap valve (3) which was identified formerly as the directional valve, and the hydrostat valve (5) (relief valve) is self-evident after an examination and with reference to the illustrations.

Thoroughly wash all parts in solvent. Inspect valve body bores and all other parts for scoring, pitting and ridging. The hydrostat valve (5) which is also the relief valve is factory adjusted with shims (12) to relieve the system pressure at 1500 psi for early models; or 1700 psi for late models. Do not attempt to reset the system unloading pressure. Hydrostat valve assembly (5) comprises the items shown bracketed in Fig. O760 but is available only as a complete assembly.

155B. Early production tractors were equipped with an overload valve (192—Fig. O762) which is no longer used. This valve, adjusted at the factory to unseat at 1700 psi, was installed to relieve any sudden shock loads imposed on the system. Remove this valve when encountered, and replace it with a pipe plug.

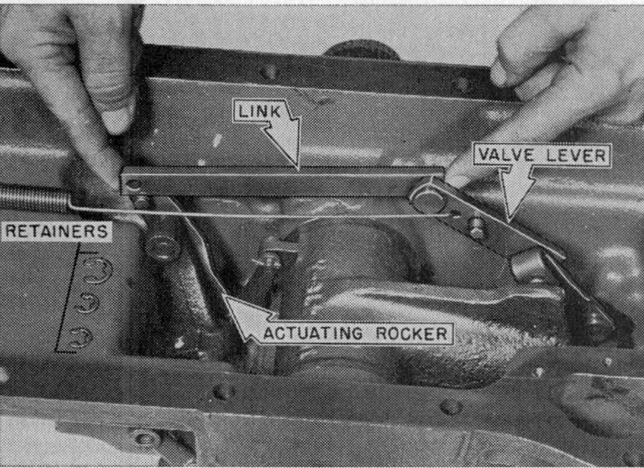

Fig. O759D—Remove actuating rocker-to-valve lever link as shown.

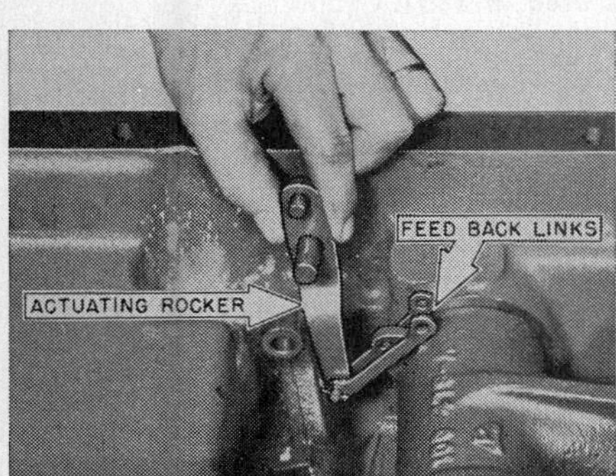

Fig. O759E — With return spring and pin from feed-back links removed, the actuating rocker and feed-back links can be removed as shown.

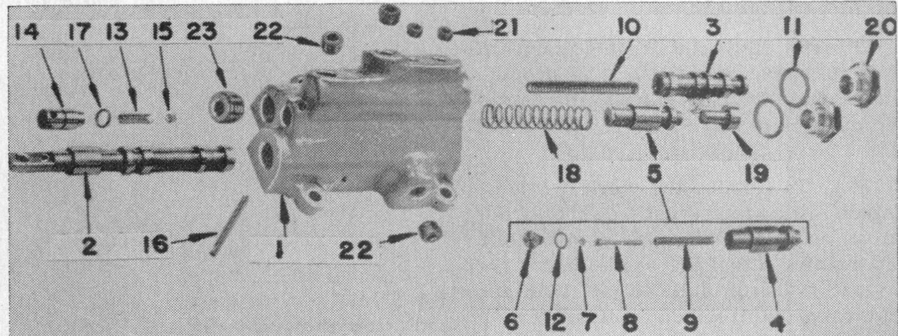

Fig. O760—Components of internal control valve which is bolted to upper face of hydraulic cylinder and carried in the hydraulic housing (16) shown in Fig. O759. For service procedure refer to paragraph 155.

1. Valve body
2. Pilot valve spool
3. Directional valve
5. Hydrostat valve
6. Plunger plug for 5
7. Ball, 5/32" for 5
8. Spring guide for 5
9. Plunger spring for 5
10. Directional valve spring
11. Plug washer ¾"
12. Shim for 5
13. Check valve spring
14. Check valve body
15. Ball, 9/32"
16. Roll pin, 3/16"
17. Sealing ring
18. Spring for valve 5
19. Spacer
20. Snap valve plug
21. Pipe plug ⅛"
22. Pipe plug ¼"
23. Pipe plug ½"

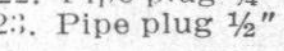

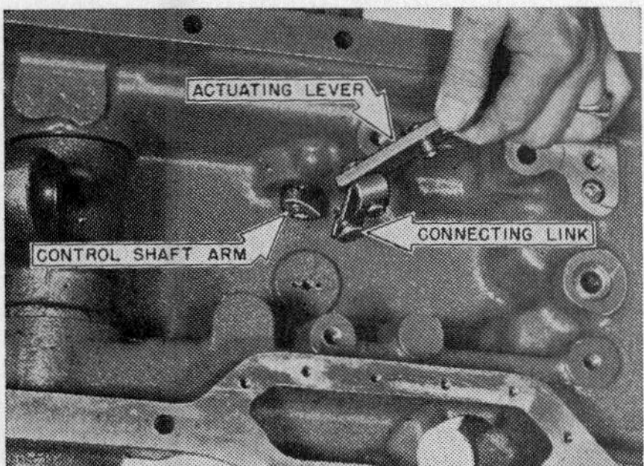

Fig. O759F—With rocker-to-valve lever link and rocker and feed-back links removed the connecting link and actuating lever can be removed as shown.

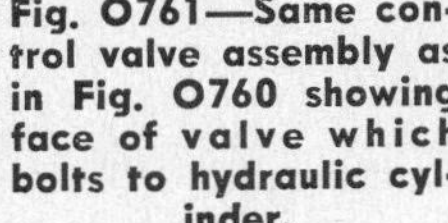

Fig. O761—Same control valve assembly as in Fig. O760 showing face of valve which bolts to hydraulic cylinder.

1. Control valve body
47. "O" ring, ¾" I. D.
49. "O" ring, ½" I. D.
91. "O" ring, 9/16" I. D.
164. Linkage adjusting rod
166. Valve actuating lever
172. Torsion spring for 166

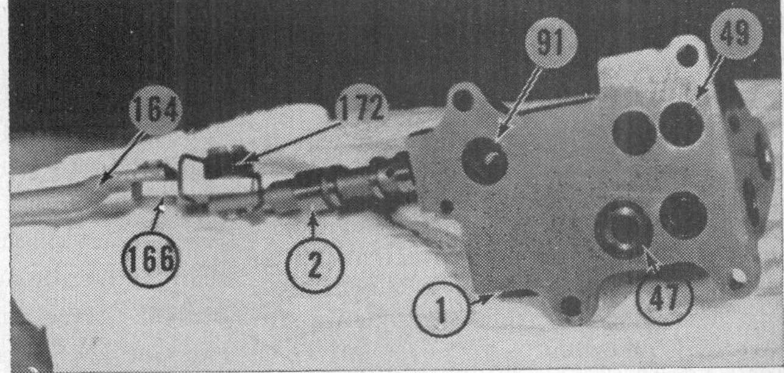

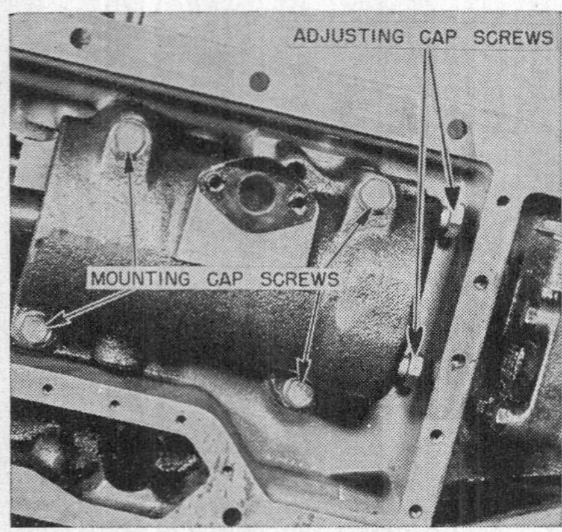

Fig. O761A — With hydraulic cylinder pushed forward and mounting cap screws tightened to 45-50 ft.-lbs. torque, adjust set screws on end of cylinder until they are snug against housing wall and tighten the lock nuts.

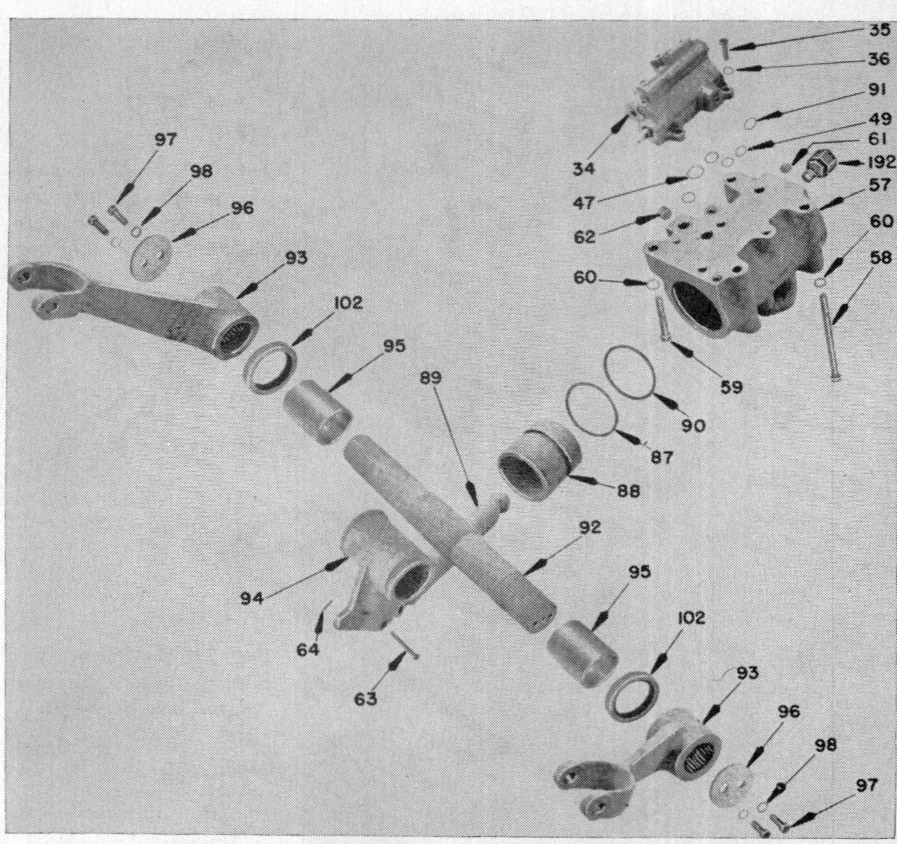

Fig. O762—Hydraulic cross shaft (92), cylinder (57) and internal control valve (34) elements of manual selector lift system which are mounted on the hydraulic housing shown at (16) in Fig. O751. In double feed-back system the arms (93) are retained by snap rings and a screw is used at (63).

34. Internal valve assembly
47. "O" ring, ¾" I.D.
49. "O" ring, ½" I.D.
57. Hydraulic cylinder
62. Pipe plug, ⅜"
87. Back-up washer
88. Hydraulic piston
89. Connecting rod
90. Piston "O" ring
91. "O" ring, 9/16" I.D.
92. Cross shaft
93. Lift arm
94. Rocker arm
95. Bushing for 92
96. Plate
102. Oil seal
192. Relief valve (obsolete)

156. **WORK CYLINDER AND PISTON.** The hydraulic (work) cylinder (57—Fig. O759 or O762) is mounted on the underside of the hydraulic housing. Work cylinder piston, equipped with an "O" ring and leather back-up ring, can be removed as follows: After removing hydraulic housing from tractor and reservoir oil pan from hydraulic housing, remove clevis pin or retainer attaching connecting rod to rocker arm. Remove connecting rod and piston.

"O" ring (90) and leather back-up ring (87) are mounted in one groove. Install the "O" ring nearest the top of the piston.

156A. To remove the work cylinder, detach linkage adjusting rod from valve actuating lever. Remove four cap screws attaching cylinder to hydraulic housing, and lift off cylinder and valve control body assembly.

When reinstalling the cylinder and control valve body assembly, use a new "O" ring Fig. O761 between cylinder and hydraulic housing.. Tighten the four attaching cap screws just lightly; then push cylinder assembly as far forward as it will go, and tighten the cap screws firmly.

The two adjusting screws on the front side of later cylinders should be turned outward until their heads apply a slight pressure on the hydraulic housing.

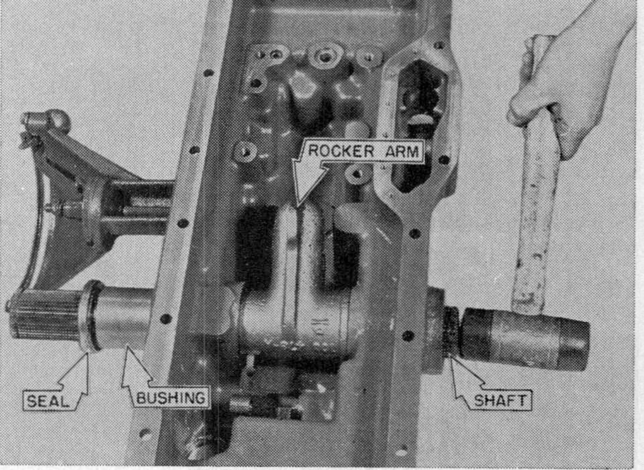

Fig. O762A — Drift cross shaft from either side with a soft faced hammer as shown. Notice that seal and bushing are forced out on opposite side. Seal will be damaged and will require renewal.

157. **LIFT ARMS CROSS SHAFT.** Refer to Fig. O762. The lift arms cross shaft (92) is supported in renewable bushings (95). Bushings are presized and will require no final sizing if carefully installed.

To renew the bushings, it will be necessary to remove the cross shaft as follows: Remove hydraulic housing from tractor, and reservoir oil pan from hydraulic housing. Remove clevis pin (63) or retainer attaching connecting rod (89) to rocker arm (94). Remove both lift arms (93) from cross shaft. Lift arms have a blind spline for correct assembly of arms to cross shaft. Bump cross shaft (92) out of rocker arm and hydraulic housing. One bushing and one oil seal for the cross shaft will be removed when bumping cross shaft out of hydraulic housing.

Fig. O762B — With draft control shaft removed the oil seal shown can be removed. Reinstall seal with lip facing inward.

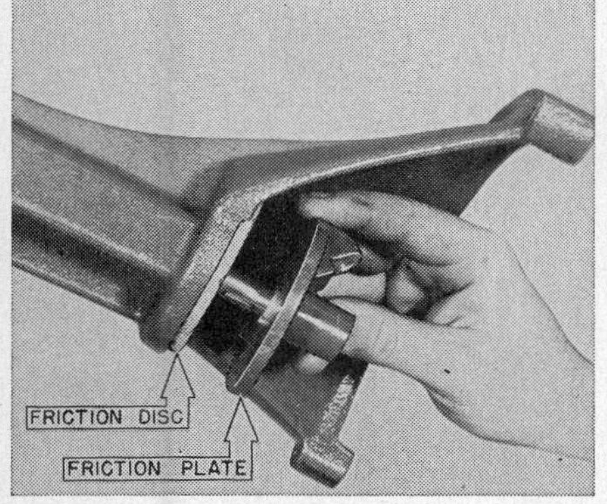

Fig. O762C—View showing removal of friction plate and disc. Note that friction disc is between plate and control quadrant.

Fig. O762D — Install cross shaft bushing until outer end is flush with bottom of oil seal counterbore. Install oil seal with lip pointing inward and drive seal into position using lift arm as a driver.

Fig. O762E—Right side of hydraulic housing showing parts to be removed when extra external control valve Fig. O765A is added to regular hydraulic system. Refer to paragraph 160.

17. Cap screw
23. Pipe plug
42. Hydraulic passage cover
45. Pipe plug (testing)

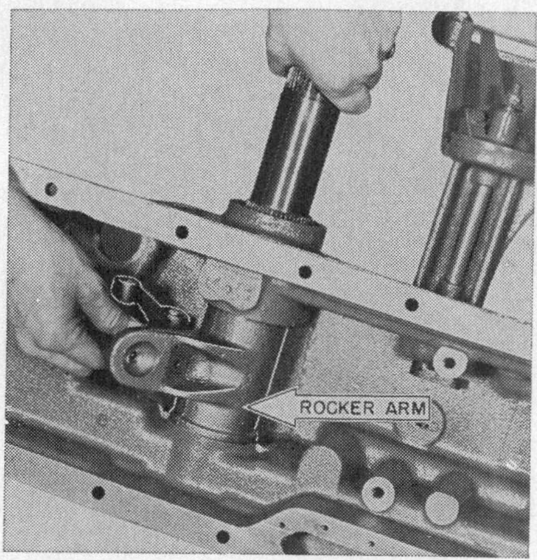

Fig. O763—With rocker arm positioned in housing, slide cross shaft into position from side without bushing or seal. Be sure to align the blind splines. Install remaining cross shaft bushing using a piece of 2-inch pipe as a driver, then install oil seal using lift arm as a driver.

158. **FILTER.** Hydraulic lift system is provided with a bypass filter which has a renewable element. Filter is located externally.

REMOTE UNIT OVERHAUL

Extra equipment, available for attachment to the main hydraulic system, includes a remote or external work cylinder and a remote or external control valve. The latter controls the operation of the remote cylinder.

160. **REMOTE CONTROL VALVE.** The remote control valve which is used to operate either single or double acting remote cylinders is a combination four-way directional flow valve with a built-in manually adjustable (flow control) valve, a relief valve and a check valve.

The control valve shown in Fig. O765A is installed on the right-hand side of the hydraulic housing after removing the hydraulic passage cover (42—Fig. O762E), two pipe plugs (23) and cap screws (17).

160A. Removal and disassembly of the remote control valve is self-evident after an examination of the unit and with reference to Figs. O762E and O765A.

The hydrostat valve (4) which is also the relief valve is factory adjusted with shims (10) to relieve system pressure at 1500 (or 1700) psi. Do not attempt to reset the system unloading pressure. Hydrostat valve is available only as a complete assembly and is the same part as used in the internal valve.

After installing the remote control valve, bleed the hydraulic system as outlined in paragraph 146.

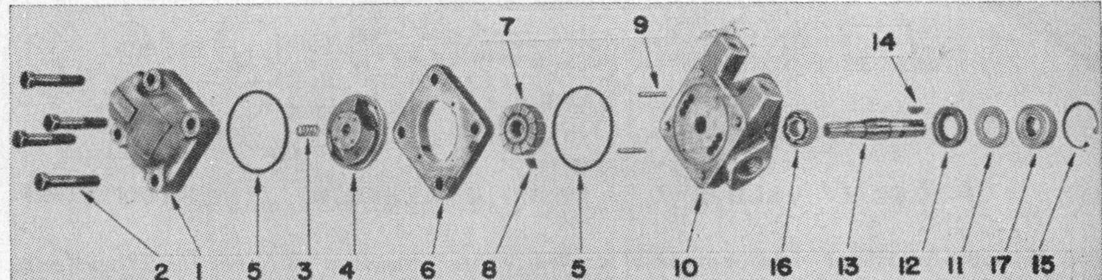

Fig. O765—This optionally available, engine mounted Vickers hydraulic pump is serviced in the same manner as the transmission mounted pump of the regular system. Refer to paragraph 153.

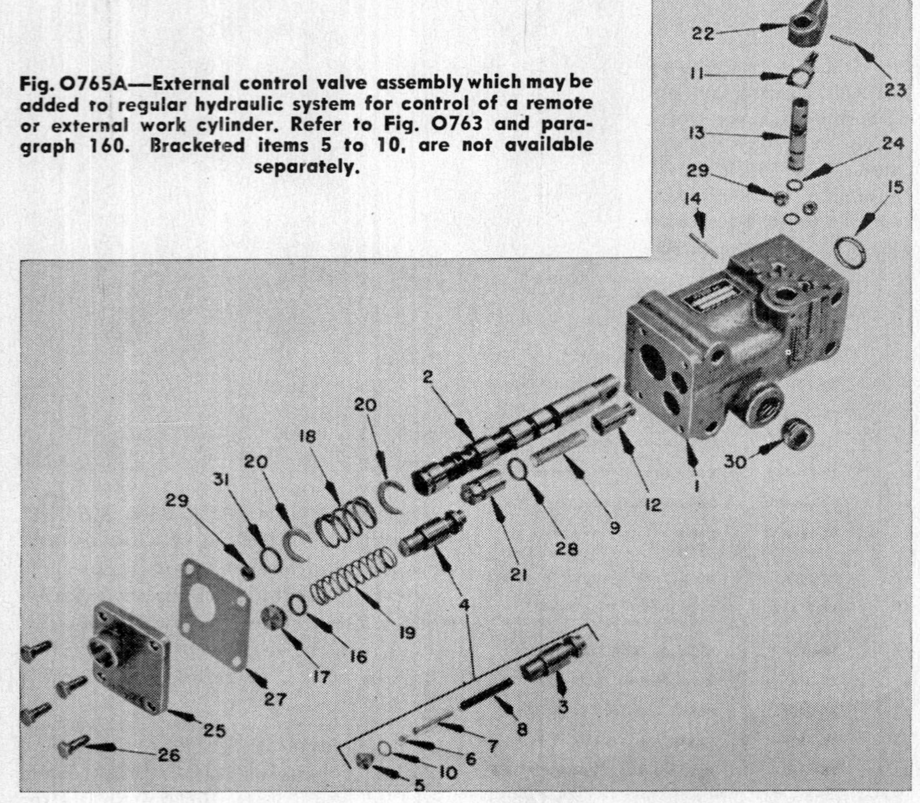

Fig. O765A—External control valve assembly which may be added to regular hydraulic system for control of a remote or external work cylinder. Refer to Fig. O763 and paragraph 160. Bracketed items 5 to 10, are not available separately.

1. Valve body
2. Control valve spool
4. Plunger assembly
5. Plunger plug
6. Plunger ball
7. Plunger spring guide
8. Plunger spring
9. Control valve spring
10. Plunger plug shim
11. Spring for valve 13
12. Check valve
13. Manual adjusting (flow control) valve
15. Sealing ring, ¾" I.D.
16. Sealing ring, ½" I.D.
17. Retainer for spring 19
18. Control valve spring
19. Flow control spring
20. Retainer for spring 18
21. Retainer
22. Adjusting knob
24. Sealing ring, 5/16" I.D.
25. Valve spool cap
28. Sealing ring, 7/16" I.D.
29. Pipe plug, ⅛"
30. Pipe plug, ⅜"
31. Sealing ring, 9/16" I.D.

Bracketed items 3-5, etc., not available separately

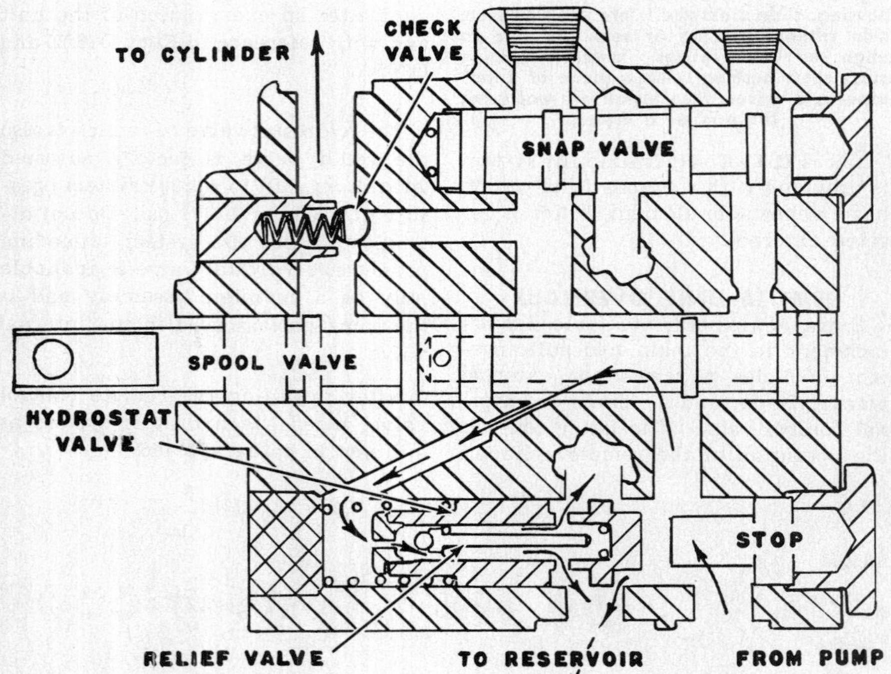

Fig. O765B—Sectional view of internal control valve (attached to underside of hydraulic housing) used on standard internal work cylinder system. An exploded view of this valve is shown in Fig. O760. Also available and shown in Fig. O765A is an external valve with manual flow (volume) control for operation of a single or double acting remote work cylinder.

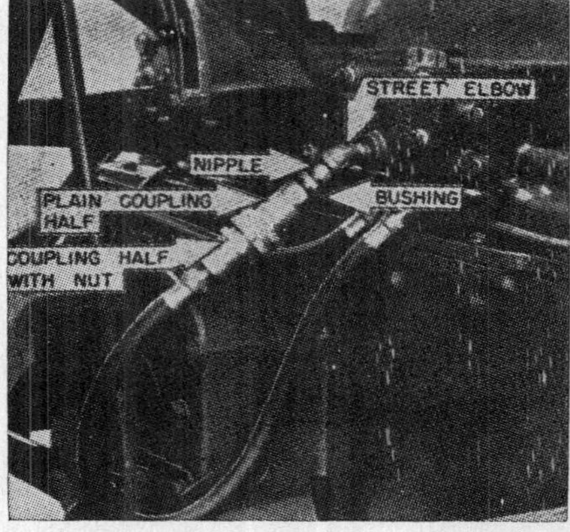

Fig. O766—External control valve installed on side of manual selector system hydraulic housing. An exploded view of this valve is shown in Fig. O765A.

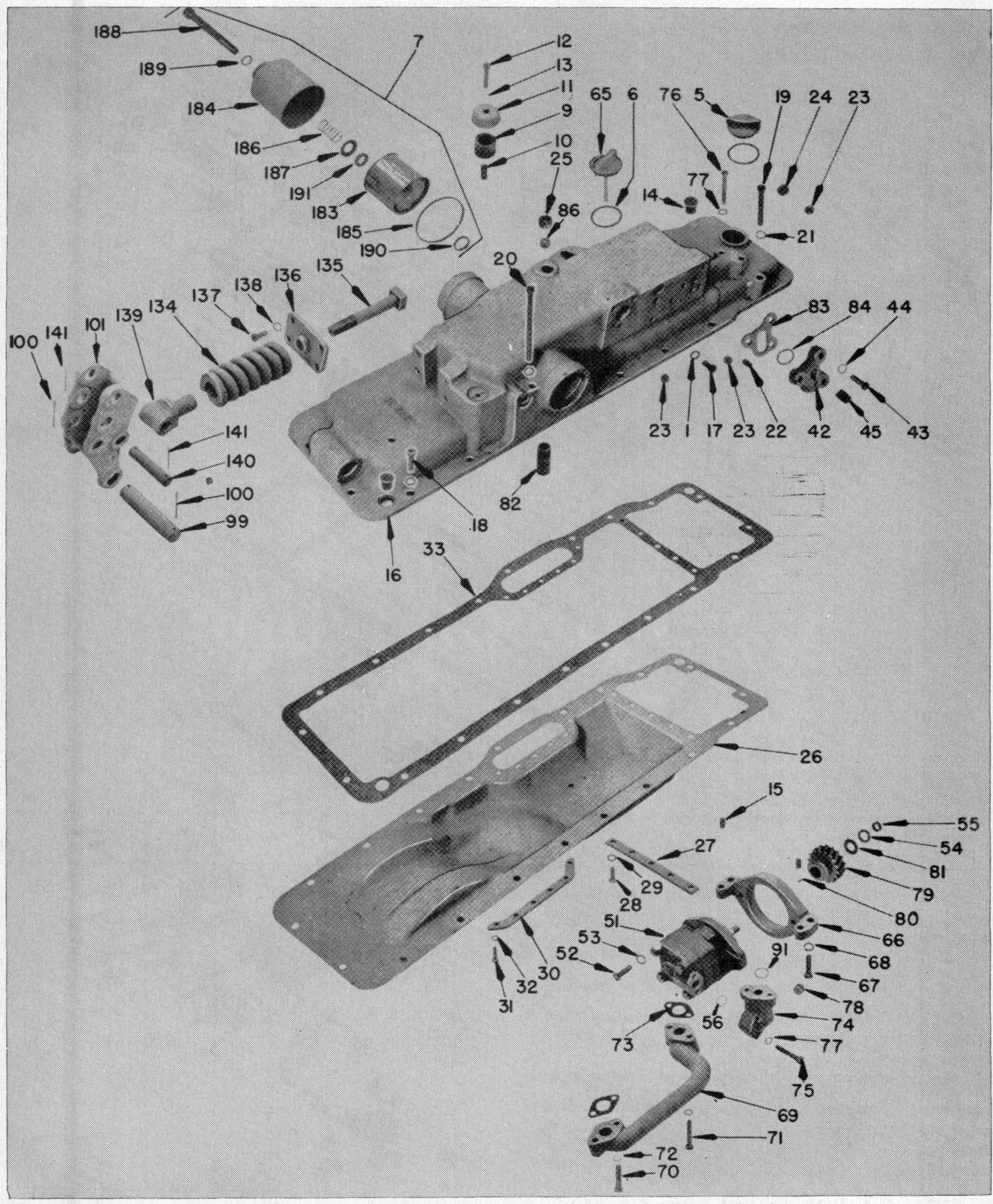

Fig. O767—Manual selector system. Some of the components attached to the upper surface of hydraulic housing (16) which is also the transmission top cover. View of double feed-back system is shown in Fig. O767A.

5. Filler cap
9. Breather element
10. Breather tube
16. Hydraulic housing
26. Reservoir oil pan
27. Pan reinforcement, front
30. Pan reinforcement, left
42. Passage cover
51. Hydraulic pump
65. Dip stick
66. Pump bracket
69. Inlet adapter
74. Outlet adapter
79. Pump gear
82. Filter outlet pipe
101. Lift rocker
134. Draft control spring
135. Adjusting bolt for 134
136. Spring guide
139. Spring yoke
183. Filter cartridge
184. Filter body
185. Filter gasket

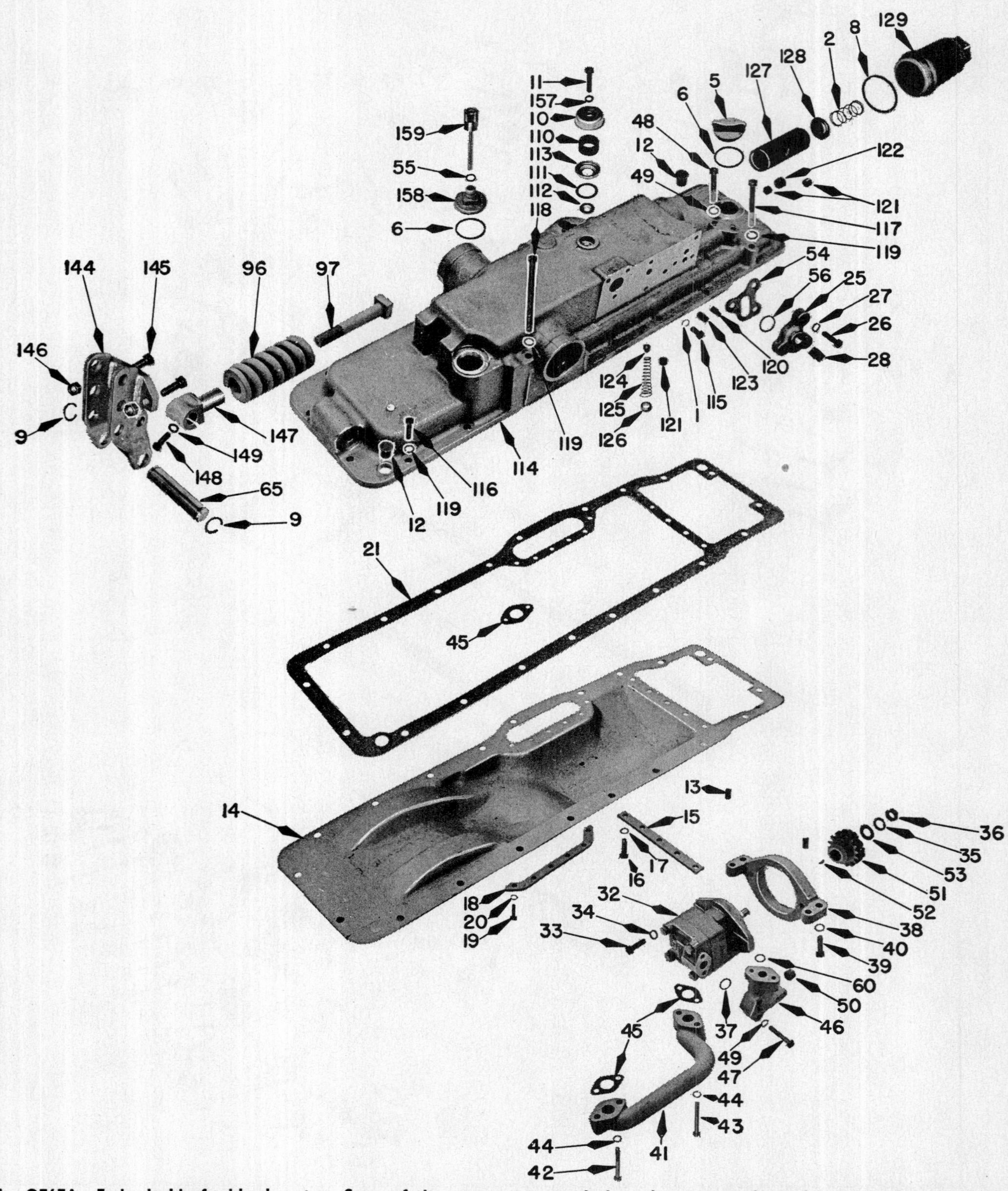

Fig. O767A—Early double feed-back system. Some of the components attached to the upper surface of hydraulic housing (114) which is also the transmission top cover. Late models are similar. View of manual selector system shown in Fig. O767.

5. Transmission filler cap
10. Breather cap
12. Dowel pin
14. Reservoir oil pan
15. Oil pan reinforcement
18. Oil pan reinforcement
21. Gasket
25. Cover
32. Hydraulic pump
38. Bracket
41. Inlet adaptor
46. Outlet adaptor
51. Pump gear
54. Passage cover plate
65. Pivot pin
96. Draft control spring
97. Adjusting bolt
110. Breather element
112. Breather baffle
114. Hydraulic housing
127. Filter element
129. Filter body
144. Rocker assembly
147. Spring yoke
158. Filler cap
159. Dip stick

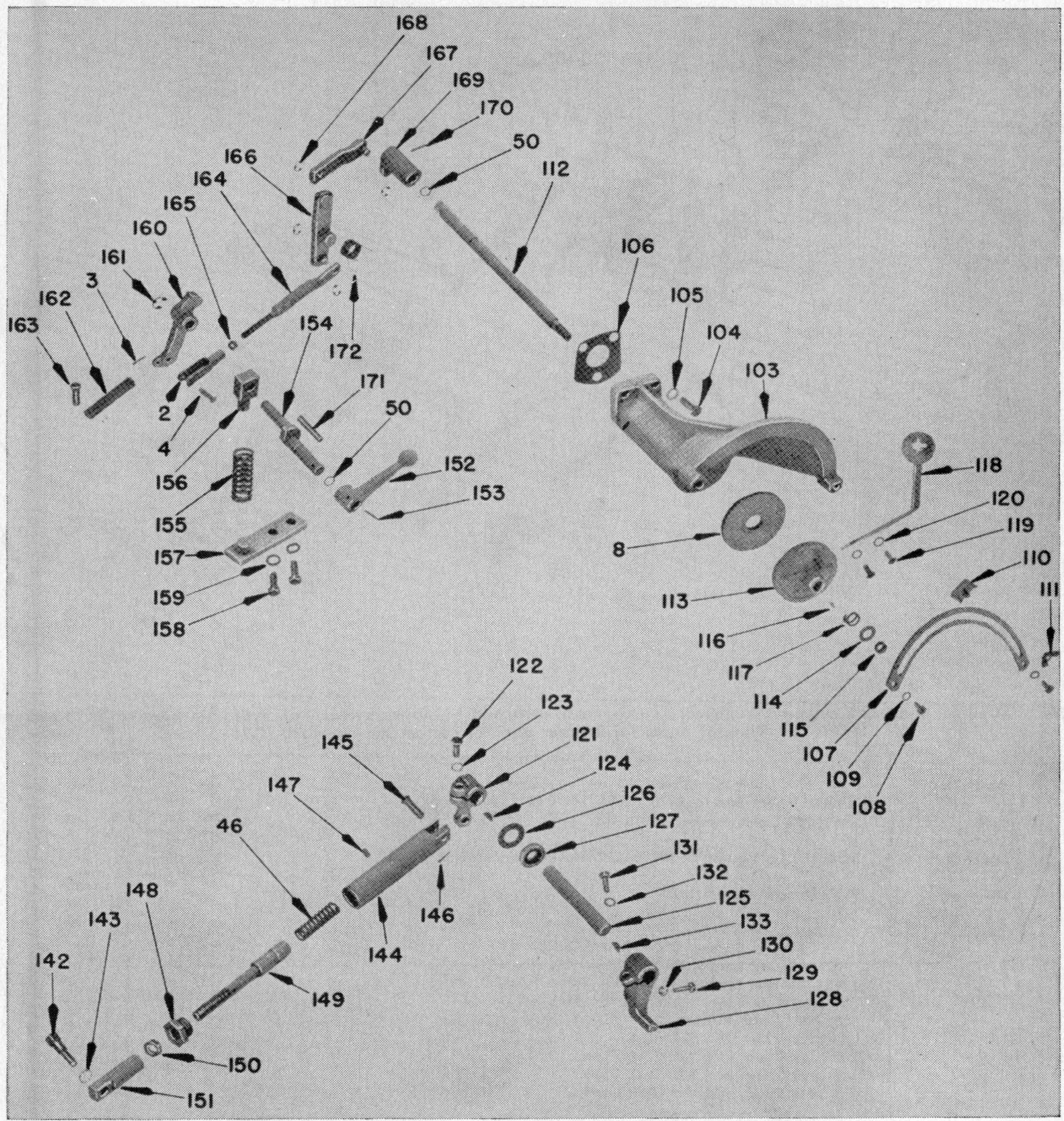

Fig. O768—Manual selector system. Additional components attached to upper surface of hydraulic housing shown at (16) in Fig. O767. View of parts for double feed-back system is shown in Fig. O768A.

2. Linkage adjusting rod end
8. Friction disc
46. Adjusting plunger spring
103. Control quadrant
107. Lever retainer
110. Lever stop
112. Hydraulic control shaft
118. Control lever
121. External draft lever
125. Draft control shaft
127. Oil seal
128. Draft internal lever
142. Draft pick-up stud
144. Front end for link
148. External link nut
149. Link adjusting plunger
151. External link rear end
152. Selector lever
154. Selector lever shaft
155. Over-center spring
156. Spring upper pivot
157. Spring support
160. Cam follower
162. Spool return spring
164. Linkage adjusting rod
166. Valve actuating lever
167. Connecting link
169. Control shaft arm
172. Torsion spring for 166

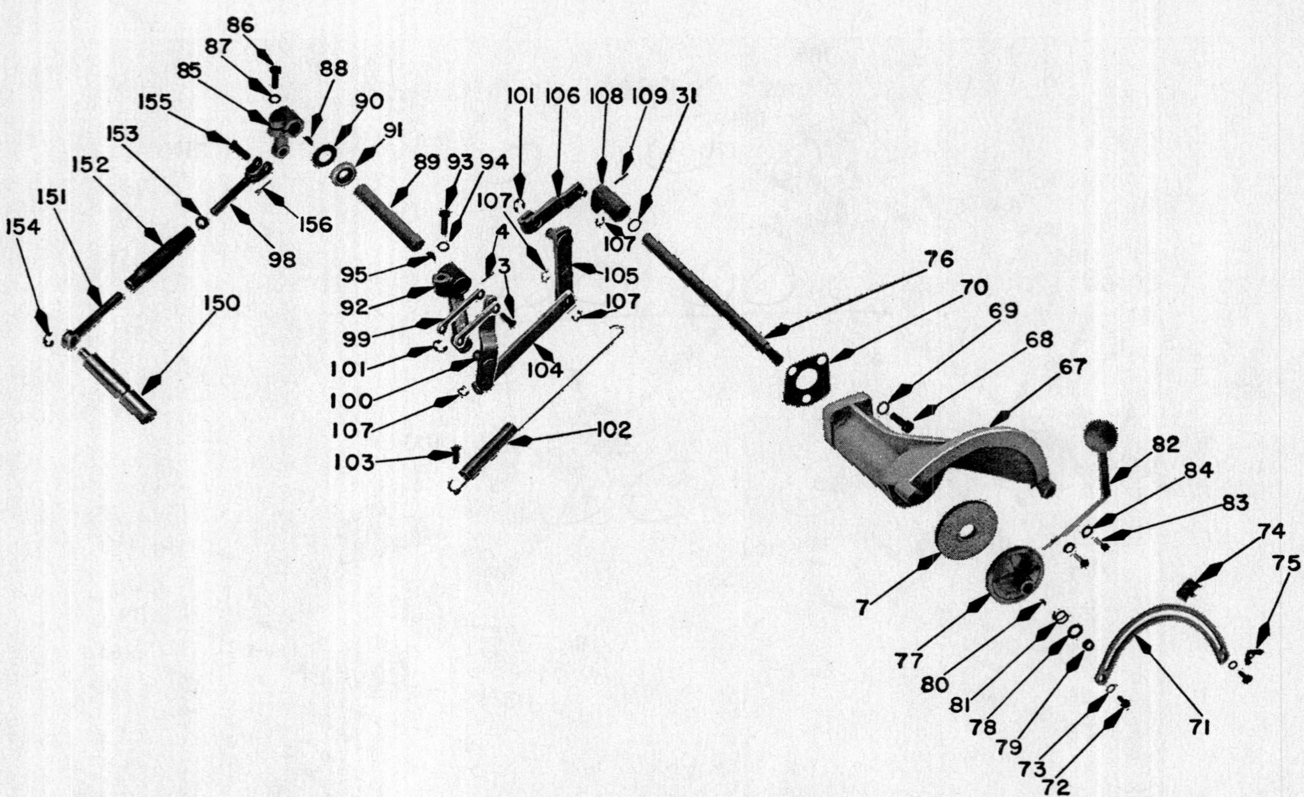

Fig. O768A—Double feed-back system. Additional components attached to upper surface of hydraulic housing (114) shown in Fig. O767A. View of parts for manual selector system shown in Fig. O768.

7. Friction disc
67. Control quadrant
70. Gasket
71. Control lever retainer
74. Control lever stop
77. Friction plate
82. Control lever
76. Control shaft
85. External draft control lever
89. Draft control shaft
91. Oil seal
92. Internal draft control lever
99. Feed-back link
100. Rocker arm assembly
102. Valve spool return spring
104. Rocker to valve lever link
105. Valve actuating lever
106. Connecting link
108. Control shaft arm
150. Control spring yoke pin
152. Adjusting turnbuckle

APPENDIX 1

POWER STEERING SYSTEM

The power steering system on the late production Super 55 series and the current 550 series consists of a pump mounted on rear of the generator, and driven by the generator shaft and a combined control valve and power unit which is integral with, and part of, the steering shaft.

As with any hydraulic system, the maintenance of cleanliness and the avoidance of nicks and burns on any of the working parts is of the utmost importance.

LUBRICATION AND BLEEDING

1s. Fill reservoir with SAE 10W engine oil or Type "A" Automatic Transmission Fluid (approximately 1½ pints) until fluid level registers with "Full" mark on bayonet gage.

To bleed system, fill reservoir to "Full" mark on bayonet gage, start engine and turn steering gear full left and full right several times to allow air to bleed out. Recheck fluid level and if necessary, add fluid to bring level to "Full" mark on bayonet gage.

TROUBLE-SHOOTING

2s. The accompanying trouble-shooting chart lists troubles which may be encountered in the servicing of the power steering system. For those not so obvious, refer to the appropriate subsequent paragraphs.

OPERATING PRESSURE

3s. To make a pressure test of the power steering system, proceed as follows: Refer to Fig. O771 and install a shut-off valve and pressure gage in series with the high pressure line of the power steering pump. Be sure gage is between pump and shut-off valve and is capable of registering at least 1100 psi.

Fig. O771—Install pressure gage and shut-off valve as shown. Note that gage is between pump and shut-off valve. Normal operating pressure range is 950-1050 psi.

With shut-off valve open, start engine and allow oil to reach operating temperature. With working fluid warmed and engine running at high idle, close the shut-off valve. The gage should show a pressure reading of 950-1050 psi. NOTE: Do not keep shut-off valve closed more than a few seconds as a rapid temperature rise will occur and damage to pump could result.

If the gage reading is within the 950-1050 psi range the pump can be considered satisfactory and any trouble is elsewhere in the system. If the gage pressure is below 875 psi, the pump must be removed and overhauled or renewed. If gage pressure is above the specified operating pressure the relief valve is probably stuck in its bore and pump will have to be removed and overhauled.

POWER STEERING PUMP

4s. **REMOVE AND REINSTALL.** Disconnect the two hoses from pump and secure ends in a raised position to prevent oil drainage. Remove the two cap screws which retain pump to generator, then remove filler cap and bayonet gage. Rotate pump until filler neck points downward and catch oil as it drains. While oil is draining, unscrew and remove engine oil filter assembly. Now while using a rotating motion pull power steering pump from generator shaft. Use caution when removing and reinstalling pump to avoid damaging the pump drive shaft oil seal as it slides over splines on generator shaft. Check condition of the rubber oil filter gasket before installing oil filter. Reconnect hoses then fill and bleed system as outlined in paragraph 1s.

5s. **OVERHAUL.** With the pump removed as in paragraph 4s, refer to Fig. O772 and proceed as follows: Remove the two cap screws which retain reservoir and remove reservoir. Be sure not to lose the two Dyna-seal washers which are under heads of cap screws. Remove screen from filler neck. The filter can be removed at this time by removing filter retainer spring from groove in cover assembly, however; if pump is to be separated, leave filter until pump is separated, at which time filter will be free. Remove lock wire from valve cover plate, slide plate off and hold spring retainer to keep parts from flying. Now remove spring retainer and spring, then slide or jar the

TROUBLE-SHOOTING CHART

	Loss of Power Assistance	Hard Steering	Noisy Pump Operation	Power Assistance in One Direction	Erratic Steering
Generator shaft splines worn or sheared	★				
Plugged filler cap air vent	★	★	★		★
Plugged oil strainer	★	★	★		★
Oil level low	★		★		★
Air in system	★		★		★
Vanes stuck in rotor slots	★				★
Relief valve stuck open	★				
Pump intake partially blocked		★	★		★
Return line leaking air			★		★
Faulty drive shaft seal			★		
Generator drive belt slipping or worn	★	★			★
Faulty or broken centering springs				★	★
Flow control valve sticking	★	★			★
Low pump pressure	★	★			

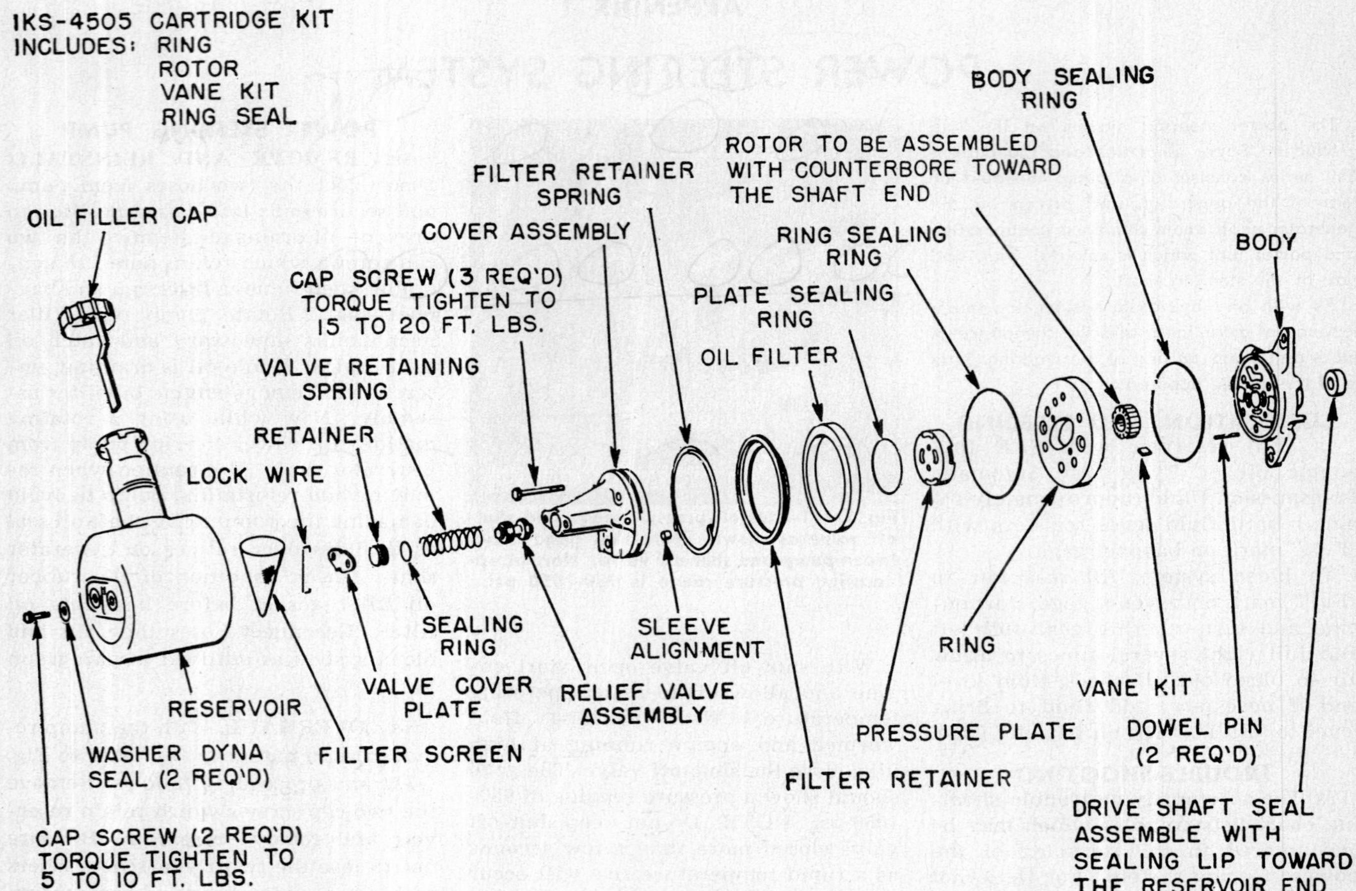

Fig. O772—Exploded view of the power steering pump used on all Oliver tractors. Rotor, vanes and ring are purchased as an assembly.

flow control and relief valve assembly from its bore. Remove the three cap screws retaining pump cover to pump body and separate the pump assembly. Use a brass rod and tap pressure plate from pump cover and remove "O" ring from groove in counterbore of pump cover. Remove alignment sleeve. DO NOT attempt to remove the two steel balls in the pump cover as they are permanently installed.

Lift the pump ring from dowels of pump body and remove "O" ring from groove in outer diameter. Lift off rotor and vane assembly and remove vanes from rotor. Pull dowels from pump body if they appear bent or damaged. Drive shaft seal can be removed from pump body by drifting it out with a brass rod from inside of body. Remove "O" ring from rear face of pump body.

With pump disassembled, clean all parts in a suitable solvent and inspect as follows: Check pump ring, rotor and vanes for wear or scoring. If ring, rotor or vanes need renewing all the parts will have to be renewed as a unit. Pay particular attention to the inside diameter of pump ring and mating surface of vanes. Vanes should fit rotor slots snugly yet be free to slide. Inspect the machined surfaces of the pump body, pump cover and pressure plate for scoring or other wear. If depth of scoring does not exceed 0.005 in pressure plate and 0.0015 in pump body, they can be reconditioned by lapping or using a surface grinder. Be sure the small sensing hole in pump body is clear and clean.

The flow control and relief valve assembly can be disassembled for cleaning by unscrewing the hex head retainer. The relief valve is pre-set at the factory and the Oliver Corporation recommends that if the valve does not maintain the recommended pressure range the entire unit be replaced. However, in case of emergency, the pressure can be regulated by adding or subtracting shims under the relief valve spring.

6s. When reassembling use new "O" rings and proceed as follows: Use a suitable driver and install drive shaft seal with lip facing reservoir end of pump. Coat all parts with SAE 10W engine oil or Type "A" Automatic Transmission fluid, then with dowel pins and new "O" ring installed in pump body, position rotor on pump body with counterbore in rotor toward front of pump. Install new "O" ring on pump ring. Place pump ring over dowels of pump body so the cap screw holes in ring align with threaded cap screw holes in pump body and the small sensing hole in pump body aligns with mating hole in pump ring. Install vanes in rotor with rounded edges outward and align rotor with opening in oil seal. Grease the face of the pressure plate which contacts rotor and position the pressure plate over dowels. Grease will aid in holding rotor while installing pump on tractor. Install alignment sleeve in pump ring and a new "O" ring in pressure plate counterbore of pump cover. Install filter retaining spring, retainer (with flat side toward retaining spring), and filter on pump cover. Now install pump cover over pressure plate, pump ring and alignment sleeve. Insert cap screws and tighten to 15-20 ft.-lbs.

Flow control valve can be cleaned if necessary by using Crocus cloth, however, DO NOT round valve edges or attempt to polish bore. With flow control and relief valve assembled, install valve with hex head outward (toward rear of pump) as shown in Fig. O773. Insert spring and retainer with new "O" ring. While holding these parts in position, slide valve cover plate over retainer and install lockwire.

Place screen in filler neck with large open end upward. Position reservoir over pump cover so filler neck points upward when pump is held in mounting position. Install the two reservoir cap screws and Dyna-seals but do not tighten cap screws at this time. After determining that reservoir is correctly installed and alignment for cap screws is satisfactory, push the reservoir into position by hand. DO NOT use the cap screws to pull reservoir into position. To do so will deform reservoir. Tighten cap screws to 5-10 ft.-lbs. torque.

Check rotation of pump assembly by using a substitute drive shaft. Pump should turn freely by hand and if any tendency to bind exists, correct condition before installing pump on tractor.

Install pump on tractor as outlined in paragraph 4s then fill and bleed system as outlined in paragraph 1s.

Be sure generator belt is adjusted to allow ¼-inch deflection when checked midway between generator pulley and crankshaft pulley.

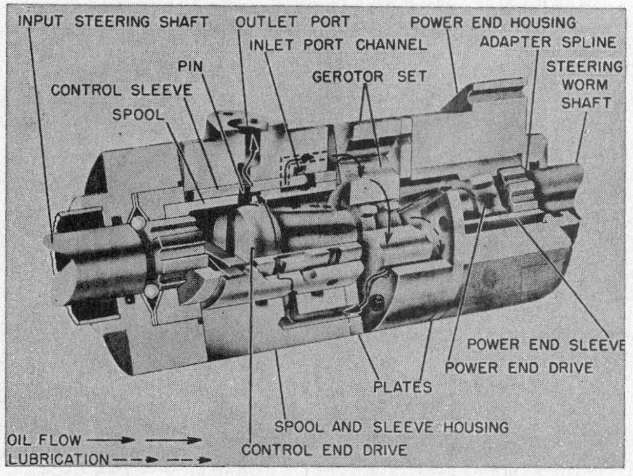

Fig. O773A — Schematic view of the power steering unit (hydraulic motor) used on the late Super 55 and the current 550 series tractors. Refer to Fig. O776 for parts breakdown.

POWER STEERING UNIT

7s. **REMOVE AND REINSTALL.** Removal of the power steering unit requires that the hood, fuel tank, battery and battery tray and the rear panel assembly be removed as shown in Fig. O774. To accomplish this proceed as follows: Remove the four hood retaining bolts and remove hood. Disconnect the oil pressure warning buzzer from its mounting. Disconnect wire from fuel gage sending unit and on diesels, disconnect the excess fuel return line. Shut off fuel and remove fuel line, then remove fuel tank mounting bolts and lift off fuel tank.

Remove steering wheel retaining nut and using a suitable puller, remove steering wheel. CAUTION: Do not drive on end of steering shaft. To do so will damage thrust bearing races in the power steering unit. These races are thin and brittle. Remove battery or batteries. Remove knob from light switch, then remove the switch retaining ring and push switch inward. Remove the self tapping screws from instrument panel cover. Disconnect tachourmeter cable at gage. Disconnect the ball joint link and the governor (or injection pump) control rod from the bellcrank mounted on bottom of battery tray support, then unbolt and remove battery tray support. Unbolt control operating rod support from steering gear housing. Disconnect choke control, or injection pump stop cable at front attachment and unthread cable from hole in rear tank support. Unclip and disconnect tail light wires at front lower left side of rear panel assembly. On diesel models, disconnect cable from glow plug and unthread from hole in rear panel assembly, then disconnect primary lead wires and battery cables from starting motor solenoid. Remove cap screws retaining rear panel assembly to tractor frame and slide the assembly up over the steering column and set aside as shown in Fig. O774.

8s. With rear panel assembly removed, place scribe marks on both Pitman arms to simplify reinstallation.

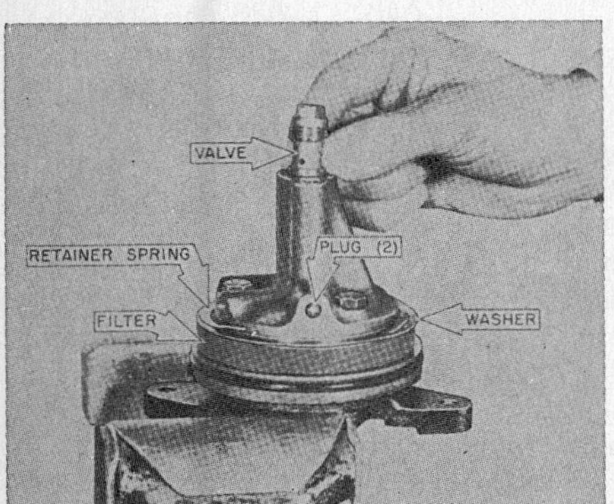

Fig. O773 — When installing flow control and relief assembly, the hex head retainer must be toward rear of pump as shown.

Fig. O774—View showing rear panel assembly removed. Remove Pitman arms from sector shafts before removing steering gear assembly.

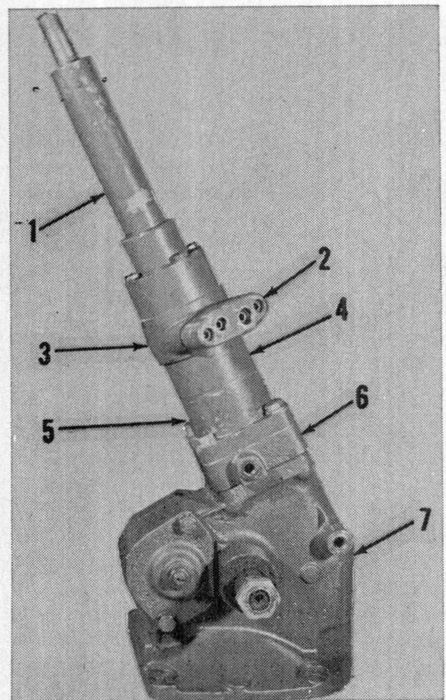

Fig. O775—View showing steering gear assembly removed. Note position of port block. Unit has no oil drain plug.

1. Tube and cap assembly
2. Port block
3. Spool and sleeve housing
4. Gerotor set
5. Power end housing
6. Adaptor
7. Steering gear housing

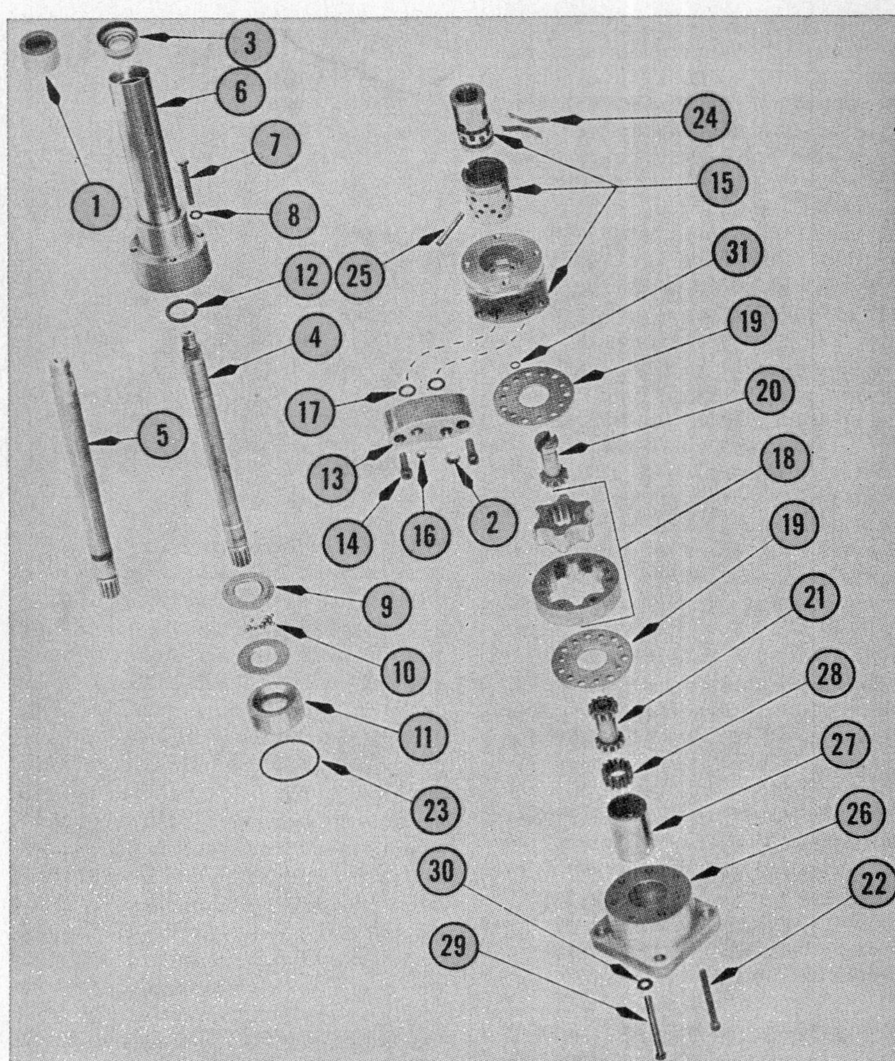

Fig. O776—Exploded view of the power unit of the power steering gear assembly. Shaft (4) with splined upper end used on models prior to serial number 100 085-AS. Shaft (5) with tapered end and keyway used on models 100 085-AS and up. Bushing (1) substituted for bearing (3) at same time.

1. Bushing
2. Flared connector
3. Bearing
4. Steering shaft
5. Steering shaft
6. Tube and cap assembly
9. Race
10. Steel balls (3/16")
11. Spacer
12. Quad ring (seal)
13. Port block
15. Spool and sleeve assembly
16. Flared connector
17. "O" ring
18. Gerotor set
19. Plate
20. Control end drive
21. Power end drive
23. "O" ring
24. Centering springs
25. Pin
26. Power end housing
27. Power end sleeve
28. Adaptor spline
29. Special socket head screw
30. Lead washer
31. "O" ring

Remove nuts retaining Pitman arms to sector shafts and pull arms from shafts. Disconnect hydraulic lines from the port block of power steering unit. Cap or tape open ends of hydraulic lines and secure lines with the ends higher than level of power steering pump reservoir to prevent fluid drainage. Remove cap screws from base of steering gear housing and be careful not to lose the spacers which are under the Pitman arm stops. Also note the position of the right and left hand Pitman arms stops. Lift power steering unit from tractor. See Fig. O775. Reinstall by reversing the removal procedure and keep in mind that when positioning the rear panel assembly on series 550 the starter safety switch actuating rod must enter its guide hole.

9s. **OVERHAUL.** With the power steering unit removed as outlined in paragraphs 7s and 8s proceed as follows: Use scribe marks as needed so unit can be reassembled in the same relative position. Refer to Fig. O776 and remove cap screws retaining power end housing and adaptor to steering gear housing. Remove cap screws retaining tube and cap assembly to power unit and remove tube and cap assembly. Invert tube and cap assembly and remove spacer spool and lower thrust bearing race. Push steering shaft slightly upward until balls are released, then remove the 13 balls. Remove the top thrust bearing race and pull steering shaft from bottom end of tube and cap assembly. Remove the sleeve and spool assembly from the spool and sleeve housing, then remove the pin and six centering springs and separate the spool and sleeve.

Invert unit and remove the seven socket head screws and note the position of the special socket head screw. Note: This special socket head screw may be one which is the same length as the other six with a groove in the threaded end to accommodate a small "O" ring; or, it may be a screw that is considerably shorter than the other six screws.

Now while holding parts together, again invert unit and remove parts as follows: Lift off spool and sleeve housing, upper plate and control end drive from Gerotor (rotor & ring) set. Lift off Gerotor set, lower plate, power end drive, adaptor spline and power end sleeve from power end housing.

Port block assembly is retained to spool and sleeve housing by two socket head screws. If necessary, the flared brass connector fittings can be removed from port block assembly by tapping same and using a cap screw and nut as a puller. With unit completely disassembled refer to table of Specifications then, wash all parts in a suitable solvent and carefully inspect all parts for undue wear or damage. Pay particular attention to the thrust bearing races (9) and balls (10). Check splines for undue wear and all parts for nicks and burrs. Be sure pin (25) is not bent or worn and that centering springs (24) are in good condition.

10s. When reassembling, use all new "O" rings and proceed as follows: Insert the special socket head screw in its proper hole (directly under the stamped model mark on the power end housing) in the power end housing and be sure the lead washer is under head of screw if so equipped. Then insert one other socket head screw on opposite side and stuff counterbores with paper to hold screws in position. Invert power end housing and install power end sleeve into power end housing with the spline in the inside diameter of sleeve downward. Now install adaptor spline, power end drive, lower plate, and Gerotor set. Postion control end drive in the rotor of the Gerotor set so that slot in upper end aligns with any of the tooth valleys (not with teeth) as shown in Fig. O776C. With control end drive so

Fig. O776A — Check tooth clearance of Gerotor set with a feeler gage as shown above. Clearance should be between 0.001-0.005.

Power Steering Unit Specifications

Torque output @ 1000 psi	1100 in. lbs.
Operating pressure—maximum	1000 psi
Operating speed	69 rpm
Gear ratio	11.68:1
Ring & rotor tooth clearance—maximum	0.001-0.005
Ring thickness	0.8629-0.8630
Rotor thickness	0.8620-0.8621 min. 0.8618
Body bore sleeve diameter	1.6603-1.6604
Control sleeve outside diameter	1.6598-1.6599
Control sleeve & body bore clearance—maximum	0.0008
Control sleeve inner bore diameter	1.2600-1.2601
Spool outside diameter	1.2598-1.2599
Spool & control sleeve clearance—maximum	0.0004
Sleeve pin hole	0.2500-0.2505
Pin outside diameter	0.2498-0.2500
Pin & sleeve clearance—maximum	0.0017
Power end housing sleeve bore	1.3745-1.3755
Power end sleeve outside diameter	1.3715-1.3735
Power end housing & sleeve clearance—maximum	0.006
Power end drive backlash in rotor	0.003-0.004
Control end drive backlash in rotor	0.001-0.002
Control end drive slot	0.2501-0.2505
Plates	Must be within 0.001 parallel on housing. Plates are reversible and interchangeable but must be flat within 0.00002. Plate thickness is 0.164-0.170 but may be lapped to a maximum of 0.005 undersize.
Centering springs	Must return to a height of 7/32" after being depressed to 1/16". Springs for the S-105-D unit are not interchangeable with previous units.

Fig. O776B — With each spring depressed to a height of 1/16-inch on a flat surface it should return to a minimum height of 7/32-inch when released.

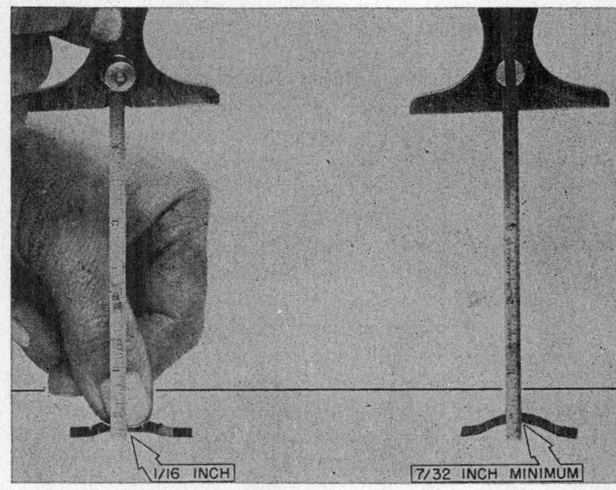

Fig. O776C—Be sure slot of control end drive is aligned with a tooth valley as shown. If slot is aligned with high point of a tooth the unit will operate in reverse.

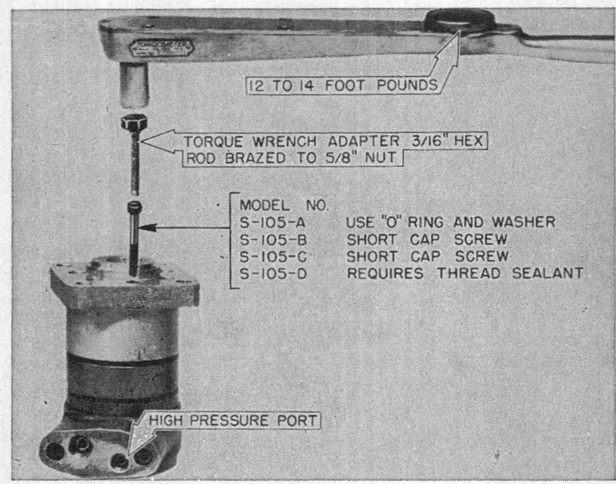

Fig. O777—Tighten socket head screws to 12-14 ft.-lbs. using an Allen wrench adapted for use with a torque wrench. Note also the model numbers and the corresponding socket head screws used.

Fig. O778A — View of adapter plate showing location of quad ring and "O" ring seals. Install adapter plate so filler plug hole is on right side and to the rear of center.

positioned, install upper plate and spool and sleeve housing. NOTE: If the special socket head screw (29) is long type be sure the small "O" ring is fitted into the counterbored screw hole in the spool and sleeve housing.

Later special socket screws may not be fitted with "O" ring (31—Fig. O776) in which case threads of special socket head screw (29) should be coated with grade "C" Loctite sealant or equivalent. If Loctite sealant is used the Oliver Corp., recommends waiting 12 hours before operating steering system. Refer to Fig. O777.

Port block should be at the two o'clock position. Now install the remaining five socket head screws and tighten all screws evenly and to a torque of 12-14 ft.-lbs. This will require a special Allen wrench and a suggested method of making one is to cut off the short shank of a regular Allen wrench and weld on a nut of sufficient size to accommodate a socket. Fig. O777 shows such a tool.

Place spool in sleeve and install pin, then install centering springs. With spool and sleeve assembled, position unit in housing so that pin fits into the slot in the control end drive. With parts installed, the spool and sleeve assembly should be flush with the top of the spool and sleeve housing.

Install steering shaft into tube and cap assembly and position top race. Place the 13 balls in cap and maneuver steering shaft until balls fit into groove of shaft and bevel of race, then pull shaft toward top of tube and cap assembly until balls are locked into position. Install lower race and spacer, with counterbore of spacer toward thrust bearing. Bolt tube and cap assembly to power steering unit, then bolt power unit to steering gear housing after making certain adaptor plate is positioned with oil filler hole toward right side.

Reinstall unit on tractor as outlined in paragraphs 7s and 8s. Fill and bleed system as outlined in paragraph 1s.

STEERING GEAR

The steering gear portion of the power steering unit, except for detail differences, is very similar to the steering gear unit used on models without power steering. The primary difference being that the steering shaft has been shortened and the upper end splined to accept the power unit, and the bearing cup for the wormshaft upper bearing is located in the adapter instead of the tube and cap assembly. In addition to the bearing cup the adapter also is fitted with a quad ring oil seal and "O" ring. See Fig. O778A.

11s. **WORMSHAFT END PLAY.** To adjust wormshaft end play remove the rear panel assembly as outlined in paragraph 7s. Remove both Pitman arms. Remove cap screws which retain power unit and adaptor to steering gear housing and lift power unit from steering gear housing. Hydraulic lines need not be disconnected.

Use spacers under the cap screws just removed, or obtain shorter cap screws, and bolt adaptor to steering gear housing. With adaptor secured in place, adjust wormshaft to zero end play by adding or subtracting shims located between adaptor plate and steering gear housing.

Shims are available in thicknesses 0.002, 0.005, 0.010 and 0.030.

12s. **BACKLASH ADJUSTMENT.** Sector mesh adjustment can be made with steering gear on tractor and is done in the same manner as previous models. Refer to paragraph 6B for procedure.

13s. **GEAR UNIT R&R.** Remove and reinstall the steering gear unit as outlined in paragraphs 7s and 8s.

14s. **GEAR UNIT OVERHAUL.** Except for minor differences which are obvious, the procedure for overhauling the gear unit on models equipped with power steering is the same as that for models without power steering. Refer to paragraphs 8 through 8D.

Note: In cases where Pitman shaft double gear (11—Fig. O778B) fails on models with power steering, a unit with 1¼-inch wide teeth is available to replace the original which has teeth 1-inch wide.

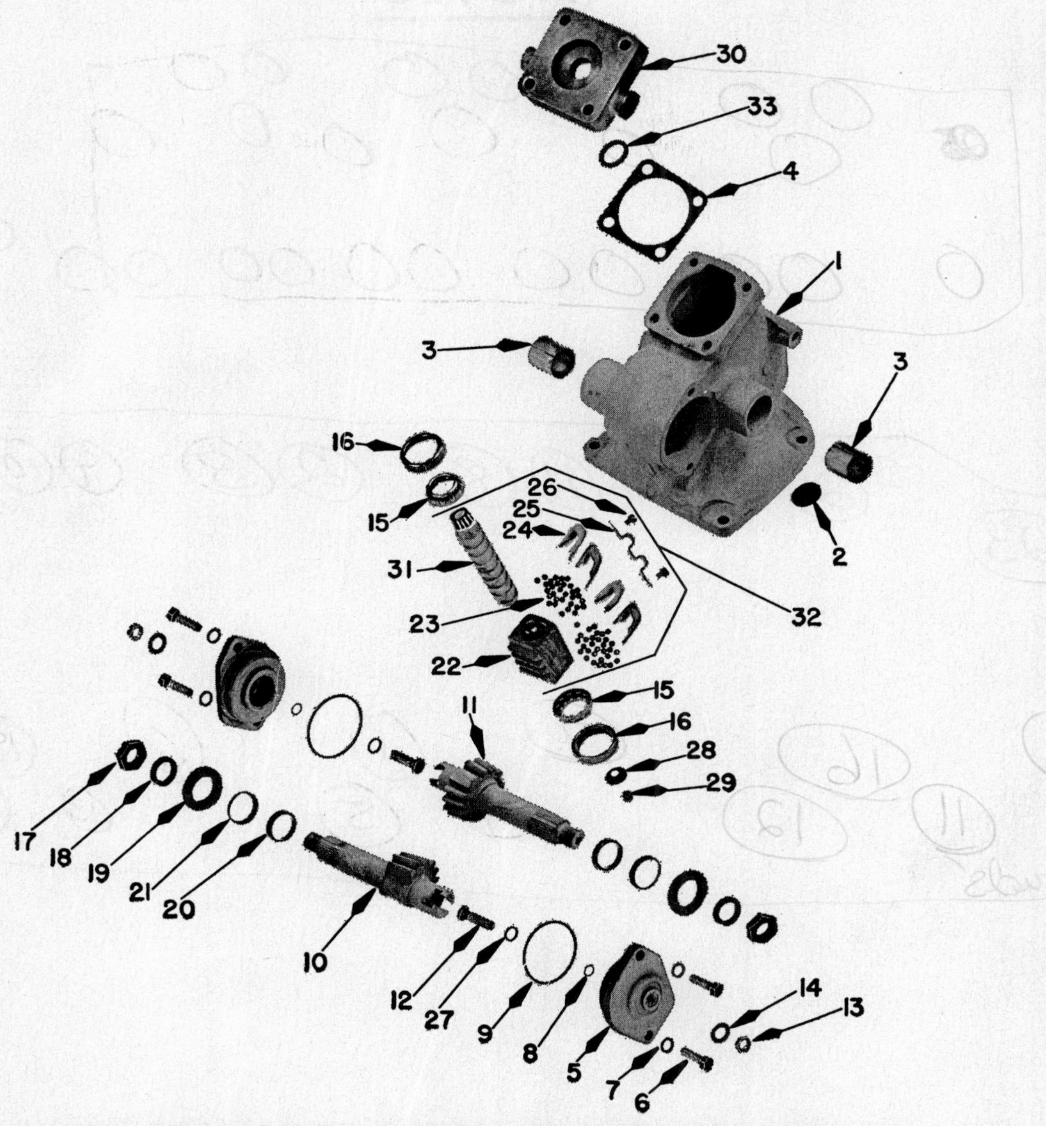

Fig. O778B—Exploded view of the gear portion of the power steering gear assembly. Unit is similar to non-power steering units except that wormshaft (31) has been shortened and the upper end splined.

1. Housing
2. Expansion plug
3. Bushing
4. Shim
5. Cover
8. Packing
9. "O" ring
10. Pitman arm shaft (single)
11. Pitman arm shaft (double)
12. Lash adjuster
13. Jam nut
14. Washer
15. Bearing cone
16. Bearing cup
17. Nut
18. Lock washer
19. Dust seal
20. Packing
21. Packing retainer
22. Ball nut
23. Steel balls
24. Ball guide
25. Ball guide retainer
27. Shim
28. Bearing retainer
29. Retainer eyelet
30. Adaptor
31. Wormshaft
32. Wormshaft and ball nut assembly
33. Quad ring (seal)

NOTES

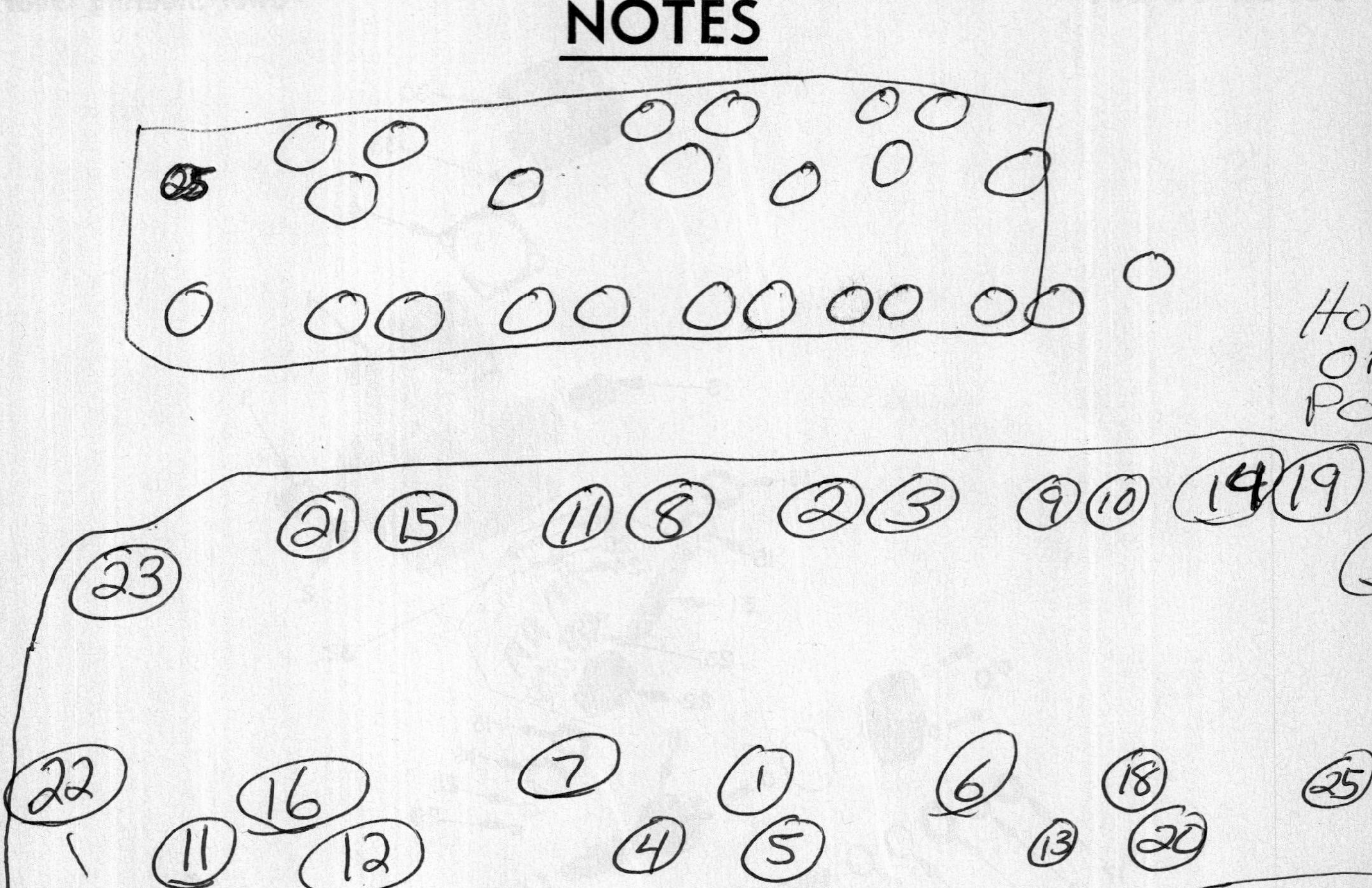

25
Ho
O
Pa
21
15
11
8
2
3
9
10
14
19
23
22
16
11
12
7
4
1
5
6
13
18
20
25
Studs